Essentials of Chemical Engineering

Essentials of Chemical Engineering

Editor: Lewis Allison

NY RESEARCH PRESS

New York

Published by NY Research Press
118-35 Queens Blvd., Suite 400,
Forest Hills, NY 11375, USA
www.nyresearchpress.com

Essentials of Chemical Engineering
Edited by Lewis Allison

International Standard Book Number: 978-1-63238-643-4 (Hardback)

Cataloging-in-Publication Data

Essentials of chemical engineering / edited by Lewis Allison.
    p. cm.
Includes bibliographical references and index.
ISBN 978-1-63238-643-4
1. Chemical engineering. 2. Chemistry, Technical. 3. Engineering. I. Allison, Lewis.
TP155 .E87 2019
660--dc23

Contents

Preface

This book on chemical engineering elucidates on the concepts and theories fundamental to this field of study. Chemical Engineering is a branch of engineering that uses the principles of applied physics, chemistry, life sciences and other scientific fields for production, use and transformation of chemicals, materials and energy to serve various engineering purposes. There has been rapid progress in this field and its applications are finding their way across multiple industries such as biotechnology, control engineering, plant design, etc. This book offers information about the essential topics of chemical engineering while also discussing the progress made in modern theory and principles of the field. It elucidates new techniques and their applications in a multidisciplinary manner. This book traces the progress of this field and highlights some of its key concepts. For all readers who are interested in chemical engineering, the case studies included in this book will serve as an excellent guide to develop a comprehensive understanding.

All of the data presented henceforth, was collaborated in the wake of recent advancements in the field. The aim of this book is to present the diversified developments from across the globe in a comprehensible manner. The opinions expressed in each chapter belong solely to the contributing authors. Their interpretations of the topics are the integral part of this book, which I have carefully compiled for a better understanding of the readers.

At the end, I would like to thank all those who dedicated their time and efforts for the successful completion of this book. I also wish to convey my gratitude towards my friends and family who supported me at every step.

Editor

1

A Comparison of the Capital Costs of a Vanadium Redox-Flow Battery and a Regenerative Hydrogen-Vanadium Fuel Cell

Mark Moore*, Robert Counce, Jack Watson and Thomas Zawodzinski

Department of Chemical Engineering, University of Tennessee, USA

Abstract

The capital costs of a Regenerative Hydrogen-Vanadium Fuel Cell and a Vanadium Redox-Flow Battery are compared for grid level energy storage. The bulk of the capital costs for a Vanadium Redox-Flow Battery lie in the costs of the vanadium electrolyte, while the Regenerative Hydrogen-Vanadium Fuel Cell presents a potential for savings by eliminating the need for half of the vanadium electrolyte required by a Vanadium Redox-Flow Battery. It was found that the Regenerative Hydrogen-Vanadium Fuel Cell would cost $57 less per kWh than the Vanadium Redox-Flow Battery, with savings garnered from the elimination of half of the electrolyte somewhat mitigated by the costs of the catalyst and air compressor required. If the capital costs are annualized through straight line depreciation, and the operation costs are included, the Vanadium Redox-Flow Battery is $5 per kWh less per year than the Regenerative Hydrogen-Vanadium Fuel Cell.

Keywords: Flow battery; Economics; Regenerative hydrogen-vanadium fuel cell; Capital costs comparison

Introduction

The goal of this paper is to estimate and compare the capital cost of a regenerable hydrogen-vanadium battery (RHVB) with an all-vanadium redox-flow battery (VRB) for grid-scale applications [1,2]. As more and more renewable power production is added to the grid the need increases for large-scale storage alternatives. The potential of the redox flow battery (RFB) for use in grid scale energy storage is well documented [3-6]. Revenue streams for RFBs are somewhat complex, including peak shaving, load leveling, energy reserve and grid stabilization capabilities to improve the performance of the utility grid and deferral of investments for additional generation capacity [7]. In a series of papers Banham-Hall and others establish the technical viability of these potential revenue streams for VRBs integrated into a system of renewable power generation [7-9]. Using grid-based prices and other relevant information, Fare and others showed the value of VRBs for frequency regulation to be about $1500/kW [10]. The combination of renewable energy production and energy storage enables the system to behave more like a conventional power generation systems [7-9]. The all-vanadium redox flow battery (VRB) is currently the leading battery alternative. For bulk energy storage the Vanadium Redox-Flow Battery (VRB) has a distinct advantage over other types of flow batteries. Vanadium cations have four different oxidation states, allowing vanadium to be used in both the anolyte and the catholyte. This is advantageous because any cross contamination of ionic species through the membrane does not present major difficulties. Vanadium, however, is more expensive that many other electrolyte sources for flow batteries and represents a significant portion of the capital costs of building a VRB [2]. A possible solution to this is to create a hybrid battery/fuel cell, designing the anode half-cell to function as in a fuel cell, and the cathode half-cell to function as in a flow battery. The chemical equations are [1]:

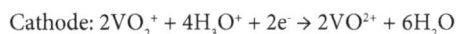

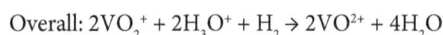

Cathode: $2VO_2^+ + 4H_3O^+ + 2e^- \rightarrow 2VO^{2+} + 6H_2O$

Anode: $H_2 + 2H_2O \rightarrow 2H_3O^+ + 2e^-$

Overall: $2VO_2^+ + 2H_3O^+ + H_2 \rightarrow 2VO^{2+} + 4H_2O$

This concept was demonstrated with a regenerative hydrogen-vanadium battery (RHVB) by Yfit et al. [1]. A RHVB has the potential for lower capital costs by eliminating the need for half of the vanadium

required in a VRB. In addition, since the electrolyte is only required for the reaction in the cathode half-cell, metal ions with only 2 oxidation states may be used as an alternative. Metal ions that provide 2 or more moles of electrons for every mole of ions when they are oxidized could further reduce the required capital cost.

In this paper the capital cost of a RHVB is estimated and compared with a VRB. The contributions to the total capital costs of a VRB can be seen in Figure 1. The relevant information about the RHBV can be found in Tables 1 and 2. In this costing study it is assumed that there is little difference in the stack components between the RHVB and the VRB. The differences in the stacks are in the reactions taking place in the anode half-cell, the electrical potential generated by the reactions in each cell, and the need for a catalyst in the regenerative battery. The VRB requires two liquid tanks for the anolyte and catholyte, and 2 pumps to move the electrolyte solutions. The RHVB battery requires just one tank and pump but also requires storage for the hydrogen as well as a compressor.

Methods

The stack component costs in the EPRI report *Vanadium Redox Flow Batteries: An In Depth Analysis* were largely used for this analysis [5]. With the addition of cost of a catalyst ink and Application of $65/m^2$ from James et al. [11]. Using the design details in the EPRI report for the total electrode area needed for a 1 MW, 6 MW hr battery, the number of cells required for the battery was calculated. This was then multiplied by the cost per cell component provided by the report. The voltage efficiency used was calculated from data provided by Che-Nan Sun in experiments conducted at Oak Ridge National Laboratory on a VRB [12]. This efficiency was used for both the RHVB and the VRB for simplicity purposes. The flow rates of the vanadium electrolyte and the

*Corresponding author: Mark Moore, Department of Chemical Engineering, University of Tennessee, USA, E-mail: mmoore76@vols.utk.edu

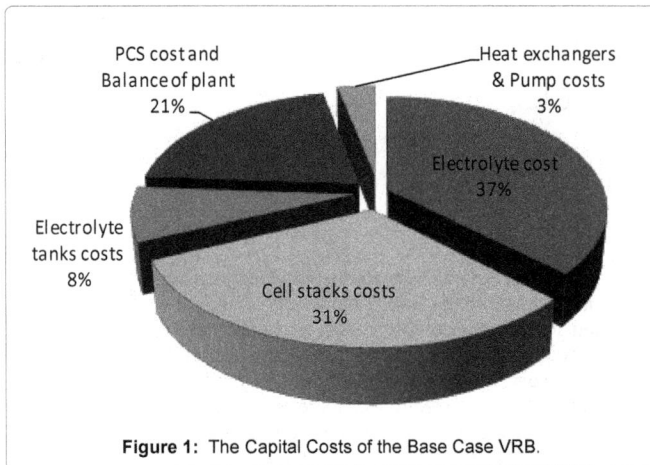

Figure 1: The Capital Costs of the Base Case VRB.

hydrogen were estimated by calculating the moles of electrons oxidized per second by one cells, then multiplied by the number of cells in the stacks. This estimate is the molar flow rate and molar concentration required, and is used to estimate the size of the pumps and the compressor required using the methods found in a textbook by Ulrich and Vasudevan [13]. The pressure drop of the vanadium electrolyte through the cell stacks was estimated by using an empirical correlation in "Understanding Vanadium Redox-Flow Batteries" by Blanc and Rufer that uses a hydraulic resistance calculated from computer simulations using the finite element method [14]. It was assumed that there is no loss to the power produced by the batteries due to species depletion as the electrolyte and the hydrogen flows through the stack.

The volume of vanadium required was estimated by multiplying the previous calculation of moles of electrons oxidized per time of the entire stack by the charge or discharge time. The method of hydrogen storage is assumed to be with the use of an adsorbent. It is important to note that the capital costs of this storage method used here are from the USDOE FreedomCAR targets [15]. The 2010 target is $133 per kg of hydrogen, the 2017 target is $67 per kg of hydrogen, and the cost used to calculate the capital cost for this paper was $100 per kg hydrogen. While the use of adsorbent storage reduces the need for high pressure storage of hydrogen, it is still necessary to use a pressurized vessel, albeit at a lower pressure than without an adsorbent. A compressor is then necessary for the hydrogen exiting the stacks during the charging process. Heat is required to desorb the hydrogen, and at steady state operation it is possible that the heat generated by the cell stacks or the compressor could be used. An external source of heat would be required at start up, however. During the charge cycle the hydrogen flowing from the stacks contains water, requiring that this water be removed from the stream or that a method of removing the water from the storage tank be found. The capital cost calculations in this paper do not reflect the costs associated with heating the adsorption materials or removing the water for the hydrogen stream or the tank. The costs of the power conditioning system and the control system for both systems were assumed to be the same and were taken from the EPRI report [3].

The annual operational costs associated with the fixed capital can be seen in Table 3. These costs are $42.77 per kWh for a RHVB and $51.02 per kWh for a VRB. Other operation costs are assumed to be the similar for the two battery systems with the exception of the costs to run the pumps and compressors. The cost of electricity is assumed to be $0.10 per kWh, and it is also assumed that the battery runs a full cycle a day (charge and discharge) 328 days a year. With these assumptions,

the costs of electricity annually for the RFB are $0.79 per kWh while the costs of electricity annually for the RHVB are $16.80 per kWh.

Results

The results of the capital cost analysis can be seen in Tables 4 and 5. The total cost per year, using straight line depreciation for the capital costs over a 20 year lifespan, would be about $70 per kWh for the VRB and $75 per kWh for the RHVB. The precious metal catalyst required for the RHVB constitutes the difference in the capital cost between the cell stacks. In addition, the higher electrical potential available to the vanadium battery cells allowed for an overall smaller stack size than the RHVB, reducing costs. The VRB uses two pumps, while the RHVB uses a pump for the vanadium electrolyte and a compressor for the hydrogen. The capital cost of the compressor is much greater than the costs of the pumps, adding to the costs of the RHVB in comparison to the VRB. In addition, the costs for the electricity to run the compressor are much greater than the costs for running the pumps. The savings in the capital costs associated with the regenerative battery is for the vanadium and its storage, the regenerative battery only requiring vanadium as a catholyte, while the VRB requires vanadium for the anolyte as well. Because of the lower electrical potential of the cells in the regenerative battery, a higher current is required to sustain the power requirement of 1 MW. This necessitates the use of more electrolytes in the regenerative battery, mitigating some of the savings in the purchase costs of vanadium.

Conclusion

Overall the VRB is about $5 per kWh per year cheaper than the RHVB. The capital costs are for batteries with the specific energy and power capacities detailed in Figure 1. A sensitivity analysis by Zhang et al. for a VRB can be used to determine how these costs will change when the energy and power capacities are adjusted [16]. With a fixed energy capacity (stored electrical energy) a VRB has a power capacity sensitivity index of 0.4881, which represents the rate of change of the capital costs with respect to the power capacity. This rate of increase in the capital costs would be higher with a RHVB as the power capacity is determined by the size of the stacks, which has a higher cost in the RHVB due to the catalyst. If the power capacity is held constant

Category	Value
Stoichiometry	Cathode: $2VO_2^+ + 4H_3O^+ + 2e^- \leftrightarrow 2VO^{2+} + 6H_2O$ Anode: $H_2 + 2H_2O \leftrightarrow 2H_3O^+ + 2e^-$
Power Capacity	1,000 kW
Energy Capacity	6,000 kWh
Overall Efficiency	0.73
Open Circuit Electrical Potential per Cell	1.1 Volts
Cross Sectional Area of Cell	236 cm^2
Current Density	604 mA/cm^2

Table 1: Design Details for Hydrogen-Vanadium Regenerative Battery.

Component	Value
Stoichiometry	Cathode: $2VO_2^+ + 4H_3O^+ + 2e^- \leftrightarrow 2VO^{2+} + 6H_2O$ Anode: $V^{2+} \leftrightarrow V^{3+} + e^-$
Power Capacity	1,000 kW
Energy Capacity	6,000 kWh
Overall Efficiency	0.73
Open Circuit Voltage per Cell	1.3 Volts
Cross Sectional Area of Cell	1 m^2
Current Density	604 mA/cm^2

Table 2: Design Details for the Vanadium Redox-Flow Battery.

Capital-related cost item	Fractions of fixed capital
Maintenance and repairs	0.06
Operating supplies	0.01
Overhead, etc.	0.03
Taxes and insurance	0.03
General	0.01
Total	**0.14**

Table 3: Annual Expenses Proportional to Fixed Capital.

Component	Cost
Total Cost of Stack	$9.00
Pump Costs	$4.11
Cost of Compressors	$50.00
Cost of Electrolyte Tank	$30.00
Cost of Adsorption Tank	$1.64
Cost of Vanadium	$59.18
Fuel Cell Balance of Plant	$97.42
PCS, Transformer, etc.	$54.12
Total Cost	$305.47

Table 4: Capital Costs of RHVB in $/kWh.

Component	Cost
Total Cost of Stacks	$4.92
Pump Costs	$8.23
Cost of Electrolyte Tanks	$60.00
Total Cost Vanadium	$139.76
Fuel Cell Balance of Plant	$97.42
PCS, Transformer, etc.	$54.12
Total Costs	$364.44

Table 5: Capital Costs of VRB in $/kWh.

the sensitivity index for capital costs due to cycle time (representing total energy capacity) and vanadium costs are 0.6101 and 0.3337, respectively, for a VRB. These rates of change for the capital costs would be less for a RHVB because the main driving force for these costs is for the vanadium, with the RHVB using half the mass of vanadium as a VRB.

In order for the RHVB to be more cost effective than the VRB more cost reductions must be found. Possibilities for cost reductions are:

1) Eliminating the need for the catalyst by operating the battery at higher temperatures

2) Reducing the pressure in the H_2 storage tank

3) Replacing vanadium with a lower cost redox material

It would be necessary to ensure that the vanadium does not precipitate from the solution at high temperatures, however. For the purposes of this study the costs of the compressor and pressurized tank were only calculated for a hydrogen storage pressure of 10 bar, a more thorough analysis may provide a cost savings in this regard. A more affordable electrolyte could significantly reduce the costs of the regenerative battery. Because of the need for a metal with only 2 oxidation states and a reduced chance of cross contamination through the membrane, it is possible a number of other metals would present a cost savings over vanadium.

References

1. Yfit V, Hale B, Matian M, Mazur P, Brandon NP (2013) J Electrochem Soc 160: A56.

2. Moore M, Counce R, Watson J, Zawodzinski T, Kamath J (2012) A Step-by-Step Design Methodology for a Base Case Vanadium Redox-Flow Battery. Chem Eng Ed 46: 239-250.

3. Weber AZ, Mench MM, Meyers JP, Ross PN, Gostick JT, et al. (2011) Redox flow batteries: a review. J Appl Electrochem 41: 1137-1164.

4. Skyllas-Kazacos M, Chakrabarti MH, Hajimolana SA, Mjalli FS (2011) Progress in Flow Battery Research and Development. J Electrochem Soc 158: R55-R79.

5. Eckroad S (2007) Vanadium Redox Flow Batteries: An In-Depth Analysis. EPRI, Palo Alto, CA.

6. Kear G, Shah AA, Walsh FC (2012) Development of the all-vanadium redox flow battery for energy storage: a review of technological, financial and policy aspects. Int J Energy Res 36: 1105-1120.

7. Banham DD (2012) Flow Batteries for Enhancing Wind Power Integration. IEEE Trans 27: 1690-1697.

8. Banham-Hall DD, Taylor GA, Smith CA, Irving MR (2011) Frequency control using Vanadium redox flow batteries on wind farms. IEEE Power and Energy Society General Meeting 1-8.

9. Banham-Hall DD, Smith CA, Taylor GA, Irving MR (2012) International Universities Power Engineering Conference, UPEC, IEEE Computer Society.

10. Fares RL, Meyers JP, Webber ME (2014) A dynamic model-based estimate of the value of a vanadium redox flow battery for frequency regulation in Texas. Appl Energy 113: 189-198.

11. James BD, Kalinoski JA, Baum KN (2010) Mass Production Cost Estimation for Direct H2 PEM Fuel Cell Systems for Automotive Applications. Directed technologies, Arlington VA.

12. Sun CN, Mench M, Zawodzinski T (2013) Comparing Limits of Performance in Aqueous and Non-aqueous Redox Flow Batteries. 224th ECS Meeting.

13. Ulrich G, Vasudevan P (2004) Process (Ulrich) Publishing. Durham, NH.

14. Blanc C, Rufer A (2010) Understanding the Vanadium Redox Flow Batteries, Paths to Sustainable Energy. InTech

15. He C, Dansai S, Brown G, Bollepalli S (2005) Electrochem Soc Interface 31.

16. Zhang M, Moore M, Watson J, Zawodzinski T, Counce R (2012) Capital Cost Sensitivity Analysis of an All-Vanadium Redox-Flow Battery. J Electrochem Soc 159: A1183-A1188.

Positive and Negative Aspects of Electrode Reactions of Hydrogen Evolution and the Influence of a Constant Magnetic Field

Marek Zieliński*

Department of Inorganic and Analytical Chemistry, Faculty of Chemistry, University of Lodz, Poland

Abstract

Nowadays, the application of electrode reactions, including those concerning hydrogen evolutions, attracts great attention in industry and in power engineering worldwide. Thus, new ways of increasing the efficiency of electrode processes are of particular interest. One of them may involve a Constant Magnetic Field (CMF). The magnetic field influences both electrons and ionized atoms leading to dynamic effects (e.g. electrolyte movement in the layer adjacent to the electrode resulting from the Lorentz force). Investigations carried out with the CV method (Cyclic Voltammetry) proved that the reaction rate constant of hydrogen production increases under the influence of CMF. Alloys well adsorbing hydrogen such as Co-Mo, Co-W, Co-Mo-W (their composition was selected by the EDX method – Energy Dispersive X-ray Analysis) and alloys well absorbing hydrogen such as Co-Pd was obtained. CMF catalyzed the increase of the reaction rate of hydrogen generation, hydrogen probably being the main ecological energy source of the future. This finding was confirmed by SEM (Scanning Electron Microscopy). The influence of CMF on so-called hydrogen corrosion of metals was also established.

Keywords: Hydrogen; Alloys; Constant magnetic field; Lorentz force

Introduction

Reactions of hydrogen formation may be beneficial for the natural environment and for industrial applications. However, they may also lead to some undesirable results. It is common knowledge that hydrogen is supposed to become a basic fuel by the end of the twenty first century. Nowadays, the source of hydrogen production is first and foremost petroleum (50%), natural gas (30%), and coal (15%). Only 0.5% of hydrogen is obtained from water electrolysis even though water belongs to renewable energy sources, along with the wind, solar radiation and thermal water [1].

Electrolytic water decomposition has been known since 1800 (Nicholson, Garlishe):

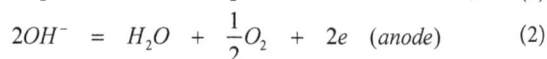

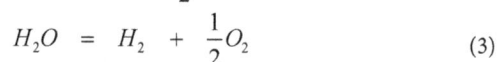

$$2H_2O \; + \; 2e \; = \; H_2 \; + \; 2OH^- \quad (cathode) \qquad (1)$$

$$2OH^- \; = \; H_2O \; + \; \frac{1}{2}O_2 \; + \; 2e \quad (anode) \qquad (2)$$

$$H_2O \; = \; H_2 \; + \; \frac{1}{2}O_2 \qquad (3)$$

High – temperature water electrolysis increases the process efficiency by decreasing the value of Gibbs free energy (ΔG), increasing the rate and number of reactions proceeding on the electrodes. It allows applying a higher current density, decreasing heat demand (by the amount of evaporation heat, as water vapor does no longer evaporate).

Hydrogen is an ecological fuel (the combustion product is water) and is easily ignited, contributing to more effective combustion. Hydrogen is easier and cheaper to store than electric energy. Its resources are unlimited. At present, its only drawback is the fact that it diffuses through metals, though this may also be turned into an advantage and used for storage.

Hydrogen can also be evolved by electrolysis through adsorption of its atoms on metal surface [2]:

$$H^+ \; + \; M \; + \; e \; \rightarrow \; MH_{ads} \quad (acidic \; solutions) \quad (4)$$

$$H_2O \; + \; M \; + \; e \; \rightarrow \; MH_{ads} \; + \; OH^- \quad (neutral \; and \; alkaline \; solutions)$$
$$\qquad (5)$$

where: M - metal atom, H_{ads} - adsorbed hydrogen atoms.

The majority of adsorbed hydrogen atoms are subjected to electrochemical desorption, and the generated molecules of H_2 dissolve in the solution or are removed from it in the form of gas bubbles. However, part of the adsorbed hydrogen atoms may permeate to the metal and become absorbed (Figure 1) [2]:

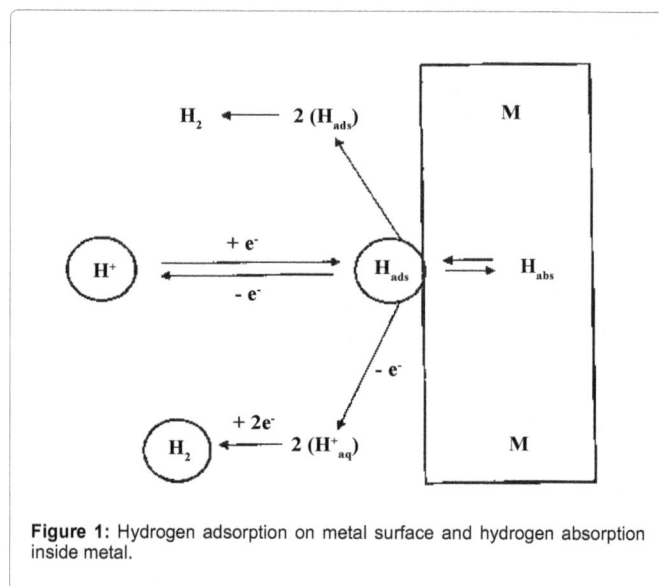

Figure 1: Hydrogen adsorption on metal surface and hydrogen absorption inside metal.

*Corresponding author: Marek Zieliński, Department of Inorganic and Analytical Chemistry, Faculty of Chemistry, University of Lodz, Poland
E-mail: zielmark@chemia.uni.lodz.pl

$$MH_{ads} \rightarrow MH_{abs} \qquad (6)$$

where: MH_{abs} - hydrogen atom absorbed inside metal.

Hydrogen may originate from the gaseous phase as well. Molecules of H_2, H_2S and others are capable of so-called dissociative chemisorption. Forces of chemisorptive interaction of molecules (H_2, H_2S) with transition metal surface (Fe, Cr, Ni, Mo, W) overcome the mutual attraction of atoms in molecules, resulting in dehydrogenation and adsorption of hydrogen atoms on metal surface as follows:

$$H_2 + 2M \rightarrow 2MH_{ads} \qquad (7)$$

$$M_nX + 2M \rightarrow MH_{(n-1)}X + MH_{ads} \qquad (8)$$

where: X - anion

It was established that effective metal thickness involved in the process of adsorption is composed of several atomic layers [3,4]. Further on, a reaction according to Equation (6) may proceed.

Metallic systems capable of hydrogen absorption have found various applications in the synthesis of many transition metal combinations and alloys as catalysts in hydrogenation reactions, as membranes serving for hydrogen separation from other gases through diffusion, and for hydrogen isotope separation (the source of high purity hydrogen obtained through thermal decomposition of hydrides) [5-9]. Metallic hydrides are an object of common interest as they are used in hydrogen storage. A metal that easily absorbs hydrogen both from the gaseous phase and electrochemically from solutions is palladium (it may dissolve a hydrogen volume 850 times greater than its own volume) [10-13]. Hydrogen may be present in palladium in two phases, preserving the crystalline structure of the pure metal but characterized by an increased value of the lattice constant. At small hydrogen concentrations, a solid hydrogen solution in palladium, called the α-phase, is generated. When the quantity of absorbed hydrogen increases, phase β, which is a non-stoichiometric palladium hydride, commences to form. A particular role in the process of absorption is played by hydrogen located directly under the surface of palladium (a thickness of several hundred atomic layers), displaying properties different from those of the absorbed hydrogen. In this layer, the concentration of hydrogen exceeds its concentration in phases α and β [14].

Compared to pure palladium, its alloys display slightly different properties in the process of hydrogen absorption. This may be explained by electron effects (changes in the electron structure after alloy formation) or by geometrical effects (changes in crystalline network parameters) [15]. Based on the electrochemical properties connected with hydrogen absorption such as the potential of absorbed hydrogen oxidation peak and the potential of α–β phase transition, it is possible to determine the volumetric composition of palladium alloys by electrochemical method [16]. The catalytic activity of metals in the reaction of hydrogen formation may be determined from the value of exchange current density (j_o). The higher the energy of the bond M – H, the more a given metal (M) is involved in the process of hydrogen generation catalysis and the smaller is the overpotential of its production. There is a relationship between the overpotential of hydrogen formation (η) and free enthalpy of hydrogen adsorption (H_{ads}) on various metals. The greater the enthalpy, the lower the overpotential of hydrogen formation is noticed for a given metal. The relationship between the overpotential of hydrogen formation (η)

and current density (j) may be defined with an equation connected to exchange current density (j_o):

$$\eta = [(R \cdot T)/(\alpha \cdot n \cdot F)]\log(j/j_0) \qquad (9)$$

where: R – gas constant [8,314 J/(K·mol)], j – limiting current density [A/cm²], j_o– exchange current density [A/cm²], T – temperature [K], α– transfer coefficient [-],

n – number ofelectrons participating in the reaction, F – Faraday constant [96485 C/mol].

Exchange current density (j_o) is a measure of hydrogen formation rate on metal (a metal electrode). The relationship between exchange current density logarithm (log j_o) in the process of hydrogen formation and free enthalpy of hydrogen adsorption on electrode surface (H_{ads}), i.e. the force of the metal-hydrogen bond made it possible to plot a so-called volcanic curve (Figure 2).

Depending on the type of metal, the stability of M–H bonds varies. This is demonstrated in the values of free enthalpy of hydrogen atom adsorption. Hence, it is concluded that the kinetics and mechanism of hydrogen formation on metallic electrodes is influenced by the type of metal. At the top of the curve, there are noble metals (Pt, Pd), being the most active in the process of hydrogen formation. Hydrogen is strongly adsorbed on the surface of such metals, which are characterized by the low overpotential of hydrogen formation (<0.2 V). Apart from the H - H interactions, hydrogen bonds with metal are formed. Transition metals are on the ascending part of the curve (Co, Ni, Fe) they form weak metal-hydrogen bonds, whereas metals strongly adsorbing hydrogen and being difficult to melt (Mo, W, Ti) are on the descending part of the curve. Those metals are characterized by average values of hydrogen formation overpotential (0.2–0.6 V) and by average values of exchange current of hydrogen formation reaction on those metals. There is no pure metal which would completely fulfill the electro-catalytic requirements. Thus, it is essential to search for alloy materials that would have more beneficial properties than noble metals. Metals and alloys adsorbing a lot of hydrogen (Co-Mo, Ni-Mo, Co-W, Ni-W) are characterized by high electrocatalytic activity in the process of hydrogen formation [17]. The adsorbed hydrogen considerably alters the electron structure of the electrode material [18]. According to the Brewery and Engel valence bond theory, catalytic activity in the process of hydrogen formation increases linearly as a function of metal-hydrogen bond force for alloys formed from transition metals. Furthermore, another theory of hydrogen formation on transition metal alloys is based on the electron structure of alloys in terms of the

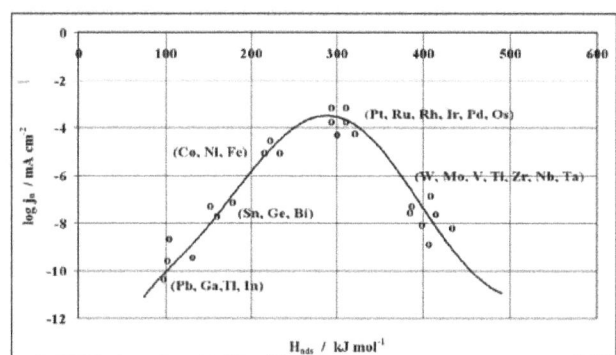

Figure 2: Relationship between the logarithm of exchange current density (log j_o) in the process of hydrogen formation on various metals and free enthalpy of hydrogen adsorption (H_{ads}).

direction of electron transfer between atoms located in the alloy [19]. The combination of various d metals in an alloy may surpass platinum in respect of catalytic activity in the process of hydrogen production. Co-Mo, Ni-Mo, Co-W, and Ni-W alloys, where the pairing up effect provides a greater number of available d orbitals and ensures better proton adsorption, surface adherence and surface transport. It was proved that the highest electrocatalytic activity is displayed by alloy combinations of metals from the dgroup characterized by minimal entropy and easier overlapping of d orbitals, which are formed in intermetallic compounds of the highest symmetry (e.g. $MoCo_3$, WCo_3, $MoNi_3$, WNi_3).

While striving for the generation of the greatest quantity of hydrogen possible through various reactions and its safe storage, and despite the fact that it is regarded as the ecological energy source of the future, it needs to be remembered that hydrogen production may be noxious as well.

Both atomic and molecular hydrogen has a harmful influence on all metals and alloys. Several atomic parts of hydrogen per million parts of metal or alloy increase its embrittlement. The metal microstructure decreases and the value of stresses increase under the influence of hydrogen. So-called hydrogen embrittlement is ascribed to atomic hydrogen diffusion in a metal or alloy and its accumulation in spaces or internal surfaces, where molecular hydrogen in the form of gas is produced. The higher the hydrogen concentration, the greater the pressure of gaseous molecular hydrogen, contributing to void expansion and cracking. Atomic hydrogen evolving on metal or alloy surfaces is adsorbed. Diffusing adsorbed hydrogen may evolve as molecular hydrogen in internal voids or on the boundary 'inclusions – matrix'. In the case of high hydrogen penetration rate, the pressure generated by molecular hydrogen may cause the formation of bubbles (so-called hydrogen cracks).

The influence of a constant magnetic field on alloy electrodeposition was investigated earlier [20-22]. The impact of the Magneto-Hydrodynamic Force (MHD) on the convection of the solution in a three-electrode layer during metal and alloy deposition was established as well [23,24]. Furthermore, the influence of a constant magnetic field on hydrogen desorption on the surface of metallic electrodes was also examined [25,26]. In this case, the hydrogen desorption rate increased whereas the size of hydrogen bubbles decreased. The study presented concerned the influence of a constant magnetic field on the kinetics of electrodeposition reaction of two-component alloys (Co-Mo, Co-W, Co-Pd) and three-component alloys (Co-Mo-W) and the concurrent processes of hydrogen formation [20,27,28].

Experimental

Cyclic Voltammetry (CV), Scanning Electron Microscopy (SEM) and Energy Dispersive X-ray Analysis (EDX) were used. A three-electrode system was used consisting of the following components: a working (gold, disk) electrode of 0.1 cm2 surface area, an auxiliary platinum mesh electrode of much larger area and a saturated calomel electrode as a reference electrode. The measuring system was located between the pole pieces of an N–S pole electromagnet (Figure 3).

The Constant Magnetic Field (CMF) applied for investigation of magnetic induction B was from 0 to 1.2 T. Vector B was parallel to the surface of the working chamber. The electrolyte applied (galvanic bath) (depending on the alloy) was composed of: 0.27 mol/dm³ $CoSO_4 \cdot$ $7H_2O$; 0.07 mol/dm³ $Na_2MoO_4 \cdot 2H_2O$; 0.06 mol/dm³ $Na_2WO_4 \cdot 2H_2O$; 0.06 mol/dm³$(NH_4)_2$ PdCl4; 0.6 mol/dm³ sodium citrate ($Na_3C_6H_5O_7 \cdot$ $2H_2O$); 0.05 mol/dm³ H_3BO_3; 0.025 mol/dm³ EDTA ($C_{10}H_{14}O_8N_2Na_2 \cdot$ $2H_2O$) and 0.1 mol/dm³H_2SO_4 (Table 1).

The alloys were deposited for 3500s. Investigations by the CV method were performed on an apparatus composed of the ER-2505 laboratory electromagnet with pole piece N and S, measurement vessel with a three-electrode electrochemical system, a device controlling the electromagnet PZP – 80, an SZP stable electromagnet current source, a hallotrone sensor of constant magnetic field, a TH-26 hallotroneteslometer, a CMR-02 digital temperature regulator, a set for electrochemical measurements ATLAS (9933 Electrochemical Interface, 9923 Frequency Response Analyzer) and a computer (with POL-99 and IMP-99). SEM investigations were carried out using an electron scanning microscope Vega 5135 MM produced by Tescan. In the EDX method, an X-ray micro-analyzer EDX Link 300 ISIS, produced by Oxford Instruments, was used to investigate the chemical composition of the alloy (in terms of quality and quantity).

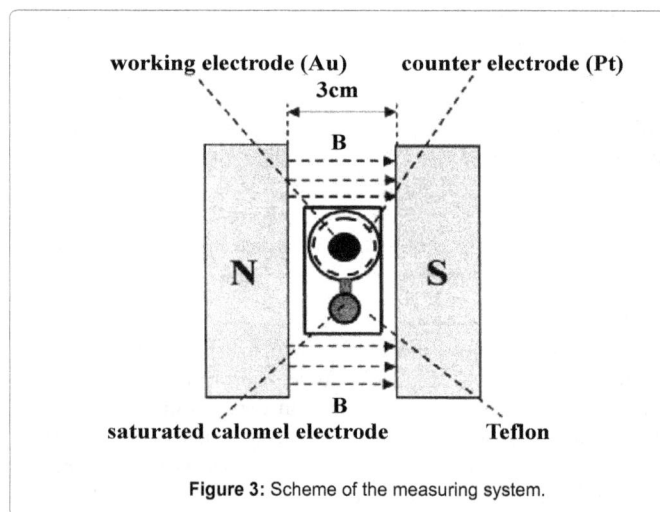

Figure 3: Scheme of the measuring system.

Chemical compound	Molar mass(g/mol)	Producer	Concentration in solution (mol/dm³)	Alloys Co-Mo	Co-W	Co-Mo-W	Co-Pd
$CoSO_4 \cdot 7H_2O$	281.12	Fluka (Germany)	0.27				
$Na_2MoO_4 \cdot 2H_2O$	241.98	POCH S.A., Gliwice (Poland)	0.07		-		-
$Na_2WO_4 \cdot 2H_2O$	329.86	BHD Chemicals Ltd (England)	0.06	-			-
$(NH_4)2PdCl_4$	284.22	POCH S.A., Gliwice (Poland)	0.06	-	-	-	
$C_6H_5O7Na_3 \cdot 2H_2O$ (sodium citrate)	294.12	POCH S.A., Gliwice (Poland)	0.6				
$C_{10}H_{14}O_8N_2Na_2 \cdot 2H_2O$ (EDTA)	372.24	ZOCH Lublin (Poland)	0.025				
H_3BO_3	61.84	POCH S.A., Gliwice (Poland)	0.05				
H_2SO_4 (95%, d=1.84 g/cm3)	98.08	"Chempur", Piekary Śląskie (Poland)	0.1				

Table 1: The chemical compounds used for alloys electrochemical preparing.

Results and Discussion

Studying the kinetics of electrochemical redox processes exposed to CMF, it was observed that a constant magnetic field may be an additional parameter controlling electrochemical processes. Currently, scientists are also interesting in investigating the processes of electrochemical hydrogen production. It was found that CMF increased the hydrogen production rate at a temperature of T=298 K. Applying Cyclic Voltammetry (CV), curves with characteristic current peaks of hydrogen formation rate on alloys deposited on golden electrodes were obtained. CFM was applied in the range of magnetic induction (B) from 0 to 1200 mT. The constant of hydrogen formation reaction rate (k) were calculated, and the function was plotted (Figure 4).

Increased magnetic induction (B) led to higher constants of the hydrogen formation reaction rate (k). CMF affected both electrons and ionized atoms. Dynamic effects occurred, one of which was volumetric movement of the medium (movement of the electrolyte). Movement of the electrolyte in a diffusion layer under the influence of the Lorentz force (F) defined with the general equation:

$$F = j \times B \qquad (11)$$

where: j - limiting current density vector; B—magnetic induction vector,

as well as other MagnetoHydrodynamic Effects (MHD) more frequently activated the surface of the electrode by removing hydrogen bubbles (H_2) more rapidly. Therefore, hydrogen bubbles were smaller and new hydrogen ions (H^+) could discharge more promptly on the surface of the electrode. Both natural and MHD convections macroscopically stir the bulk electrolyte, whereas growth and detachment of gas bubbles induce microscopic convections on the electrode surface. In addition to macroscopic MHD forced convection, micro-MHD convection among gas bubbles might be induced by the sophisticated electric field. When the gas bubbles are produced very actively at more than 50 mA/cm^2, the mass transfer process can be significantly enhanced by the superimposition of magnetic field (the conditions of the experiment of 200mT). The unique convection induced adjacent to the bubble nucleation and growth site probably helps agitating the electrolyte near the electrode to drastically decrease the supersolubility of dissolved H2 gas in a magnetic field. As mentioned in the Introduction, alloys (e.g. cobalt with molybdenum or tungsten) work better as electrodes in the processes of hydrogen formation than single metals. Molybdenum and tungsten are metals that strongly adsorb hydrogen, and moreover they

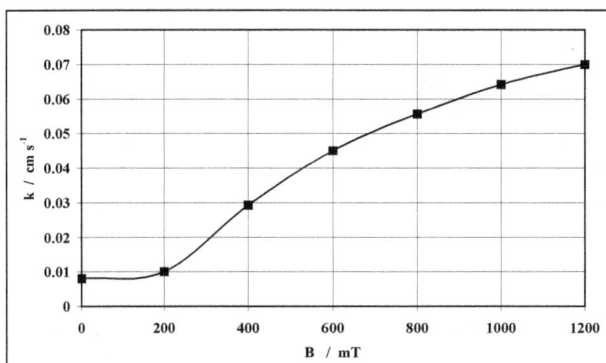

Figure 5: Relationship between current density (j) in hydrogen formation on golden electrodes coated with Co-Mo or Co-Mo-W and magnetic induction (B).

are mechanically stable. Two-component alloys of Co-Mo and Co-W, which are characterized by a low overpotential of hydrogen formation (η), displayed high activity in these processes. The three-component alloy of Co-Mo-W, characterized by an even lower overpotential of hydrogen formation than two-component alloys, was obtained at the potential of -1.17 V$_{SHE}$. With the EDX method, it was established that in the obtained alloy of Co-Mo-W there was 84.65% w/w of cobalt, 11.90% w/w of molybdenum and 3.45% w/w of tungsten. The process was carried out without a magnetic field. When electrodes covered with such alloys were transferred to CMF and subjected to the processes of hydrogen formation by the CV method, the relationship of current density (j) and magnetic induction (B) was established (Figure 5). The increase in the alloy of component, resulting in lower overpotential of hydrogen formation (η) (e.g. Pd, W, Mo), causes an increase the hydrogen evolution. Reduction overpotential of hydrogen formation (η), is a consequence higher energy of metal-hydrogen bond (e.g. W-H), which catalyzes the process of evolution of hydrogen [higher values the exchange current density (j$_o$)].

Equation (12) defining this relationship is as follows: [20]:

$$j \approx 0{,}63\pi^{1/6} \cdot A^{-1/6} \cdot \rho^{-1/3} \cdot D^{8/9} \cdot v^{-2/9} \cdot (nFc)^{4/3} \cdot B^{1/3} \quad (12)$$

where: j – current density [mA/cm2], n – number of electrons participating in the reaction, A – area of the working electrode [m^2 or cm^2], D – diffusion coefficient [cm^2/s], v – kinematic viscosity [m^2/s or St=cm^2/s], c – concentration of electroactive ions in the electrolyte [mol/dm^3], B – magnetic induction [T or mT], ρ– electrolyte density [g/cm^3], F – Faraday constant [96485 C/mol].

The influence of CMF was also displayed by the decreasing thickness of a diffusion layer (δD) which may be defined by the following equation (13) [20]:

$$\delta_D \approx 1{,}59(\rho \cdot r \cdot v^{2/3} \cdot D^{1/3})^{1/3} \cdot (n \cdot F \cdot c \cdot B)^{-1/3} \quad (13)$$

where: r - radius of the working disc electrode [mm].

Subsequently, an alloy of Co-Pd was obtained electrochemically. One of its components, namely palladium, located at the top of the volcanic curve (Figure 2), is a noble metal of a low overpotential of hydrogen evolution, and is very active in the process.

The Co-Pd alloy was obtained at a temperature of T=298 K, at the potential of U=-1.05VS$_{HE}$. Based on the EDX method it was established that after a redox reaction and without a magnetic field, a

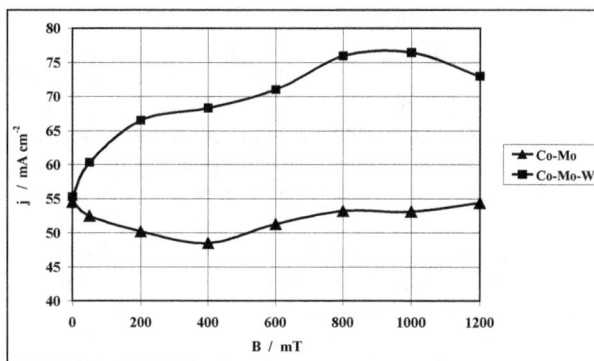

Figure 4: Relationship between the constant of hydrogen formation reaction rate (k) and magnetic induction (B).

solution with 0.27 mol/dm3 of Co^{2+} and 0.06 mol/dm^3 of Pd^{2+} yielded an alloy of Co-Pd composed of 75.2% w/w of cobalt and 24.8% w/w of palladium. Apart from adsorbing hydrogen, the alloy displayed the capability of hydrogen absorption which could be a perfect method of its storage. Based on the peak of absorbed hydrogen oxidation, the quantity of this element could be determined in the alloy of Co-Pd using the CV method. Furthermore, the energy of adsorption changed with the potential applied. Thus, it may be concluded that controlling the electrode potential, the alloy may be coated with hydrogen up to a defined extent, or its total desorption may be achieved. SEM observations (at magnification x100) proved that CMF contributed to an increase in stresses in the structure when the alloy Co-Pd was formed, and some change in the surface microstructure occurred (Figure 6).

As previously mentioned, the increase of absorbed hydrogen in the Co-Pd alloy resulted in a higher alloy embrittlement (decreasing its mechanical strength). Faster hydrogen penetration into the alloy increased its internal pressure resulting in the formation of bubbles and cracks. This occurred most intensively in the palladium subsurface layer, where hydrogen concentration was even higher than in phases α and β, in which most of the hydrogen is adsorbed. For better visualization of CMF influence on hydrogen formation reactions in the Co-Pd alloy, Co-Pd alloy electrodeposition was performed at a potential of U=- 1.05 V_{SHE}, with magnetic inductions of B=0, B=600 mT and B=1200 mT, but at a temperature of T=323 K. Photographs were taken using a SEM at a magnification of X500 (Figure 7).

It may be observed that both the temperature and CMF accelerated the reaction of hydrogen formation. Obviously, this had a negative impact on the surface of the material (so-called hydrogen corrosion), but the advantage of the applied conditions was the increase of the hydrogen volume evolution. As hydrogen is intended to be the main ecological energy source of the future, every possibility of increasing its resources certainly represents a positive aspect of the actions undertaken.

Conclusions

A constant magnetic field (CMF) may be a parameter increasing the efficiency of the formation of hydrogen, which is believed to be the main ecological energy source of the future. This is confirmed by an

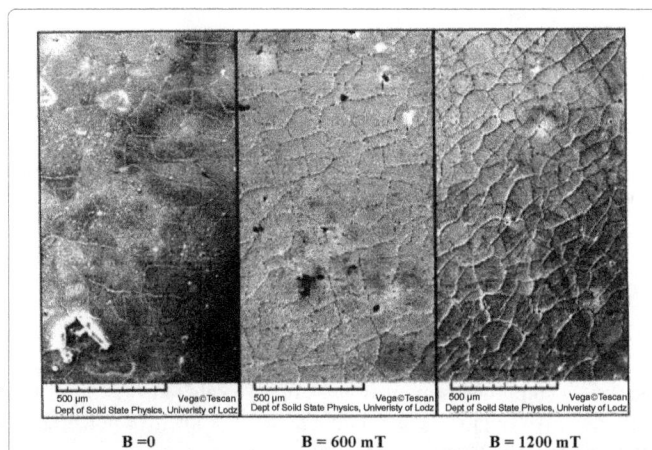

Figure 7: SEM images of Co-Pd alloys (Co=75.2% and Pd=24.8%) obtained in CMF at different magnetic induction at T=323 K. The photographs were taken using a Vega 5135 MM scanning electron microscope produced by Tescan (magnification X500).

increase in the reaction rate constant (k) under the influence of CMF as determined with the CV method. Appropriately selected compositions of galvanic solutions and alloys well adsorbing hydrogen (Co-Mo, Co-W, Co-Mo-W), and additionally absorbing hydrogen (Co-Pd), determined with the EDX method, resulted in an increased reaction rate of hydrogen formation under the influence of CMF. While conducting investigations at two different temperatures, SEM imaging proved that, either temperature or CMF increase the reaction rate of hydrogen formation. This was observed the formation of bubbles and cracks of various sizes on the surface of the alloys.

Acknowledgement

This work was supported by the Lodz University.

References

1. David Hirsch (2003) Decarbonization of fossil fuels: Hydrogen production by the solar thermal decomposition of natural gas using a vortex – flow solar reactor.

2. Andrez Czerwiński (1994) The adsorption of carbon oxides on a palladium electrode from acidic Solution, J Electroanal Chem 379: 487-493.

3. DAJ Rand, R Woods (1972) Determination of the surface composition of smooth noble metal alloys by cyclic voltammetry. J Electroanal Chem Interfac Electrochem 36: 57-69.

4. DAJ Rand, R Woods (1973) Electrosorption characteristics of thin layers of noble metals electrodeposited on different noble metal substrates. J ElectroanaL Chem Interfac Electrochem 44: 83-89.

5. Z Jusys, H Massong, H Baltruschet (1999) A New Approach for Simultaneous DEMS and EQCM: Electro-oxidation of Adsorbed CO on Pt and Pt-Ru. J Electrochem Soc 146: 1093-1098.

6. N Furuya, T Yamazaki, N Shibata (1997) High performance Ru-Pd catalysts for CO2 reduction at gas-diffusion electrodes. J Electroanal Chem 431: 39-41.

7. JM Leger (2005) Preparation and activity of mono- or bi-metallic nanoparticles for electrocatalytic reactions. Electrochim Acta 50: 3123-3129.

8. T Iwasita (2002) Electrocatalysis of methanol oxidation. Electrochim Acta 47: 3663-3674.

9. SS Gupta, J Datta (2006) A comparative study on ethanol oxidation behavior at Pt and Pt Rh Electrodeposits. J Electroanal Chem 594: 65-72.

10. I Kleperis, G Wójcik, A Czerwiński, I Skowroński, M Kopczyk et al. (2001) Electrochemical behavior of metal hydrides. J Solid State Electrochem 5: 229-249.

11. FA Lewis (1982) A survey of hydride formation and the effects of hydrogen contained within the metal lattices. Plat Met Rev 26: 20-27.

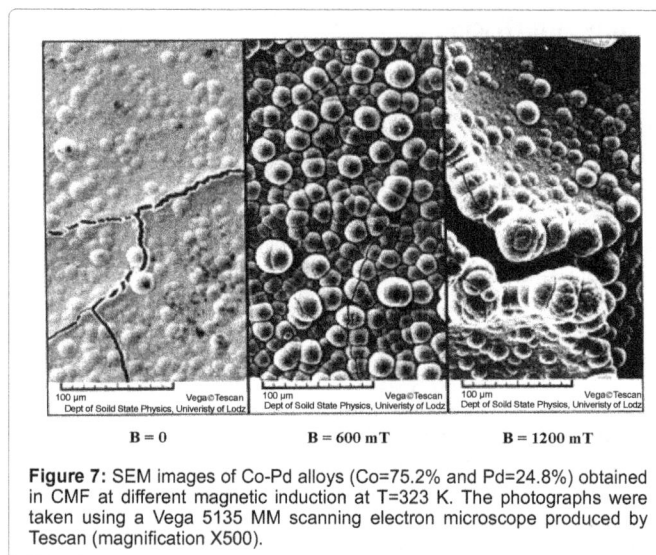

Figure 6: SEM images of Co-Pd alloys (Co=75.2% and Pd=24.8%) obtained with and without a constant magnetic field at a temperature of T=298 K. The images were made using a Vega 5135 MM scanning electron microscope produced by Tescan (magnification X100).

12. Y Sakamoto, K Kajihara, E Ono, K Baba, TB Flanagan (1989) Hydrogen solubility inpalladium-vanadium alloys. Zeitschrift für Physikalische Chemie 165: 67-81.

13. Y Sakamoto, FL Chen, M Ura, TB Flanagen (1995) Thermodynamic properties for solution of hydrogen in palladium-based binary alloys. Ber Bunsenges Phys Chem 99: 807-820.

14. RV Bucur, F Bota (1982) Galvanostatic desorption of hydrogen from palladium layers-2. the transfer process. Electrochim Acta 27 (1982) 521-528.

15. S Thiebaut, A Bigot, JC Achard, B Limacher, D Leroy et al. (1995) Structural and thermodynamic properties of the deuterium-palladium solid solutions systems: De-[Pd(Pt), Pd(Rh), Pd(Pt, Rh)]. J Alloys Comp 231: 440-447.

16. M Łukaszewski, A Żurowski, M Grdeń, A Czerwiński (2007) Correlations between hydrogen electrosorption properties and composition of Pd-noble metal alloys. Electrochem. Commun 9: 671-676.

17. MM Jaksic (1984) Electrocatalysis of hydrogen evolution in the light of the brewer-engel theory for bonding in metals and intermetallic phases. Electrochim Acta 29: 1539-1550.

18. I Paseka (1995) Evolution of hydrogen and its sorption on remarkable active amorphous smooth Ni-P(x) electrodes. Electrochim Acta 40: 1633-1640.

19. I Paseka, I Felicka (1997) Hydrogen evolution and hydrogen sorption on amorphous smooth Me-P(x) (Me=Ni, Co and Fe-Ni) electrodes. Electrochim Acta 42: 237-242.

20. M Zieliński, E Miękoś (2008) Influence of constant magnetic field on the electrodepositionof Co–Mo–W alloys. J Appl Electrochem 38: 1771-1778.

21. M. Zieliński (2013) Effects of constant magnetic field on the electrodeposition reactions and cobalt-tungsten alloy structure. Mat Chem Phys 141: 370-377.

22. M. Zieliński (2013) Influence of Constant Magnetic Field on the Electrodeposition of Cobalt and Cobalt Alloys. Int J Electrochem Sci 8: 12192-12204.

23. H Matsushima, T Nohira, I Mogi, Y Ito (2004) Effects of magnetic fields on iron electrodeposition. Surf Coat Technol 179: 245-251.

24. K Msellak, IP Chopart, O Ibara, O Aaboubi, I Amblard (2004) Magnetic field effects on Ni–Fe alloys codeposition. J Magn Magn Mater 281: 295-304.

25. Bund A, Kuehnlein HH (2005) Role of Magnetic Forces in Electrochemical Reactions at Microstructures. J Phys Chem B 109: 19845-19850.

26. A Bund, S Kohler, HH Kuehnlein, W Plieth (2003) Magnetic field effects in electrochemical reactions. Electrochim Acta 49: 147-152.

27. IA Koza, S Mullenhoff, M Uhlenmann, K Ecklert, A Gebert et al. (2009) Desorption of hydrogen from an electrode surface under influence of an external magnetic field – in– situ microscopic observations. Electrochem Commun 11: 425-429.

28. IA Koza, M Uhlenmann, A Gebert, L Schultz (2008) Desorption of hydrogen from theelectrode surface under influence of an external magnetic field. Electrochem Commun 30: 1330-1333.

3

Control of Morphology and Acidity of SAPO-5 for the Methanol-To-Olefins (MTO) Reaction

Kazusa Terasaka, Hiroyuki Imai* and Xiaohong Li

Faculty of Environmental Engineering, The University of Kitakyushu, 1-1 Hibikino, Wakamatsu, Kitakyushu, Fukuoka, Japan

Abstract

Silicoaluminophosphate (SAPO) zeotype materials, a family of zeolites with micropores, have moderate acid strengths compared with conventional aluminosilicate zeolites; furthermore, their acid amounts can be tuned by the incorporation of Si species into the aluminophosphate (AlPO) framework. The conversion of methanol to light olefins including ethene, propene, and butenes (isobutene, 1-butene, and 2-butenes), methanol-to-olefins (MTO) reaction, is carried out over zeolites as an acid catalyst. In the MTO reaction, the enhancement of the diffusivity of reactants and products and the tuning of the acidity of zeolites are crucial keys to the improvement of the catalyst life due to the suppression of the coke deposition in the pores. In the present study, we have focused on the improvement of the catalytic performance of SAPO-5 materials with the AFI structure with large micropores of 0.73 nm apertures in the MTO reaction. Highly crystalline SAPO-5 with different morphologies and acidities were readily synthesized by merely varying the concentration of the starting gel. The employment of a highly concentrated starting gel with a H_2O/Al ratio of 5 led to the formation of smaller-sized SAPO-5 crystallites with a larger amount of mild acid sites compared with SAPO-5 synthesized with the conventional compositions with the H_2O/Al ratio of 50. The catalytic performance of the synthesized SAPO-5 materials as an acid catalyst was evaluated in the MTO reaction. The crystallite morphology as well as the acid amount scarcely affected the initial activity and product distribution, while the catalyst life was considerably affected. The decrease in the crystallite size of SAPO-5 led to improving the catalyst life due to the improvement of the resistance to the coke deposition.

Keywords: AFI-type silicoaluminophosphate; SAPO-5; Crystallite size; Acidity; Methanol-to-olefins (MTO) reaction; Gel concentration

Introduction

Accompanying an increase in worldwide energy consumption, the utilization of carbon resources alternative to crude oil, such as biomass, coal and natural gas, has been highly desired for the production of chemicals. Methanol is one of attractive materials as a sustainable feedstock for the production of increasingly-demanded hydrocarbons in industrial processes because methanol can be industrially manufactured by the conversion of syngas obtained through the gasification of various carbon resources, and be readily converted to hydrocarbons over solid acid catalysts. Thus, the conversion of methanol to light olefins including ethene, propene and butenes has attracted much attention in recent decades [1]; so-called "methanol-to-olefins (MTO) reaction". The MTO reaction is generally carried out over zeolites or zeotype materials with strong Brønsted acid sites and micropores as well as internal cavities. Among the zeolite catalysts, silicoaluminophosphate zeotype material SAPO-34 with the CHA structure, which is composed of a 3-dimensional pore system with 8-membered ring (8-MR) openings (0.38 nm × 0.38 nm) [2], is well-known as an excellent catalyst for the MTO reaction; it provides high yields of light olefins with a long catalyst life due to moderate acid strength and the limitation of the size of the products imposed by the small pores [1,3-7].

It has been proposed that the formation of hydrocarbon pool species, alkylated aromatic compounds such as hexamethylbenzene, in the cages of SAPO-34 should be required for the production of light olefins during the MTO reaction [8-11]. Meanwhile, the formation of the bulky aromatic compounds in the cage leads to the formation of coke through the further polymerization to cause the rapid deactivation due to the regulation of the diffusion of reactants as well as products [12]. Therefore, improving the resistance to the coke deposition is required for the utilization of zeolite catalysts for a longer period. The catalyst life is strongly influenced by the crystallite

size of zeolites. In fact, the catalyst life of SAPO-34 can be improved by decreasing the crystallite size without varying the acid amount due to the improvement in the mass transfer inside the pores of the catalyst [7,13]. The synthesis of smaller-sized SAPO-34 compared with SAPO-34 synthesized by the conventional method has been attained by using various approaches, such as dry-gel method and microwave-assisted heating [14-22]. These smaller-sized SAPO-34 catalysts exhibited a longer catalyst life compared with the conventional SAPO-34 catalyst in the MTO reaction. A facile method for preparing the nano-sized crystallites of ZSM-5 was developed by enhancing the nucleation by mixing the starting gel under moderate conditions before the crystallization [23]. Furthermore, the control of the crystallite size of ZSM-5 can be attained by merely changing the water content in the starting amorphous gel. The nano-sized ZSM-5 exhibited an excellent catalytic performance in the hexane cracking and higher resistance to the coke deposition compared with conventional micro-sized ZSM-5 catalysts. The facile method with controlling the gel concentration is expected to be applied to the synthesis of a variety of zeolite catalysts to form smaller-sized crystallites for improving catalytic performances.

Recently, silicoaluminophosphate zeotype material with the AFI topology, SAPO-5, has been used as a model catalyst for investigating detailed mechanisms of the MTO reaction. SAPO-5 consists of a

---

*Corresponding author: Hiroyuki Imai, Faculty of Environmental Engineering, The University of Kitakyushu, 1-1 Hibikino, Wakamatsu, Kitakyushu, Fukuoka 808-0135, Japan, E-mail: h-imai@kitakyu-u.ac.jp

1-dimensional pore system with 12-membered ring (12-MR) openings (0.73 nm × 0.73 nm) [2], whose diameter is similar to that of the cavities of SAPO-34. Thus, bulky compounds would be capable of diffusing in and out of the catalyst through the large pore openings. In fact, comparing SAPO-5 with SAPO-34 in the MTO reaction, large-sized alkenes were preferentially formed as an intermediate molecule in the micropores of SAPO-5 to produce mainly butenes, in particular isobutene, through the decomposition of the intermediate alkenes [24,25]. By contrast, ethene and propene were selectively produced through the formation of bulky aromatic compounds over SAPO-34 [26]. Since even light olefins can be distinguished by the size of micropore aperture, SAPO-5 has an advantage over SAPO-34 in respect of the production of butenes from methanol. Furthermore, the absence of cavities and intersections in SAPO-5 facilitates the estimation of the space available for the formation of intermediates and products in the MTO reaction. In addition to the control of the product distribution, the large pore openings of SAPO-5 would be advantageous to suppressing the coke deposition due to lower diffusion barriers of molecules in comparison with zeolites with small pores. Meanwhile, the diversification of SAPO-5 by tuning the acidity and/or the crystallite morphology and size has not been well-studied for improving the catalytic performance in the MTO reaction. It is expected that the investigation of the catalytic properties of SAPO-5 with different physicochemical properties would lead to the promotion of the potential of SAPO-5 as an acid catalyst in the MTO reaction.

In the present study, we report on the synthesis of SAPO-5 catalysts with different morphologies as well as acidities. In particular, we aimed to develop a facile synthesis method for decreasing the crystallite size of SAPO-5. Then, the synthesized SAPO-5 catalysts were applied as an acid catalyst to the MTO reaction in order to investigate the catalytic performance of the SAPO-5 catalysts. In addition, we also investigated the effects of the crystallite morphology and acidities on the catalytic performance on the basis of the activity, the product distribution, and the catalyst life.

Experimental

Synthesis of SAPO-5 materials with different morphologies

The SAPO-5 materials with different morphologies were hydrothermally synthesized with water, Al(O^{i}Pr)$_3$ (Kanto Chem. Co.) as an Al source, orthophosphoric acid (H$_3$PO$_4$, 85%, Kanto Chem. Co.) as a P source, fumed silica (Aerosil 200, Aerosil) as a Si source and tetraethylamine (TEA, Tokyo Chem. Inc.) as an organic-structure-directing agent (OSDA). Al(O^{i}Pr)$_3$ was added to an aqueous solution containing H$_3$PO$_4$ and TEA to form a white suspension. Fumed silica was added to the mixture. The molar composition of the prepared amorphous gel was 1 Al$_2$O$_3$: 1 P$_2$O$_5$: 0.2 SiO$_2$: 1.0 TEA: 10-100 H$_2$O. The prepared gel was transferred into a Teflon-lined stainless steel vessel and hydrothermally treated at 473 K for 2 days with tumbling at 30 rpm. The obtained product was recovered by filtration, washed with deionized water, and dried at 363 K. Then, the final product of SAPO-5 was obtained by calcination of the as-synthesized sample at 823 K for 10 h for removing the OSDA. For the purpose of varying crystallite sizes of SAPO-5, 5 wt% SAPO-5 product, which was synthesized with a H$_2$O/Al ratio of 10, as a seed was added to the initial amorphous gel.

Characterization

XRD patterns were collected on a SmartLab (Rigaku) instrument using a Cu-Kα X-ray source (45 kV, 200 mA). Nitrogen adsorption-desorption measurements were conducted at 77 K on a BELSORP-mini II (MicrotracBEL Corp.) instrument. Prior to the measurement, the

sample was evacuated at 623 K for 2 h. The BET (Brunauer-Emmett-Teller) specific surface area was calculated from the adsorption data. External surface area was estimated by the t-plot method. Field-emission scanning microscopic (FE-SEM) images of the samples were obtained on an S-5200 microscope (Hitachi) operating at 1.0 kV. Elemental analyses of the samples were performed on an inductively coupled plasma-atomic emission spectrometer (ICP-AES, PerkinElmer). Ammonia temperature-programmed desorption (NH$_3$-TPD) profiles of the samples were recorded on a BELCAT (MicrotracBEL Corp.) apparatus. Typically, the sample was pretreated at 773 K in He (50 mL min^{-1}) for 1 h and then was cooled to 373 K. Then, 10% NH$_3$ in He was allowed to make contact with the sample for 30 min. Subsequently, the sample was evacuated to remove weakly adsorbed NH$_3$ for 15 min at 373 K. Finally, the sample was heated up to 773 K at a ramping rate of 10 K min^{-1} with the He flow (50 mL min^{-1}) passed through the reactor. A mass spectrometer was used to monitor desorbed NH$_3$ (m/e=16).

Methanol-to-Olefins (MTO) reaction

The MTO reaction was carried out in a 0.25 in. (OD) quartz tubular flow microreactor loaded with 100 mg of 50/80 mesh zeolite pellets without a binder. The catalyst was centered at a quartz reactor in a furnace. The catalyst was pretreated at 500°C for 1 h in the stream of N$_2$ prior to the reaction, and then the reactor was cooled to 450°C. The initial partial pressure of methanol was set at 2.6 kPa. N$_2$ gas was used as a carrier. The catalyst weight to the flow rate ratio (W/F$_{-methanol}$) was 67.5 g$_{cat}$ h (mol$_{-methanol}$)$^{-1}$, which corresponded to the weight hourly space velocity of methanol (WHSV) of 0.47 h^{-1}.

The reaction products were analyzed with an on-line gas chromatograph (Shimadzu GC-2014) with a flame ionization detector and a capillary column (HP-PLOT 30 m × 0.53 mm, 6 μm film thickness). The selectivities to the products were calculated based on the carbon numbers.

Results and Discussion

Synthesis of SAPO-5 with different morphologies

Figure 1 shows XRD patterns of the samples obtained through the hydrothermal treatment with the molar ratio of H$_2$O/Al varied from 5 to 50 in the starting gel and the following calcination. All the samples exhibited the XRD pattern typical of the AFI structure with high crystallinity, and showed almost the same XRD pattern although a small peak derived from another phase appeared at 22° in the pattern of the sample synthesized with the H$_2$O/Al ratio of 5. It is indicated that the concentration of the starting gel hardly affected the crystallinity of the silicoaluminophosphate zeotype material with the AFI topology. The crystallinity of the samples was further investigated by evaluating the intensities of the XRD diffraction lines attributed to the (100), (210), (002), and (102) planes of the AFI structure. The relative intensities of the diffraction lines of the samples are summarized in Figure 2. As the H$_2$O/Al ratio in the starting gel increased, the intensity of the diffraction line derived from the (002) plane increased in comparison with those from the (100), (210), and (102) planes. Moreover, the larger increase in the relative intensity of the (002) diffraction line to the (210) diffraction line was observed in comparison with that of (002) to (102). Since the 1-dimensional (1-D) pore system of the AFI structure is arranged along the c-direction [2], it is suggested that the pore length of SAPO-5 can be shortened through the crystallization with highly concentrated starting gels.

The morphology of the samples was evaluated by FE-SEM observations (Figure 3). All the samples, except for the sample

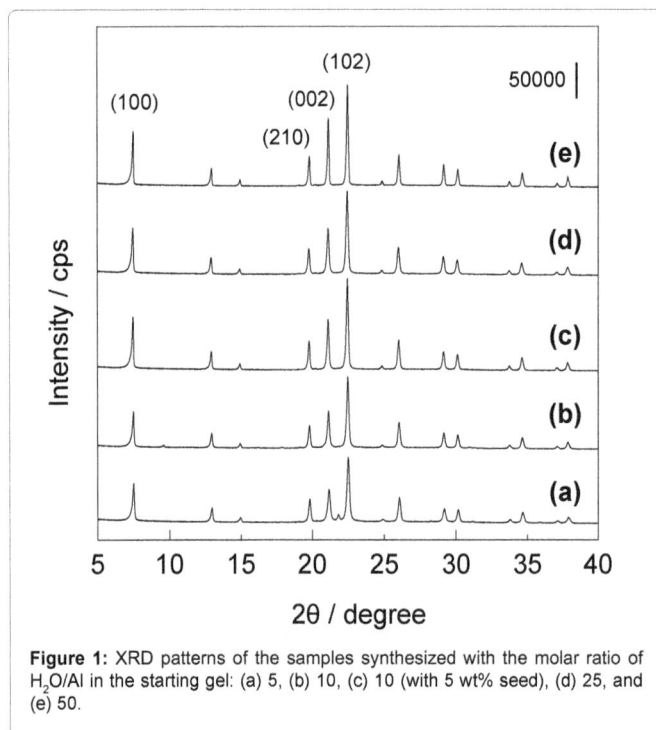

Figure 1: XRD patterns of the samples synthesized with the molar ratio of H_2O/Al in the starting gel: (a) 5, (b) 10, (c) 10 (with 5 wt% seed), (d) 25, and (e) 50.

5 crystallites in comparison with the crystallites synthesized in the absence of the seed crystallites even when there is no difference in the gel concentration.

The physicochemical properties of the samples synthesized with the various water contents are summarized in Table 1. The samples synthesized with the H_2O/Al ratios set at 5, 10, 25 and 50 are denoted by S(5), S(10), S(25) and S(50), respectively. The BET specific surface area of the samples was independent of the gel concentration although the external surface area was well-relevant to the crystallite size of the samples; the external surface area was increased with the decrease in the crystallites size. Since smaller-sized crystallites were agglomerated to form larger-sized crystallites, the differences in the BET surface areas may be derived from the heterogeneity of the crystallite size in the large agglomerates.

All the samples, except for S(50), contained almost the same compositions of Si, Al, and P; moreover, the Si/Al ratios in the samples were similar to those in the corresponding starting gel. By contrast, when the water content was increased to the H_2O/Al ratio of 50, the Si content in the sample was increased; simultaneously, the Al and P contents were relatively decreased. Roldán et al. have reported that in the synthesis of SAPO-5, Si species are incorporated into the AlPO framework through the substitution of one Si^{4+} for one P^{5+} in the starting gel with high pH, while silica islands can be formed as a result of the substitution of two Si^{4+} for one P^{5+} and one Al^{3+} in the starting gel with low pH [30]. Thus, it is suggested that Si species were incorporated into the AlPO framework in replace of P species, independent of the water content in the range of the H_2O/Al ratio of 5-25 due to the dissolution of Si species in the concentrated gel. In the case of low concentration of starting gels, pH in the gel would be decreased to decrease the solubility of Si species, leading to the promotion of the formation of Si-O-Si networks through the substitution of two Si^{4+} for one P^{5+} and one Al^{3+}.

NH_3-TPD measurements were performed in order to study the acidity of the samples. The results are shown in Figure 4 and Table 1. Two large peaks were overlapped in the NH_3-TPD profiles of all the samples; a peak was observed at around 440 K and the other at around 523 K (Figure 4). The low-temperature peak corresponds to NH_3 adsorbed on non-acidic -OH groups and NH_4^+, which forms by the reaction of NH_3 with Brønsted acid sites (BASs), and does not correspond to NH_3 adsorbed on catalytically active BASs and Lewis acid sites (LASs) [31,32]. On the other hand, the high-temperature peak corresponds to the NH_3 desorption from catalytically active BASs and LASs. The high-temperature peak of the product increased with the decrease in the water content in the starting gel, and the temperature where the peak top was observed was almost unchanged. The addition of the seed crystals to the starting gel also resulted in virtually increasing the high-temperature peak as S(10) was compared with S(10)-seed; simultaneously, a shoulder peak was observed at around 623 K, indicating that strong acid sites were newly formed compared with the products synthesized in the absence of seed crystals. Furthermore, the area ratio of the high-temperature peak to the low-temperature peak decreased with the decrease in the water content in the gel. The acid amounts of the products estimated from the high-temperature peak are listed in Table 1. In SAPO molecular sieves, the acid sites are generated by the incorporation of Si species into the AlPO framework [33-35]. In the case of the replacement of P species alone in the AlPO framework by Si species, Si species should be surrounded by Al species to give rise to relatively weaker BASs. By contrast, the substitution of two Si^{4+} for neighboring P^{5+} and Al^{3+} is prone to form silica islands, and stronger acid sites are generated at the boundaries of silica islands. Simultaneously, the formation of silica islands decreases

synthesized with the H_2O/Al ratio of 5 in the starting gel, were composed of hexagonal crystallites. In the synthesis with the H_2O/Al ratio of 50 in the starting gel, the huge crystallites were formed with mostly around 16 µm in diameter and 8 µm in length. The crystallite size was decreased with the decrease in the water content in the gel; the average sizes were found to be 4 µm in diameter and 0.7 µm in length for the H_2O/Al ratio of 25. By contrast, the crystals grew along the c-direction through the crystallization with the H_2O/Al ratio of 10 although the crystallite diameter was decreased to be 3 µm. When the sample synthesized with the H_2O/Al ratio of 10 was employed as a seed for the synthesis with the H_2O/Al ratio of 10, the crystal growth along the c-direction was suppressed and the hexagonal plates with around 2 µm in diameter and 0.1-0.3 µm in length were formed. Interestingly, the hexagonal-cylindrical crystallites of the seed was no longer observed after the crystallization of the gel containing the seed, indicating that the seed crystals were mostly dissolved during the hydrothermal treatment. The further decrease in the water content drastically changed the crystallite morphology; large agglomerates composed of plate-like crystallites were formed (Figures 3a and 3b). Considering our findings above, the SAPO-5 materials with the different morphologies can be readily synthesized by merely changing the water content in the starting gel under the present hydrothermal conditions. It has been reported that the higher alkalinity favored the nucleation of a lot of particles due to the easy-dissolution of silicon and aluminium sources in the solution, forming the smaller-sized zeolite [27,28]. In addition, the increase in the concentration of the starting gel leads to increasing a nucleation density in the early stages of the crystallization to make the crystallite size reduced [29]. Thus, decreasing the water content in the starting gel, which corresponds to the increasing the gel concentration, would lead to the enhancement of the nucleation in the early stage of the crystallization due to accelerating the dissolution of the silicon and aluminium sources in the increased pH solution, resulting in the formation of the smaller-sized SAPO-5 crystallites. Furthermore, the nucleation followed by the crystallization may be promoted on the surface of the seed crystallites to form the smaller-sized SAPO-

Sample	H₂O/Al in gel	S.A._-BET[a] /m² g⁻¹	S.A._-EXT[a] /m² g⁻¹	Si/Al[b]	P/Al[b]	Si/(Al+P+Si)[b]	Acid amount[c] /mmol g⁻¹
S(5)	5	304	36	0.11	0.85	0.054	0.367
S(10)	10	289	12	0.11	0.87	0.056	0.301
S(10)-seed	10 (with seed)	267	23	0.11	0.91	0.052	0.366
S(25)	25	245	3.6	0.10	0.92	0.050	0.296
S(50)	50	278	7.9	0.13	0.83	0.069	0.238

[a]Estimated by N₂ adsorption-desorption method
[b]Measured by ICP-AES
[c]Estimated from NH₃-TPD measurement

Table 1: Physicochemical properties of the samples synthesized with various water contents.

the amount of the acid sites in SAPO materials in comparison with the amount of the acid sites derived from Si species surrounded by Al species [36-39]. Thus, in the highly concentrated gels, Si species would be homogeneously incorporated into the framework with surrounded by Al species to generate relatively weak acid sites. On the other hand, in the low-concentrated gels, relatively strong acid sites would be generated by the formation of silica islands due to the incorporation of larger amounts of Si species. Indeed, S(50), which was synthesized with the lowest-concentrated gel, contained the highest Si content and the smallest acid amount among the samples (Table 1). Except for S(50), the acid amount of the samples was strongly dependent on the gel concentration, indicating that there are some differences in the Si distribution in the framework among the samples although all the samples contained similar Si amounts to those in the starting gels. However, since the acid sites attributed to the high-temperature peak were classified as weak BASs and the Si content in the samples was low, it is difficult to clarify the slight differences in the acid strengths derived from the Si distribution in the samples in detail. It is probable that the seed would also contribute to the homogeneous distribution of Si species in preference to the formation of silica islands due to the enhancement of the nucleation, resulting in the generation of the acid sites derived from Si species surrounded by Al species.

Catalytic performance of SAPO-5 catalysts in the methanol-to-olefins (MTO) reaction

The catalytic properties of the prepared SAPO-5 samples as an acid catalyst were evaluated in the MTO reaction. The results are shown in Figures 5 and 6 and Table 2. Figure 5 shows the methanol conversion over the prepared SAPO-5 catalysts as a function of time on stream. All the catalysts exhibited 100% methanol conversion in the initial stage of the reaction (at 3 min). A fast deactivation occurred over S(50) in the initial stage of the reaction, and the methanol conversion was decreased along with the reaction time. By contrast, S(5) exhibited the longest catalyst life among the catalysts; complete methanol conversion was kept for 5 h after the reaction started. The catalyst life was increased in line with the following order: S(5)>S(10)-seed>S(10)>S(25)>S(50), which is consistent with the order of the acid amounts estimated from the NH₃-TPD profiles (Table 1). In addition, the order of the catalyst life is opposite to that of the crystallite sizes of the catalysts, except for S(5): S(50)>S(25)>S(10)>S(10)-seed. In the MTO reaction, the effects of acidic properties and the crystallite size of zeolite catalysts on the catalyst life has been investigated in detail. The polymerization of carbonaceous species for the coke deposition in/on zeolites can be enhanced on strong acid sites, in spaces with high acid densities, and in long micropores. It has been reported that the decrease in the Si content, corresponding to the decrease in the acid density, of SAPO-34 leads to improving the coking resistance to increase the catalyst life [7]. Furthermore, the crystallite size is a strongly important factor for improving the catalyst life compared with the acid density; the decrease in the crystallite size leads to the increase in the catalyst life [7,37].

According to the findings reported above, it is indicated that in the present study the catalyst life was affected by the crystallite size of the catalysts; the smaller-sized catalyst can exhibit the longer catalyst life due to the suppression of the coke deposition by the reduction of diffusion barriers inside the micropores of the catalyst. The small amount of the acid sites of the products (0.24 mmol g⁻¹- 0.37 mmol g⁻¹) would hardly influence the catalyst life even if the acid amount was varied. The conversion of methanol was 100% over all the catalysts, except for S(50), for reaction time of 1 h (Table 2). All the catalysts produced propene as the main-product followed by butenes. The formation of dimethyl ether (DME) was observed over S(50), resulting from the decrease in the active sites for the conversion of DME to hydrocarbons due to the deactivation. The slight differences in the behavior of the product distributions were observed; the ratio of propene to butenes over the catalysts followed the order: S(5) > S(10)-seed > S(10) > S(25) > S(50). In the MTO reaction, the product distribution is strongly dependent on the acid strength of zeolite catalysts. Indeed, comparing SAPO-5 with SSZ-24, which isomorphous aluminosilicate zeolite with stronger acid strength, in the MTO reaction, key reaction steps of alkene and arene cycles can display different sensitivities to acid strength [24]. By contrast, it has been reported that the product distribution is irrelative to the amount of the acid sites [40]; furthermore, the initial product distribution is independent of the crystal size of zeolite [13]. The NH₃-TPD measurement revealed that there was no significant difference in the acid strength among the catalysts since the high-temperature peaks of all the catalysts were observed at similar temperatures (Figure 4). Considering that the ratios of propene to butenes in the early stage of the reaction time were affected by the acid strength of the catalysts, fine tunings of the acid strength by controlling the Si distribution in the framework should be required for optimizing the product distribution.

Changes in the selectivities to light olefins, paraffins and DME as a function of time on stream are shown in Figure 6. All the catalysts produced mainly propene followed by butenes during the reaction. After the deactivation started, the production of DME was observed, independent of the catalysts. The selectivity to DME was drastically increased with the reaction time, while the selectivities to the other products were gradually decreased, indicating that the conversion of DME to hydrocarbons was suppressed by the deactivation due to the coke deposition on the acid sites for the transformation to hydrocarbons. During the reaction with the complete methanol conversion, no marked change in the selectivity to propene was observed. By contrast, the selectivity to butenes was gradually increased with the reaction time, and that to ethene was decreased, resulting in the decrease in propene/butenes ratios and the increase in propene/ethene ratios over all the catalysts with the reaction time. Dahl et al. have proposed that the product distribution was affected by the increase in diffusion barriers generated by the coke deposition inside the micropores of SAPO-34 during the MTO reaction; propene/ethene ratios were decreased with the reaction time [41]. In the MTO reaction over SAPO-5, the alkene cycle is a main process for the formation

Sample	Conv. / %	Selectivity/C-%							
		C2=	C3=	C4=	C5<	C1	C2-C4	aroma[b]	DME
S(5)	100	9.4	46.6	11.3	10.4	4.1	14.9	3.3	0
S(10)	100	9.0	45.9	15.5	11.7	4.5	8.5	4.8	0
S(10)-seed	100	12.7	47.8	13.6	7.5	4.4	10.7	3.1	0
S(25)	100	10.4	43.5	18.6	9.2	5.7	9.5	3.2	0
S(50)	97.1	8.5	43.9	24.4	12.9	3.1	6.2	1.0	0.02

[a]Reaction conditions: cat., 100 mg; P(MeOH), 2.6 kPa; W/F=67.5 g h mol[-1].
[b]aroma=Aromatic compounds.

Table 2: The results of the MTO reaction for reaction time of 1 h at 450°C[a].

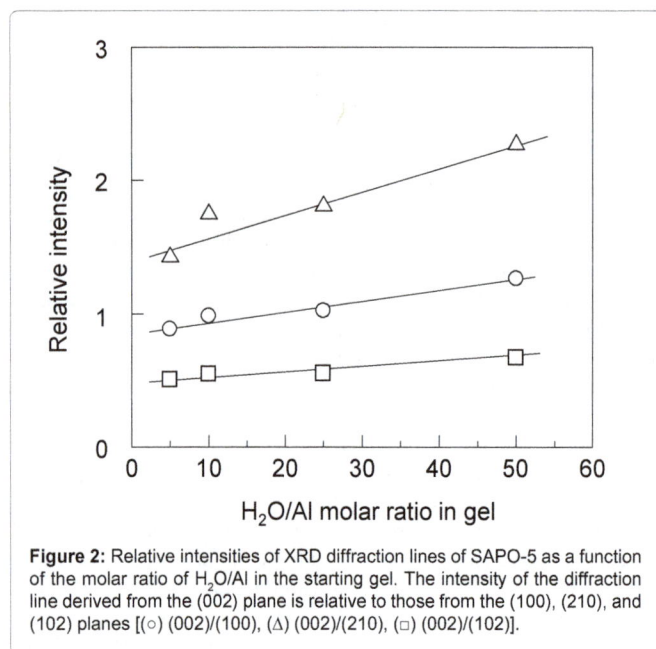

Figure 2: Relative intensities of XRD diffraction lines of SAPO-5 as a function of the molar ratio of H_2O/Al in the starting gel. The intensity of the diffraction line derived from the (002) plane is relative to those from the (100), (210), and (102) planes [($\circ$) (002)/(100), ($\triangle$) (002)/(210), ($\square$) (002)/(102)].

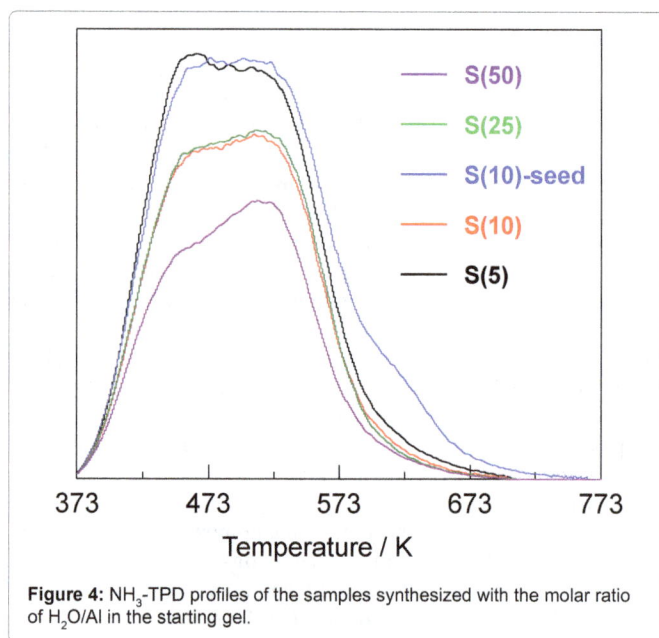

Figure 4: NH_3-TPD profiles of the samples synthesized with the molar ratio of H_2O/Al in the starting gel.

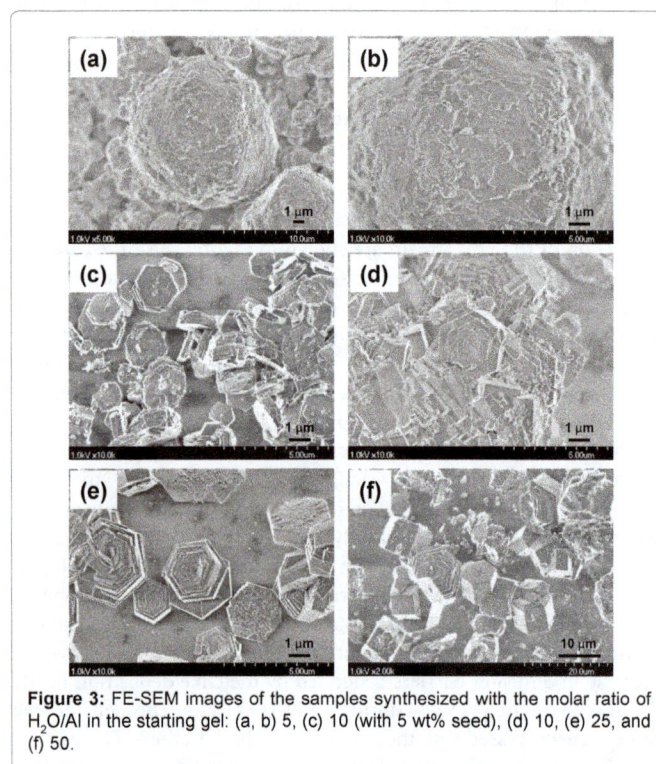

Figure 3: FE-SEM images of the samples synthesized with the molar ratio of H_2O/Al in the starting gel: (a, b) 5, (c) 10 (with 5 wt% seed), (d) 10, (e) 25, and (f) 50.

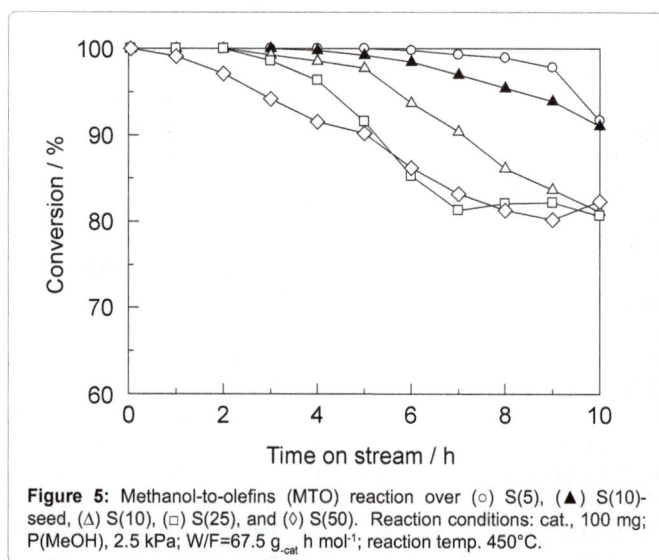

Figure 5: Methanol-to-olefins (MTO) reaction over ($\circ$) S(5), ($\blacktriangle$) S(10)-seed, ($\triangle$) S(10), ($\square$) S(25), and ($\diamond$) S(50). Reaction conditions: cat., 100 mg; P(MeOH), 2.5 kPa; W/F=67.5 g$_{-cat}$ h mol[-1]; reaction temp. 450°C.

of hydrocarbons to produce selectively propene and butenes, while propene/butenes ratios are affected by the methanol conversion; the propene/butens ratios were increased with the increase in the methanol conversion as well as the reaction temperature [24,25]. In the present study, the micopore openings of SAPO-5 would be too large to generate enough diffusion barriers even when the coke deposition occurred inside the micropores; furthermore, the weak acid strength of SAPO-5

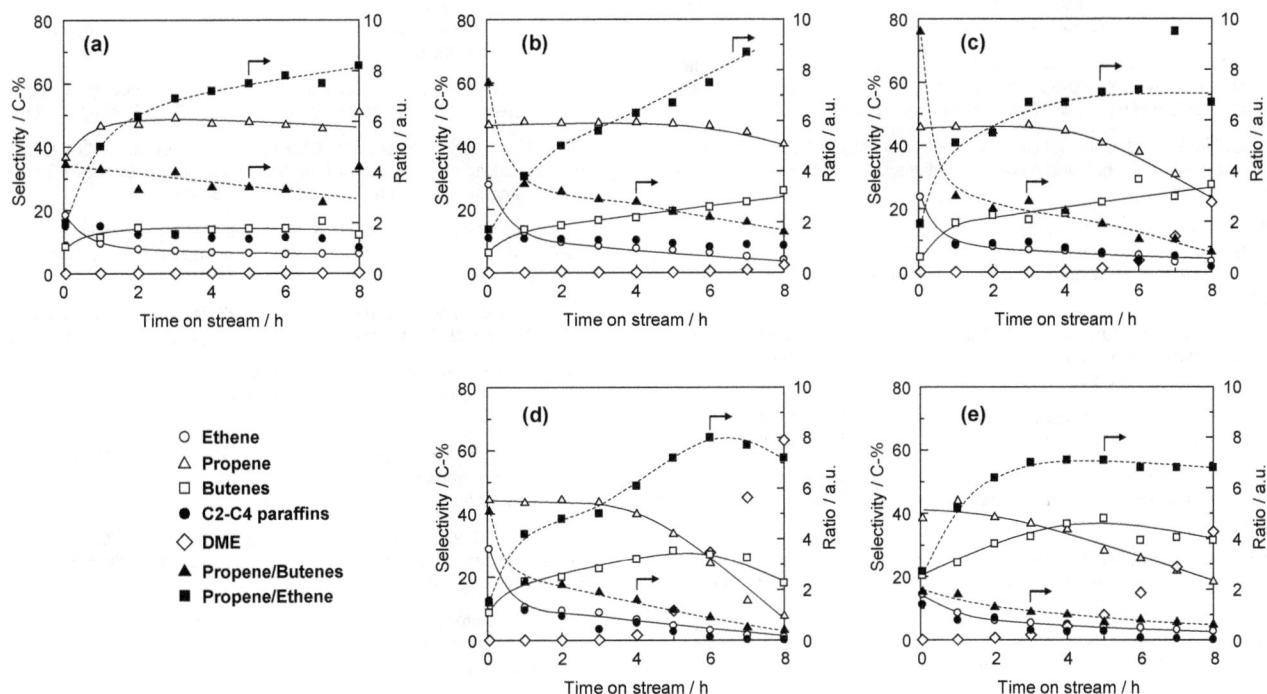

Figure 6: The product distributions in the Methanol-to-olefins (MTO) reaction over (a) S(5), (b) S(10)-seed, (c) S(10), (d) S(25), and (e) S(50). Reaction conditions: cat., 100 mg; P(MeOH), 2.5 kPa; W/F=67.5 g$_{-cat}$ h mol^{-1}; reaction temp. 450°C.

as evidenced by NH$_3$-TPD measurement would cause the alkene cycle, resulting in the predominant production of propene and butenes in comparison with ethene. In addition, it is assumed that decrease in the acid sites of SAPO-5 by the coke deposition during the MTO reaction led to decreasing the propene/butenes ratios with reaction time.

Conclusion

SAPO-5 samples with different morphologies and sizes were synthesized by tuning the water content in the starting gel, that is, the gel concentration. The gel concentration also affected the acidity of the SAPO-5 catalyst; the acid amount of the catalyst was increased with the decrease in the water content in the gel although the acid strength was mostly independent of the gel concentration. The use of the highly concentrated gel resulted in the formation of the highly-crystalline and small-sized SAPO-5 catalyst with the large amount of the acid sites. The SAPO-5 catalyst synthesized with the highly concentrated gel showed a higher stability of the catalytic activity in comparison with that with the low-concentrated gel in the MTO reaction. In addition, the SAPO-5 catalyst produced predominantly propene followed by butenes with the complete conversion of methanol.

References

1. Stöcker M (2010) Methanol to Olefins (MTO) and Methanol to Gasoline (MTG): Zeolites and Catalysis. Wiley-VCH, Weinheim.

2. Baerlocher Ch, McCusker LB: Database of Zeolite Structures.

3. Stöcker M (1999) Methanol-to-hydrocarbons: catalytic materials and their behavior. Micropor Mesopor Mater 29: 3-48.

4. Stöcker M (2005) Gas phase catalysis by zeolites. Micropor Mesopor Mater 82: 257-292.

5. Yuen LT, Zones SI, Harris TV, Gallegos EJ, Auroux A (1994) Product selectivity in methanol to hydrocarbon conversion for isostructural compositions of AFI and CHA molecular sieves. Micropor Mater 2: 105-117.

6. Barger P (2002) Methanol to Olefins (MTO) and beyond: Zeolites for Cleaner Technologies. Imperial College Press, London.

7. Wilson S, Barger P (1999) The characteristics of SAPO-34 which influence the conversion of methanol to light olefins. Micropor Mesopor Mater 29: 117-126.

8. Haw JF, Song W, Marcus DM, Nicholas JB (2003) The mechanism of methanol to hydrocarbon catalysis. Acc Chem Res 36: 317-326.

9. Arstad B, Kolboe S (2001) The reactivity of molecules trapped within the SAPO-34 cavities in the methanol-to-hydrocarbons reaction. J Am Chem Soc 123: 8137-8138.

10. Haw JF, Marcus DM (2005) Well-defined (supra)molecular structures in zeolite methanol-to-olefin catalysis. Top Catal 34: 41-48.

11. Song W, Fu H, Haw JF (2001) Selective Synthesis of Methylnaphthalenes in HSAPO-34 Cages and Their Function as Reaction Centers in Methanol-to-Olefin Catalysis. J Phys Chem B 105: 12839-12843.

12. Kvisle S, Fuglerud T, Kolboe S, Olsbye U, Lillerud KP, et al. (2008) Handbook of heterogeneous catalysis. Wiley-VCH, Weinheim.

13. Chen D, Moljord K, Fuglerud T, Holmen A (1999) The effect of crystal size of SAPO-34 on the selectivity and deactivation of the MTO reaction. Micropor Mesopor Mater 29: 191-203.

14. Lin S, Li J, Sharma RP, Yu J, Xu R (2010) Fabrication of SAPO-34 crystals with different morphologies by microwave heating. Top Catal 53: 1304-1310.

15. Hirota Y, Murata K, Tanaka S, Nishiyama N, Egashira Y, et al. (2010) Dry gel conversion synthesis of SAPO-34 nanocrystals. Mater Chem Phys 123: 507-509.

16. Álvaro-Muñoz T, Márquez-Álvarez C, Sastre E (2013) Enhanced stability in the methanol-to-olefins process shown by SAPO-34 catalysts synthesized in biphasic medium. Catal Today 215: 208-215.

17. Álvaro-Muñoz T, Márquez-Álvarez C, Sastre E (2012) Use of different templates on SAPO-34 synthesis: effect on the acidity and catalytic activity in the MTO reaction. Catal Today 179: 27-34.

18. Pengfei W, Dexing Y, Jie H, Jing'an X, Guanzhong L (2013) Synthesis of SAPO-34 with small and tunable crystallite size by two-step hydrothermal crystallization and its catalytic performance for MTO reaction. Catal Today 212: 62e1-62e8.

19. Yang M, Tian P, Wang C, Yuan Y, Yang Y, et al. (2014) A top-down approach to prepare silicoaluminophosphate molecular sieve nanocrystals with improved catalytic activity. Chem Commun (Camb) 50: 1845-1847.

20. Álvaro-Muñoz T, Sastre E, Márquez-Álvarez C (2014) Microwave-assisted synthesis of plate-like SAPO-34 nanocrystals with increased catalyst lifetime in the methanol-to-olefin reaction. Catal Sci Technol 4: 4330-4339.

21. Yang G, Wei Y, Xu S, Chen J, Li J, et al. (2013) Nanosize-enhanced lifetime of SAPO-34 catalysts in methanol-to-olefin reactions. J Phys Chem C 117: 8214-8222.

22. Li Z, Martínez-Triguero J, Concepción P, Yu J, Corma A (2013) Methanol to olefins: activity and stability of nanosized SAPO-34 molecular sieves and control of selectivity by silicon distribution. Phys Chem Chem Phys 15: 14670-14680.

23. Bellón JM, Contreras LA, Pascual G, Bujan J (1999) Neoperitoneal formation after implantation of various biomaterials for the repair of abdominal wall defects in rabbits. Eur J Surg 165: 145-150.

24. Erichsen MW, Svelle S, Olsbye U (2013) The influence of catalyst acid strength on the methanol to hydrocarbons (MTH) reaction. Catal Today 215: 216-223.

25. Erichsen MW, Svelle S, Olsbye U (2013) H-SAPO-5 as methanol-to-olefins (MTO) model catalyst: Towards elucidating the effects of acid strength. J Catal 298: 94-101.

26. Hereijgers BPC, Bleken F, Nilsen MH, Svelle S, Lillerud KP, et al. (2009) Product shape selectivity dominates the Methanol-to-Olefins (MTO) reaction over H-SAPO-34 catalysts. J Catal 264: 77-87.

27. Persson E, Schoeman BJ, Sterte J, Otterstedt JE (1994) The synthesis of discrete colloidal particles of TPA-silicalite-1. Zeolites 14: 557-567.

28. Brar T, France P, Smirniotis PG (2001) Control of Crystal Size and Distribution of Zeolite A. Ind Eng Chem Res 40: 1133-1139.

29. An T, An J, Yang H, Li G, Feng H, et al. (2011) Photocatalytic degradation kinetics and mechanism of antivirus drug-lamivudine in TiO2 dispersion. J Hazard Mater 197: 229-236.

30. Roldán R, Sánchez-Sánchez M, Sankar G, Romero-Salguero FJ, Jiménez-Sanchidrián C (2007) Influence of pH and Si content on Si incorporation in SAPO-5 and their catalytic activity for isomerisation of n-heptane over Pt loaded catalysts. Micropor Mesopor Mater 99: 288-298.

31. Niwa M, Katada K (1997) Measurements of acidic property of zeolites by temperature programmed desorption of ammonia. Catal Surv Jpn 1: 215-226.

32. Suzuki K, Aoyagi Y, Katada N, Choi M, Ryoo R, et al. (2008) Acidity and catalytic activity of mesoporous ZSM-5 in comparison with zeolite ZSM-5, Al-MCM-41 and silica-alumina. Catal Today 132: 38-45.

33. Borade RB, Clearfield A (1994) A comparative study of acidic properties of SAPO-5, SAPO-11, SAPO-34 and SAPO-37 molecular sieves. J Mol Catal 88: 249-266.

34. Sastre G, Lewis DW, Richard C, Catlow A (1997) Modeling of silicon substitution in SAPO-5 and SAPO-34 molecular sieves. J Phys Chem B 101: 5249-5262.

35. Li Z, Martínez-Triguero J, Yu J, Corma A (2015) Conversion of methanol to olefins: Stabilization of nanosized SAPO-34 by hydrothermal treatment. J Catal 329: 379-388.

36. Wei Y, Zhang D, He Y, Xu L, Yang Y, et al. (2007) Catalytic performance of chloromethane transformation for light olefins production over SAPO-34 with different Si content. Catal Lett 114: 30-35.

37. Wang L, Guo C, Yan S, Huang X, Li Q (2003) High-silica SAPO-5 with preferred orientation: synthesis, characterization and catalytic applications. Micropor Mesopor Mater 64: 63-68.

38. Sinha AK, Sainkar S, Sivasanker S (1999) An improved method for the synthesis of the silicoaluminophosphate molecular sieves, SAPO-5, SAPO-11 and SAPO-31. Micropor Mesopor Mater 31: 321-331.

39. Yang C, Gong C, Peng T, Deng K, Zan L (2010) High photocatalytic degradation activity of the polyvinyl chloride (PVC)-vitamin C (VC)-TiO2 nano-composite film. J Hazard Mater 178: 152-156.

40. Zhu Q, Kondo JN, Ohnuma R, Kubota Y, Yamaguchi M, et al. (2008) The study of methanol-to-olefin over proton type aluminosilicate CHA zeolites. Micropor Mesopor Mater 112: 153-161.

41. Dahl IM, Wendelbo R, Andersen A, Akporiaye D, Mostad H, et al. (1999) The effect of crystallite size on the activity and selectivity of the reaction of ethanol and 2-propanol over SAPO-34. Micropor Mesopor Mater 29: 159-171.

4

Expendable Railroad Sleepers, of Modern Fuels for Power Stoker fired Boilers

Eugeniusz Orszulik*

Central Mining Institute, Pl. Gwarków 1, 40–116 Katowice, Poland

Abstract

The paper presents a concept of producing energy on the basis of modern alternative fuels with the expendable railroad sleepers to be burnt in power stokerfired boilers. The thermal energy contained in water vapour and hot water will be utilized in producing in combination, of electrical energy, and for heating of cubature objects. There have been presented the properties of alternative fuels obtained:

- During the combustion of flammable mixture no I (waste - code 17 02 01 wood - from expendable railroad sleepers with hard coal) met the class 5 (highest) emission value requirements with regard to PN-EN 303-5: 2012 standard. Other flammable mixtures, no II and III meet the requirements of class 3 (lowest) value with regard to the PN-EN 303-5 standard. The limit for such a low class was the emission value of dust. The laboratory boiler used for testing was not equipped with a device for the separation of dust. In case of equipping the boiler with dust removal device, satisfactory emission results in accordance with the accepted standards of emissions would be obtained.

- From combustion of flammable mixtures have been qualified for development in land areas at a depth of 0.3 - 15 m bgl. (23):

  - Group A: (a) Land immovable of the area protected on the basis of the Water Law regulations. (b) Areas subject to protection on the basis of the nature protection regulations, if maintaining the current level of land pollution does not pose a threat to human health or environment - for these areas the concentrations resulting from the actual state maintain standards.

  - Group B: Land classified as farming land with the exception of lands under ponds and under ditches, forests as well as lands with a high amount of trees and bushes, wasteland as well as built-up and urbanised land with the exception of industrial, mining and communications areas;

  - Group C: Industrial, mining and communications areas for the flammable mixture no III - waste - code 17 02 01 wood - from expendable railroad sleepers with cereal straw. Flammable mixtures: no I - waste - code 17 02 01 wood.

  - From railroad sleepers with hard coal and no II - waste - code 17 02 01 wood - from railroad sleepers with lignite, were qualified to apply to land development in the areas of : - Group C: industrial, mining and communications areas, developed at the depth of 0.3 - 15 bgl.

- Using the basic component - wood - from railroad sleepers as composite of alternative fuels will:

  - Enable to save primary fuels: hard coal and lignite.

  - Not cause the excess of permissible emission standards of dust and gaseous substances to the air.

  - Enable the development of slag and ash obtained from the combustion process in groups of lands from: Groups A, B, and C.

- Using alternative fuel composites in energetics, containing wood - from expendable railroad sleepers will not require construction changes of boilers and will provide the meeting of conditions for thermal conversion of wastes in accordance with the requirements given in the Regulation of the Minister of Economy.

  Composites of alternative fuels containing wood - from expendable railroad sleepers will be qualified based on the Minister of Environment Regulation and could be qualified as a part of thermal energy conversion of municipal sewage sludge.

Keywords: Wood; Railroad sleepers; Fuel; Combustion; Boiler; Emission

Introduction

Due to the continuous increase of the demand for fossil fuels to generate heat and electricity, one should reach for new energy sources [1-10]. Waste - code 17 02 01 wood (from railroad sleepers) - constitutes a great nuisance for the environment [11-13]. Known waste wood from railroad sleepers treatment methods do not fully solve the problem of the recovery or disposal. The choice of disposal technology of this type of waste is associated with meeting legal regulations [3,14-19]. Methods known from the literature [7,20-25] have not developed

***Corresponding author:** Eugeniusz Orszulik, Central Mining Institute, pl. Gwarków 1, 40 – 116 Katowice, Poland, E-mail: e.orszulik@gig.katowice.pl

a way of combustion or co-combustion of from railroad sleepers in stoker-fired boilers .

In the present work, [26] a test has been conducted, based on managing stabilized from railroad sleepers by their thermal disposal in the D10 process [15] by co-combustion with fuels i.e. hard coal. The research was aimed at the possibility of using alternative fuels composites containing wood from railroad sleepers as fuel for energy boilers equipped with stoker burners. Hard coal used in the tests, their extraction and acquisition is dependent on the economic situation [7,27].

Research Methodology

The assumptions of applying new modern energy fuels for combustion in stoker-fired boilers are:

- Reduction of energy acquisition costs.

- Partial substitution of alternative fuel produced of non-hazardous waste, obtained from expendable railroad sleepers, and unsuitable for further recovery, for such a primary fuel as hard coal.

- Ecological effect relying on reduction of:

- Emission of gas and dust substances into the atmosphere, relative to commonly used coal. Production of modern alternative energy fuels whose quality will be comparable with the hard coal commonly used.

In the test was a used expendable wooden railroad sleeper from refurbished railway. Regulation of the Minister of Environment of 27 September 2001 on Waste Catalogue, Law Journal 01.112.1206 [11] expendable wooden railroad sleepers has code 17 02 01 wood.

The purpose of the research was to determine the thermal properties, the measurements of the emissions of dust and gaseous substances emitted to the atmosphere and pollutants to the ground (waste slag and ash) arising during and after the process of co-combustion of composites of alternative fuels containing waste wood from discarded railway sleepers as a fuel to power boilers. The following flammable mixtures were selected to testing:

- Waste - code 17 02 01 wood (from railroad sleepers) with hard coal.

- Waste - code 17 02 01 wood (from railroad sleepers) with lignite.

- Waste - code 17 02 01 wood (from railroad sleepers) with wood chips.

- Waste - code 17 02 01 wood (from railroad sleepers) with cereal straw.

Measurements and tests were carried out in accordance with regulations in force [1,2,28-30]. The scope of the study included the following measurements:

Tests and measurements were carried out on a test post, which was equipped with a 25 kW laboratory water boiler. The boiler was equipped with a mechanical fuel feeder, which was directly connected with a dust burner. The test post was equipped with measuring and control apparatus for continuous and periodic measurement:

- Of the temperature of the water supplying the boiler.

- Of the temperature of the water coming back from the boiler.

- Of water pressure in the boiler.

- Of the flow of water through the boiler.

- Of exhaust gas analysis i.e. O_2, CO_2, CO, NO_2, SO_2.

- Of the analysis of the associated organic carbon content.

- Of the analysis of the content of hydrogen fluoride and hydrogen chloride.

- Of the concentration of dust in exhaust gas.

- Of polluting substances emitted to the ground in order to determine the applicability of the products of combustion in the form of slag and ash into a group of types of land which could be developed [31].

The boiler had a feeder for flammable mixture, the PID controller to control the work of the boiler from heating water temperature in the system.

At the time of tests, the control of combustion process of flammable mixtures and heat generation was based primarily on the maintenance of thermodynamic parameters through adequate air supply to the combustion and the analysis of the composition of the exhaust gas leaving the combustion chamber of the boiler for a specified boiler load. Figure 1 shows the layout of the measuring and control system of the test post.

Flammable mixtures used in testing

The composition of flammable mixtures was determined on the basis of the obtained research of energy properties of fuels, biomass and waste:

- Hard coal - sortment's sortment coal fine MII.

- Lignite - sortment's sortment coal fine II.

- Cereal straw (biomass).

- Waste - code 17 02 01 wood (from expendable railroad sleepers) after drying.

Properties of used fuels, biomass and wood (from railroad sleepers) after drying as ingredients of flammable mixtures for thermal and environmental testing were presented in Table 1 [32-36].

The composition of flammable mixtures was determined on the basis of the obtained research of energy properties of fuels, biomass and waste:

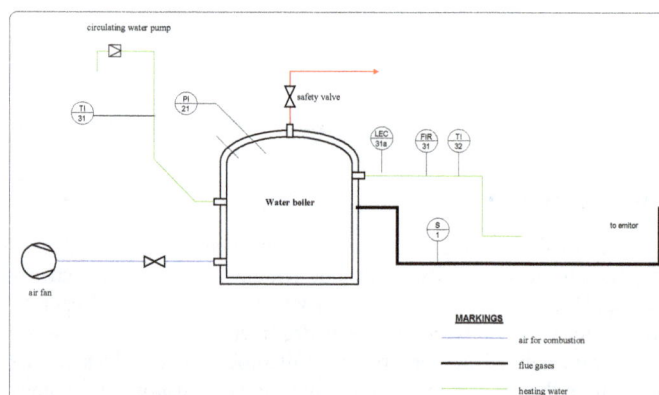

Figure 1: The post for testing the water boiler during the combustion of flammable mixtures.

Number	Specification	Unit	Fuel			
			Hard coal	Lignite	Cereal straw	Waste-code 17 02 01 wood
1	Calorific value[1]	kJ/kg	22033	8133	13218	13860
2	Sulphur content[2]	%	1.02	0.47	0.39	0.04
3	Ash content[3]	%	7.21	11.07	6.55	2.65
4	Moisture Content[4]	%	20.22	52.16	19.86	31.18
5	Chlorine Content[5]	%	0.095	0.046	0.254	0.10

The method to determine:

[1]PN – ISO 1928 - Solid mineral fuels – Determination of gross calorific value by the calorimetric method, and calculation of net calorific value.

[2]PN – ISO 334 - Solid mineral fuels – Determination of total sulphur – Eschka method.

[3]PN-ISO 1171:2002: "Solid mineral fuels - Determination of ash.

[4]PN-80/G-04511: "Solid fuels. Determination of moisture.

[5]PN – ISO 587 - Solid mineral fuels – Determination of total chlorine using Eschka mixture.

Table 1: Properties of used fuels for thermal and environmental tests.

- Hard coal - sortment's sortment coal fine MII.
- Waste - code 17 02 01 wood - from railroad sleepers.
- Lignite had the highest moisture content (52.16%) and the lowest cereal straw (19.86%).
- Cereal straw had the highest chlorine content (0.254%), and the lowest wood chips (0.10%).

Waste - code 17 02 01 wood - from expendable railroad sleepers were characterized by the calorific value of 13860 kJ/kg, which was higher than the limit set out in the Regulation of the Minister of Economy [18]. For this reason, the waste should not be stored at dumping site.

The composition of flammable mixtures was determined after the established criteria:

- The average calorific value should amount to minimum 8.0 MJ/kg.
- Sulphur content should amount to 0.6%.
- Ash content should amount to 15%.
- Chlorine content should amount to 5%.
- Moisture content should amount to 20%.

The criteria were established on the basis of the coals currently used for power generation and heating, mostly lignite in dust boilers.

Table 2 presents the results obtained on the basis of the results of the tests of fuels, biomass and wood - from expendable railroad sleepers compositions of flammable mixtures, and in Table 3 are presented the properties of flammable mixtures used for thermal and environmental tests.

Flammable mixtures: No I, II and III meet the established criteria.

The best energy properties have the flammable mixture No I - waste - 17 02 01 wood - from expendable railroad sleepers with hard coal. The mixture has a high calorific value (20094 kJ/kg), low ash content of 12.50%, but contains a high sulphur content of 0.9%.

The worst energy properties have the flammable mixture No II - waste - 17 02 01 wood - from expendable railroad sleepers with lignite. It has a very low calorific value (8112 kJ/kg), low ash content of 13.55% and sulphur of 0.43% and a high moisture content of 46.71%. Properties of the mixture No II are comparable to lignite burned in

BOT power plants.

Chlorine content in flammable mixtures No I, II and III is very low, which will require the maintenance of the combustion process to ensure the temperature of exhaust gases above 850°C for at least 2 seconds [4,37].

Determining the energy efficiency of the boiler during the combustion of flammable mixtures

Boiler energy efficiency [38] at the time of the combustion of flammable mixtures was determined by definition for the tested combustible mixture, taking into account the supplied energy flow as the chemical energy contained in the mixture:

$$\eta = \frac{Q_N}{Q_z} \qquad [a]$$

Where: Q_N - flow of the energy brought out, kW, Q_z - flow of supplied energy, kW.

Boiler efficiency determined by the formula [a] refers to a specific point of the boiler heat load (at the moment of measurement). Stream energy brought out from water boiler:

$$Q_N = \overset{0}{m}(h_2 - h_1)$$

Where: $\overset{0}{Q_N}$ - determined thermal useful power, kW

$\overset{0}{m}$ - Hot-water mass flow, kg/s

h_2 - Enthalpy of water at an average temperature of outflow, t_2, kJ/kg

h_1 - Enthalpy of water at an average temperature on inflow, t_1, kJ/kg

The stream of energy supplied to the boiler:

$$Q_z = \overset{0}{m}\left[\left(H_{(N)} + h_F\right)/(1 - l_u) + J_{(N)A}\right]$$

Where - flow of supplied energy, kW

$\overset{0}{m}$ - Flow of fuel mass, kg/s

$H_{(N)}$ - The calorific value of fuel at reference temperature, t_r, kJ/kg

h_F - Fuel enthalpy $h_F = c_F\left(t_F - t_r\right)$, kJ/kg

c_F - Proper heat capacity of fuel, kJ/kgK

t_F - Fuel temperature, °C

t_r - Reference temperature, °C

Mixture	Fuel	Waste-code 17 02 01 wood
No I	Hard coal-80% of the mass	20% of the mass
No II	Lignite-20% of the mass	80% of the mass
No III	Cereal straw-50% of the mass	50% of the mass

Table 2: The composition of the flammable mixtures for thermal and environmental tests.

Number	Specification	Unit	Flammable mixture		
			No I	No II	No III
1	Calorific value	kJ/kg	20094	8112	13315
2	Sulphur content	%	0.90	0.62	0.43
3	Ash content	%	12.50	14.74	13.55
4	Moisture content	%	21.02	46.71	25.64
5	Chlorine content	%	0.070	0.049	0.018

Table 3: Properties of the flammable mixtures for thermal and environmental testing.

l_u - Non-burnt fuel flow mass to flow of supplied fuel mass ratio

$$l_u = m_{Fu} / m_{Fo}$$

$J_{(N)A}$ - Enthalpy of combustion air $J_{(N)A} = \mu_A C_{pA} \left(t_A - t_r \right)$

, kJ/kg

C_{pA} - Proper heat capacity of air, kJ/(kgK)

μ_A - Air mass to fuel mass ratio

t_A - Air temperature at the border of the balance cover, °C

Study of solid products - Slag and ash obtained after combustion of flammable mixtures

Solid waste obtained after combustion of flammable mixtures - slag and ash - have undergone the tests of the chemical composition for the content of trace elements. Marked quantities of heavy metals in the analysed samples were compared to the standards of the quality of the soil or land contained in the Regulation of the Minister of Environment [2] with regard to (current and planned land development) groups of types of land where they can be used.

The Results of the Tests and Measurements from Co-Combustion of Flammable Mixtures

Combustion of the flammable mixture no I

Study of the combustion of flammable mixture no I - wood - from expendable railroad sleepers with hard coal has been carried out for a heat load laboratory boiler of 85 ± 5%.

Thermal-emission tests were aimed at determining the energy efficiency of the boiler and the designation of emission values occurring during the combustion of flammable mixtures no I. Thermal-emission test results for the flammable mixture no I are presented in Table 4.

Thermal and emission values of the boiler obtained during combustion of flammable mixture no I were compared with the limit emission values contained in standard PN-EN 303-5: 2012 [1] (Table 5):

Emission values obtained during the tests of the combustion of flammable mixture no I (waste - code 17 02 01 wood - From expendable railroad sleepers with hard coal) meet the class 5 (highest) emission value requirements with regard to PN-EN 303-5: 2012 standard [1].

Solid product tests - slag and ash - obtained from the combustion process of the flammable mixture number I were aimed at designation of factors qualifying its choice for use as a raw material useful in the construction industry or farming.

Table 6 shows the comparison of the concentrations of trace elements obtained in testing. Obtained concentrations were compared to the standards of the quality of the soil or land contained in the Regulation of the Minister of Environment [2].

Test results presented in table 6 for the tested sample of the waste - slag and ash obtained after the combustion of flammable mixture no I, contain values higher than the limit values for soil and land quality standards with regard to surface land assigned to group A and B [2]. The sample meets the requirements of land development in areas of Group C - industrial, mining and communication areas, developed at a depth of 0.3 - 15 m bgl. [2].

Test of combustion of flammable mixture no II

Testing the combustion of flammable mixture no II - 17 02 01 wood - from expendable railroad sleepers with lignite has been carried out for a heat load laboratory boiler of 85 ± 5%.

Thermal-emission tests were aimed at determining the energy efficiency of the boiler and the designation of emission values occurring during the combustion of flammable mixture no II. Thermal-emission test results of the flammable mixture no II were shown in table 7.

At the time of the thermal - emission test the following values were obtained (table 8):

Thermal and emission values of the boiler obtained during combustion of flammable mixture no II were compared with the limit emission values contained in standard PN-EN 303-5: 2012 [1] (table 9):

Emission values obtained during the tests of the combustion of flammable mixture no II (waste - code 17 02 01 wood - from expendable railroad sleepers with lignite) meet the class 3 (lowest) emission value requirements with regard to PN-EN 303-5: 2012 standard [1].

Solid product tests - slag and ash - obtained from the combustion process of the flammable mixture no II were aimed at the designation of factors qualifying its choice for use as a raw material useful in the construction industry or farming. Table 10 provides a comparison of the concentration values of trace elements in combustion tests of the flammable mixture no II with the standards of the soil or land quality contained in the Regulation of the Minister of Environment [2].

Test results presented in table 10 for the tested sample of the waste - slag and ash obtained after the combustion of flammable mixture no II, contain values higher than the limit values for soil and land quality standards with regard to surface land assigned to group A and B [2]. The tested sample meets the requirements of land development in areas of Group C - industrial, mining and communication areas, developed at a depth of 0.3-15 m bgl. [2].

Test of combustion of flammable mixture no III

Testing the combustion of flammable mixture no III - 17 02 01 wood - from expendable railroad sleepers with cereal straw has been carried out for a heat load laboratory boiler of 85 ± 5%.

Thermal - emission tests were aimed at determining the energy efficiency of the boiler and the designation of emission values occurring during the combustion of flammable mixtures no III.

Thermal - emission test results of the flammable mixture no III were shown in table 11.

At the time of the thermal-emission study the following values were obtained (Table 12):

Thermal and emission values of the boiler obtained during combustion of flammable mixture no III were compared with the limit emission values contained in standard PN-EN 303-5: 2012 [1] (Table 13):

Emission values obtained during the tests of the combustion of flammable mixture no III (waste - code 17 02 01 wood - from expendable railroad sleepers with cereal straw) meet the class 3 (lowest) emission value requirements with regard to PN-EN 303-5: 2012 standard [1].

Table 14 provides a comparison of the concentration values of trace elements in the solid product - slag and ash - obtained in combustion tests of the flammable mixture no III with the standards of the soil or land quality contained in the Regulation of the Minister

Water temperature, °C		Water flow, kg/h.	Amount of generated energy, kW	Water pressure in the boiler, bar	Amount of consumed fuel, kg/h	Coefficient of air excess, λ	Boiler load %	Boiler efficiency %
Supply	Return							
46.70	65.45	21.51	0.5	0.1	0.15	1.38	86	79.65

Content of combustible parts in ash and slag: 42.09%

Unit	O_2 %	CO_2 %	SO_2 */	NO_2 */	CO */	Dust */	OGC */	HCl	HF	
mg/m³	11.3	9.5	1105.7	146.3	457	30	0.15	13	p.o.	
kg/h			0.012	0.002	0.005	0.003	0.00000	0.00014	p.o.	

*/- by the oxygen content of 10% in the exhaust gas in the contractual conditions at a temperature of 273 K and a pressure of 101.3 kPa

Table 4: Thermal-emission test results obtained during combustion of flammable mixtures no I.

Specification	Unit	Result	Limit values in accordance with PN-EN 303-5		
Boiler efficiency in accordance with PN-EN 303-5	%	81			
			Class		
			3	4	5
The concentration of dust and gaseous substances emitted into the air: - Dust - OGC - Sulphur dioxide - Nitrogen dioxide - Carbon monoxide	mg/m³ mg/m³ mg/m³ mg/m³ mg/m³	30 0.15 1106 146 457	150 80 - - 3000	60 30 - - 1000	40 20 - - 500

Table 5: Thermal and emission values of the boiler obtained during combustion of flammable mixture no I.

Element	Land"			The value determined for the flammable mixture no I
	Group A	Group B	Group C	
	Depth in m bgl			
		0-0.3	0-2.0	
	mg/kg s.m.			
	I Metals			
Arsenic As	20	20	60	<2
Barium Ba	200	200	1000	692
Cadmium Cd	1	4	15	<2
Cobalt Co	20	20	200	6
Chrome Cr	50	150	500	35
Copper Cu	30	150	600	26
Molybdenum Mo	10.	10	250	<2
Nickel Ni	35	100	300	24
Lead Pb	50	100	600	29
Tin, Sn	20	20	350	2
Zinc Zn	100	300	1000	92

"By the Regulation of the Minister of Environment.

Table 6: Limit values for concentrations in the soil or land, compared with the results obtained from the analyses of the test sample of the solid product-ash and slag-from combustion of the flammable mixture no I.

Water temperature, °C		Water flow, kg/h	Amount of generated energy, kW	Water pressure in the boiler, bar	Amount of consumed fuel, kg/h	Coefficient of air excess, λ	Boiler load %	Boiler efficiency %
Supply	Return							
43.70	61.48	43.06	1.01	0.1	0.45	1.33	80	62.22
Content of combustible parts in ash and slag,%		42.09%						

Table 7: Thermal-emission test results obtained during combustion of flammable mixture no II.

Unit	O_2 %	CO_2 %	SO_2 */	NO_2 */	CO */	Dust */	OGC */	HCl	HF
mg/m³	13.0	7.8	211.8	134.8	1098.6	122	0.89	8	p.o.
kg/h			0.002	0.001	0.010	0.003	0.00014	0.00008	p.o.

*By the oxygen content of 10% in the exhaust gas in the contractual conditions at a temperature of 273 K and a pressure of 101.3 kPa.

Table 8: Thermal-emission test results

Specification	Unit	Result	Limit values in accordance with PN-EN 303-5		
Boiler efficiency in accordance with PN-EN 303-5	%		62		
			Class		
			3	4	5
The concentration of dust and gaseous substances emitted into the air:					
- Dust	mg/m³	122	150	60	40
- OGC	mg/m³	1	80	30	20
- Sulphur dioxide	mg/m³	212	-	-	-
- Nitrogen dioxide	mg/m³	135	-	-	-
- Carbon monoxide	mg/m³	1099	3000	1000	500

Table 9: Thermal and emission values of the boiler obtained during combustion of flammable mixture no II.

Element	Land[ ]			The value determined for sample: Flammable mixture no II
	Group A	Group B	Group C	
	Depth in m bgl			
		0-0.3	0-2.0	
	mg/kg s.m.			
	I Metals			
Arsenic As	20	20	60	<2
Barium Ba	200	200	1000	5
Cadmium Cd	1	4	15	194
Cobalt Co	20	20	200	<2
Chrome Cr	50	150	500	11
Copper Cu	30	150	600	53
Molybdenum Mo	10	10	250	19
Nickel Ni	35	100	300	85
Lead Pb	50	100	600	<2
Tin, Sn	20	20	350	37
Zinc Zn	100	300	1000	29

[ ]by the Regulation of the Minister of Environment dated 9 September 2002 on the soil and ground quality standards - Dz. U. 02.165.1358 (Journal of Laws 02.165.1358).

Table 10: Limit values for concentrations in the soil or land, compared with the results obtained from the analysis of the test sample of the solid product-ash and slag-from combustion of the flammable mixture no II.

Water temperature, °C		Water flow, kg/h	Amount of generated energy, kW	Water pressure in the boiler, bar	Amount of consumed fuel, kg/h	Coefficient of air excess, λ	Boiler load %	Boiler efficiency %
Supply	Return							
45.12	63.12	14.56	0.33	0.1	0.16	1.25	83	61.8
Content of combustible parts in ash and slag,%		10.33%						

Table 11: Thermal-emission test results obtained during combustion of flammable mixture no III.

Unit	O₂ %	CO₂ %	SO₂[ ]	NO₂[ ]	CO[ ]	Dust[ ]	OGC[ ]	HCl	HF
mg/m³	12.8	7.6	637	217	966	565	2.5	3	p.o.
kg/h			0.002	0.001	0.010	0.003	0.00000	0.00005	p.o.

[ ]By the oxygen content of 10% in the exhaust gas in the contractual conditions at a temperature of 273 K and a pressure of 101.3 kPa.

Table 12: Thermal-emission study results.

Specification	Unit	Result	Limit values in accordance with PN-EN 303-5		
Boiler efficiency in accordance with PN-EN 303-5	%		62		
			Class		
			3	4	5
The concentration of dust and gaseous substances emitted into the air:					
- Dust	mg/m³	565	150	60	40
- OGC	mg/m³	3	80	30	20
- Sulphur dioxide	mg/m³	637	-	-	-
- Nitrogen dioxide	mg/m³	217	-	-	-
- Carbon monoxide	mg/m³	966	3000	1000	500

Table 13: Emission values obtained during the tests of the combustion of flammable mixture no III.

Element	Land"			The value determined for Sample: Flammable mixture no III
	Group A	Group B	Group C	
	Depth in m bgl			
		0-0.3	0-2.0	
	mg/kg s.m.			
I Metals				
Arsenic As	20	20	60	<0.19
Barium Ba	200	200	1000	<0.19
Cadmium Cd	1	4	15	<0.047
Cobalt Co	20	20	200	<0.19
Chrome Cr	50	150	500	<0.19
Copper Cu	30	150	600	<0.19
Molybdenum Mo	10	10	250	0.37
Nickel Ni	35	100	300	<0.47
Lead Pb	50	100	600	<0.47
Tin, Sn	20	20	350	<0.19
Zinc Zn	100	300	1000	<0.19

"by the Regulation of the Minister of Environment dated 9 September 2002 on the soil and land quality standards - Dz. U. 02.165.1358 (Journal of Laws 02.165.1358).

Table 14: Limit values for concentrations in the soil or land, compared with the results obtained from the analysis of the test sample of the solid product-ash and slag-from combustion of the flammable mixture no III.

of Environment [2].

Results of tests contained in Table 9 for the tested sample of waste - slag and ash obtained after combustion of flammable mixture no III - waste - code 17 02 01 wood - from expendable railroad sleepers with cereal straw showed values lower than the limit values for soil and land quality in relation to surface grounds assigned to group A, B and C [2]. The sample meets the requirements for application in development of land:

- Group A: (a) Land immovable of the area protected on the basis of the Water Law regulations. (b) Areas subject to protection on the basis of the nature protection regulations, if maintaining the current level of land pollution does not pose a threat to human health or environment - for these areas the concentrations resulting from the actual state maintain standards.

- Group B: land classified as agricultural land except the land under ponds and ditches, forest land, land with a high amount of woods and bushes, waste land as well as built-up and urbanized land, with the exception of industrial land, mining areas, and communication areas;

- Group C: industrial, mining and communication areas; Developed to a depth of 0.3-15 m bgl. [2].

Summary

The paper presents a concept of producing energy on the basis of modern alternative fuels with the expendable railroad sleepers to be burnt in power stokerfired boilers. The thermal energy contained in water vapour and hot water will be utilized in producing in combination, of electrical energy, and for heating of cubature objects. There have been presented the properties of alternative fuels obtained, and the concept of their utilization in the process of emissions of natural environment. The paper presents research methodology thermal properties, the measurements of the emissions of dust and gaseous substances emitted to the atmosphere and pollutants to the ground arising during, flammable mixtures used in testing on the basis of the obtained research of energy properties of fuels, biomass and waste and the energy efficiency of the boiler during the combustion of flammable mixtures: hard coal, lignite, cereal straw and waste - wood (from railroad sleepers).

References

1. PN-EN 303-5 (2012) Heating boilers for solid fuels, hand and automatically stocked, with a nominal heat of up to 500 kW-Terminology, requirements, testing and marking.

2. Regulation of the Minister of Environment (2002) on the soil and land quality standards.

3. Regulation of the Minister of Environment (2011) on standards for emissions from installations-Journal of Laws 11.95.558.

4. Regulation of the Minister of Economy, Labour and Social Policy (2003) amending the regulation concerning the requirements on the process of thermal transformation of waste. Journal of Laws 03.1.2.

5. Regulation of the Minister of Environment (2010) concerning detailed technical conditions of qualifying part of energy recovered from thermal transformation of municipal wastes-Journal of Laws 110.117.788.

6. Regulation of the Minister of Economy (2008) concerning the detailed scope of obligation to obtain and present for the discontinuance of certificate, acquaintance of compensatory payment, purchasing electric energy and heat generated from renewal sources and the obligation to confirm the data on the amount of electric energy generated in a renewal energy source-Journal of Laws 08.156.969.

7. Czop M, Kajda-Szcześniak M (2013) Methods of disposal of sewage sludge. Archives of Waste Management and Enviromental Protection 15: 83-92.

8. Eduljee G (1994) Organic micro pollutants emissions from waste incineration. In: Hester RE, Harrison RM (ed.)-Waste incineration and the environment, Royal Society of Chemistry, London.

9. Karekez S, Coelhob ST (2004) Renewable energy-traditional biomass vs. Modern biomass. Energy Policy 32: 711–714.

10. Demirbas A (2005) Potential applications of renewable energy sources, biomass combustion problems in boiler power systems and combustion related environmental issues. Progress in energy and combustion Science 31: 171-192.

11. Regulation of the Minister of Environment (2001) on waste catalogue-Journal of Laws 01.112.1206.

12. Środa K, Kijo-Kleczkowska A, Otwinowski H (2013) Methods of disposal of sewage sludge. Archives of Waste Management and Environmental Protection 15: 33-50.

13. Środa K, Kijo-Kleczkowska A, Otwinowski H (2012) Thermal disposal of sewage sludge. Inżynieria Ekologiczna 28: 67-81.

14. Law (2001) Environmental Law-Journal of Laws 01.62,627 consolidated text Journal of Laws 08.25.150 dated 23.01.2008

15. Law (2012) on waste-Journal of Laws 13.21.

16. Regulation of the Minister of Environment (2012) on permissible substance

degree in the air-Journal of Laws 12.1031.

17. Regulation of the Minister of Environment (2010) on reference values for some substances in the air, Journal of Laws 10.16.87.

18. Regulation of the Minister of Economy and Labour (2005) on criteria and procedures of permissions for waste collection at particular types of landfills, Journal of Laws 05.186.1553.

19. Regulation of the Minister of Environment (2002) on the soil and land quality standards, Journal of Laws 02.165.1358.

20. Fieducik J, Gawroński (2010) A Sewage sludge drying and combustion as a method of its utilization based on the example of incineration plant in Olsztyn. Zeszyty Naukowe Politechniki Rzeszowskiej. Budownictwo i Inżynieria Środowiska 57: 147–154.

21. Girczys J, Rećko K (2001) The possibilities of mutual management of sewage sludge and waste coal slime. Inżynieria i Ochrona Środowiska 4: 107-116.

22. Pająk T (2003) Combustion and co-combustion of sewage sludge-basic determinants. Przegląd Komunalny 1: 35-38.

23. Pająk T, Wielgosiński G (2003) Contemporary technologies of sewage sludge drying and combustion-criteria and determinants of technology choice. II Międzynarodowa i XIII Krajowa Konferencja Naukowo-Techniczna n.t. Nowe spojrzenia na osady ściekowe. Odnawialne źródła energy i Częstochowa 491-500.

24. Wandrasz JW, Kozioł M, Landrat M, Ścierski W, Wandrasz AJ (2000) Possibility of sludge co-combustion from sewage-treatment plants jointly with coal in stoker boilers. Gospodarka Paliwami i Energią 8: 10–15

25. Wielgosiński G (2002) Combustion, co-combustion and drying of sewage sludge. Przegląd Komunalny 1:10-15.

26. GIG documentation of research and development work- No 582 0777 3-323.

27. Ochrona Środowiska 2000-2010. Główny Urząd Statystyczny. Informacje i Opracowania Statystyczne. Warszawa 2010.

28. Regulation of the Minister of Economy and Labour (2005) on criteria and procedures of permissions for waste collection at particular types of landfills, Journal of Laws 05.186.1553.

29. Regulation of the Minister of Environment (2008) on requirements concerning measurements on volume of emission and collected water-Journal of Laws 08.206.1291.

30. PN-Z-04030-7 (1994) Measurement of dust concentration and flow in waste gases using the gravimetric method. Flow, gas physical parametres, dust emission measurements.

31. Regulation of the Minister of Environment (2008) on types of results of measurements carried out because of installation or device exploitation as well as other data terms and ways of their presentation. Journal of Laws.08.215.1366.

32. PN-ISO 1928 (2002) Solid mineral fuels-Determination of gross calorific value by the calorimetric method, and calculation of net calorific value.

33. PN-ISO 334 (1997) Solid mineral fuels-Determination of total sulphur-Eschka method.

34. PN-ISO 1171 (2002) Solid mineral fuels-Determination of ash.

35. PN-ISO 587 (2000) Solid mineral fuels-Determination of total chlorine using Eschka mixture.

36. Regulation of the Minister of Economy (2002) concerning other waste types than dangerous and types of installations and devices where their thermal transformation is allowed, Journal of Laws 02.18.176.

5

Experimental Determination of Bubble Size in Solution of Surfactants of the Bubble Column

Maedeh Asari[1] and Faramarz Hormozi[2]*

[1]*Faculty of Engineering- Department of Chemical Engineering, Islamic Azad University, Shahrood Branch, Shahrood, Iran*
[2]*Faculty of Chemical, Gas and Petroleum Engineering, Semnan University, Semnan, Iran*

Abstract

This paper focuses on the effect of surfactants on the bubble size. Bubble size in Sodium dodecyl sulfate / water system were investigated at various superficial gas velocities (0.13, 0.26 and 0.5 cm/s). On the other hands, Bubble diameter was determined for different values of Sodium dodecyl sulfate surfactant concentration. Surfactant concentration in water were 0.05, 0.02 and 0.1 vol.%. Tap water and aqueous solutions with surfactants (anionic, non-ionic and zwitterionic) are used as liquid phases. The bubbles size in this phase is determined at C_s=0.05%vol and u_g=0.13 cm/s. The bubbles are generated into a small-scale bubble column making of Plexiglas with height of 1.2 m. High speed photography techniques are used to measure the bubble size. The experimental results were shown that bubble diameter in Sodium dodecyl sulfate /water system is larger than other systems. In solution of Sodium dodecyl sulfate, Sauter mean bubble diameter (Location A and D) decreases when superficial gas velocity increased.

Keywords: Bubble column; Surfactant; Bubble size

Introduction

Bubble column reactors are widely used in chemical and biochemical processes such as oxidation, chlorination, polymerization, hydrogenation, synthetic fuels by gas conversion processes, fermentation and wastewater treatment. Bubble columns can be employed in many mass transfer processes [1]. However, the lack of a more complete knowledge on the bubble column fluid dynamic behavior in its various regimes causes several operational difficulties and design uncertainties, which include poor predictions of the mean bubble diameter, gas hold up and interfacial area [2,3]. A bubble column reactor is basically a cylindrical vessel with a gas distributor at the bottom [4]. The interfacial area available for mass transfer is the most important design parameter defined by gas holdup and bubble size which in turn are affected by the operating conditions, the physic-chemical properties of the two phases, the gas sparger type and the column geometry [5]. Bubble column are preferred to be two-phase contactors for their ease of operation, maintenance and absence of moving parts, yet they have complex hydrodynamics characteristics [6]. Knut [7] studied dynamic simulation of 2D bubble column and shown that two dimensional dynamic simulation of the flat bubble column is feasible, applying state-of the art dynamic turbulence models. Surfactant designates a substance that exhibits some superficial or interfacial activity. Different methods have been employed for bubble dimension evaluation [8]. Gas bubbles in transparent fluids can be photographed and measured, usually using image-analysis [9]. This is the simplest technique but cannot be used with opaque media such as those found in fermentation systems. Statistical models [10] are required to calculate bubble –size distributions from the measured chord lengths. Several authors studied bubble size and interfacial phenomena in different types of bubble column reactors. Colella et al. [11] studied the interfacial mechanisms focusing on the coalescence and breakage phenomena of bubbles in three different bubble columns. They investigated the influences of gas superficial velocity and different hydrodynamic configurations on bubble size distribution in the bubble columns. Lehr and Mewes [12] evaluated the bubble sizes in two-phase flows. They predicted the bubble size distribution in bubble columns including the formation of large bubbles at high superficial gas velocities. Schäfer et al. [13] discussed the influence of operating conditions and physical properties of gas and liquid phase on initial and stable

bubble sizes in a bubble column reactor under industrial conditions. Akita and Yoshida [14] determined the bubble size distribution using a photographic technique. The gas was sparged through perforated plates and single-orifice using various liquids (water, aqueous and pure glycol, methanol, and carbon tetrachloride). It has been reported in the literature that with increase in surfactant concentration, coalescence time increases [15]. Sardeing et al. [16] reported that in superficial gas velocities between 1.5×10^{-4}-2×10^{-4} m/s, bubble diameter was in surfactant solutions between 1 mm-8 mm. In these studies we have also analysed the influence of SDS surfactant concentration and the gas flow-rate upon the bubble diameter in bubble column. On the other hand, the bubble size distribution has been studied in ionic, nonionic and zwitterionic surfactants on the bubble column. Yang and Maa [17] studied ionic surfactants (SDS, SDBS) in water systems and showed that the coalescence time increased with increasing surfactant concentration.

Experimental Setup and Technique

The schematic diagram of the modified bubble column is shown in Figure 1. It consists of an air compressor, (1), a rotameter, (2), an halegon lamp, (3), etc. Four different surfactant (Tween20, [CAS No:9005-64-5], Triton X-100, [CAS No:9002-93-1], Cocamidopropy Betaine, [CAS No:61789-40-0], SDS(Sodium dodecyl sulfate), [CAS No:151-21-3]) have been used in the present work. All of surfactants were purchased from Merck Company (Germany). Bubble size is reported at ambient conditions [atmospheric pressure 25 (± 0.5)°C]. The gas from the compressed air line passed through calibrated rotameter. The photographic method, used in this study to determine

*Corresponding author: Faramarz Hormozi, Faculty of Chemical, Gas and Petroleum Engineering, Semnan University, Semnan, Iran
E-mail: fhormozi@semnan.ac.ir

the bubble size of the two-phase mixture, has been developed using a rectangular bubble column (20 cm×5 cm×120 cm). The liquid column heights during the operation were 45 cm. To determine profiles of ellipsoid, bubble was monitored over distance ca. 1m and was using professional video recorder. The photographs were taken by a digital camera (Casio Exilim (EX-F1)) taken along the height of the column, from bottom to top. The digital photographs were processed and enhanced by using Image Processing MATLAB Software that enabled to distinguish clearly the bubble boundaries. The diameters of the bubbles were determined from photographs of the operating column, 5, 20, 30 and 40 cm above the gas distributor. The images were taken at three axial positions for different operating conditions. The 2d picture shapes of the bubbles were approximated by ellipsoid whose maximum and minimum axes were automatically computed by the software program used for image analysis [18,19]. The third dimension was calculated with the assumption that the bubbles are symmetric around the minimum axes. From the known values of maximum and minimum axes, an equivalent ellipsoid bubble diameter was calculated by the following equation [20]:

$$d_{be} = (d^2_{b,max} d_{b,min})^{1/3} \tag{1}$$

where $d_{b,max}$ and $d_{b,min}$ are the ellipsoid maximum and minimum axes of bubble. The distributions were obtained by sorting the equivalent diameters of bubbles into different uniform classes. At a particular operating condition, the bubble picture taken from different locations of the column are shown in Figure 2. The Sauter mean bubble diameter (d_{32}) is defined as the volume-to-surface mean bubble diameter [21]:

$$d_{32} = \frac{\sum_{i=1}^{N} n_i d_{Bi^3}}{\sum_{i=1}^{N} n_i d_{Bi^2}} \tag{2}$$

where n_i is the number of bubbles of diameter d_{Bi}.

Between 1000 and 3000 bubbles were counted for determination of the size distribution, using 30 photographs.

Bubble Size Distribution

Bubble coalescence and breakup play a significant role in determining bubble size distribution. Coalescence was found to take place when more than about a half of the projected area of the following bubble was overlapped with that of the leading bubble at the critical distance. In contrast, the breakup occurred in the case the overlapping

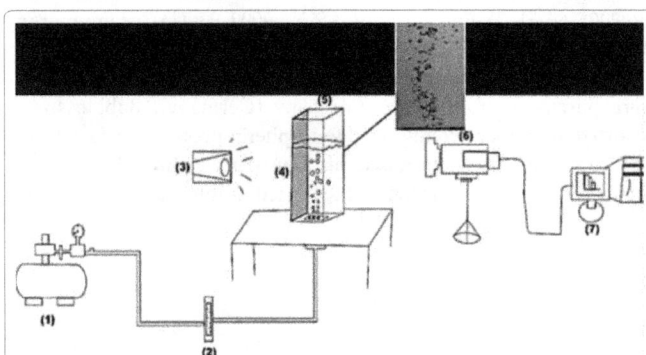

Figure 1: Gas-liquid experimental set-up. (1): Air compressor (2): Rotameter (3): Halogen lamp (4): Plexiglas plate (5): Bubble column (6): Camera (7): Image processing.

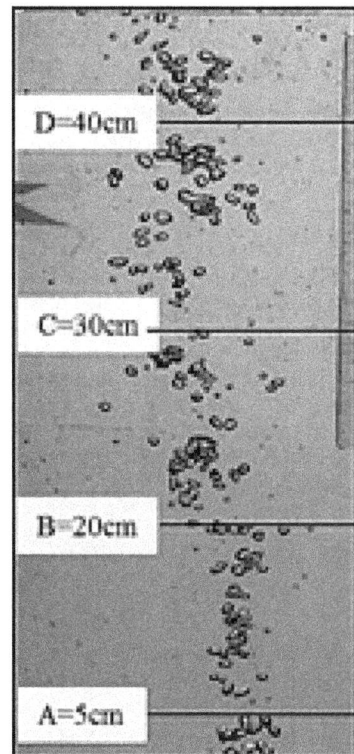

Figure 2: Photograph taken from different location of column in Tween20/water system at C_s=0.05% vol and u_g=0.13 cm/s Location A , B, C and D.

was less than about a half of the projected area of the following bubble. Thus, when the leading bubble is larger than the following one, the latter has a tendency to coalesce. In contrast, in the case of the smaller size of the leading bubble, the following bubble tends to breakup. Coalescence is significantly influenced by the physical properties of the liquid. Analysis of bubble size in bubble columns must distinguish between bubble-size distribution just after bubble formation at the sparger and size distribution further away from the distributor [18]. Two basic methods – photography and probe techniques – exist for determining bubble size, however; they do not lead to identical results. Both methods are subject to certain limitations in view of the marked bubble selection that may occur (i.e., not all bubble sizes can be detected). In particular, any measurement method only leads to realistic results if the flow is homogeneous (i.e., a narrow bubble-size distribution is found). As yet, no method can be recommended for the measurement of large bubbles in the heterogeneous flow regime.

Results and Discussion

Effect of superficial gas velocity upon bubble size in SDS +water system

First, there is general observation that applies to all solutions. For example, regardless of type and presence of chemical added, the average bubble radius decreases with increasing of gas flow rate. Figure 3 show bubble size distribution for SDS-water system in regions A and D. As the gas flow rate increases the gas holdup and kinetic energy increases which increase turbulent intensity, bubble- bubble interactions, velocity of bubbles and the probability of coalescence which is because of as increasing collision frequency between bubbles with increase in gas flow rate (Figure 3).

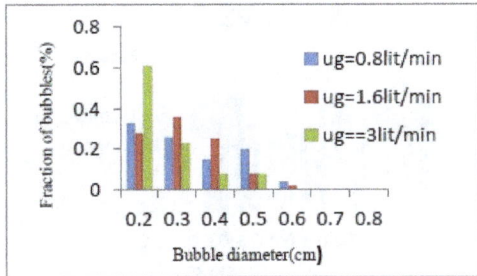

(a)

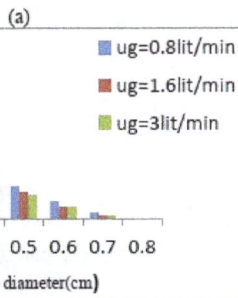

(b)

Figure 3: Effect of gas flow rate upon bubble size in SDS/water system at C_s=0.02% vol. (a) Location A; (b) Location D.

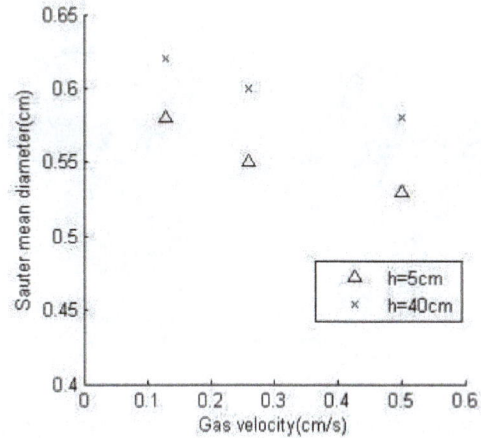

Figure 4: Influence of gas flow rate upon Sauter mean bubble diameter in SDS/water system at C_s=0.02% vol.

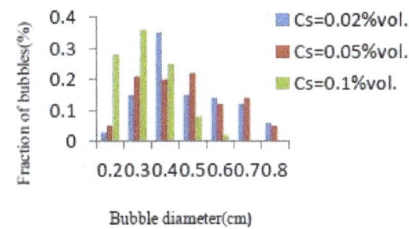

(a)

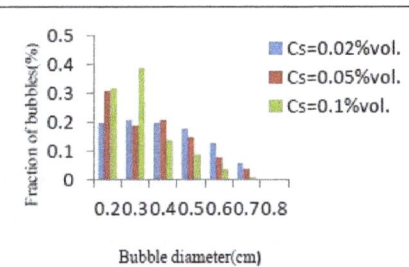

(b)

Figure 5: Influence of SDS concentration upon bubble diameter at u_g=0.13cm/s; (a) Location A. (b) Location D.

The probability of coalescence is higher in region D but the bubble size decreases with increasing superficial gas velocity in A and D location. This is due to bubble break- up with increasing gas flow rate. Also as the superficial gas velocity increases, the Sauter mean bubble diameter decreases (Figure 4). For u_g greater than 0.13 cm/s smallest bubbles are obtained in solution of lowest static surface tension. The rate of coalescence decreases with the gas flow rate increasing (Figure 4).

Effect of SDS concentration upon bubble size

One of the parameters that effect bubble size, is surfactant concentration. Effect of various SDS concentrations (0.02, 0.05 and 0.1% vol) at u_g=0.13 cm/s on bubbles diameter is shown in Figure 5. SDS addition to pour water decreased the bubbles diameter. Further, surfactant concentration enhancement decreased the of bubbles diameters by decreasing the surface tension and buoyancy force. Sardeing et al. [16] used various surfactants and investigated that bubbles diameter decreased about 30% (as an average value). The bubble size distribution in an emulsification processes is a result of the competition between opposite processes, bubble breakage and bubble-bubble coalescence. It was shown experimentally that the bubble size rapidly decreases with an increase of SDS concentration [22]. Sample photographs of the bubble populations shown in Figure 6. They clearly showed that as the SDS concentration increases, the bubble populations will become smaller in size. Sauter mean bubble diameter (d_{vs}) decreases due to SDS concentration increasing (Figure 7).

Effect of ionic, non-ionic and zwitterionic surfactants on bubble size

Presence of surfactants has a great effect on the bubble diameters. The bubble size distribution was obtained in four axial locations A (of height 0.05 m), B (of height 0.2 m), C (of height 0.3 m) and D (of height 0.4 m) from the bottom of the column (Figure 2). Typical results

for these four locations are presented in Figure 8. It is seen that the bubble size in location D are greater than location A, B and C (Figure 2). The average bubble size in location C and B are almost the same. All calculations regarding goodness of fit have been performed by MATLAB software. Bubble diameter increased with increasing the distance from the bottom of the column due to the coalescence of smaller bubbles. The coalescence bubbles of location A go up due to their buoyancy and accumulate in location B, C and D. Also the bubble number flux varies in different locations due to the same reason. That bubble number flux decreases in location C and D over location A and B is result of an increase in bubble size due to coalescence. As shown in Figure 8, there is no significant variation of bubble size in location

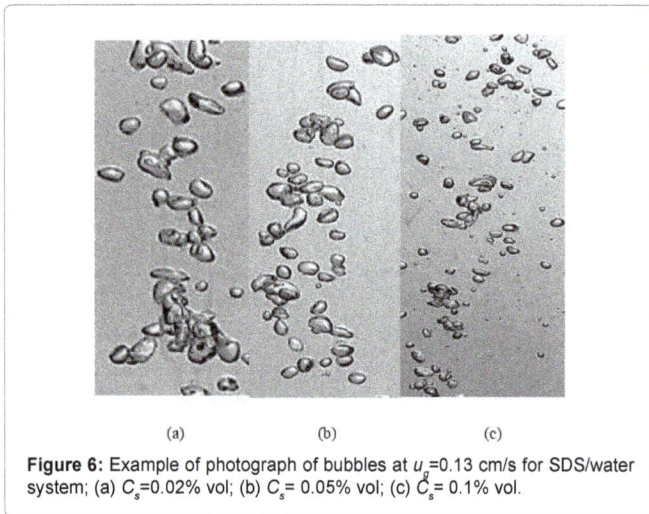

Figure 6: Example of photograph of bubbles at u_g=0.13 cm/s for SDS/water system; (a) C_s=0.02% vol; (b) C_s= 0.05% vol; (c) C_s= 0.1% vol.

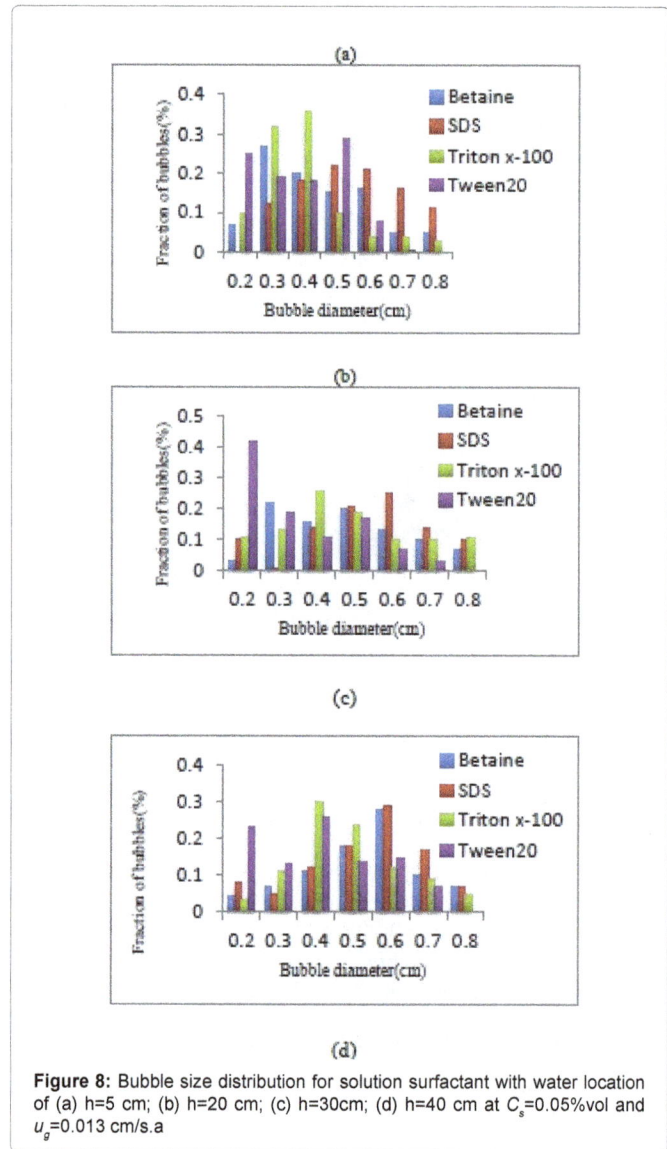

Figure 7: Effect of SDS concentration upon sauter mean diameter at ug=0.13 cm/s.

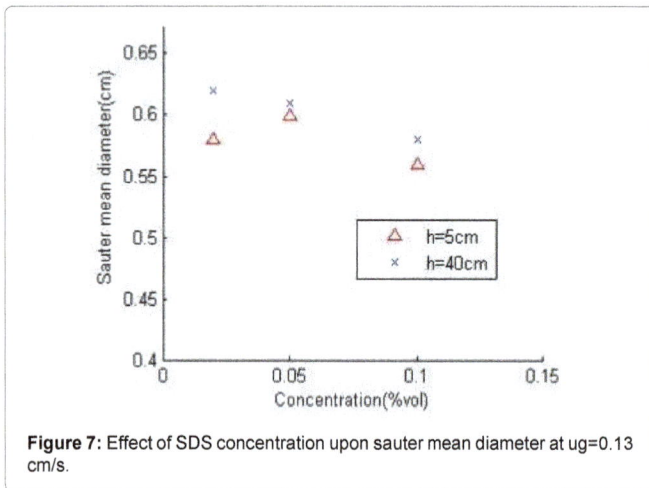

Figure 8: Bubble size distribution for solution surfactant with water location of (a) h=5 cm; (b) h=20 cm; (c) h=30cm; (d) h=40 cm at C_s=0.05%vol and u_g=0.013 cm/s.a

B and C. The bubble size in location A is much smaller than other locations due to a break-up.

Figure 8 shows the relation between the detached bubble diameter and fraction of bubbles for the different surfactants. Whatever the liquid properties are, the bubble diameters vary between 0.2 and 0.8 for ug=0.13cm/s (d<0.2 cm, effective force is surface tension and bubbles are spherical). For this gas flow rate, the order below is observed:

$$d_{SDS} > d_{Betaine} > d_{Triton\ X-100} > d_{Tween20}$$

Sauter mean bubble diameter was investigated in four different axial positions (A, B, C and D).Typical profile of d_{32} as a function of height above sparger are show in Figure 9. The sauter mean bubble diameter varies with axial location due to coalescence effect whereas the variation of gas holdup is due to variation of bubble number flux. The values of d_{32} obtained in the range of 0.4-0.65. Increasing in height increases d_{32} at all type of surfactants. The value of d_{32} in SDS+water system is more than other systems.

Conclusion

Effect of surfactant on the bubble size in rectangular bubble column has been studied. In order to obtain bubble size distribution about 1000-3000 bubbles were analyzed. The evaluation of bubble size distribution in different location of the column and the influence gas

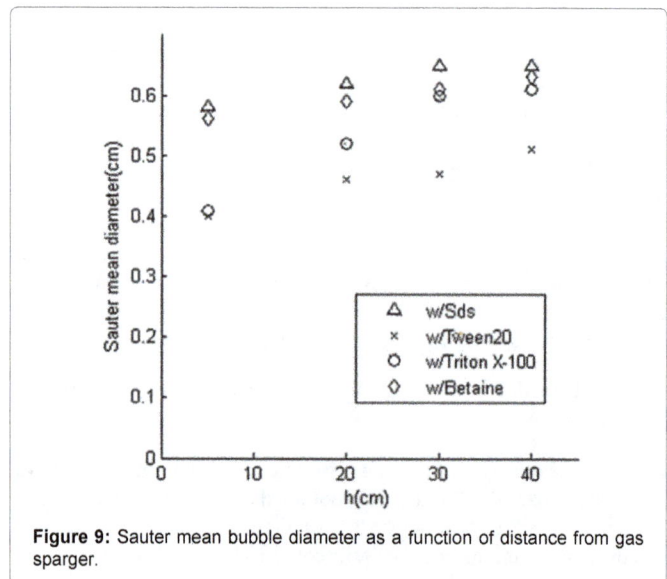

Figure 9: Sauter mean bubble diameter as a function of distance from gas sparger.

flow rate and SDS concentration were pointed out. The measurements were done using photographic techniques. The bubble size in bubble column increased with increasing distance from the bottom of the column due to coalescence. The bubble diameter in SDS+ water system were bigger than other systems (Betaine, Triton X-100 and Tween20 at C_s=0.05% vol and u_g=0.013 cm/s). When gas flow rate increase (SDS +water system), an increase in the number of small bubbles was also observed, and Sauter mean bubble diameter also decreased due to breakage bubbles. The Sauter mean bubble diameter decreases, when SDS concentration increasing.

Nomenclature

$d_{b,max}$ major axis of the projected ellipsoid(m)

$d_{b,min}$ minor axis of the projected ellipsoid(m)

d_{32} sautor mean diameter bubble(m)

ug superficial gas velocity (m/s)

h distance from gas distributor(m)

d diameter(m)

C_s surfactant concentration

References

1. Ariyapadi S, Balachandar R, Berruti F (2004) Effect of surfactant on the characteristics of a droplet-laden jet. Chemical Engineering and Processing: Process Intensification 43: 547-553.

2. Al-Masry WA, Ali EM, Aqeel YM (2005) Determination of bubble characteristics in bubble columns using statistical analysis of acoustic sound measurements. Chemical Engineering Research and Design 83: 1196-1207.

3. Lage PLC, Esposito RO (1999) Experimental determination of bubble size distributions in bubble columns: prediction of mean bubble diameter and gas hold up. Powder Technology 101: 142–150.

4. Kantarci N, Borak F, Ulgen KO (2005) Review Bubble column reactors. Process Biochemistry 40: 2263-2283.

5. Camarasa E, Vial C, Poncin S, Wild G., Midoux N, et al. (1999) Influence of coalescence behaviour of the liquid and of gas sparging on hydrodynamics and bubble characteristics in a bubble column. Chemical Engineering and Processing: Process Intensification 38: 329–344.

6. Waheed AAM, Emad MA, Yehya MA (2006) Effect of antifoam agent on bubble characteristic in bubble columns based on acoustic sound measurements. Chemical Engineering Science 61: 3610-3622.

7. Knut B (2005) Dynamic simulation of a 2Dbubble column. Chemical Engineering Science 60: 5294–5304.

8. Gómez-Díaz D, Navaza JM, Sanjurjo B (2008) Interfacial area evaluation in a bubble column in the presence of a surface-active substance Comparison of methods. Chemical Engineering Journal 144: 379-385.

9. Machon V, Pacek AW, Nienow AW (1997) Bubble size in electrolyte and alcohol solutions in a turbulent stirred vessel. Chemical Engineering Science 75: 339-348.

10. Clark NN, Turton R (1998) C Chord length distributions related to bubble size distributions in multiphase flows. International Journal multiphase Flow 14: 413-424.

11. Colella D, Vinci R, Bagatin M, Masi E, Abu Bakr (1999) A study on coalescence and breakage in three different bubble columns. Chemical Engineering Science 54: 4767-4777.

12. Lehr F, Mewes D (2001) A transport equation for the interfacial area density applied to bubble columns. Chemical Engineering Science 56: 1159-1166.

13. Schäfer R, Merten C, Eigenberger G (2002) Bubble size distribution in a bubble column reactor under industrial conditions. Experimental Thermal and Fluid Science 26: 595-604.

14. Akita K, Yoshida F (1974) Bubble size, interfacial area and liquid-phase mass transfer coefficient in bubble columns. Ind. Eng. Chem. Process Des Dev 13: 84-91.

15. Ghosh P, Juvekar VA (2002) Analysis of the drop rest phenomenon. Chemical Engineering Research and Design 80: 715-728.

16. Sardeing R, Painmanakul P, Hébrard G (2006) Effect of surfactants on loquid-side mass transfer coefficient in gas-liquid systems: A first step to modeling. Chemical Engineering Science 61: 6249-6260.

17. Yang YM, Maa JR (1984) Bubble coalescence in dilute surfactant solution. Journal of Colloid and Interface Sci 98: 120–125.

18. Majumder SK, Kundu G, Mukherjee D (2006) Bubble size distribution and gas-liquid interfacial area in a modified downflow bubble column. Chemical Engineering Journal 122: 1-10.

19. Polli M, Stanislao MD, Bagatin R, Abu Bakr E, Masi M (2002) Bubble size distribution in the sparger region of bubble columns. Chemical Engineering Science 57: 197-205.

20. Couvert A, Roustan M, Chatellier P (1999) Two-phase hydrodynamic study of a rectangular airlift loop reactor with an internal baffle. Chemical Engineering Science 54: 5245-5252.

21. Asgharpour M, Mehrnia MR, Mostoufi N (2011) Effect of surface contaminants on oxygen transfer in bubble column reactors. Biochemical Engineering Journal 49: 351-360.

22. García-Abuín A, Gómez-Díaz D, Losada M, Navaza JM (2012) Bubble column gas-loquid interfacial area in a polymer + surfactant + water system. Chemical Engineering Science 75: 334-341.

6

Antimicrobial Activity of Bimetallic Cu/Pd Nanofluids

Aashit Kumar Jaiswal[1]*, Mayank Gangwar[2,3], Gopal Nath[2] and RR Yadav[1]

[1]*Department of Physics, University of Allahabad, Allahabad, Uttar Pradesh, India*
[2]*Department of Microbiology, Institute of Medical Sciences, Banaras Hindu University, Varanasi, Uttar Pradesh, India*
[3]*Department of Pharmacology, Institute of Medical Sciences, Banaras Hindu University, Varanasi, Uttar Pradesh, India*

Abstract

A series of copper/palladium bimetallic nanostructures based nanofluids has been prepared with the aim of investigating antimicrobial activity. Synthesized nanofluids were characterized by UV-visible spectroscopy, X-ray diffraction, scanning electron microscopy and acoustic spectroscopy to determine their optical absorbance, structure, surface morphology and particle size distribution, respectively. Prepared nanofluids were tested for their antimicrobial activity using the agar disc diffusion method and their minimum inhibitory concentration (MIC) values were calculated by micro-dilution method. Results of antimicrobial activity revealed that prepared nanofluids possessed a good antibacterial activity against microbial species. In summary, the application of these bimetallic nanofluids as antimicrobial agent will be very valuable for biomedical and industrial applications.

Keywords: Nanofluids; Bimetallic; Antibacterial activity; MIC

Introduction

Nanofluids (the dispersion of nano-sized particles in fluids) have been extensively used in a wide variety of applications [1-7]. In recent years, nanofluids have been found to possess excellent antimicrobial activity, largely due to the ability of small sized particles being able to penetrate the living microbe-cells and causing chemically induced internal damage [1,6,7]. In this regard, although both nano-sized organic and inorganic particles have been extensively investigated, inorganic nanomaterials are considered to be better antimicrobial agents [6] as organic materials become unstable at high temperatures or pressures. Therefore, the use of inorganic materials has emerged up as novel antimicrobial agents. Among inorganic metallic nanomaterials, antibacterial properties of silver and gold nanoparticles are comprehensively studied for their potential applications in food packaging, in disinfection of water and in the infection control in medicine [7].

Due to improved catalytic property, bimetallic nanoparticles are of greater interest in comparison of monometallic nanoparticles [8-14]. By virtue of their unique properties, they have been applied in sensors, catalysts, electronic devices and optics. Depending on preparation conditions, bimetallic nanoparticles can be in alloy form or core-shell structure. Among various bimetallic nanoparticles, considerable interest has been paid to the preparation of the bimetallic nanoparticles of Cu since Cu exhibits excellent features like electrical conductivity and chemical activity [10-12]. Few reports are available on synthesis of Cu/Pd nanoparticles using different techniques [13-15]. However, there is no literature on the bactericidal properties of Cu/Pd bimetallic nanofluids (BMNFs). Also among the different antimicrobial agents, silver and gold nanoparticles have been extensively studied and used since ancient times to fight infections and prevent spoilage. Cu/Pd nanofluids are cost effective in comparison to silver and gold. Hence, we prepared a series of Cu/Pd nanofluids by varying their molar ratios and the antimicrobial properties of the nanofluids (NFs) have been studied. Antimicrobial activities of the prepared nanofluids were evaluated using the agar disc diffusion method against standard strains viz. *E. coli*, *P. aeruginosa*, *E. faecalis*, *S. aureus* and against the yeasts *C. albicans*, *C. tropicalis*, *C. neoformans* and their MIC values were calculated by micro-dilution method. The antibacterial activity measurements for copper-palladium nanofluids are performed first time. Results of antibacterial activity were found to depend on the concentration of copper and palladium in nanofluids.

Experimental

A series of bimetallic Cu/Pd nanostructures based nanofluids were prepared via a facile method [10] by taking the three different molar ratios - 1/20, 1/1 and 20/1 of Cu/Pd bimetals at nanoscale. A flow chart for the synthesis is shown in Figure 1. Trisodium citrate (147 mg) was used as a complexing agent and polyvinyl pyrrolidone (PVP) (500 mg) as a protecting agent. Whole synthesis was carried on at room temperature. The synthesized Cu:Pd in ratio 1:20, 1:1 and 20:1, were labelled as S1, S2 and S3, respectively. The structural properties and crystallite size of the material was analyzed with X-ray diffractometer (X-Pert PRO PANalytical) using monochromatized Cu K_a radiation (λ=1.54059 Å). Thin film was used as target material in the XRD measurement. The morphology of the thin film was investigated with scanning electron microscope (SEM, LEO-0430). Particle size of Cu/Pd bimetallic nanoparticles and their distribution in the nanofluids were determined with a Matec Applied Sciences acoustic particle sizer (APS-100). APS-100 determines particles size and their distribution

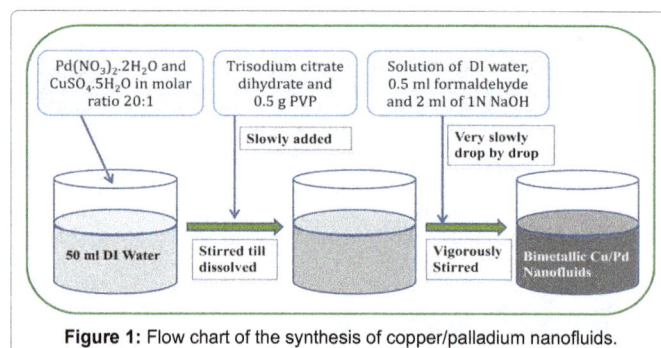

Figure 1: Flow chart of the synthesis of copper/palladium nanofluids.

*Corresponding author: Aashit Kumar Jaiswal, Department of Physics, University of Allahabad, Allahabad-211 002, Uttar Pradesh, India
E-mail: ajaiswal386@gmail.com

from the acoustic attenuation data using the software based on Epstein and Carhart theory [16]. The UV-Vis spectra of the NFs were recorded by Ultrospec 4000 UV/Visible Spectrophotometer.

Antimicrobial (antibacterial and antifungal) activities of NFs were evaluated by measuring the zone of inhibition in the agar disc as per standard protocol [17]. The disc diffusion method was used to screen the antibacterial activity and antifungal activity. In this study, different bacterial strains viz. *Escherichia coli* (ATCC 35218), *Pseudomonas aeruginosa* (ATCC 27853), *Enterococcus faecalis* (clinical isolate), *Staphylococcus aureus* (ATCC 25323) were used to test the antibacterial activity and different strains of candida viz. *Candida albicans* (ATCC 90028), *Candida tropicalis* (ATCC 750), *Cryptococcus neoformans* were tested for the antifungal properties of BMNFs. All cultures were preserved at Department of Microbiology, Institute of Medical Sciences, BHU, Varanasi, India. For antimicrobial Susceptibility Test: Muller Hinton agar (MHA) plates were prepared by pouring 15 ml of molten media into sterile petriplates. The fresh grown bacteria were suspended in sterile saline to achieve concentration of 10^7 cfu/ml. This suspension was spread on the surface of plates. The different concentrations of nanofluids were loaded on 6 mm sterile disc and were placed on the surface of medium. The plates were kept for incubation at 37°C for 24 hr for bacteria and 48 hr at 25°C for fungal agents. At the end of incubation, inhibition zones were examined around the discs, were measured with transparent ruler in millimetres.

Further, MIC values were also determined for the standard solution of ciprofloxacin (CPF) and nanofluid samples by micro-dilution method [18,19] using serially diluted (2 fold) test compounds according to guidelines of National Committee for Clinical Laboratory Standards (NCCLS) [20]. MIC was determined by dilution of nanofluids with different concentration. Standardized inoculums of bacteria/fungus were added in each well of microtiter plate. The nanofluids were serially diluted in specific well and were then incubated at 35°C for 24 hr for bacterial growth and at 25°C for fungal growth for 48 hr. The lowest concentration (highest dilution) of the nanofluid which inhibits the visible bacterial growth of microorganism was regarded as MIC.

Results and Discussion

Figure 2a displays the XRD pattern of the bimetallic nanoparticles which indicates that diffraction peaks at 2θ=39.6°, 45.7° and 67.8° correspond to the (111), (200) and (220) planes of Pd (JCPDS card no. 00-05-0681) of the fcc lattice, respectively. Other peaks located at 2θ=43.6° and 74.4° can be assigned to the reflections of corresponding (111) and (220) planes, respectively of Cu (JCPDS card no. 00-04-0836). The average crystallite size estimated from the Debye–Scherrer relation was 3 nm. The UV-vis absorption spectra of S1, S2 and S3 are presented in Figure 2b. It can be clearly seen that there is no absorption peak above the 300 nm in case of S1 which indicates the reduction of Pd(II) [21]. The peak observed at 650 nm for sample S3 shows the formation of copper nanoparticles due to more proportionality of Cu in comparison to Pd. Figure 2c shows the particle size distribution determined with the principle of ultrasonic spectroscopy using acoustic particle sizer (APS-100). The APS-100 measures the ultrasonic attenuation (dB/cm) at different frequencies of the ultrasonic waves (1-100 MHz) of nanofluids with high accuracy. These measurements are commonly referred to as ultrasonic attenuation spectroscopy. The APS-100 simultaneously measures the velocity of the waves at different frequencies. The measured ultrasonic attenuation spectra are converted into particle size distribution data using the software based on Epstein and Carhart theory [16]. This theory incorporates the intrinsic absorption, visco-internal dissipation losses, thermal

dissipation losses and scattering losses of ultrasonic energy interacting with the nanofluids. It is obvious from the Figure 2c that the particle size distributions are not very much different for S1, S2 and S3. It is clear from figure that particle distributions of S1 and S2 are in the range of 10 to 20 nm, while particle distribution of S3 is in the range of 10 to 15 nm. For S1 and S2, distribution curve overlaps. However, it is slightly different for S3. It seems due to more concentration of copper in the ratio. Figure 2d-2f shows SEM micrographs of the S1, S2 and S3, respectively. Figure 2d shows at random distribution of the particles. Figure 2e shows hexagonal shaped nanostructures. Figure 2f shows the mixed shaped nanostructures. Thus, a varied composition of Cu/Pd bimetallic system was observed on their optical and surface morphological features. The surface morphology of S2 may be helpful for its application for the development of sensors as it possesses special surface structure [22].

Table 1 shows results of antimicrobial activity of nanofluids. It is obvious from the Table 1 that antibacterial activity of S3 is found to be higher against both Gram positive and negative bacteria than those of S2. The antibacterial activity of S1 and S3 is equal against the Gram negative bacteria - *P. aeruginosa*. Interestingly, S3 shows higher antibacterial activity against *S. aureus* and *E. coli* than S2; however, lesser activity is recorded *E. faecalis*. The response of antibacterial activity is depicted in Figure 3. Although the difference in antibacterial activity of S1 and S3 is insignificant, the bacterial inhibition potential was displayed significantly increased as against the standard solution of well-known antibiotic drug, CPF. The zone of bacterial growth

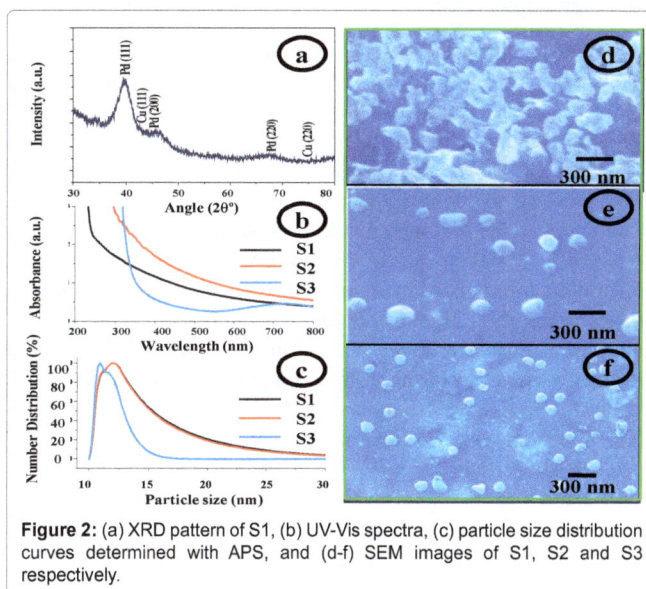

Figure 2: (a) XRD pattern of S1, (b) UV-Vis spectra, (c) particle size distribution curves determined with APS, and (d-f) SEM images of S1, S2 and S3 respectively.

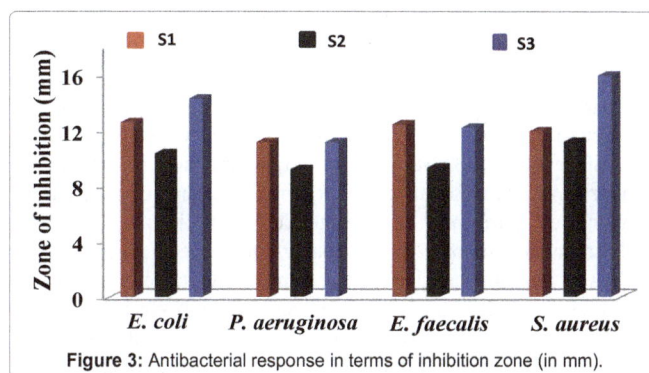

Figure 3: Antibacterial response in terms of inhibition zone (in mm).

(a) In terms of inhibition zone (in mm)						
Microbial species						
	Gram negative bacteria		Gram positive bacteria		Fungi		
Nanofluids	E. coli	P. aeruginosa	E. faecalis	S. aureus	C. albicans	C. tropicalis	C. neoformans
S1	12.54 ± 0.47	11.12 ± 0.33	12.41 ± 0.68	11.91 ± 0.43	-	-	-
S2	10.31 ± 0.19	9.16 ± 0.66	9.28 ± 0.11	11.16 ± 0.66	-	-	-
S3	14.24 ± 0.21	11.12 ± 0.33	12.17 ± 0.18	15.91 ± 0.43	-	-	-
Ciprofloxacin	28. 06 ± 1.30	24.76 ± 0.76	23.7 ± 1.05	29.93 ± 0.49	-	-	-
Amphotericin B	-	-	-	-	18.07 ± 0.28	16.22 ± 0.40	19.51 ± 0.52
(b) In terms of MIC values (µg/ml)						
Microbial species						
	Gram negative bacteria		Gram positive bacteria		Fungi		
Nanofluids	E. coli	P. aeruginosa	E. faecalis	S. aureus	C. albicans	C. tropicalis	C. neoformans
S1	46.98	93.97	46.98	93.97	-	-	-
S2	93.97	187.96	375.9	93.98	-	-	-
S3	46.98	93.97	93.98	46.98	-	-	-
Ciprofloxacin	6.25	6.25	3.12	6.25	-	-	-
Amphotericin B	-	-	-	-	0.5	0.5	0.5

Table 1: Antimicrobial activity of nanofluids (a) in terms of inhibition zone (in mm) and (b) in terms of MIC values (µg/ml) against various microbial species.

inhibition on culture dishes was ranged from 9-16 mm by application of S1, S2 and S3 NFs at very low concentration, i.e., 46.98, 93.97 and 46.98 µg/ml, respectively. It is revealed from these data that selected samples of BMNFs have mild antibacterial property. This experiment further exhibited substantially increased state of MIC values in S2 sample in comparison to S1 and S3 samples as well as CPF. Antifungal characteristics of these nanofluids were found completely absent as against the amphotericin B, a well-known antifungal drug.

Although a few antibacterial inducing mechanisms have been postulated by exposure of some mono and/or bimetallic nanoparticles [6,23], but none of the mechanism was established as central mechanism of mortality of different strains of bacteria. Xiu et al. [23] suggested that metal ions, released from metallic nanoparticles, strongly bound to cell wall of the bacteria and easily pass through membrane into the protoplasm, and thus damage the single stranded DNA which ultimately leads to cell death. We may also speculate the similar mechanism of action of metallic nanofluids at certain ratio and concentration. Their large surface area of nanoparticles provides better contact with the microorganisms and hence, provides better fusion with the bacterial cell membranes leading to membrane damage and cell death. Hence, this laboratory is pioneer to report antibacterial property of some selected bimetallic nanofluids at certain ratios. The smaller is the particle, the greater is its surface area to volume ratio. This enhances its biological and chemical activity by increasing the contact area of the bimetal with a microorganism. The use of nanoscale bimetals allows achieving hundred time decreased concentration and at the same time increase in antimicrobial properties.

In the present investigation in order to improve antibacterial properties, we have synthesized bimetallic nanoparticles based nanofluids. Previously, Wu et al. [24] tried to synthesize combined nanoparticles for this purpose with advanced antibacterial properties. On perusal of Table 1, it is obvious that S3 has Cu:Pd in the ratio 20:1. Since Cu nanoparticles have larger antibacterial activity in comparison to Pd nanoparticles, therefore S3 has greater antibacterial activity in comparison to S1 in which Cu:Pd ratio is 1:20. The lowest antibacterial activity of S2 was probably due to the equal proportion of Cu and Pd nanoparticles (1:1) present in bimetallic nanofluid. In S1, Pd nanoparticle proportion is greater and antibacterial activity is larger than that of S2. It shows that antibacterial activity due to Pd nanoparticles is significant but smaller in comparison to Cu addition.

Antibacterial activity of Pd nanoparticles is not reported in the literature. Lowest value of antibacterial activity in S2 may be understood due to influence of shape of the particles and effective concentration of Cu/Pd bimetals. Further investigations are required for substantial increase of antibacterial characteristics of Cu/Pd nanofluids by modulating the molar ratios of nanoparticles considering their applied use in pharmaco-therapeutic industry and nanodrug delivery system. The different antimicrobial agents, such as Ag and Au nanoparticles extensively studied by the others, are very costly. Cu/Pd BMNFs are cost effective in comparison to these novel metal nanoparticles.

Conclusion

In the present study, a facile approach was used for one-pot synthesis of polyvinylpyrrolidone (PVP) stabilised copper/palladium bimetallic nanostructure based nanofluids with small range distribution (~10-20 nm) at room temperature. The results on the investigations of the nanofluids demonstrate that these nanofluids inhibited the growth of bacteria at very low concentrations. Also the influence of Pd nanoparticles and morphology of the Cu/Pd nanostructure is significant for the antimicrobial activity of nanofluids which is important for pharmaceutical industries. Thus, the application of these nanofluids as antibacterial agent will be very valuable for biomedical and industrial applications.

Acknowledgements

The authors are thankful to Professor Shanti Sundaram, Department of Biotechnology, University of Allahabad, India for UV-Vis measurements and to Professor Neeraj Khare, Department of Physics, IIT Delhi, India for XRD measurement. The authors also thank Dr. KP Singh, Department of Zoology, University of Allahabad for useful discussion. AKJ acknowledges the Department of Science and Technology, New Delhi, India (Project no.: SR/S2/CMP-0038/2011) for the financial support.

References

1. Taylor R, Coulombe S, Otanicar T, Phelan P, Gunawan A, et al. (2013) Small Particles, Big Impacts: A Review Of The Diverse Applications Of Nanofluids. J Appl Phys 113: 011301.

2. Loulijat H, Zerradi H, Dezairi A, Ouaskit S, Mizani S, et al. (2015) Effect of Morse potential as model of solid–solid inter-atomic interaction on the thermal conductivity of nanofluids. Adv Powder Tech 26: 180-187.

3. Zerradi H, Ouaskit S, Dezairi A, Loulijat H, Mizani S (2014) New Nusselt number correlations to predict the thermal conductivity of nanofluids. Adv Powder Tech 25: 1124-1131.

4. Loulijat H, Zerradi H, Mizani S, Achhal EM, Dezairi A, et al. (2015) The behavior of the thermal conductivity near the melting temperature of copper nanoparticle. J Mol Liq 211: 695-704.

5. Zerradi H, Mizani S, Loulijat H, Dezairi A, Ouaskit S (2016) Population balance equation model to predict the effects of aggregation kinetics on the thermal conductivity of nanofluids. J Mol Liq 218: 373-383.

6. Kon K, Rai M (2013) Metallic Nanoparticles: Mechanism of Antibacterial Action and Influencing Factors. J Comp Clin Path Res 2: 160-174.

7. Moritz M, Geszke-Moritz M (2013) The Newest Achievements In Synthesis, Immobilization And Practical Applications Of Antibacterial Nanoparticles. Chem Eng J 228: 596-613.

8. Toshima N, Yonezawa T (1998) Bimetallic Nanoparticles-Novel Materials for Chemical and Physical Applications. New J Chem 22: 1179-1201.

9. Jo C, Lee J, Jang Y (2005) Electronic and Magnetic Properties of Ultrathin Fe-Co Alloy Nanowires. Chem Mater 17: 2667-2671.

10. Jaiswal AK, Singh S, Singh A, Yadav RR, Tandon P, et al. (2015) Fabrication Of Cu/Pd Bimetallic Nanostructures With High Gas Sorption Ability Towards Development Of LPG Sensor. Mat Chem Phys 154: 16-21.

11. Chen S, Zhang H, Wu L, Zhao Y, Huang C, et al. (2012) Controllable Synthesis Of Supported Cu–M (M = Pt, Pd, Ru, Rh) Bimetal Nanocatalysts And Their Catalytic Performances. J Mater Chem 22: 9117-9122.

12. Shih ZY, Wang CW, Xub G, Chang HT (2013) Porous Palladium Copper Nanoparticles for the Electrocatalytic Oxidation of Methanol in Direct Methanol Fuel Cells. J Mater Chem A1: 4773-4778.

13. Zhang L, Hou F, Tan Y (2012) Shape-tailoring of CuPd nanocrystals for enhancement of electro-catalytic activity in oxygen reduction reaction. Chem Commun (Camb) 48: 7152-7154.

14. Lo SHY, Chen TY, Wang YY, Wan CC, Lee JF, et al. (2007) A Mechanism Study On The Synthesis Of Cu/Pd Nanoparticles With Citric Complexing Agent. J Phys Chem. C 111: 12873-12876.

15. Mishra G, Singh D, Yadawa PK, Verma SK, Yadav RR (2013) Study Of Copper/ Palladium Nanoclusters Using Acoustic Particle Sizer. Plat Met Rev 57: 186.

16. Epstein PS, Carhart RR (1953) The Absorption Of Sound In Suspensions And Emulsions. I. Water Fog in Air. J Acoust Soc Am 25: 553-565.

17. Sharma AK, Gangwar M, Tilak R, Nath G, Sinha ASK, et al. (2012) Comparative In Vitro Antimicrobial And Phytochemical Evaluation Of Methanolic Extract Of Root, Stem And Leaf Of Jatropha Curcas Linn. Pharm Journal 4: 34-40.

18. Wiegand I, Hilpert K, Hancock REW (2008) Agar and Broth Dilution Methods to Determine the Minimal Inhibitory Concentration (MIC) Of Antimicrobial Substances. Nature Protocol 3: 163-175.

19. Gangwar M, Kumar D, Tilak R, Singh TD, Singh SK, et al. (2011) Qualitative Phytochemical Characterization And Antibacterial Evaluation Of Glandular Hairs Covering of Mallotus phillippinensis Fruit Extract. J Pharma Res 4: 4214-4216.

20. National Committee for Clinical Laboratory Standards NCCLS (2000) Methods for dilution antimicrobial susceptibility tests for bacteria that grow aerobically: approved standards. NCCLS document M7-A5. NCCLS. Wayne, PA, USA.

21. Yonezawa T, Imamura K, Kimizuka N (2001) Direct Preparation and Size Control of Palladium Nanoparticle Hydrosols by Water-Soluble Isocyanide Ligands. Langmuir 17: 4701-4703.

22. Rawal I (2015) Facial Synthesis of Hexagonal Metal Oxide Nanoparticles for Low Temperature Ammonia Gas Sensing Applications. RSC Adv 5: 4135-4142.

23. Xiu ZM, Zhang QB, Puppala HL, Colvin VL, Alvarez PJJ (2012) Negligible Particle-Specific Antibacterial Activity of Silver Nanoparticles. Nano Lett 12: 4271-4275.

24. Wu P, Xie R, Imlay K, Shang JK (2010) Visible-Light-Induced Bactericidal Activity Of Titanium Dioxide Codoped With Nitrogen And Silver. Environ Sci Tech 44: 6992-6997.

7

Preparation and Characterization of Polyester Composites Reinforced with Bleached, *Diospyros perigrina* (Indian persimmon) Treated and Unbleached Jute Mat

Chowdhury Kaiser Mahmud[1]*, Md. Asanul Haque[1], AM Sarwaruddin Chowdhury[1], Mohammad Abdul Ahad[2] and Md. Abdul Gafur[3]

[1]*Department of Applied Chemistry and Chemical Engineering, Faculty of Engineering and Technology, University of Dhaka, Bangladesh*

[2]*Department of Chemistry, Faculty of Science, University of Chittagong, Chittagong, Bangladesh*

[3]*Department of Pilot Plant and Product Development Center, Bangladesh council of Scientific and Industrial Research, Dhaka, Bangladesh*

Abstract

In this research work, jute mat reinforced polyester matrix composites have been developed by cold compression molding technique with varying process parameters, such as fiber condition (untreated, bleached and Indian persimmon juice treated). The developed jute fiber reinforced composites were then characterized by tensile test, flexural test, water absorption test, IR spectroscopy, thermal test and scanning electron microscopy. The results shows that tensile strength, flexural strength and value of elastic modulus increases in case of treated and bleached jute mat in compared to untreated fiber. The highest change in tensile strength, flexural strength and elastic modulus has been observed for bleached jute mat-UPR composite which means that the bleached jute mat-UPR composite is stronger than the other two. The water absorption property for the bleached jute mat-UPR composite is the lowest than the other two. The bulk density value shows that the bleached jute mat-UPR composite is denser than the other two. The thermo gravimetric analysis and thermo mechanical analysis also show that the bleached jute mat-UPR composite is more thermally and thermo mechanically stable. Microscopic observation suggests that the fracture behavior is brittle in nature and maximum fiber pull out has been occurred for unbleached jute mat-UPR composite.

Keywords: Jute mat reinforced polyester matrix composite; Tensile test; Flexural test; Water absorption test; IR spectroscopy; Thermal test; Scanning electron microscopy

Introduction

Natural fiber-reinforced composites have attracted great research and economic interests because of their outstanding 'green' characteristics compared with glass fiber-reinforced composites [1]. But the superior resistance of glass fibers to environmental attack made glass-fiber-reinforced polymers more attractive for marine products and in the chemical and food industries [2-5]. On the other hand, the manufacture, use and removal of traditional composite structures usually made of glass, carbon and aramid fibers are considered critically because of growing environmental conscious [6,7]. Scientists have been faced with the difficult challenge of lowering the human impact on their surroundings while maintaining or even bettering the quality of life. The potential environmental and economic benefits of using natural renewable resource based materials have therefore, led to increased interest in the development of novel bio-composite materials. As a result, lingo cellulose fibers which are a natural fiber have been identified as a potential substitute for commonly applied man made synthetic fibers in the preparation of composites due to their lightweight, low cost, no hazardousness, biodegradability, renewability and above all environmental friendly characteristics [8-14]. However, some of the infirmities of natural fiber such as wet ability, non-compatibility with some polymeric matrices, low thermal stability, and hydrophilic nature of fiber surface, poor adhesion with the matrices and swelling and maceration of the fibers due to moisture absorption have prevented complete replacement of synthetic fibers. But the scientific and industrial path of Fiber Reinforced Polymer (FRP) composites was successful primarily because these materials was able to offer structural efficiency and strength to weight ratios over traditional materials such as metals, plastics and wood. Cellulose fibers from kenaf, hemp, ramie, flax, sisal, coir and jute are also being used as reinforcement and they have some advantages over synthetic fibers such as low abrasion, multi

functionality, low density, low cost, unlimited availability and no waste disposal problems which encourage their applications in composite. One of the vastly available natural fibers is jute which is frequently used as reinforcing agent because it is non-abrasive, has low density and high strength. Jute appears to be a promising material because it is relatively inexpensive and commercially available in the required form. It has higher strength and modulus than plastic and is a good substitute for conventional fibers in many situations. Jute composite can be used as particleboard, ceiling, blocks for building construction and furniture.

There are several key factors that serve as a motivation for me in the development of bio-composite materials. Firstly, bio-composite materials offer high strength and stiffness at lighter weights than most traditional materials such as metals, glass fiber reinforced polymer (GFRP) composites and wood products. The seemingly endless supply of agricultural resources to serve as feedstock for bio-composite materials also allows for lower costs and makes them less subject to economic fluctuations. From an environmental and sustainability standpoint, bio based composites from renewable resources are superior to traditional FRP composites as they do not contribute to the depletion of the

***Corresponding author:** Chowdhury Kaiser Mahmud, Department of Applied Chemistry and Chemical Engineering, Faculty of Engineering and Technology, University of Dhaka, Bangladesh
E-mail: kaisermahmudacce@gmail.com.

world's natural resources, require less energy and chemicals to produce [15-18], emit less greenhouse gases and toxins during their production, use and disposal [16,17] and have the potential to be biodegradable, recyclable or used for energy harvesting upon incineration [15,17].

In this research work, several experimental works were done on three different composites as well as comparative studies on the experimental works were done to find the best reinforcing agent for making a convenient composite.

Materials and Methods

The raw materials for preparing the samples were unsaturated polyester resin (UPR), Monomer (Styrene), Jute mat, juice of *Diospyros perigrina* (Indian persimmon) and Hardener (Methyl ethyl ketone peroxide). The three different types of composite specimen are then prepared by the following way-

First the jute mat (bleached, unbleached) were cut into 31-21 cm pieces, then they are dried. After that Indian persimmon juice is hand-lay up on bleached jute mat which is again dried in an oven. Finally three types (bleached, unbleached and treated) of three layered composite is produced by open molding technique. Then different properties are measured for these three types of composite like Bulk Density, Water absorption, Tensile properties, Flexural properties, Hardness, Thermal, Thermo-mechanical, Chemical, Fracture properties etc.

Bulk density was calculated using following formula,

$$D = \frac{W_s}{V}$$

Where, D=Density of the specimen in kg/m^3,

Ws=Weight of the specimen in kg and V=Volume of the specimen in m^3.

For water intake measurement, the specimen was prepared according to ASTM-D570-98.The test specimen was 76 mm length, width above 25 mm.

The total water absorption was calculated following the rules given below.

Absorption %=100 (W$_s$-W$_d$)/ W$_d$

Where, W$_d$= Conditioned weight of the specimen, W$_s$=Saturated weight of the specimen after submersing in distill water.

Tensile specimen was prepared according to ASTM-638M-91a.

Tensile strength = Applied loadCross sectional area of the load bearing area

σ = PA

= PA in Pa

Tensile strain is calculated according to ASTM D-638M - 91a.

Tensile strain = $\frac{l-l_0}{l_0}$

Where, l_0= Original length of the sample.

l= Length of the material after stretching

Flexural specimen was prepared according to ASTM D790M, 3 point loading.

The Flexural strength may be calculated for any point of the load deflection by means of the following equation-

S = 3PL / 2bd^2

Where,

S = stress in the outer fibers at midspan in MPa, P = load at a given point on the load – deflection curve in N, L = support span in mm,

b = width of specimen tested in mm,

d = depth of tested specimen in mm.

Flexural strain may be calculated for any deflection using the following equation-

$$\varepsilon_f = \frac{6Dd}{L^2}$$

Where, ε_f = Main strain in the outer surface in mm/mm, D= Maximum deflection of the center of the beam in mm, L= Support span in mm, d= Depth in mm

Microhardness of the samples was measured by using following formula:

$$HV = 1.854\frac{F}{d^2}(approximately)$$

Where, F= Load in Kgf, d = Arithmetic mean of the two diagonals, d$_1$ and d$_2$ in mm, HV= Vickers hardness

Leeb Rebound Hardness value is derived from the energy loss of a defined impact body after impacting on a metal sample, similarly to the Shore scleroscope.

Thermal analysis of the specimen includes a group of techniques where some physical property of the sample is monitored under controlled conditions with variation of temperature at a programmed rate. When the mass change is monitored the results, which indicated chemical reactions, are called Thermo gravimetry (TG). When heat absorption monitored, the result indicate crystallization, phase change etc. as well as reactions. This is called Differential Thermal Analysis (DTA). Together, they are a powerful method of analysis.

SEM analyses have been carried out in the Centre for Advanced Research in Sciences (CARS) situated in university of Dhaka. This is a JEOL JSM-6490LA Scanning Electron Microscopy for image analysis. For SEM analysis, the fracture surface of the three different types of composite need to be analyzed. First the non-conducting samples were treated with powdered platinum for making the sample conductive and then set on a carbon tape for analysis.

IR test was done by using Shimadzu IR prestige-21 Fourier Transform Infrared Spectroscopy (FTIR) technique.

Thermo-mechanical analyses of composites were carried out by using an S II-TMA/SS6300 analyzer.

Results and Discussion

From Figure 1 we can clearly see that for bleached jute mat-UPR (Unsaturated Polyester Resin) composite, the value of Bulk Density, Water absorption, Tensile strength, Elastic modulus, Flexural strength , Leeb hardness value is the highest among the three different types of composite. For Indian persimmon treated composite these properties are moderate in nature. The Indian persimmon treated composite has the lowest percentage of elongation Figure 1a-1c, which means that it has the least ductility among the three different types of composite. From Figure 1d-1h we also see that the unbleached jute mat-UPR

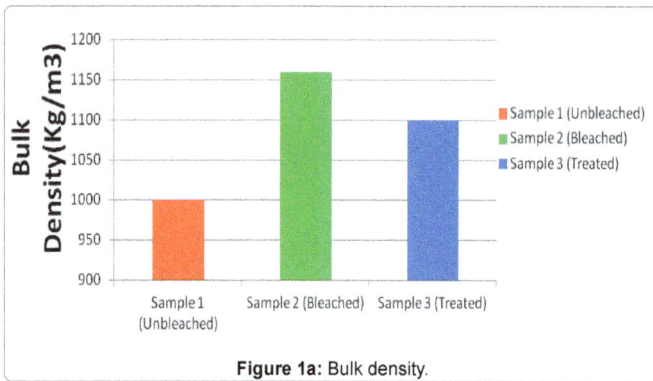

Figure 1a: Bulk density.

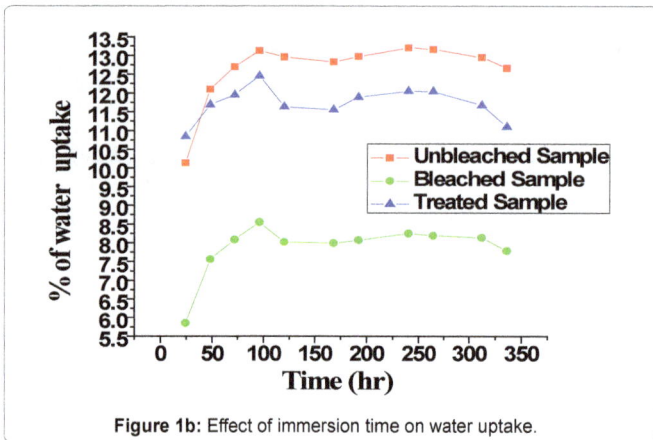

Figure 1b: Effect of immersion time on water uptake.

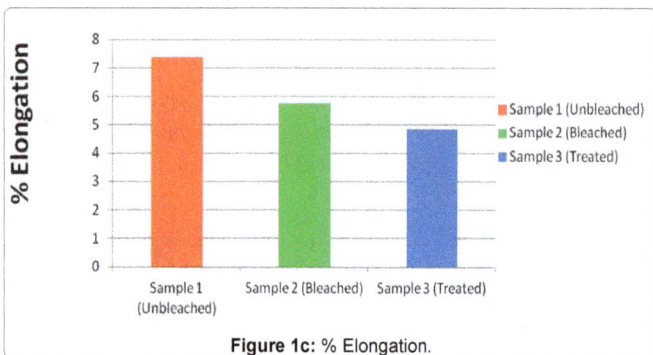

Figure 1c: % Elongation.

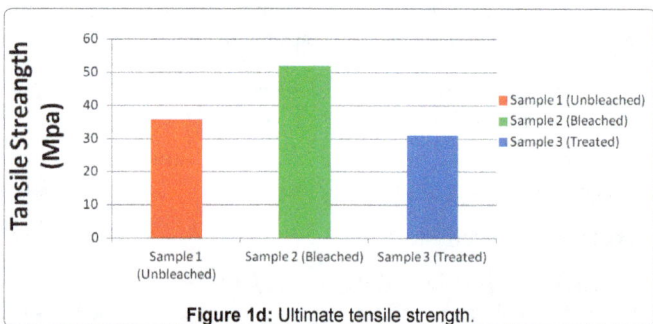

Figure 1d: Ultimate tensile strength.

Thermal Analysis (DTA) and Differential Thermo Gravimetry (DTG) curves of bleached, treated and unbleached jute mat-UPR composite respectively. The top (blue) one is the TG, bottom (red) one is the DTG and middle (black) one is the DTA curve. The major degradation occurs in one stage. The lighter substances remove initially and then heavier material removed. Initially, the composites loss their weight due to moisture removal. Figure 2d shows the overall Thermo Gravimetric (TG) curves of bleached, treated and unbleached jute mat-UPR composite. The top (blue) one is for bleached comp., bottom (green) one is for unbleached comp. and middle (red) one is for treated comp. The TG curve reveals that the onset temperature of unbleached, treated and bleached jute mat-UPR composite is 333.4°C, 313.5°C and 307.7°C respectively which means the unbleached jute mat-UPR composite is

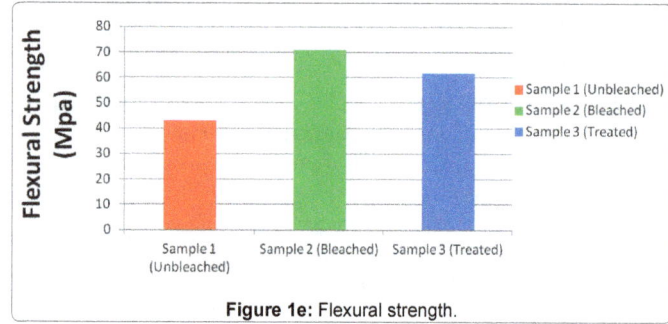

Figure 1e: Flexural strength.

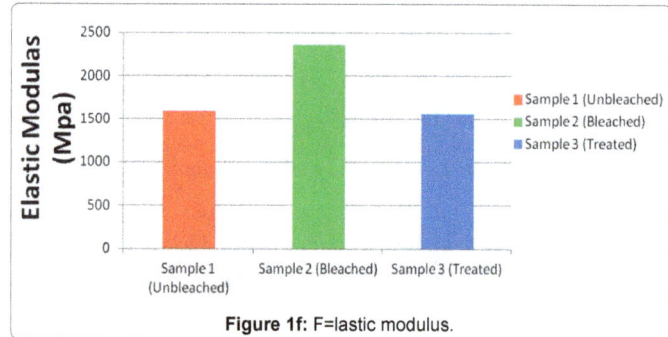

Figure 1f: F=lastic modulus.

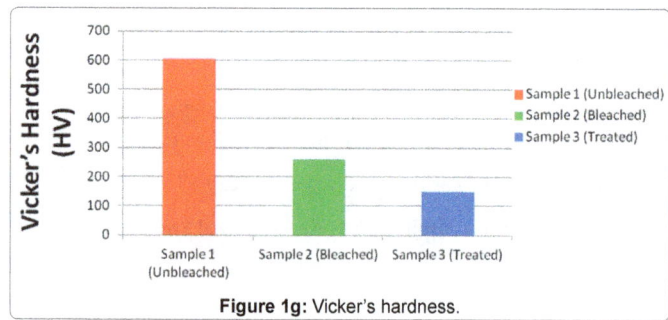

Figure 1g: Vicker's hardness.

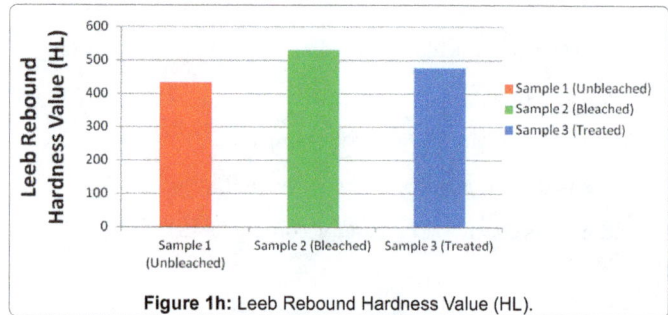

Figure 1h: Leeb Rebound Hardness Value (HL).

composite has the highest Vicker's hardness value among the three different types of composite. From the above discussion we can say that the unbleached jute mat-UPR (Unsaturated Polyester Resin) composite is harder and stiffer than the other two composite (bleached and treated one).

Figure 2a-2c shows the Thermo Gravimetric (TGA), Differential

and bleached jute mat-UPR composite is 333.4°C, 313.5°C and 307.7°C respectively which means the unbleached jute mat-UPR composite is more thermally stable than the other two. Figure 2f shows the Differential Thermo Gravimetric (DTG) curves of bleached, treated and unbleached jute mat-UPR composite. The top (green) one is for unbleached comp., bottom (red) one is for treated comp. and middle

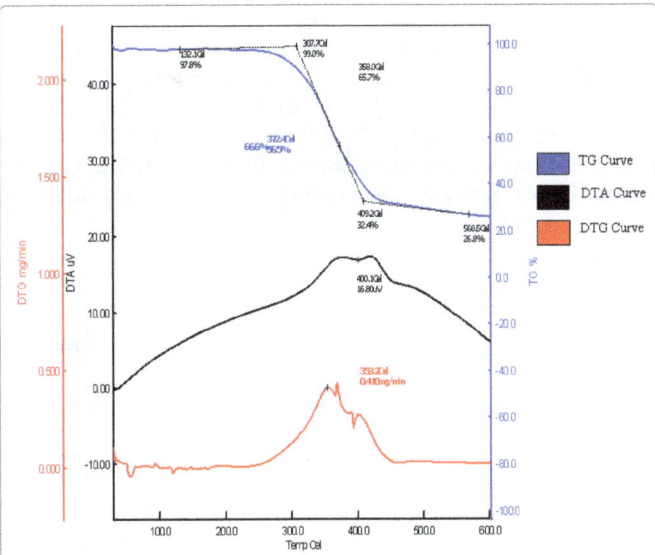

Figure 2a: TG, DTG and DTA curves of bleached jute mat + UPR composite.

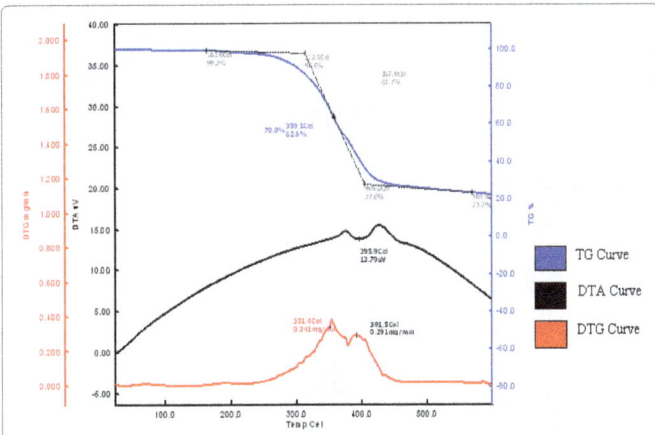

Figure 2b: TG, DTG and DTA curves of treated jute mat + UPR composite.

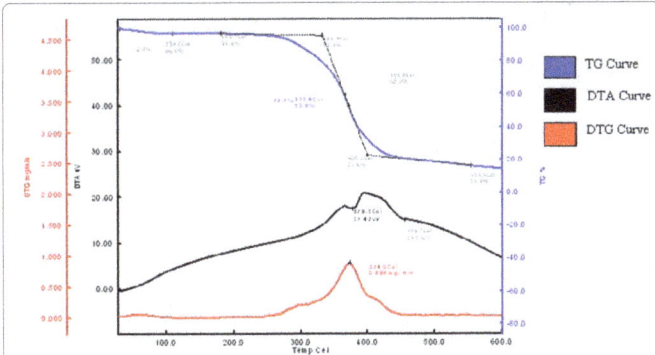

Figure 2c: TG, DTG and DTA curves of unbleached jute mat + UPR composite.

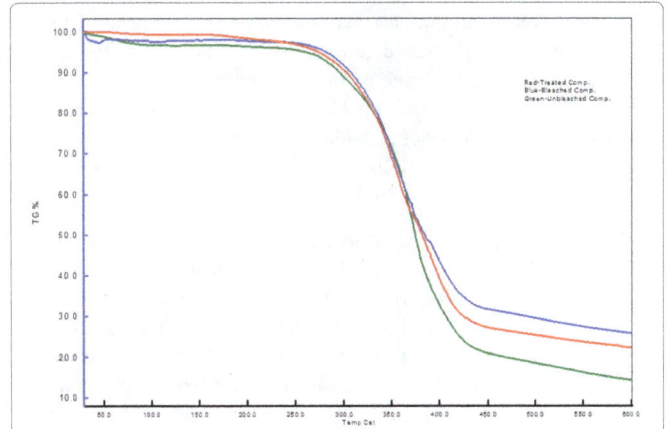

Figure 2d: Thermo Gravimetric Analysis of differently treated jute mat + UPR composite.

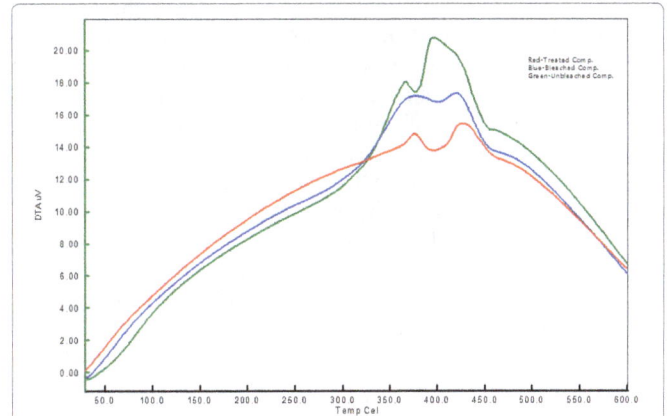

Figure 2e: DTA curves of differently treated jute mat + UPR composite.

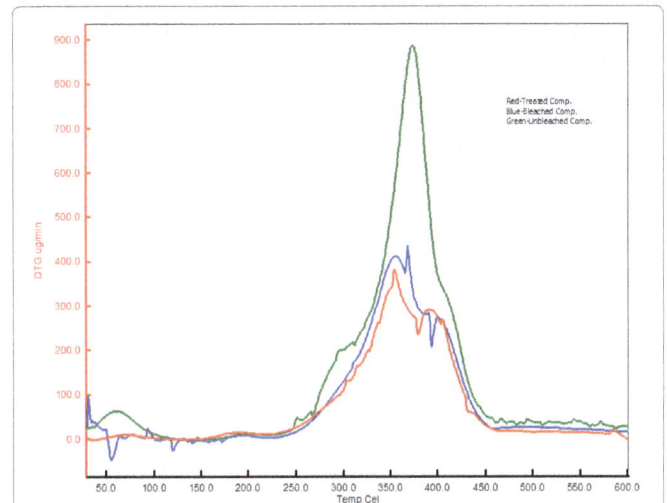

Figure 2f: DTG curves of differently treated jute mat + UPR composite.

more thermally stable than the other two.

Figure 2d-2e shows the Thermo Gravimetric Analysis (TGA) curves of bleached, treated and unbleached jute mat-UPR composite. The top (blue) one is for bleached comp., bottom (green) one is for unbleached comp. and middle (red) one is for treated comp. Then the TG curve reveals that the onset temperature of unbleached, treated

(blue) one is for bleached comp. Then the DTG curve reveals that the maximum degradation rate of unbleached, treated and bleached jute mat-UPR composite is 0.886 mg/min at 374°C, 0.341 mg/min at 351.4°C and 0.410 mg/min at 353.2°C respectively which also means the unbleached jute mat-UPR composite is more thermally stable than the other two.

Figure 2e shows the Differential Thermal Analysis (DTA) curves of bleached, treated and unbleached jute mat-UPR composite. The top (green) one is for unbleached comp., bottom (red) one is for treated comp. and middle (blue) one is for bleached comp. The DTA curve yields one endothermic peak for treated and bleached jute mat-UPR composite which is 395.9°C and 400.1°C respectively. The DTA curve yields two endothermic peak for unbleached jute mat-UPR composite is 378.3°C and 455.7°C respectively.

Figure 3a shows the Thermo Mechanical Analysis (TMA) and Differential Thermo Mechanical Analysis (DTMA) curves of treated jute mat-UPR composite. The top (blue) one is for DTMA, bottom (green) one is for TMA. The TMA curve shows that the onset of contraction occurs in the range of 24.6 to 51.6°C, the value of which is 7.53 μm. From DTMA curve we can see that there are two stages of contraction occurs at a rate of 0.719 μm/°C at 63.6°C and at a rate of 0.847 μm/°C at 96.2°C. Figure 3b shows the Thermo Mechanical Analysis (TMA) and Differential Thermo Mechanical Analysis (DTMA) curves of bleached jute mat-UPR composite. The top (blue) one is for DTMA, bottom (green) one is for TMA. The TMA curve shows that the onset of contraction occurs in the range of 23.9 to 61°C, the value of which is 5.20 μm and there is also an expansion occurs in the range of 79.9 to 109.9°C the value of which is 0.55 μm. The maximum contraction occurs at 67°C, the value of which is 9.41 μm. From DTMA curve we can see that there is one contraction occurs at a rate of 0.705 μm/°C at 66.9°C. Figure 3c shows the Thermo Mechanical Analysis (TMA) and Differential Thermo Mechanical Analysis (DTMA) curves of unbleached jute mat-UPR composite. The top (blue) one is for DTMA, bottom (green) one is for TMA. The TMA curve shows that the onset of contraction occurs in the range of 28.1 to 58.5°C, the value of which is 6.68 μm. The maximum contraction occurs at 66.9°C, the value of which is 16.05 μm. From DTMA curve we can see that there is one contraction occurs at a rate of 1.111 μm/°C at 66.6°C

Figure 3d shows the Thermo Mechanical Analysis (TMA) curves of bleached, unbleached and treated jute mat-UPR composite. The top (green) one is for bleached jute mat-UPR composite, bottom (red) one is for treated jute mat-UPR composite and middle (black) one is for unbleached jute mat-UPR composite. The TMA curve shows that for bleached, unbleached and treated jute mat-UPR composite, the onset of contraction occurs at 23.9°C, 28.1°C and 24.6°C respectively, the value of which is 8.20 μm, 6.68 μm and 7.53 μm respectively which means that bleached one is thermo mechanically unstable than the other two composite. Figure 3e shows the Differential Thermo Mechanical Analysis (DTMA) curves of bleached, unbleached and treated jute mat-UPR composite. The top (black) one is for unbleached jute mat-UPR composite, bottom (green) one is for bleached jute mat-UPR composite and middle (red) one is for treated jute mat-UPR composite. From DTMA curve we can see that there are two stages of contraction occurs for treated jute mat-UPR composite at a rate of 0.719 μm/°C at 63.6°C and at a rate of 0.847 μm/°C at 96.2°C. For bleached and unbleached jute mat-UPR composite we can see that there is one contraction occurs at a rate of 0.705 μm/°C at 66.9°C and 1.111 μm/°C at 66.6°C respectively.

Figure (S-1-S-3) shows the FTIR spectrum of the treated, bleached and unbleached jute mat-UPR composite. The intense peak for

ester group appeared at wavenumber of 1728 cm⁻¹ for bleached and unbleached jute mat-UPR composite. For treated jute mat-UPR composite ester group appeared at wavenumber of 1724.36 cm⁻¹. The C=O absorption band of saturated aliphatic esters with intense appearance is presenting the 1750-1725cm⁻¹ region. The more intense peak for C-O appeared at wavenumber of 1041.56,1043.49 and 1039.63 cm⁻¹ for bleached, treated and unbleached jute mat-UPR composite

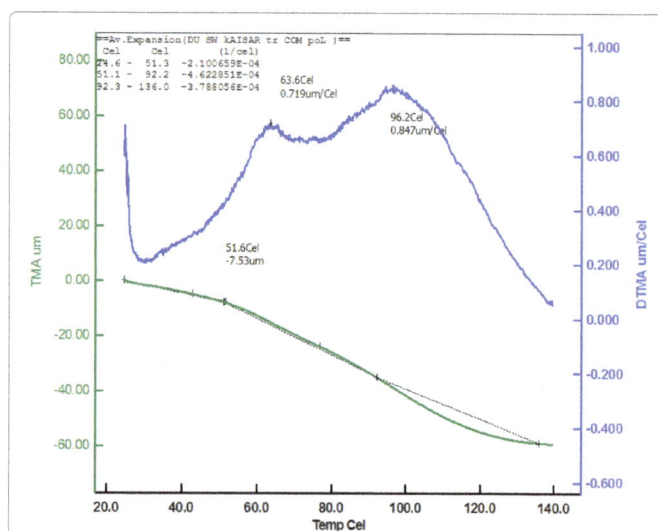

Figure 3a: TMA and DTMA curves of treated jute mat + UPR composite.

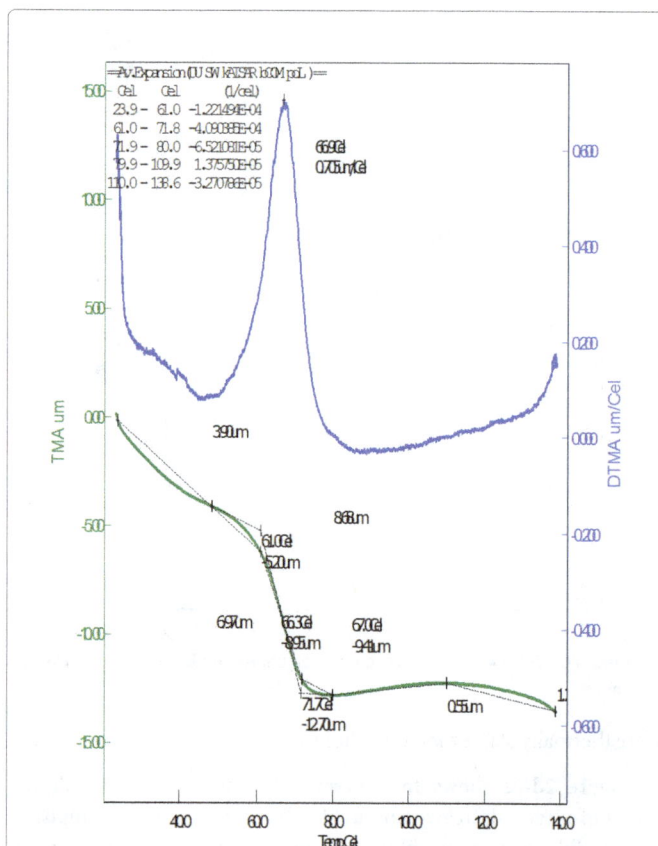

Figure 3b: TMA and DTMA curves of bleached jute mat + UPR composite.

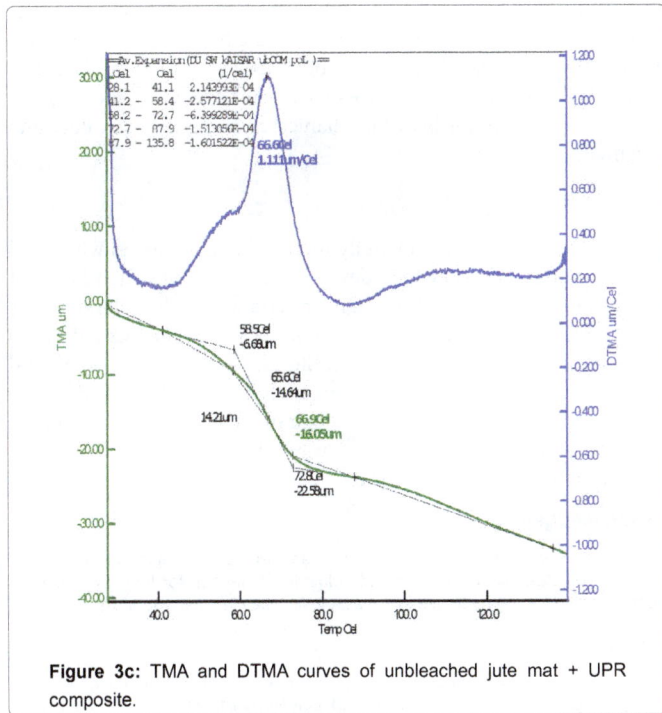

Figure 3c: TMA and DTMA curves of unbleached jute mat + UPR composite.

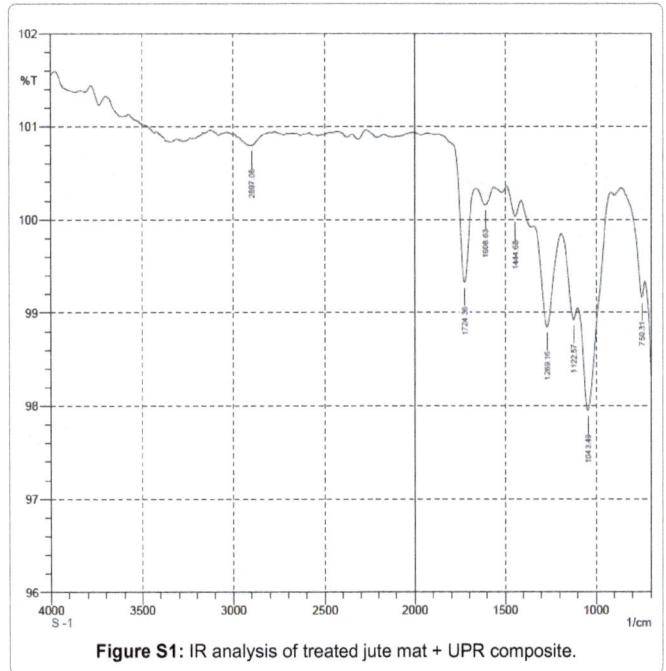

Figure S1: IR analysis of treated jute mat + UPR composite.

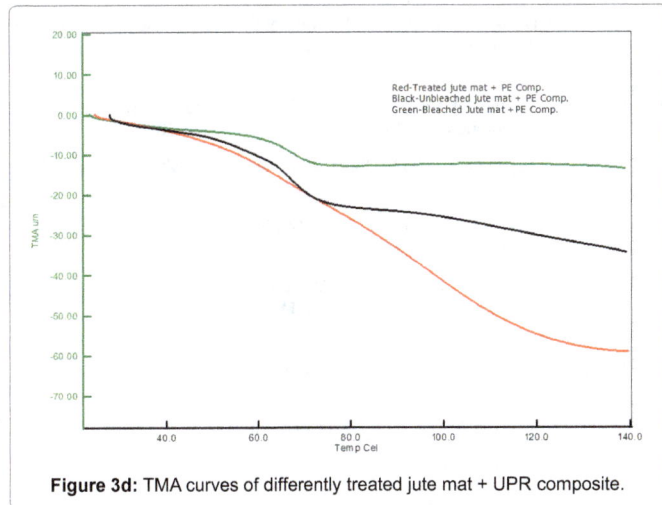

Figure 3d: TMA curves of differently treated jute mat + UPR composite.

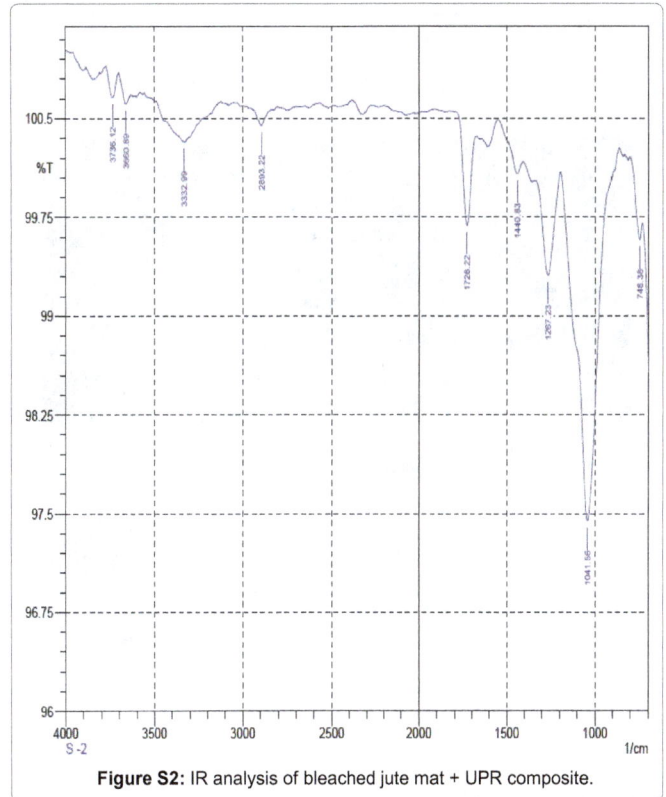

Figure S2: IR analysis of bleached jute mat + UPR composite.

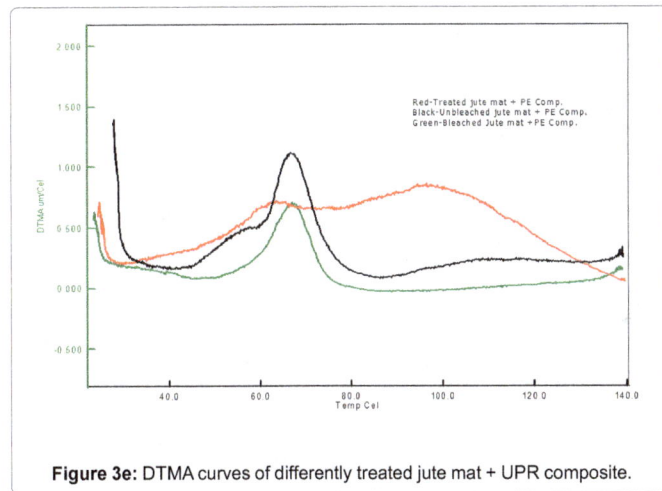

Figure 3e: DTMA curves of differently treated jute mat + UPR composite.

which should be found in the range of 1300-1000 cm^{-1} in the case for either the C-O bond in cellulose or for C-O bond of ester in polyester. Another intense peak is found at wavenumber of 3660 and 3610 cm^{-1} for bleached and unbleached jute mat-UPR composite which is for alcohol group in cellulose.

Scanning electron microscopic (SEM) investigation of the fracture surface of the jute composites is performed to study interfacial properties between jute fiber and UPR matrix. SEM images of fracture

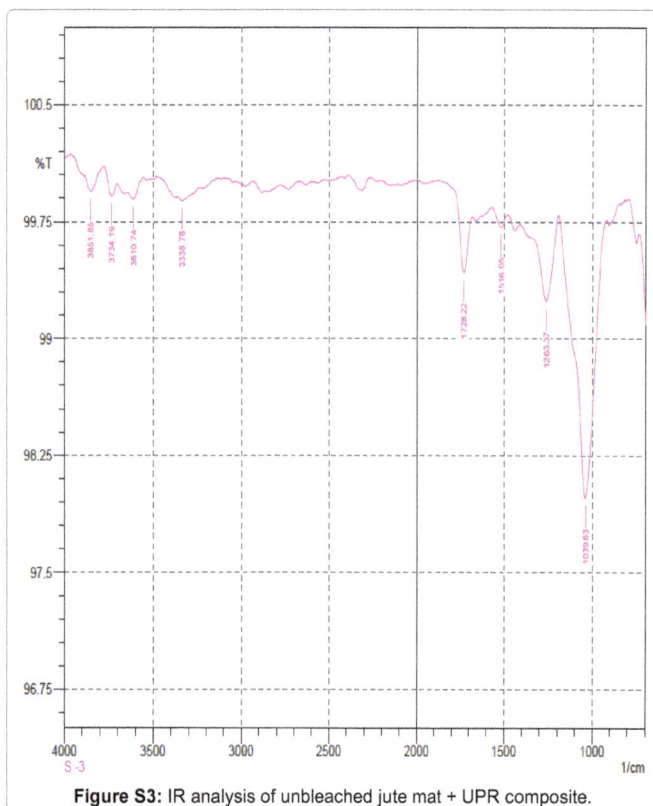
Figure S3: IR analysis of unbleached jute mat + UPR composite.

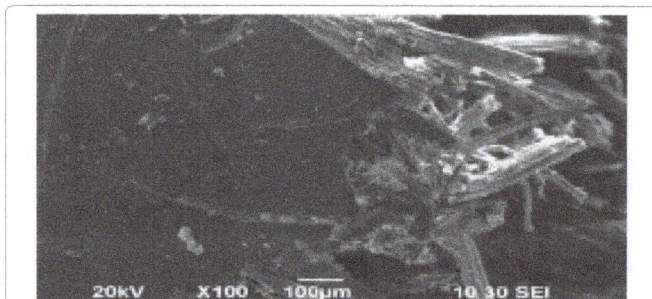
Figure 4a: Fracture surface of bleached jute mat –UPR composite shows fibre pull out and poor fibre matrix adhesion (100 magnification).

Figure 4b: Fracture surface of unbleached jute mat-UPR composite shows fibre pull out and poor fibre matrix adhesion (100 magnification).

surface of (a) Bleached jute mat + UPR composite (Figure 4a) and (b) Unbleached jute mat + UPR composite (Figure 4b) are presented in Figure 4a and 4b The Figure clearly indicates that there is considerable difference in the fiber-matrix interaction between the bleached and unbleached jute mat + UPR composites. SEM observation suggests that the fracture behavior is brittle in nature. Fiber pull-out phenomena are observed for all cases but for unbleached jute mat + UPR composite

fiber pull-out is observed in bundle form whereas in bleached jute mat + UPR composites agglomeration of the fibers into bundle form is relatively prevented. Between the bleached samples and unbleached samples (Figure 4a and 4b) shows significant change of morphology and this is effective for better mechanical bonding between fiber and polymer matrix.

Conclusions

The result of the present study reveals that composites with good strength could be successfully developed by using bleached jute mat as the reinforcing agent. From characterization of three layered jute mat-UPR show high strength but composite's SEM analysis indicating a strong adhesion between lignocellulose bleached jute mat and polyester. So from the thesis work we concluded that bleached fiber was the best one for reinforcement although medium performance Indian persimmon treated jute fiber can be used where there is a cost concern involved with the bleaching agent.

Acknowledgement

The authors acknowledge 'Pilot plant and Product development center' of "Bangladesh Council of Scientific and Industrial Research" for their excellent co-operation by allowing us to use the available facilities in their laboratory.

References

1. Yan ZL, Zhang JC, Lin G, Zhang H, Ding Y, et al. (2013) Fabrication process optimization of hemp fiber-reinforced polypropylene composites. Journal of reinforced plastics and composites 32: 1504-1512.

2. Wambua P, Ivens J, Verpoest I (2003) Natural Fibers: Can they replace Glass in Fiber Reinforced Plastics? Composites Science and Technology 63: 1259-1264.

3. Sharifah HA, Martin PA, Simon TC, Simon RP (2005) Modified polyester resins for natural fiber composites. Compos Sci Technol 65:525-535.

4. Bledzki AK, Gassan J (1999) Composites reinforced with cellulose based fibers. Prog Polym Sci 24: 221-274.

5. Marion P, Andreas R, Marei HM (2003) Study of wheat gluten plasticization with fatty acids. Polym 44: 115-122.

6. Sebe G, Cetin NS, Hill CAS, Hughes M (2000) RTM Hemp Fiber-Reinforced Polyester Composites. Applied Composite Materials 7: 341-349.

7. Chawla KK, Bastos AC (1979) The Mechanical Properties of Jute Fiber and Polyester/Jute Composites. Mechanical Behaviour of Materials 3: 191-196.

8. Jiang L, Hinrichsen G (1999) Flax and cotton fiber reinforced biodegradable polyester amide. Die Angew Makromol Chem 268:13-17.

9. Fan L, Du Y, Hung R, Wang Q, Zhang L (2005) Preparation and Characterization of Alginate/Gelatin Blend Fibers. Journal of Applied Polymer Science 96: 1625-1629.

10. Idicula M, Joseph K, Thomas S (2010) Mechanical Performance of Short Banana/ Sisal Hybrid Fiber Reinforced Polyester Composites. Journal of reinforced plastics and composites 29: 12-29.

11. Khan MA, Khan RA, Zaman HU, Alam MN, Houqe MA (2009) Effect of Surface Modification of jute with Acrylic Monomers on the Performance of polypropylene.A Composites. Journal of Reinforced Plastics and Composites.

12. Khan RA, Khan MA, Sultana S, Khan MN, Shubra QTH, et al. (2009) Mechanical, Degradation and Interfacial Properties of Synthetic Degradable Fiber Reinforced Polypropylene Composites. Journal of Reinforced Plastics and Composites 49:466-476.

13. Shokoohi S, Arefazar A, Khosrokhavar R (2008) Silane Coupling Agents in polymer- based Reinforced Composites: A Review. Journal of Reinforced Plastics and Composites 27:473-485.

14. Kuswaha P, Kumar R (2009) Effect of Silanes on Mechanical properties of Bamboo Fiber-epoxy Composites. Journal of Reinforced Plastics and Composites.

15. Saheb DN, Jog JP (1999) Natural Fiber Polymer Composites: A Review. Advances in Polymer Technology 18: 351-363.

16. Schloesser TP (2004) Natural Fiber Reinforced Automotive Parts. In: Natural Fibers, Plastics and Composites. Wallenburger FT, Weston NE (eds) Kluwer Academic Publishers, Norwell, MA, USA.

17. Joshi SV, Drzal LT, Mohanty AK, Aurora S (2004) Are Natural Fiber Composites Environmentally Superior to Glass Fiber Reinforced Composites? Composites Part A: Applied Science and Manufacturing 35: 371-376.

18. Patel M, Narayan R (2005) How Sustainable are Polymers and Biobased Products? The Hope, the Doubts and the Reality. In: Natural Fibers, Biopolymers and Biocomposites. Mohanty AK, Mishra M, Drzal LT (eds) CRC Press, Boca Raton, USA.

8

Experimental Measurements of Octanol-Water Partition Coefficients of Ionic Liquids

Montalbán MG[1], Collado-González M[2], Trigo R[1], Díaz Baños FG[2] and Víllora G[1]*

[1]Department of Chemical Engineering, University of Murcia, Murcia, Spain
[2]Department of Physical Chemistry, University of Murcia, Murcia, Spain

Abstract

Interest in ionic liquids (ILs) has increased due to their promising use as "green solvents" because of their negligible vapor pressure. However, their solubility in water could lead to their dispersion into the environment through liquid effluents, generating an important toxicological effect in soils and seawater. One of the most relevant parameters related to the assessment of environmental risk is the octanol-water partition coefficient (K_{ow}). With this parameter is possible to estimate some ecosystem risk factors such as bioaccumulation, sorption to soils and sediments and toxicity in fish by the usage of experimental correlations. Shake-flask and slow-stirring methods are the most currently used methods for determining the K_{ow} of a chemical compound. The former has the disadvantage that equilibrium might not be reached quickly, while the slow-stirring method is not always suitable for ILs, since some of them may decompose after continuous contact with water. We have developed a combined version of both methods. Here, we present measurements of the K_{ow} of twenty-four ILs at 30°C, using the three experimental methods. The types of anion and alkyl chain length of the cation are among the parameters studied. The K_{ow} of ILs used in this study range between 0.0017 and 3.6567 at 30°C. The K_{ow} of ILs studied is lower than in commonly used industrial solvents.

Keywords: Ionic liquids; Octanol-water partition coefficient; Hydrophobicity; Hydrophilicity; Ecosystem risk factor

Introduction

Ionic liquids (ILs) are low melting point organic salts, most of which are liquids at room temperature. In the past decade they have generated a huge interest in research and industrial fields for their capacity to serve as chemical and biochemical reaction media. In addition, ILs are of interest because they constitute a new group of polar and non-aqueous solvents, whose most important advantage is their negligible vapor pressure [1]. It is mainly for this reason that they are considered "green solvents" compared to conventional volatile organic compounds (VOCs). Due to their high chemical and thermal stability, ILs can resist high temperatures. However, for their potential applications, ILs are mainly valued because of the possibility of modulating their physical and chemical properties, such as melting point, viscosity, density, hydrophobicity and polarity by selecting the anion, the cation or substituent present in their specific structure. The number of different combinations of anions and cations that can be used to form potential ILs is enormous.

Due to the non-volatile character of ILs, they do not contribute to atmospheric pollution [2]. However, despite their significant degree of solubility in water, their effect in this respect has not been studied in depth [3-5]. Moreover, because of their high stability, ILs could represent serious pollutants in aqueous waste streams or accidental spills. For this reason, it is very important to quantify this effect by means of toxicological parameters such as the octanol-water partition coefficient (K_{ow}) that we study in this work. The K_{ow} is known to be one of the quantitative physical properties that best correlate with biological activity because the water–saturated octanol system is considered a reasonable model of the physico-chemical environment in living organisms [6,7]. K_{ow} classifies ILs according to their hydrophobicity or hydrophilicity, the latter characteristic being closely linked to the lipophilicity of a chemical compound and this parameter constitutes an essential physicochemical property in medicinal chemistry. For instance, it plays a crucial role in the absorption, distribution, metabolism and excretion (ADME) characteristics of drugs [8]. In fact, K_{ow} is used to model blood/lipid partition in toxicology to understand the tendency of a compound to cross biological membranes [9]. Therefore, K_{ow} is the key parameter for use in experimental correlations to estimate some parameters related to bioconcentration [10-13] and toxicity in fish [14,15], as well as sorption to soils [16-18].

Octanol and lipids have similar molecular structures containing both polar and hydrophilic oxygen at the end of a long hydrophobic alkyl chain and also similar physical properties. For this reason, correlations between natural ecosystems and K_{ow} are highly useful [19]. A high number of empirical correlations have been developed to estimate bioconcentration (BCF) and bioaccumulation (BAF) factors, soil sorption coefficients (K_{OC}), and toxicity. Bioconcentration is the process that causes an increased chemical concentration in an aquatic organism compared to that observed in water, due to the absorption of chemicals by different metabolic routes. The bioconcentration factor, BCF, is the equilibrium ratio of the chemical's concentration in the organism, in µg/kg lipid, to the concentration in the water, in µg/L, when exposure is only to the chemical in the water [20]. Pollutants also enter the biota through the food chain, which is referred to as biomagnification. The sum of bioconcentration and biomagnification is referred to as bioaccumulation [21]. The soil sorption coefficient, K_{OC}, describes soil-water partitioning. The K_{OC} is the ratio of the mass of a chemical adsorbed per unit weight of organic carbon in a soil to the concentration of the chemical in a liquid phase. Toxicity is typically reported in terms of mortality to various species quantified by the

---

*Corresponding author: Villora G, Faculty of Chemistry, Regional Campus of International Excellence "Campus Mare Nostrum", University of Murcia, P.O. Box 4021, Campus of Espinardo, E-30071, Murcia, Spain, E-mail: gvillora@um.es

LC_{50} in µmol/L, which is the concentration lethal to half of the species population [22]. Examples of equations based on experimental data relating K_{ow} with BCF, K_{OC} and LC_{50} are shown in Table 1 [23,24].

The activity of a compound, a_i, in the water-rich phase and the octanol rich-phase will be, by definition, the same ($a_i^{(w)} = a_i^{(o)}$) when the equilibrium is reached. Because $a_i = \gamma_i x_i$, where γ_i is the activity coefficient and x_i is the mole fraction in the water rich (^{w}) and octanol rich (^{o}) phases, respectively, the following expression can be deduced:

$$\frac{x_i^o}{x_i^w} = \frac{\gamma_i^w}{\gamma_i^o} \qquad (1)$$

If the test compound solution is very dilute and pressure and temperature are constant, concentrations and mole fractions will be proportional. In addition, if the compound is extremely dilute in both phases, so-called "infinite dilution", activity coefficients can be considered not to change with small variations in the concentrations. As a result, the Nernst distribution law is followed:

$$K_{ow} = \frac{C_i^o}{C_i^w} \qquad (2)$$

where K_{ow} is the octanol-water partition coefficient and C_i are sufficiently dilute concentrations in both phases [3,6].

Ideally, concentrations of the same solute species are measured in the two liquid phases when determining K_{ow} values. This may be difficult for species such as acids or salts, in which the solute tends to dissociate more in the aqueous phase than in the octanol-rich phase. As shown in Figure 1, ILs $[M_{v+}X_{v-}]$ tend to have a greater tendency to dissociate in the water-rich phase [3].

When solute dissociation is expected, the K_{ow} is calculated as the ratio between the concentration of the undissociated and dissociated species of the salt in the octanol-rich phase and in the water-rich phase. For this work, concentrations were measured in each phase using UV-Vis spectroscopy, which detects the imidazolium, pyridinium and ammonium group on the cation (whether dissociated or undissociated). Therefore, the K_{ow} values reported are given by Eq.3:

$$K_{ow} = \frac{[M_{v+}X_{v-}]^{octanol} + [M^{Z+}]^{octanol}}{[M_{v+}X_{v-}]^{water} + [M^{Z+}]^{water}} \qquad (3)$$

It is important to recognize that octanol and water are not completely immiscible. At 25°C, the solubility of water in octanol is approximately 0.275 mole fractions, but the solubility of octanol in water is 7.45×10^{-5} mole fraction [25]. Since the solubility of water in octanol is high, the K_{ow} cannot be represented simply as the ratio of saturation concentrations of a solute dissolved in separate solutions of octanol and water. The mutually saturated octanol and water phases affect the partitioning of the solute between the two solvents.

Many different experimental methods (direct and indirect) exist for determining K_{ow} of a chemical compound, but two methods are the most commonly used: shake-flask and slow-stirring, both direct methods. In the traditional shake-flask method, octanol and water are mutually saturated for three days. A sufficiently dilute solution of the test compound and water-saturated octanol is brought into contact with the same quantity of saturated water and shaken for about five minutes to achieve equilibrium. Next, both phases are separated by centrifugation and the test chemical concentrations are measured in each phase. This method has speed as an advantage and the disadvantages of the possible formation of microdroplets after centrifugation and the fact that equilibrium is not guaranteed, then the evaluation of partition equilibrium can be done by monitoring the absorbance after the phase separation [3,26-28]. Slow-stirring method is similar to the previous method. However, instead of rigorously shaking both phases, they are stirred slowly for an extended period of time (45 days, approximately). After that, each phase is analyzed in the same way as in the shake-flask method. The main advantage of this method is to avoid the emulsification due to the reduction of the stagnant diffusion layer between the two phases. Nevertheless, care must be taken with the length of time that the phases are in contact since some ILs may decompose after continuous contact with water [3,29].

Our research group is especially interested in working with ILs hence this paper is focused on the determination of the K_{ow} for a group of ILs. Information about the toxicological risk of ILs is limited in the literature. Our research group has already measured several K_{ow} values for some imidazolium-based ILs using our own method, which will be explained below [30]. Other authors, for example, Ropel et al. [3], Deng et al. [7,31] and Ventura et al. [32] used the slow-stirring method to measure K_{ow} for imidazolium, pyridinium, ammonium and pyrrolidinium-based ILs. Others like Kaar et al. [33], Zhao et al. [34] and Lee and Lee [9] determined K_{ow} values for imidazolium and pyrrolidinium-based ILs using the shake-flask method. The results reported by these publications for ILs have been inconsistent due to different experimental methods and range of IL concentrations established in the experiments. However, some tendencies can be extracted. For instance, it is clear that more hydrophobic anions such as [NTf$_2$] lead to higher K_{ow} values and, K_{ow} values increase with alkyl chain length in ILs with the same anion. Some studies have developed an indirect way to determine K_{ow} from the ratio of IL solubility in water and in octanol but this value does not exactly correspond with the definition of K_{ow} [35,36]. Other authors have preferred to use computational simulations [37,38] or DFT calculations (LFER parameters) [39] to determine K_{ow} for ILs. Recently, a robust and automated method for measurement of K_{ow} has been established. It consists of a syringe pump with a selection valve, a holding column, a silica capillary flow-cell and an in-line spectrophotometer. Distribution of the drug between the aqueous and octanol phases occurs by the oscillation movement of the syringe pump piston. The system has been applied to the determination of the K_{ow} of some common drugs, achieving high precision with only one-phase measurement [40]. Other new experimental methodology to obtain K_{ow} has been developed by the use of a bubble column set-up in combination with headspace concentration measurement [41]. Finally, a K_{ow} prediction method based on the lipophilicity estimation of ILs by chromatographic methods was used by Stepnowski et al. [42], Ranke et al. [43] and Studzińska et al. [44]. Due to the quantity of ionic ILs synthesized and the lack of information data about their toxicological

Correlation	Reference
Log (BCF)=0.85log (K_{ow})-0.70	[23]
Log (BCF)=0.79log (K_{ow})-0.40	[24]
Log (BCF)=log (K_{ow}) − 1.32	[10]
For Log (K_{ow}) < 5; Log (BCF)=0.5	[13]
Log (BCF)=0.791log (K_{ow})-0.729	[15]
Log (K_{oc})=0.544log (K_{ow})+1.377	[16]
Log (K_{oc})=0.989log (K_{ow})-0.364	[17]
Log (K_{oc})=0.679log (K_{ow})+0.663	[18]
Log (1/LC_{50})=0.854log (K_{ow})-1.74[a]	[15]
Log (1/LC_{50})=0.629log (K_{ow})-0.489[a]	[15]
Log (1/LC_{50})=0.89(± 0.03)log (K_{ow})-1.75(± 0.05)[b]	[14]

Table 1: Correlations of K_{ow} with ecosystem risk parameters. [a]For guppies, fathead minnow, rainbow trout and medaka. [b]for guppies, fathead minnow and rainbow trout.

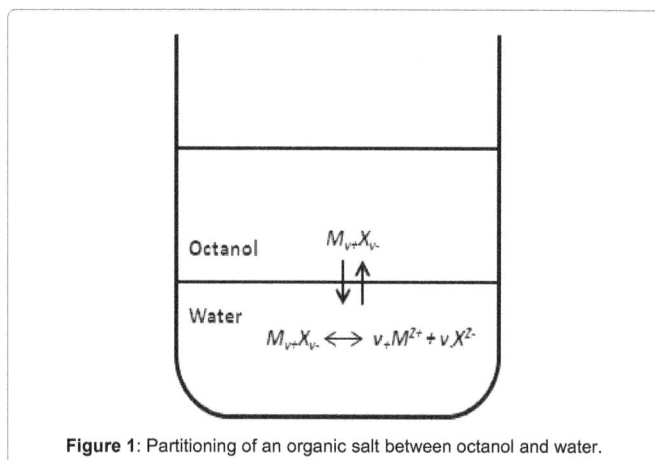

Figure 1: Partitioning of an organic salt between octanol and water.

risk, more studies are required to ensure the safe and environmental friendly use of ILs by researchers and industries.

In the present paper, we measured the K_{ow} of dilute samples of several imidazolium, pyridinium and ammonium-based ILs. The K_{ow} values previously reported for some ILs have shown huge discrepancies, as can be seen below. For this reason, we have used three different experimental methods and compared the results obtained, which has not been done before now. Furthermore, the effect of the anion and cation alkyl chain length on the K_{ow} values is discussed.

Experimental

Experimental method

The K_{ow} values of ILs were measured using three different methods, two of them explained in the previous section (the shake-flask and slow-stirring methods) and the other one a combined version of the same which our research group developed in a previous paper [30]. Before to carry out each experiment, both water and octanol were mutually saturated by stirring for three days or more. The experimental setup for the slow-stirring method and the combined method is very similar, both consisting of a 22 mL glass vial with an open-top screw cap sealed with a silicone/Teflon septum. In the slow-stirring method and our combined method, approximately 10 mL of distilled, deionised water, presaturated with octanol was added to the glass vial. A 12-gauge Teflon tube was introduced into the vial, reaching just below the water surface. Approximately 10 ml of octanol-IL "stock" solution, consisting of octanol presaturated with water and containing a known concentration of IL (1 mM), was added to the vial. In the combined method, the vial was shaken vigorously for 5 minutes and then maintained at 30 ± 0.5°C in a thermostatic bath without stirring for 13 days. In the slow-stirring method, the vial was not shaken intensely, but was maintained in the same bath at 30 ± 0.5°C with a 1 cm Teflon coated magnetic stir bar which provided slow-stirring for 45 days. In both methods, samples were taken from the octanol-rich phase by penetrating the silicone septum with a stainless-steel needle fitted with a glass syringe. Samples were withdrawn from the water-rich phase by inserting the needle through the Teflon tubing directly into the aqueous phase to prevent octanol contamination. In both cases, samples of each phase were taken from the vials until three measurements coincided confirming that the concentrations in both phases had stabilized.

In the shake-flask method, the apparatus consists of a 12 mL centrifuge tube with cap. Approximately 5 mL of distilled, deionised water, presaturated with octanol and 5 mL of octanol-IL "stock" solution, consisting of octanol presaturated with water and containing a known concentration of IL (1 mM) was added to the centrifuge tube. The tube was shaken vigorously for about five minutes and then the octanol and water phases were separated by centrifugation. Samples were taken from the octanol-rich phase and the water-rich phase with a glass syringe and analyzed directly.

In all three methods, the IL concentration in each phase was analysed by UV-VIS spectrophotometry as described in the section Analytical Method. Determinations were made at least in triplicate to ensure repeatability of the tests and the mean values are reported. The samples were diluted until the absorbance was less than 1. The initial concentration of IL in the octanol phase was less than 1.2×10^{-2} mol/L, but enough to ensure that the IL could be measured accurately. The final concentrations of IL in the octanol phase were between 1.77×10^{-5} mol/L and 3.92×10^{-3} mol/L and in the water phase between 2.75×10^{-5} and 8.38×10^{-3} mol/L. IL concentrations were kept at the dilution limit so that K_{ow} values would be independent of concentration.

In the present paper, we have studied twenty-four ILs based on imidazolium, pyridinium and ammonium salts with different substituents: 1-butyl-3-methylimidazolium hexafluorophosphate ($[bmim^+][PF_6^-]$), 1-hexyl-3-methylimidazolium hexafluorophosphate ($[hmim^+][PF_6^-]$), 1-methyl-3-octylimidazolium hexafluorophosphate ($[omim^+][PF_6^-]$), 1-butyl-3-methylimidazolium tetrafluoroborate ($[bmim^+][BF_4^-]$), 1-methyl-3-octylimidazolium tetrafluoroborate ($[omim^+][BF_4^-]$), 1-ethyl-3-methylimidazolium bis(trifluoromethylsulfonyl)imide ($[emim^+][NTf_2^-]$), 1-butyl-3-methylimidazolium bis(trifluoromethylsulfonyl)imide ($[bmim^+][NTf_2^-]$), 1-hexyl-3-methylimidazolium bis(trifluoromethylsulfonyl)imide ($[hmim^+][NTf_2^-]$), 1-methyl-3-octylimidazolium bis(trifluoromethylsulfonyl)imide ($[omim^+][NTf_2^-]$), 1-butyl-2,3-dimethylimidazolium bis(trifluoromethylsulfonyl)imide ($[bdmim^+][NTf_2^-]$), 1-ethyl-3-methylimidazolium triflate ($[emim^+][TfO^-]$), 1-ethyl-3-methylimidazolium ethylsulphate ($[emim^+][EtSO_4^-]$), 1-ethyl-3-methylimidazolium acetate ($[emim^+][CH_3COO^-]$), 1-butyl-3-methylimidazolium methylsulphate ($[bmim^+][MeSO_4^-]$), 1-butyl-3-methylimidazolium 2-(2-methoxiethoxy) ethylsulphate ($[bmim^+][MDEGSO_4^-]$), 1-methylimidazolium chloride ($[mim^+][Cl^-]$), 1-ethyl-3-methylimidazolium chloride ($[emim^+][Cl^-]$), 1-(2-hydroxi-ethyl)-3-methylimidazolium chloride ($[hemim^+][Cl^-]$), 1,2-dimethylimidazolium chloride ($[dmim^+][Cl^-]$), 1-ethylimidazolium chloride ($[eim^+][Cl^-]$), 1-hexyl-3-methylimidazolium chloride ($[hmim^+][Cl^-]$), 1-ethyl-3-methylpyridinium ethylsulphate ($[empy^+][EtSO_4^-]$), 1-butyl-3-methylpyridinium tetrafluoroborate ($[bmpy^+][BF_4^-]$) and ethylammonium nitrate (ETAN).

Materials

The structures of the ILs used in this study are listed in Table 2. The ILs $[bmim^+][PF_6^-]$ (purity>99%), $[hmim^+][PF_6^-]$ (purity>99%), $[omim^+][PF_6^-]$ (purity>99%), $[bmim^+][BF_4^-]$ (purity>99%), $[emim^+][NTf_2^-]$ (purity>99%), $[bmim^+][NTf_2^-]$ (purity>99%), $[hmim^+][NTf_2^-]$ (purity>99%), $[omim^+][NTf_2^-]$ (purity>99%), $[bdmim^+][NTf_2^-]$ (purity>99%), $[emim^+][TfO^-]$ (purity>99%), $[emim^+][EtSO_4^-]$ (purity>99%), $[emim^+][CH_3COO^-]$ (purity>95%), $[mim^+][Cl^-]$ (purity>98%), $[emim^+][Cl^-]$ (purity>98%), $[hemim^+][Cl^-]$ (purity>99%), $[dmim^+][Cl^-]$ (purity>98%), $[eim^+][Cl^-]$ (purity>98%), $[hmim^+][Cl^-]$ (purity>98%) and ETAN (purity>97%) were purchased from Iolitec (Germany). The ILs $[omim^+][BF_4^-]$ (purity>99%), $[bmim^+][MeSO_4^-]$ (purity>99%), $[bmim^+][MDEGSO_4^-]$ (purity>98%), $[empy^+][EtSO_4^-]$ (purity>99%) and $[bmpy^+][BF_4^-]$ (purity>99%) were purchased from Solvent Innovation GmbH (Cologne, Germany). The reagent octanol was purchased from Merck Eurolab (Germany).

Abbreviation	Molecular weight (g mol^{-1})	Structure
[bmim$^+$][PF$_6$]	284.18	
[hmim$^+$][PF$_6$]	312.08	
[omim$^+$][PF$_6$]	340.29	
[bmim$^+$][BF$_4$]	226.02	
[omim$^+$][BF$_4$]	282.13	
[emim$^+$][NTf$_2$]	391.31	
[bmim$^+$][NTf$_2$]	419.37	
[hmim$^+$][NTf$_2$]	447.42	
[omim$^+$][NTf$_2$]	475.47	
[bdmim$^+$][NTf$_2$]	433.39	
[emim$^+$][TfO$^-$]	260.24	
[emim$^+$][EtSO$_4$]	236.29	
[emim$^+$][CH$_3$COO$^-$]	170.21	
[bmim$^+$][MeSO$_4$]	250.32	
[bmim$^+$][MDEGSO$_4$]	338.43	

[mim⁺][Cl⁻]	118.57	
[emim⁺][Cl⁻]	146.02	
[hemim⁺][Cl⁻]	162.62	
[dmim⁺][Cl⁻]	131.58	
[eim⁺][Cl⁻]	131.58	
[hmim⁺][Cl⁻]	202.72	
[empy⁺][EtSO₄⁻]	247.32	
[bmpy⁺][BF₄⁻]	237.05	
ETAN	108.1	

Table 2: Abbreviations and structures of the studied ILs.

Analytical method

The concentration of the ILs in octanol and water was measured by UV-VIS spectrophotometry using a ThermoSpectronic UV-VIS recording spectrophotometer (Heλios α), which has a sensitivity of ± 0.001. A calibration curve was prepared for the different ILs at their maximum absorbance wavelength: 212 nm for imidazolium, 216 nm for pyridinium and 210 nm for ammonium-based ILs. The calibration curves were made in duplicate and mean values of the extinction coefficients are reported (Table 3). Samples were diluted if their concentration exceeded the calibration range.

Results and Discussion

Experimental technique evaluation

The combined experimental technique developed by our group was validated in a previous work [30] with benzaldehyde, which has a well-known $\log(K_{ow})$=1.48. Benzaldehyde shows two characteristic peaks at 250 and 200 nm in water, and at 245 and 207 nm in octanol. The wavelengths at 250 nm and 245 were selected for absorbance measurements in water and octanol, respectively. The extinction coefficient in water and octanol is shown in Table 3. $\log(K_{ow})$ of benzaldehyde obtained was 1.43, which is very close to the value given in the literature (1.48), confirming the validity of the proposed method [30].

Extinction coefficients

To determine the concentrations of IL in each phase and the resulting K_{ow} values, the IL extinction coefficients (ε) in octanol and water were obtained at the wavelength of maximum absorption of the imidazolium (λ_{max}= 212 nm), pyridinium (λ_{max}= 216 nm) and ammonium (λ_{max}= 210 nm) groups (Table 3).

The extinction coefficients of all the ILs have values of between 3000 and 6000 L mol⁻¹ cm⁻¹ except [hmim⁺][NTf₂⁻] in water and in octanol and [bmpy⁺][BF₄⁻] in octanol which were greater. The ones of [hmim⁺] [NTf₂⁻] in water and in octanol and [bmpy⁺][BF₄⁻] and ETAN in octanol

Compound	ε in water/L mol^{-1} cm^{-1}	ε in octanol/L mol^{-1} cm^{-1}
Benzaldehyde	1363.7	1303.6
[bmim$^+$][PF$_6$$^-$]	4472.5	3455.7
[hmim$^+$][PF$_6$$^-$]	2977.6	5719.0
[omim$^+$][PF$_6$$^-$]	3828.4	5100.0
[bmim$^+$][BF$_4$$^-$]	4212.3	4639.6
[omim$^+$][BF$_4$$^-$]	4207.1	4787.0
[emim$^+$][NTf$_2$$^-$]	4397.7	3996.4
[bmim$^+$][NTf$_2$$^-$]	3367.6	4904.1
[hmim$^+$][NTf$_2$$^-$]	6817.3	9380.8
[omim$^+$][NTf$_2$$^-$]	3103.3	4483.1
[bdmim$^+$][NTf$_2$$^-$]	3067.3	5168.7
[emim$^+$][TfO$^-$]	4326.7	4977.2
[emim$^+$][EtSO$_4$$^-$]	3604.8	4190.8
[emim$^+$][CH$_3$COO$^-$]	5481.9	4617.4
[bmim$^+$][MeSO$_4$$^-$]	3999.5	4777.0
[bmim$^+$][MDEGSO$_4$$^-$]	5372.7	4776.9
[mim$^+$][Cl$^-$]	4197.7	5136.6
[emim$^+$][Cl$^-$]	4477.8	4513.2
[hemim$^+$][Cl$^-$]	3741.3	4703.9
[dmim$^+$][Cl$^-$]	5808.9	5803.6
[eim$^+$][Cl$^-$]	3988.6	4562.5
[hmim$^+$][Cl$^-$]	4671.8	4368.0
[empy$^+$][EtSO$_4$$^-$]	4062.2	4961.0
[bmpy$^+$][BF$_4$$^-$]	4346.1	9246.3
ETAN	5254.5	12865

Table 3: Extinction coefficients of Benzaldehyde and ILs in water and Octanol.

are bigger than the others, perhaps as a result of different interactions with the solvents.

K_{ow} Values

The values of K_{ow} measured for each ionic liquid and obtained with the three methods indicated in Experimental Method Section are shown in Table 4. The K_{ow} values found in literature for some of the ILs have been added to compare the results, distinguishing between the K_{ow} values obtained by the shake-flask or slow-stirring method. The uncertainties reported are standard deviations of multiple tests, as detailed in the Experimental section.

The K_{ow} of a given chemical structure can also be estimated. Most methods divide the molecule into fragments or groups of atoms, each with its corresponding empirical constant and structural factor. Using experimental data, a database of the different contributing fragments and structure factors possible can be computed. We used two of these approaches to estimate the K_{ow} values of ILs, both using SMILES notation (Simplified Molecular Input Line Entry System): web page molinspiration.com [45] and Bio-Loom Software [46]. The estimated values are also shown in Table 4 [47-49].

Discussion

The ILs studied here based on imidazolium, pyridinium and ammonium, are fairly hydrophilic so their K_{ow} were expected to be low. Indeed, the experimental values shown in Table 4 are lower than those in commonly used industrial solvents (i.e., $K_{ow(ethanol)}$=0.479 [50]; $K_{ow(n\text{-}heptane)}$=31623 [51]).

On the other hand, the three methods used here to determine the K_{ow} of ILs have their advantages and drawbacks.

The shake-flask method is a classical method to measure the K_{ow} that

has been widely used in recent decades; however, several parameters need to be taken into account for the measurement to be valid. For example, pre-saturation of the two solvents with each other is essential to obtain accurate numbers for the volume of each phase. Shaking has to be carried out manually or mechanically for a sufficiently long time to homogenize the solution and, since octanol and water form an emulsion, the two phases have to be separated. This problem may be overcome by centrifugation, but, even so, micro droplets might still remain in each phase, which can introduce large errors in the final measurements, especially in the case of hydrophobic ionic compounds. The K_{ow} values obtained by the shake-flask method are generally the lowest, probable because the equilibrium had not been achieved.

The slow stirring method can avoid the measurement errors introduced by the emulsion between octanol and water, especially for hydrophobic compounds. This method is generally considered a better representation of ecological conditions, whereby oil and water are equilibrated and the IL is present in very dilute amounts. Nevertheless, it is not appropriate for all ILs, since some may decompose after continuous contact with water for a long period of time. The values obtained by the slow stirring method are generally much higher than those obtained by the other methods.

The combined version of both methods seeks to avoid the disadvantages of each of the conventional methods. As can be seen in Table 4, the values obtained with the combined method are generally intermediate with respect to the other two procedures. However, depending on the IL, different situations can be found. For some ILs the combined method provides results very similar to those obtained with the shake-flask method and, for others they are similar to the slow-stirring results. This may be due to the different rates of degradation in water and/or the different tendencies to form microemulsions, and demonstrates the difficulty involving in using the technique.

However, while the three methods deliver different results, some trends can be identified. The K_{ow} values increase with increasing alkyl chain length of the cation (Figure 2), which is consistent with the findings of other authors [3,5,7,30,37,39,52]. The ILs with a bis(trifluoromethylsulfonyl)imide anion have higher K_{ow} values and the ILs with an ammonium cation have the lowest values.

Keeping the imidazolium cation constant, it is possible to see the effect by changing the anion. The hydrophilic order for the ILs which are not completely soluble with water are [BF$_4$$^-$]>[PF$_6$$^-$]>[NTf$_2$$^-$]. These results can be explained by the greater symmetry and consequent greater hydrophobic character of [PF$_6$$^-$] compared with [BF$_4$$^-$] and of [NTf$_2$$^-$] compared with [PF$_6$$^-$], due to the inclusion of two carbon atoms in the [NTf$_2$$^-$] skeleton. At the other extreme, the hydrophilic order for the ILs completely soluble in water is [CH$_3$COO$^-$]>[Cl$^-$]. For the same cation, the ILs with the anions [MDEGSO$_4$$^-$], [MeSO$_4$$^-$], [TfO$^-$] and [EtSO$_4$$^-$] have an intermediate hydrophilicity between the two groups of anions mentioned above (Figure 3).

The insertion of a hydroxy group reduces the lipophilicity of an imidazolium chloride IL, as can be observed when [hemim$^+$][Cl$^-$] is compared with [emim$^+$][Cl$^-$].

There are some differences between the K_{ow} values calculated by the two molecular simulation methods, but the trends observed are similar and agree with the experimental results.

Conclusions

The K_{ow} of ILs is an important parameter because describes the lipophilicity of ILs and can be used to determine properties such as

Compound	K_{ow} shake-flask method	K_{ow} combined method	K_{ow} slow-stirring method	Literature values		K_{ow} estimated
				shake-flask method	slow-stirring method	
[bmim+][PF6-]	0.0200 ± 0.0062	0.0323 ± 0.0001 [30]	0.2539 ± 0.0443	0.0200 [47], 0.0041 [33], 0.0035 [9], 0.0191 [48].	0.0219 [3] 0.0037 [32]	0.0091[45], 0.0054 [46]
[hmim+][PF6-]	0.0230 ± 0.0072	0.1120 ± 0.0023	0.3130 ± 0.0019	0.0631 [48]	0.1202-0.338 [49] , 0.1380 [32]	0.0931[45], 0.0617[46]
[omim+][PF6-]	0.4663 ± 0.0338	0.2388 ± 0.0279 [30]	0.9196 ± 0.0338	0.4467 [48]	1.2106 [49], 0.4009 [32]	0.9500[45], 0.7079[46]
[bmim+][BF4-]	0.0094 ± 0.0003	0.0054 ± 0.0005 [30]	0.1303 ± 0.0186	0.0031 [34]	0.0030 [3],	0.0091[45], 0.0054[46]
[omim+][BF4-]	0.0152 ± 0.0009	0.0537 ± 0.0011 [30]	0.0746 ± 0.0085	0.2089 [48], 0.0457 [34]	0.5754-1.2190 [49],	0.9500[45], 0.7079[46]
[emim+][NTf2-]	0.0607 ± 0.0039	0.0642 ± 0.0142	0.1407 ± 0.0124	0.0661 [48]	0.0891-0.1096 [3], 0.0070 [32],	925[45], 0.4467[46]
[bmim+][NTf2-]	0.1193 ± 0.0063	0.1076 ± 0.0038 [30]	0.3350 ± 0.0190	0.0229 [9], 0.3162 [48], 1.2882 [34]	0.1096-0.6166 [3], 0.4266-0.6607 [49], 0.0210 [32],	925[45], 0.4467[46]
[hmim+][NTf2-]	0.0352 ± 0.0127	0.1336 ± 0.0679	0.7827 ± 0.1734	1.4454 [48]	1.4125-1.6596 [3],	925[45], 0.4467[46]
[omim+][NTf2-]	3.6141 ± 1.0810	3.6567 ± 0.1912 [30]	9.8147 ± 0.1258	11.2202 [48]	6.3095-11.2202 [3],	925[45], 0.4467[46]
[bdmim+][NTf2-]	0.0326 ± 0.0107	0.7252 ± 0.0591	0.7327 ± 0.2768		0.6095-1.1587 [49]	925[45], 0.4467[46]
[emim+][TfO-]	0.0084 ± 0.0016	0.0322 ± 0.0021	0.1150 ± 0.0987			0.0008[45], 0.0005[46]
[emim+][EtSO4-]	0.0069 ± 0.0033	0,0226 ± 0.0093	0.0127 ± 0.0015			0.0008[45], 0.0005[46]
[emim+][CH3COO-]	0.0098 ± 0.0008	0.0092 ± 0.0008	0.539 ± 0.0712	0.0030 [34]		0.0008[45], 0.0005[46]
[bmim+][MeSO4-]	0.0072 ± 0.0007	0.0111 ± 0.0002	0.0038 ± 0.0038			0.0091[45], 0.0054[46]
[bmim+][MDEGSO4-]	0.0476 ± 0.0363	0.0191 ± 0.0036	0.0111 ± 0.0006			0.0001[45], 1.95×10^{-5}[46]
[mim+][Cl-]	0.0643 ± 0.0230	0.0354 ± 0.0138	0.1539 ± 0.0953			0.0005[45], 2.24×10^{-5}[46]
[emim+][Cl-]	0.0849 ± 0.0049	0.096 ± 0.0242	0.1612 ± 0.0278			0.0008[45], 0.0005[46]
[hemim+][Cl-]	0.0111 ± 0.0036	0.0058 ± 0.0001	0.0125 ± 0.0047			7.85×10^{-5}[45], 0.0001[46]
[dmim+][Cl-]	0.0140 ± 0.0026	0.0242 ± 0.0022	0.0227 ± 0.0032			0.0003[45], 0.0001[46]
[eim+][Cl-]	0.0445 ± 0.0042	0.0699 ± 0.0013	0.0764 ± 0.0014			0.0013[45], 3.1623[46]
[hmim+][Cl-]	0.0213 ± 0.0064	0.0221 ± 0.0010	0.1246 ± 0.0041	0.0186 [48]		0.0931[45], 0.0617[46]
[empy+][EtSO4-]	0.0049 ± 0.0011	0.0203 ± 0.0057	0.0313 ± 0.0100			0.0019[45], 0.0007[46]
[bmpy+][BF4-]	0.0082 ± 0.0009	0.0105 ± 0.0004	0.0170 ± 0.0049			0.0215[45], 0.0076[46]
ETAN	0.0028 ± 0.0008	0.0017 ± 0.0004	0.0026 ± 0.0008			0.0005[45], 0.0003[46]

Table 4: Octanol-water partition coefficients (K_{ow}) of ILs measured with shake-flask method, combined method and slow-stirring method at 30°C. Literature values are also shown, classified into those obtained by shake-flask or slow-stirring method. Estimated K_{ow} values obtained by molinspiration.com and Bio-Loom Software are also shown.

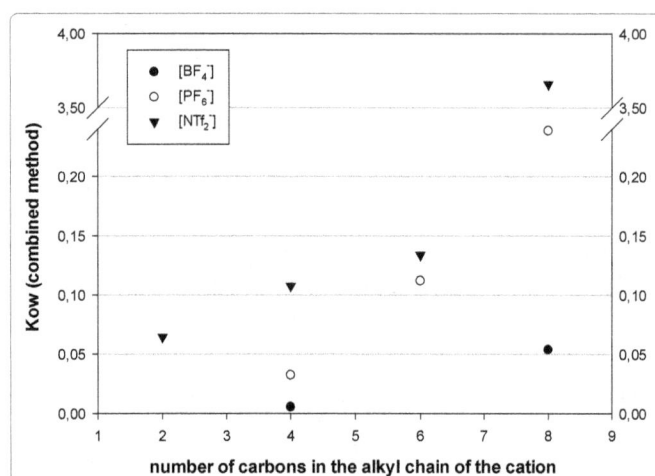

Figure 2: Octanol-water partition coefficients (K_{ow}) obtained by the combined method of ILs with different anions as a function of the alkyl chain length of the cation.

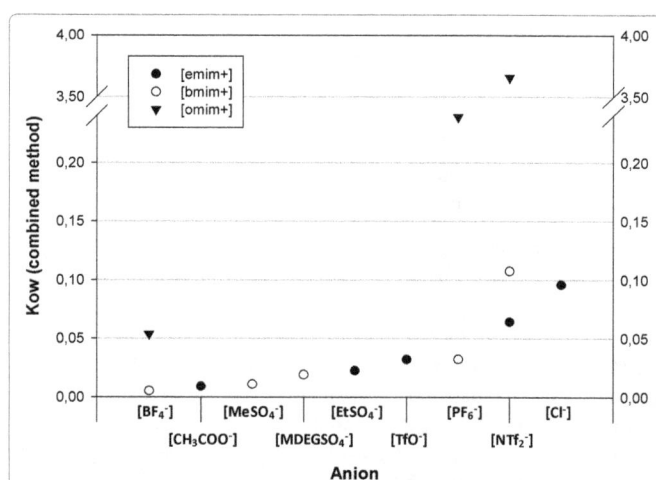

Figure 3: Octanol-water partition coefficients (K_{ow}) obtained by the combined method of ILs with three different alkyl chains ([emim+], [bmim+] y [omim+]) and different anions.

environmental effects, liquid/liquid solute partitioning and miscibility with other solvents. However, the values of the K_{ow} of ILs differ very much according to the experimental method used for their determination. It is possible that the differences in K_{ow} values obtained by different authors for the same ionic liquid might be also due to differences in ionic liquid concentrations.

The K_{ow} of ILs used in this study range between 0.0017 (ETAN) and 3.6567 ([omim+][NTf2-]) at 30°C. The ILs studied here are fairly hydrophilic, and their K_{ow} are lower than in commonly used industrial solvents.

The values of K_{ow} are lowest for the most hydrophilic ILs and increase with the cation alkyl chain length. Since all of the K_{ow} values are very small, we conclude that none of the ILs studied will accumulate or concentrate in the biota.

Acknowledgements

This work has been partially supported from the European Commission (FEDER/ERDF) and the Spanish MINECO (Ref. CTQ2011-25613 and Ref. CTQ2014-57467-R). Mercedes G. Montalbán acknowledges support from Spanish MINECO (FPI grant, BES-2012-053267).

References

1. Keskin S, Kayrak-Talay D, Akman U, Hortaçsu O (2007) A review of ionic liquids towards supercritical fluid applications. J Supercrit Fluids 43: 150-180.

2. Kabo GJ, Blokhin AV, Paulechka YU, Kabo AG, Shymanovich MP, et al. (2004) Thermodynamic Properties of 1-Butyl-3-methylimidazolium Hexafluorophosphate in the Condensed State. J Chem Eng Data 49: 453-461.

3. Ropel L, Belveze LS, Aki SN, Stadtherr MA, Brennecke JF (2005) Octanol-water partition coefficients of imidazolium-based ionic liquids. Green Chem 7: 83-90.

4. Freire MG, Carvalho PJ, Silva AM, Santos LM, Rebelo LP, et al. (2009) Ion specific effects on the mutual solubilities of water and hydrophobic ionic liquids. J Phys Chem B 113: 202-211.

5. Domanska U, Rekawek A, Marciniak A (2008) Solubility of 1-Alkyl-3-ethylimidazolium-Based Ionic Liquids in Water and 1-Octanol. J Chem Eng Data 53: 1126-1132.

6. Danielsson LG, Zhang YH (1996) Methods for determining n-octanol-water partition constants. TrAC 15: 188-196.

7. Deng Y, Besse-Hoggan P, Sancelme M, Delort AM, Husson P, et al. (2011) Influence of oxygen functionalities on the environmental impact of imidazolium based ionic liquids. J Hazard Mater 198: 165-174.

8. Han SY, Qiao JQ, Zhang YY, Lian HZ, Ge X (2012) Determination of n-octanol/water partition coefficients of weak ionizable solutes by RP-HPLC with neutral model compounds. Talanta 97: 355-361.

9. Lee SH, Lee SB (2009) Octanol/water partition coefficients of ionic liquids. J Chem Technol Biotechnol 84: 202-207.

10. Mackay D (1982) Correlation of bioconcentration factors. Environ Sci Technol 16: 274-278.

11. Mackay D, Fraser A (2000) Kenneth Mellanby Review Award. Bioaccumulation of persistent organic chemicals: mechanisms and models. Environ Pollut 110: 375-391.

12. Wang X, Ma Y, Yu W, Geyer HJ (1997) Two-compartment thermodynamic model for bioconcentration of hydrophobic organic chemicals by alga: Quantitative relationship between bioconcentration factor and surface area of marine algae or octanol/water partition coefficient. Chemosphere 35: 1781-1797.

13. Meylan WM, Howard PH, Boethling RS, Aronson D, Printup H, et al. (1999) Improved method for estimating bioconcentration/bioaccumulation factor from octanol/water partition coefficient. Environ Toxicol Chem 18: 664-672.

14. Raevsky OA, Grigor'ev VY, Weber EE, Dearden JC (2008) Classification and Quantification of the Toxicity of Chemicals to Guppy, Fathead Minnow and Rainbow Trout: Part 1. Nonpolar Narcosis Mode of Action. QSAR Comb Sci 27: 1274-1281.

15. Su LM, Liu X, Wang Y, Li JJ, Wang XH, et al. (2014) The discrimination of excess toxicity from baseline effect: effect of bioconcentration. Sci Total Environ 484: 137-145.

16. Lyman WJ, Reehl WF, Rosenblatt DH (1990) Handbook of chemical property estimation methods: environmental behavior of organic compounds. American Chemical Society.

17. Karickhoff SW (1981) Semi-empirical estimation of sorption of hydrophobic pollutants on natural sediments and soils. Chemosphere 10: 833-846.

18. Gerstl Z (1990) Estimation of organic chemical sorption by soils. J Contam Hydrol 6: 357-375.

19. Chiou CT (2003) Partition and Adsorption of Organic Contaminants in Environmental Systems. John Wiley & Sons.

20. Voutsas E, Magoulas K, Tassios D (2002) Prediction of the bioaccumulation of persistent organic pollutants in aquatic food webs. Chemosphere 48: 645-651.

21. Thomann RV (1989) Bioaccumulation model of organic chemical distribution in aquatic food chains. Environ Sci Technol 23: 699-707.

22. Allen DT, Shonnard DR (2001) Green Engineering: Environmentally Conscious Design of Chemical Processes. Pearson Education.

23. NSCEP (1975) Modeling dynamics of biological and chemical components of aquatic ecosystems. National Service Center for Environmental Publications.

24. Veith GD, Kosian P (1983) Estimating bioconcentration potential from octanol/water partition coefficients. Phys Behav PCBs Gt, Lakes Ann Arbor Sci, Ann Arbor MI 269-282.

25. Marcus Y (1990) Structural aspects of water in 1-octanol. J Solut Chem 19: 507-517.

26. NSCEP (1996) Product Properties Test Guidelines: Partition Coefficient (n-Octanol/Water), Shake Flask Method. US Environmental Protection Agency, Washington DC, USA.

27. OECD (1995) Test No. 107: Partition Coefficient (n-octanol/water): Shake Flask Method. Organisation for Economic Co-operation and Development, Paris.

28. Berthod A, Carda-Broch S (2004) Determination of liquid-liquid partition coefficients by separation methods. J Chromatogr A 1037: 3-14.

29. OECD (2006) Test No. 123: Partition Coefficient (1-Octanol/Water): Slow-Stirring Method. Organisation for Economic Co-operation and Development, Paris.

30. de los Rios AP, Hernández-Fernandez FJ, Tomas-Alonso F, Rubio M, Gomez D, et al., (2008) On the importance of the nature of the ionic liquids in the selective simultaneous separation of the substrates and products of a transesterification reaction through supported ionic liquid membranes. J Membr Sci 307: 233-238.

31. Deng Y, Besse-Hoggan P, Husson P, Sancelme M, Delort AM, et al. (2012) Relevant parameters for assessing the environmental impact of some pyridinium, ammonium and pyrrolidinium based ionic liquids. Chemosphere 89: 327-333.

32. Ventura SPM, Gardas RL, Gonçalves F, Coutinho JAP (2011) Ecotoxicological risk profile of ionic liquids: octanol-water distribution coefficients and toxicological data. J Chem Technol Biotechnol 86: 957-963.

33. Kaar JL, Jesionowski AM, Berberich JA, Moulton R, Russell AJ (2003) Impact of ionic liquid physical properties on lipase activity and stability. J Am Chem Soc 125: 4125-4131.

34. Zhao H, Baker GA, Song Z, Olubajo O, Zanders L, et al. (2009) Effect of ionic liquid properties on lipase stabilization under microwave irradiation. J Mol Catal B Enzym 57: 149-157.

35. Chapeaux A, Simoni LD, Stadtherr MA, Brennecke JF (2007) Liquid Phase Behavior of Ionic Liquids with Water and 1-Octanol and Modeling of 1-Octanol/Water Partition Coefficients. J Chem Eng Data 52: 2462-2467.

36. Domanska U, Bogel-Lukasik E, Bogel-Lukasik R (2003) 1-Octanol/Water Partition Coefficients of 1Alkyl-3-methylimidazolium Chloride. Chem Eur J 9: 3033-3041.

37. Kamath G, Bhatnagar N, Baker GA, Baker SN, Potoff JJ (2012) Computational prediction of ionic liquid 1-octanol/water partition coefficients. Phys Chem Chem Phys 14: 4339-4342.

38. Lee BS, Lin ST (2014) A priori prediction of the octanol-water partition coefficient (K_{ow}) of ionic liquids. Fluid Phase Equilibria 363: 233-238.

39. Cho CW, Preiss U, Jungnickel C, Stolte S, Arning J, et al. (2011) Ionic liquids: predictions of physicochemical properties with experimental and/or DFT-calculated LFER parameters to understand molecular interactions in solution. J Phys Chem B 115: 6040-6050.

40. Wattanasin P, Saetear P, Wilairat P, Nacapricha D, Teerasong S (2015) Zone fluidics for measurement of octanol-water partition coefficient of drugs. Anal Chim Acta 860: 1-7.

41. Heynderickx PM, Spanel P, Van Langenhove H (2014) Quantification of Octanol-water partition coefficients of several aldehydes in a bubble column using selected ion flow tube mass spectrometry. Fluid Phase Equilibria. 367: 22-28.

42. Stepnowski P, Storoniak P (2005) Lipophilicity and metabolic route prediction of imidazolium ionic liquids. Environ Sci Pollut Res Int 12: 199-204.

43. Ranke J, Müller A, Bottin-Weber U, Stock F, Stolte S, et al. (2007) Lipophilicity parameters for ionic liquid cations and their correlation to in vitro cytotoxicity. Ecotoxicol Environ Saf 67: 430-438.

44. Studzinska S, Stepnowski P, Buszewski B (2007) Application of Chromatography and Chemometrics to Estimate Lipophilicity of Ionic Liquid Cations. QSAR Comb Sci 26: 963-972.

45. http://www.molinspiration.com/services/logp.html

46. http://www.biobyte.com

47. Choua CH, Perng FS, Wong DSH, Su WC (2003) 1-Octanol/Water Partition Coefficient of Ionic Liquids. Boulder, CO, USA.

48. Lee SH (2005) Biocatalysis in Ionic Liquids: Influence of Physicochemical Properties of Ionic Liquids on Enzyme Activity and Enantioselectivity. Pohang University of Science and Technology, Pohang, Korea.

49. Gardas RL, Freire MG, Marrucho IM, Coutinho JAP (2006) Octanol-water partition coefficients of (imidazolium-based) ionic liquids. In: The 2nd National Conference on Thermodynamics of Chemical & Biological Systems, The Indian Thermodynamics Society, Veer Narmad South Gujarat University, Surat, India.

50. Leo A, Hansch C, Elkins D (1971) Partition coefficients and their uses. Chem Rev 71: 525-616.

51. Ruelle P (2000) The n-octanol and n-hexane/water partition coefficient of environmentally relevant chemicals predicted from the mobile order and disorder (MOD) thermodynamics. Chemosphere 40: 457-512.

52. Hsieh CM, Lin ST (2009) Prediction of 1-octanol-water partition coefficient and infinite dilution activity coefficient in water from the PR + COSMOSAC model. Fluid Phase Equilibria 285: 8-14.

9

Preparation of Aluminum Oxide from Industrial Waste Can Available in Bangladesh Environment: SEM and EDX Analysis

Tushar Kumar Sheel[1], Pinku Poddar[1], ABM Wahid Murad[3], AJM Tahuran Neger[2] and AM Sarwaruddin Chowdhury[1]*

[1]Department of Applied Chemistry and Chemical Engineering, Faculty of Engineering and Technology, University of Dhaka, Dhaka,
[2]Institute of Glass and Ceramic Research and Testing (IGCRT), BCSIR Dhaka, Dhaka, Bangladesh
[3]Institute of Leather Engineering and Technology, University of Dhaka, Dhaka, Bangladesh

Abstract

Aluminum oxide is an important chemical due to its many valuable properties such as hard ware resistance, good thermal conductivity, high strength and stiffness. Aluminum oxide is commonly referred to as alumina, possesses strong ionic interatomic bonding giving rise to its desirable characteristics. It can exist in several crystalline phases which all revert to the most stable hexagonal alpha phase at elevated temperatures. There are different types of Can which are used in everywhere. Most of the energy drinks are marked as canning (in which used mostly aluminum sheet). Can is a waste material which polluted our environment. We have prepared aluminum oxide from industrial Can by two methods. One is acid method and another one is alkali method. UV, thermo gravimetric (TGA), SEM and EDX analysis have been done. It has been showed that acid method is more feasible to prepare aluminum oxide.

Keywords: Aluminum oxide; Thermo gravimetric; SEM; EDX; Environment

Introduction

Aluminum wastes [1] cans recycling provide many environmental, economic and community benefits to individuals, communities, organizations, companies and industries. Recycling aluminum waste cans [2] save precious natural resources, energy, time and money. Money earned from recycling cans helps people help themselves and their communities. Zeepperfeld [3] prepared aluminum hydroxide from alkaline aluminates. Aluminum hydroxide [4-6] also recovered from aluminum containing waste solution [7].

Alpha phase alumina is the strongest and stiffest of the oxide ceramics. Its high hardness, excellent dielectric properties, refractoriness and good thermal properties make it the material of choice for a wide range of applications. High purity alumina is usable in both oxidizing and reducing atmospheres to 1925°C. Weight loss in vacuum ranges from 10^{-7} to 10^{-6} g/cm^{2}sec is over a temperature range of 1700°C to 2000°C. It resists attack by all gases except wet fluorine and is resistant to all common reagents except hydrofluoric acid and phosphoric acid. Elevated temperature attack occurs in the presence of alkali metal [8,9] vapors particularly at lower purity levels.

The composition of the ceramic body can be changed to enhance particular desirable material characteristics. An example would be additions of chrome oxide or manganese oxide to improve hardness and change color. Other additions can be made to improve the ease and consistency of metal films fired to the ceramic for subsequent brazed and soldered assembly.

Ceramic aluminas are generally produced by calcining Bayer aluminum hydroxide at temperatures high enough for the formation of α-Al_2O_3. By control of calcinations time and temperature and by the addition of mineralizes. Such as fluorine and boron, the crystallite size in the calcined product can be varied from 0.2 to 100 µm. These calcined aluminas can be categorized broadly according to their sodium content. There are two general types: those having about 0.5% Na_2O and low-soda grades with content 0.1%.

Reactive alumina is a material manufactured by dry grinding calcined alumina [10-12] to particle sizes smaller than 1 µm. The large surface area associated with very fine particles and the high packing densities obtainable considerably lower the temperatures [13] required for sintering. Tabular aluminas are manufactured by grinding, shaping, and sintering calcined alumina. The thermal treatment at 1900-2150 k causes the oxide to recrystallize into large, tabular crystals of 0.2-0.3 mm.

The model of industrial ecology emphasizes the containment and reuse of wastes generated by society as an overarching guideline for improving environmental quality. To realize this model, industry and society should work together to recover metals by recycling waste metal from all secondary sources and losing a minimum amount of material from the industrial/social system [14,15]. Aluminum oxide preparation from wastage cans by acid and alkali methods is the aim of this present work.

Experimental

Materials

We have collected three types of cans from our local market. These three types of cans are- (1) Pran juice can, (2) Tiger energy drink can, (3) Wild Brew energy drink can. Reagent grade tartaric acid, thioglycolic acid, ammonia, nitric acids are required for the determination of iron. To prepare aluminum oxide by acid method conc. HCl, NH$_4$Cl, 1:1 ammonia solution, NaOH, H$_2$O$_2$, and distilled water are required. Again conc. HCl, NaOH, H$_2$O$_2$, and distilled water are required for alkali method to prepare aluminum oxide.

***Corresponding author:** AM Sarwaruddin Chowdhury, Department of Applied Chemistry and Chemical Engineering, Faculty of Engineering and Technology, University of Dhaka, Dhaka-1000, Bangladesh
E-mail: profdrsarwar@gmail.com

Methods

Determination of iron by ammonia method: Ammonia reacts with trivalent iron in presence of thioglycolic acid produces pink color. Depending upon the intensity of pink color, iron content could be determined spectrophotometrically. The maximum molar absorptivity of the above developed color was found at 540 nm using UV visible spectrophotometer (Model: UV-2201 Shimadzu, Japan).

Preparation of standard stock solution: 0.864 g pre-dried A.R. ferric ammonium sulphate was dissolved with double distilled water containing 1 ml conc. HCl into a 100 ml volumetric flask and diluted up to the mark. It produced a solution of iron concentration of mg/ml.

Preparation of reagent:

1. Thioglycolic acid: 10 ml thioglycolic acid was dissolved to 10 ml by double distilled water into a 100 ml volumetric flask.

2. Tartaric acid: 10 g tartaric acid was dissolved with distilled water and diluted to 100 ml into a 100 ml volumetric flask.

3. Ammonia: 1:1 ammonia was used for the purpose.

4. Preparation of standard sample: 1 ml stock solution of concentration 1 mg/ml iron was taken in a 100 ml volumetric flask and diluted up to the mark by water to give a solution of 0.01 mg/ml iron.

The absorption was measured by UV visible spectrophotometer at 535 nm in a 1 cm cell against a reagent blank and the absorption reading were plotted against concentration. A linear straight line was obtained intercepting zero which was used for future sample determination.

Determination of the iron from the sample: 0.5 g sample is dissolved with 10 ml 1:1 HCl. Then it is diluted up to 100 ml. 5 ml solution is taken in a beaker and 25 ml nitric acid is added and it is heated for 20 minutes to convert the entire ferrous ion to ferric ion. After heating the solution it is up to 25 ml. 5 ml sample is taken and dissolve up to the 100 ml. from then 1 ml sample is taken to prepare 10 ml solution in which iron is determined. It is found that the percentage of iron in the Pran juice can is higher (around 60%) than the Tiger energy drink can and in the Wild Brew can (around 0.5%).

Preparation of aluminum oxide by acid method: 5 g of aluminum sheet is taken and it is reacted with 40 ml HCl with diluted 160 ml distilled water. By this reaction aluminum sheet is converted to aluminum chloride.

$$2Al \text{ (impure)} + 6HCl \text{ (aq)} \rightarrow 2AlCl_3 + 3H_2$$

This aluminum chloride solution is filtered to remove the foreign impurities and this filtrate is diluted up to the about 1000 ml. 5 g NH_4Cl is then added to the solution and boiled. After boiling, about 80 ml 1:1 HCl is added with drop wise by stirring with an electrical stirrer. At this position aluminum hydroxide is produced by the reaction of aluminum chloride with ammonium hydroxide. This aluminum hydroxide is separated by using a pressure filter and in this position sodium hydroxide is added to aluminum hydroxide and diluted it up to 500 ml and heated the solution 10-15 minutes. 5 ml H_2O_2 is added to settle the $Fe(OH)_2$ Sodium aluminates are formed which is soluble in water, and the insoluble compounds are settled down.

$$Al(OH)_3 + NaOH \rightarrow NaAlO_2 + 2H_2O$$

After setting the solution, the solution is filtered by the pickup with help of a pipette and pipette filler. In this filtered solution 1:1 HCl is added drop wise about 80 ml to complete the reaction to form aluminum hydroxide. This aluminum hydroxide is washed at least 6 times in one day to remove sodium chloride to get the pure product. After washing 6 times, the solution is filtered by the filter press. Thus we get aluminum hydroxide which is kept in the dryer at about 105°C. After drying aluminum hydroxide, it is kept in the furnace with about 800°C to form aluminum oxide.

$$2Al(OH)_3 \xrightarrow{800°C} Al_2O_3 + 3H_2O$$

This aluminum oxide is grinded in a grinder and it is passed through 100 meshes and final product is obtained (Figures 1-4).

Preparation of aluminum oxide by alkali method: 5 g of aluminum sheet is taken and it is reacted with 9 g NaOH and diluted with 250 ml distilled water. NaOH is reacted with aluminum sheet to form sodium aluminate. Then this solution is filtered to remove foreign particles. After heating (until boiling), about 3-5 ml H_2O_2 is added to this solution and it is kept for 24 hours to settle ferric oxide which is brick red in color. After settling, clear solution is picked up by the pipette filler and filtered. About 80 ml 1:1 HCl is added in the filtrate to form aluminum hydroxide.

$$NaAlO_2 + 2HCl + H_2O \rightarrow Al(OH)_3 + 2NaCl$$

Then the aluminum hydroxide is kept in a woven at 105°C temperature for drying. Again it is kept in a furnace at about 800°C temperature to form aluminum oxide. This aluminum oxide is grinded in a grinder and it is passed through 100 meshes and final product, aluminum oxide is obtained.

Results and Discussion

We have studied about three different types of cans which we have collected from our local market. At first, we analyzed the percentage of iron which is the main impurities of aluminum oxide preparation. In the experiment, we have seen that in the Pran juice can, the percentage of iron is around 60% where as in the Tiger energy drink can and in the Wild Brew can, the percentage of iron is around 0.5%. In case of Pran juice can, extraction of aluminum oxide was not as good as like as other two brands due to the higher percentage of iron. Again it was very difficult to get pure aluminum oxide from the Pran juice can. On the other hand the product yield from the Tiger energy drink can and from the Wild Brew can was higher than the Pran juice can. The product quality as well as product quantity of the Tiger energy drink can and the Wild Brew can were superior to the Pran juice can. Acid method and alkali method were used separately to prepare aluminum

Figure 1: Prepared raw materials.

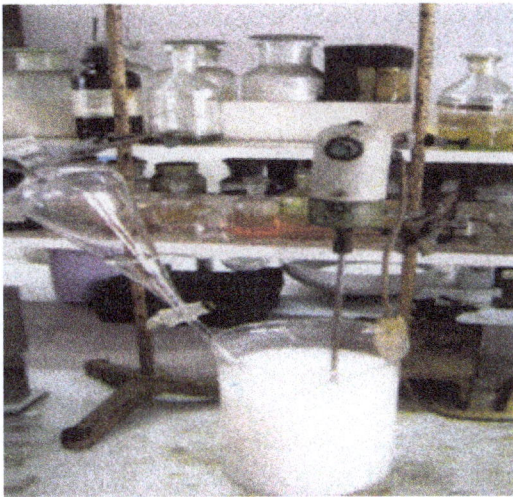

Figure 2: Preparation of Aluminum oxide.

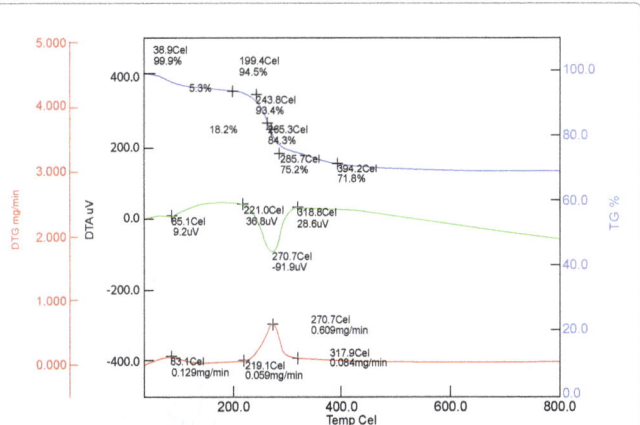

Figure 3: Aluminum hydroxide prepared in the laboratory.

Figure 4: Aluminum oxide.

oxide. We have seen that acid method was better than the alkali method because of the product quality and quantity both was higher in the acid method.

From the Figures 5-7, we have seen that about 300°C temperature is the suitable temperature to convert aluminum hydroxide to aluminum oxide. But about 800°C is the actual temperature to convert aluminum hydroxide to aluminum oxide. So, we can say that water and impurities help to convert aluminum hydroxide to aluminum oxide in less temperature. On the other hand, in case of alkali process for Tiger energy drink can about 275-300°C was needed to convert aluminum hydroxide to aluminum oxide. In the alkali process impurities were more than the acid process.

From the SEM analysis, we have seen that the surface of the grinding aluminum oxide magnifying 1000 times and the surface of the aluminum oxide was not smooth. So, we can say that it was not too much fine particles. May be impurities are responsible for the roughness of the surface (Figures 8-11).

From the analysis of EDX (Energy Dispersive X-ray) and Table 1, we have seen that 52.81% aluminum and 44.28% oxygen were present in the product. In the raw materials percentage of iron was about 0.5% but it was 0.17% in the product. Na and Cl_2 were also present in the product because those were added in washing steps. Because washing procedure was very lengthy and for this reason 1.24% Na and 1.18% Cl_2 was present in the acid method to produce aluminum oxide from Tiger energy drink can. We have also seen that K and Mn were present in the product below detection level (Figure 12).

In case of alkali method of Tiger energy drink can, from EDX analysis and Table 2, we have seen that the percentages of aluminum and oxygen were lower than the acid method. Again in the alkali method we have observed that the final product contained higher amount of Na and Cl_2 due to available NaOH was used in the method. So the product quality was not pure as like as acid method. In this method, K and Mn were absent but Fe was presenting similar to acid method (Figure 13).

Conclusion

Aluminum oxide is a material of commercial importance due to its many valuable properties such as hard ware resistance, excellent ware resistance, good thermal conductivity, excellent size and shape capability, high strength and stiffness. For the above properties it is used adsorbents, catalyst, high temperature electrical insulators, high voltage insulators, furnace liver tubes. Thermal sensors, laboratories

Figure 5: Thermogravimetric analysis of aluminum hydroxide, prepared by acid method of Tiger energy drink can. DTG=Differential Thermo Gravimetry, DTA=Differential Thermal Analysis, TG=Thermo Gravimetry.

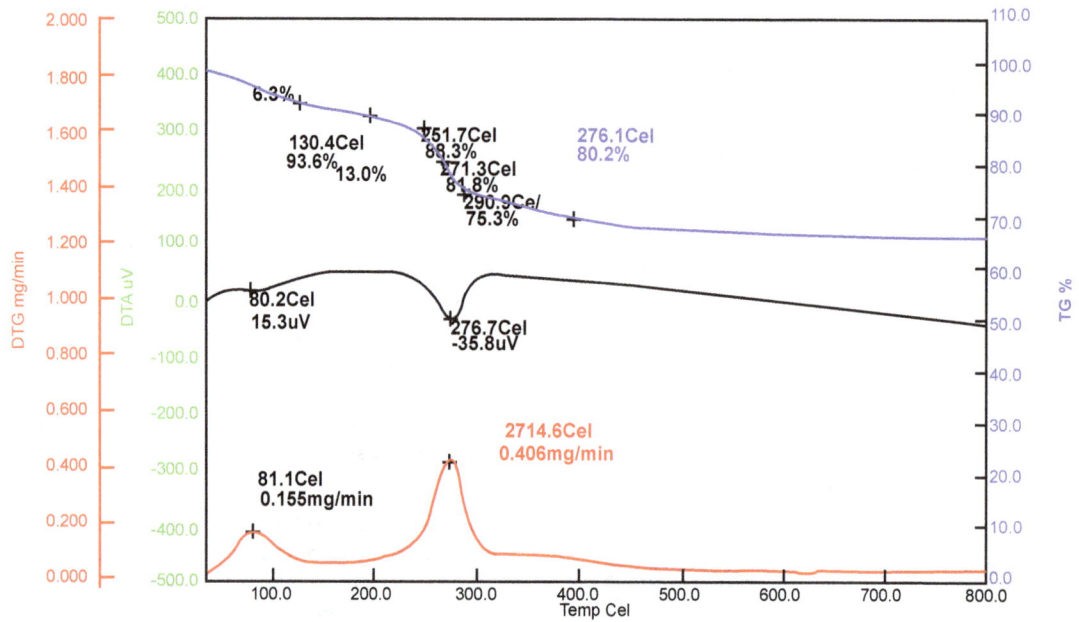

Figure 6: Thermogravimetric analysis of aluminum hydroxide, prepared by alkali method of Tiger energy drink can.

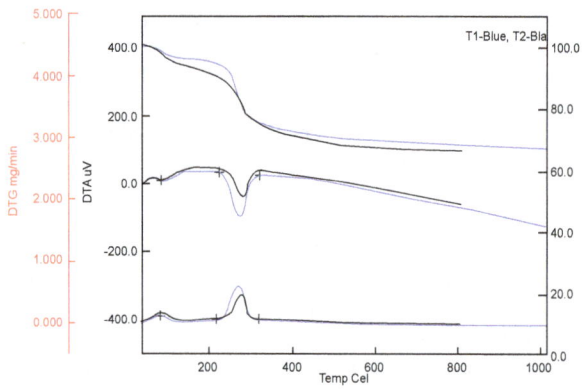

Figure 7: Comparative study of thermogravimetric analysis of two methods (acid and alkali).

Figure 9: 500 times magnification of aluminum oxide (alkali method of Tiger energy drink can).

Figure 8: 500 times magnification of aluminum oxide (acid method of Tiger energy drink can).

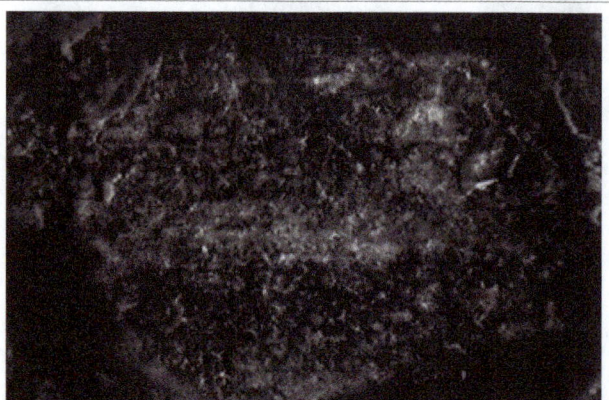

Figure 10: 1000 times magnification of aluminum oxide (acid method of Tiger energy drink can).

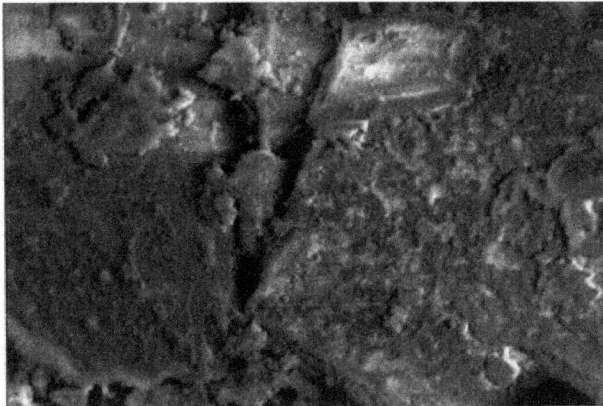

Figure 11: 1000 times magnification of aluminum oxide (alkali method of Tiger energy drink can).

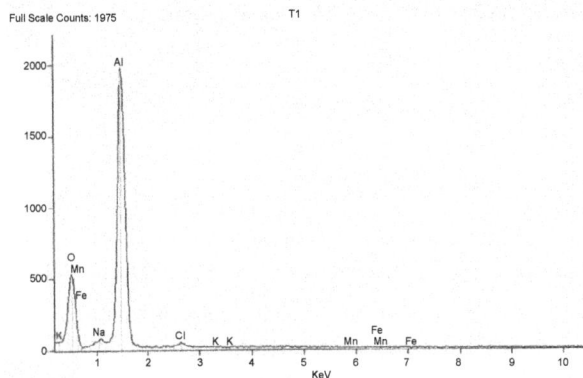

Figure 12: Spectrum of prepared aluminum oxide (acid method) by EDX.

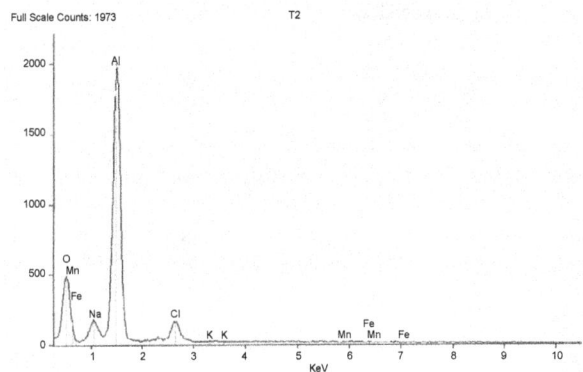

Figure 13: Spectrum of prepared aluminum oxide (alkali method) by EDX.

Element	Net Counts	Weight%	Weight% (Error)
O_2	7019	44.28	± 0.73
Na	418	1.24	± 0.17
Al	32898	52.81	± 0.51
Cl_2	444	1.18	± 0.15
K	57	0.16	± 0.12
Mn	21	0.16	± 0.29
Fe	20	0.17	± 0.34
Total		100.00	

Table 1: Quantitative analysis of aluminum oxide (acid method) which is prepared from Tiger energy drink can.

Element	Net Counts	Weight%	Weight% (Error)
O_2	6358	39.47	± 0.70
Na	2120	5.45	± 0.41
Al	33027	48.71	± 0.47
Cl_2	2609	6.18	± 0.37
K	0	0.00	-----
Mn	0	0.00	-----
Fe	24	0.18	± 0.32
Total		100.00	

Table 2: Quantitative analysis of aluminum oxide (alkali method) which is prepared from Tiger energy drink can.

instrument tubes and sample holders, grinding media and other ceramic uses. Different types of chemicals such as alum are prepared from aluminum oxide. Wastage cans were collected and aluminum oxide was prepared by acid and alkali methods. In the acid method, product quality was higher. In the alkali method, sodium chloride was the main impurities. UV- spectrometric analysis was done due to study raw materials analysis for iron. From SEM analysis it was seen that the surface of aluminum oxide is rough and from EDX result we were confirmed the purity of the product. We have also studied aluminum hydroxide by TGA and seen that 270°C temperature was needed to convert it to aluminum oxide. It was very low temperature because water and impurities were available there.

References

1. Galindo R, Padilla I, Rodríguez O, Sánchez-Hernández R, López-Andrés S, et al. (2015) Characterization of Solid Wastes from Aluminum Tertiary Sector: The Current State of Spanish Industry. Journal of Minerals and Materials Characterization and Engineering 3: 55-64.

2. Kaufman SM, Goldstein N, Millrath K, Themelis NJ (2004) The State of Garbage in America. 14th Annual Nationwide Survey of Solid Waste Management in the United States. Bio Cycle, pp: 31-41.

3. Zeppenfeld K (2003) Electrochemical procedure for production of aluminum hydroxide from alkaline aluminates. International Journal for Industry, Research and Applications 79: 754.

4. Krnel K, Drazic G, Kosmac T (2004) Degradation of AlN Powder in Aqueous Environments. Journal of Materials Research 19: 1157-1163.

5. López FA, Medina J, Gutierrez A, Tayibi H, Peña C, et al. (2004) Treatment of Aluminium Dust by Aqueous Dissolution. Revista de Metalurgia 40: 389-394.

6. Shinzato MC, Hypolito R (2005) Solid Waste of Aluminium Recycling Process: Characterization and Reuse of Its Valuable Constituents. Waste Management 25: 37-46.

7. Woodard F (2001) Method to recover aluminum hydroxide from aluminum containing waste solution. Industrial Waste Treatment Handbook.

8. Observatorio Industrial del Metal (2010) El sector del reciclaje de metales en España. Ministry of Industry. Energy and Tourism, Madrid, Spain.

9. Joint Research Center (2014) Best Available Techniques (BAT) Reference Document for the Non-Ferrous Metal Industries. Institute for Prospective Technological Studies Sustainable Production and Consumption Unit European IPPC Bureau. European Commission Final Draft, pp: 496-505.

10. Sanchez-Valente J, Bokhimi X, Toledo JA (2004) Synthesis and catalytic properties of nanostructured aluminas obtained by sol-gel method. App Catal A: General 264: 175-181.

11. Boumaza A, Favaro I, Ledion J, Sattonnay G, Brucbach JB, et al. (2009) Transition alumina phases induced by heat treatment of boehmite: an X-ray diffraction and infrared spectroscopy study. J Sol State Chem 182: 1171-1176.

12. Li J, Wang X, Wang L, Hao Y, Huang Y, et al. (2006) Preparation of alumina membrane from aluminium chloride. J Membrane Sci 275: 6-11.

13. Das BR, Dash B, Tripathy BC, Bhattacharya IN, Das SC (2007) Production of η-alumina from waste aluminium dross. Minerals Engineering 20: 252-258.

14. Allenby BR (1992) Industrial Ecology: The Materials Scientist in an Environmentally Constrained World. Materials Research Bulletin 17: 46-51.

15. Tibbs H (1992) Industrial Ecology: An Environmental Agenda for Industry. Whole Earth Review. Winter, pp: 4-19.

10

Deoxygenation of Palmitic Acid to Produce Diesel-like Hydrocarbons over Nickel Incorporated Cellular Foam Catalyst: A Kinetic Study

Lilis Hermida[1,2], H Amani[1], Ahmad Zuhairi Abdullah[1]* and Abdul Rahman Mohamed[1]

[1]*School of Chemical Engineering, Universiti Sains Malaysia, 14300 Nibong Tebal, Penang, Malaysia*
[2]*Department of Chemical Engineering, Universitas Lampung, Bandar Lampung 35145, Indonesia*

Abstract

Nickel incorporated mesostructured cellular foam (NiMCF) was studied as a catalyst for palmitic acid deoxygenation to primarily synthesize *n*-pentadecane and 1-pentadecene. The kinetic behaviour was tested in a temperature range from 280 to 300°C. The reaction was found to follow a first order kinetic model with respect to the palmitic acid with an activation energy of 111.57 KJ/Mol. In the reusability study, it was found that the average reduction in palmitic acid conversions was about 40.5%, which indicated the occurrence of catalyst deactivation during the deoxygenation. Fresh and spent catalysts were characterized by means of scanning electron microscope. Energy-dispersive X-ray spectroscopy and X-ray powder diffraction correlate their characteristics with catalytic activity and to identify the main catalyst deactivation mechanism. The catalyst deactivation was mainly due to phase transformation of metallic nickel (Ni^0) to nickel ion (Ni^{2+}) and the deposition of organic molecules on the catalyst during the deoxygenation. Regeneration of spent catalyst successfully reduced the drops in the palmitic acid conversions between the reaction cycles from 40.5% to 11.3%.

Keywords: Mesostructured cellular foam catalyst; Palmitic acid deoxygenation; Microstructure; Diesel-like hydrocarbons; Kinetics

Introduction

Exhaustion of petroleum oil or crude oil is predicted in the near future [1]. The shortage of crude oil will have global impact on the economy, culture and health of every nation as in this situation; fuel oil prices will go up. The high fuel oil price will induce other commodities or goods to be sold at relatively high prices. Diesel fuel is a kind of fuel oil that is obtained by refining crude oil in petroleum refineries. It has been predicted that the world demand for diesel fuel will grow faster than any other refined oil products in 2035 [2]. Therefore, renewable sources with related technologies should be identified as alternatives. Extensive studies on biofuels production from various renewable feedstocks have been carried out for many years. Among others, catalytic deoxygenation of fatty acids as renewable resources can be a potential technology to synthesize diesel-like hydrocarbons. During fatty acid deoxygenation, n-alkanes and alkenes will be produced through decarboxylation and decarbonylation [3] reactions, respectively;

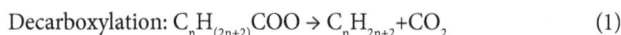

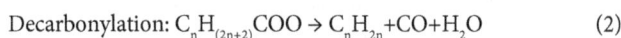

Decarboxylation: $C_nH_{(2n+2)}COO \rightarrow C_nH_{2n+2}+CO_2$ (1)

Decarbonylation: $C_nH_{(2n+2)}COO \rightarrow C_nH_{2n}+CO+H_2O$ (2)

The *n*-alkanes and alkenes are hydrocarbons that are similar to those found in crude oil derived diesel fuel [4].

Recently, extensive works have been reported on deoxygenation of various fatty acids to produce diesel-like hydrocarbons. They were carried out either in the presence of solvent or solventless condition using various catalysts such as Pd supported on mesoporous carbon [5-11] or on mesoporous silica [12-14]. Although the catalyst derived from Pd exhibited significant activity in the deoxygenation reactions, they are usually expensive. Ni-based catalysts are more practical on an industrial scale due to their availability and economic feasibility [15-18].

The use of Al_2O_3 or MgO-Al_2O_3 supported Ni catalysts for oleic acid deoxygenation in a batch reactor under solvent free and inert atmosphere conditions have been investigated [19]. Ni supported on mesostructured cellular foam (MCF) silica has been reported to be a more effective catalyst compared to Ni supported on Al_2O_3 or SBA-15 catalysts for pyrolytic decomposition of cellulose to produce H_2 [20].

This could be due to the larger pores in MCF silica support to minimize the diffusional effects of reactants as well as the products [21,22].

In our earlier work, various Ni functionalized mesostructured cellular foam (NiMCF) catalysts were used for solventless deoxygenation of palmitic acid at 300°C for 6 h in a semi batch reactor to produce *n*-pentadecane and 1-pentadecene as diesel-like hydrocarbons [23]. The catalysts were synthesized using different structures of MCF silicas prepared under various conditions. It was found that NiMCF catalyst using MCF silica prepared with 9.2 ml of tetra-ethyl-ortho-silicate (TEOS) and subsequently aged for 3 days showed the highest activity for the process. However, no attempt has been made so far to study the reaction kinetic and reusability of the NiMCF catalyst.

In the present study, kinetic of solventless deoxygenation of palmitic acid has been studied in a temperature range from 280 to 300°C. Reusability of NiMCF catalyst has also been evaluated in the deoxygenation process. The main mechanism for the catalyst deactivation has been successfully elucidated based on SEM, XRD and TGA results.

Experimental

Preparation of NiMCF catalyst

NiMCF catalyst was synthesized using MCF silica prepared with 9.2 mL of tetraethyl orthosilicate (TEOS) and aged for 3 days [23]. Nickel incorporation of into the support was achieved through deposition-

*Corresponding author: Ahmad Zuhairi Abdullah, School of Chemical Engineering, Universiti Sains Malaysia, 14300 Nibong Tebal, Penang, Malaysia
E-mail: chzuhairi@usm.my

precipitation process and subsequently reduced in hydrogen stream at 550°C for 2.5 h and cooled down under nitrogen flow.

Kinetic of solventless palmitic acid deoxygenation

Kinetic of solventless palmitic acid deoxygenation was performed in a semi-batch mode in which gaseous products (CO_2, CO, etc.) produced during the reaction were continuously removed. The deoxygenation reaction was carried out in a 250 mL three-necked flask reactor equipped with a magnetic stirring bar, reflux condenser and a tube to pass pure nitrogen flow to reaction mixture. During the reaction, nitrogen stream was used to sweep the evolved gaseous products. The reaction vessel was heated with a stirring hot plate. 6 g of palmitic acid and 15 wt% (with respect to palmitic acid) of catalyst amount were then added into the reactor. Before an experiment was started, nitrogen flow was passed through the reaction mixture for 30 min to create an inert reaction environment. Then, the reaction mixture was heated to different temperatures (280-300°C) and maintained for different reaction times (2 to 6 h) under rapid stirring. During the reaction, the gaseous products were collected in a gas-sampling bulb. At the end of the reaction, the mixture was allowed to cool to about 100°C, and then poured through a filter paper to separate the liquid products from the spent catalyst. Analysis of the liquid products was achieved using Agilent Technology 7890A gas chromatograph and then validated using a Perkin Elmer GC-MS system (Clarus 600).

Product analysis

Gas products in were identified by means of a Shimadzu C11484811134 GC system equipped with a thermal conductivity detector (TCD) and a capillary column. A mixture of standard gases with known composition was injected into the GC to identify each of the gas components based on their retention times. The mixture standard gas consisted of 30 vol% CO_2, 30 vol% CO, 30 vol% H_2 and 10 vol% Ar. It should be noted that the palmitic acid conversion and the desired product selectivity were based on the liquid-phase concentrations of the products. The gas-phase analyses were performed only for the confirmation of the presence of CO_2 and CO in the gaseous products.

Meanwhile, liquid product was collected and analyzed by means of an Agilent Technology 7890A GC system equipped with a flame ionization detector and a non-polar capillary column (GsBP-5). The detector and injector temperatures were set at 280°C and 250°C, respectively. The column temperature was set at 135°C for 1 min and was then programmed at 15°C/min to 290°C, and it was maintained constant at this temperature for 2 min. 20 µL sample was dissolved in 200 µL hexane, and then a direct injection into the gas chromatograph was carried out. The 1-pentadecane ($CH_3(CH_2)_{13}CH_3$) was used as standard to identify substances in a liquid product and to create calibration curves. The calibration curves were used to determine the concentration of substances in the liquid product. Calibration curve was generated using a series of chemical standards with known concentrations. Palmitic acid conversion, product selectivity and product yield were calculated according to Fu et al. [24] using Equations (3), (4) and (5), respectively.

$$\text{Conversion}\%, C = \frac{[C_{PA,0}] - [C_{PA,t}]}{[C_{PA,0}]} \times 100\% \quad (3)$$

$$\text{Selectivity}\%, S = \frac{[C_{PA,t}]}{[C_{PA,0}] - [C_{PA,t}]} \times 100\% \quad (4)$$

$$\text{Yield}\% = C \times S \times 100\% \quad (5)$$

where, $[C_{PA,0}]$=Concentration of palmitic acid before reaction

$[C_{PA,t}]$=Concentration of palmitic acid after reaction

$[C_{P,t}]$=Concentration of a product (n-pentadecane or 1-pentadecene) after reaction

Reusability study of NiMCF catalyst

For reusability study of the catalyst, fresh NiMCF catalyst with amount 15 wt.% (with respect to palmitic acid) was used for the solventless deoxygenation of palmitic acid at 300°C for 2, 3, 4, 5 and 6 h. After the experiment, the spent NiMCF catalyst was filtered out from the catalytic reaction mixture, washed thoroughly with dichloromethane, and then dried at 100°C overnight. Then, the washed spent catalyst was directly reused for the following deoxygenation runs using the same procedure.

Meanwhile, in the regeneration of catalyst, the washed spent catalyst was re-reduced under H_2 flow at 550°C for 2.5 h to regenerate it. After that, the spent catalyst was reused for the next deoxygenation run using the same procedure. The regenerated catalyst was characterized by means of TGA, SEM and XRD analyses. For comparison, spent NiMCF catalyst without regeneration was also characterized using the same analytical method.

Characterization of NiMCF catalyst

The SEM images were captured using a Leo Supra 50 VP field emission SEM. Before observations were made at room temperature, the samples were coated with high purity gold for electron reflection at a thickness of 20 nm by using a Polaron SC 515 Sputter Coater. Then, samples were mounted on aluminium stubs with double-sided adhesive tape for the observations carried out at a magnification of 50 kX.

X-ray diffraction (XRD) analysis was performed using a Siemens 2000X system to obtain XRD patterns of fresh, spend and regenerated catalysts in order to identify the different phases in the catalysts. The X-ray diffraction pattern was recorded using Cu-Kα radiation at 2θ angles ranging from 10-100°C. The TGA was carried out to observe the change in weight of the catalyst sample as it was heated in a certain temperature range. The thermal gravimetric analyzer unit coupled with a TG controller (TAC 7/DX) was supplied by Perkin-Elmer, USA. About 5 mg of catalyst sample was heated from 31-840°C at a heating rate of 10°C/min and an airflow rate of 25 ml/min.

Results and Discussion

Mechanisme of palmitic acid deoxygenation over NiMCF catalyst

In this study, the catalyst used was in powder form as obtained from the preparation method. Attempt to measure the particle sizes was not done bearing in mind that the catalyst particles were visibly reduced to even smaller sizes during the reaction under continuous mixing for up to 6 h. The stirring speed was set at about 250 rpm. Under such conditions, no significant external mass transfer effect was expected as generally reported in literatures [10-13].

Solventless palmitic acid deoxygenation over NiMCF catalyst in a semi batch reactor at 300°C produced mainly n-pentadecane and 1-pentadecene (Figure 1). In this reaction, n-pentadecane and 1-pentadecene were the predominant products so that the concentrations of both products were high at the end of the reaction. In the first 3 h of reaction, palmitic acid was converted

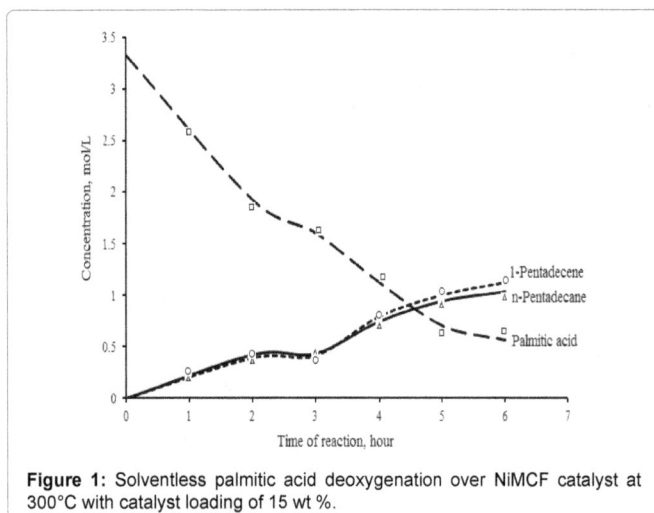

Figure 1: Solventless palmitic acid deoxygenation over NiMCF catalyst at 300°C with catalyst loading of 15 wt %.

into n-pentadecane and 1-pentadecene in which concentration of n-pentadecane was slightly higher than that of 1-pentadecene. This indicated that decarboxylation reaction was a more favoured reaction than decarbonylation reaction during this period. On the other hand, at reaction times longer than 3 h, concentrations of 1-pentadecene were usually higher than those of n-pentadecane. This observation gave the indication that decarbonylation reaction was the more dominant reaction compared to decarboxylation reaction. The lower n-pentadecane concentration was ascribed to the higher formation of CO gas that could inhibit the palmitic acid decarboxylation. CO gas which is a by-product from palmitic acid decarbonylation is poison to the catalyst in the decarboxylation reaction [25]. The activity of this catalyst was comparable with that reported by Roh et al. [19] using Ni/MgO-Al$_2$O$_3$ catalyst.

CO$_2$ and CO gases were detected in gaseous product through GC equipped with TCD detector. Meanwhile, n-pentadecane and 1-pentadecene were also detected in liquid products through GC-MS analysis. Besides n-pentadecane and 1-pentadecene, other products such as cyclopentadecane, ketone, etc. were also present at small quantities in the liquid products. This indicated that deoxygenation of palmitic acid acids over NiMCF catalyst in this study occurred not only through decarboxylation and decarbonilation reactions, but other reactions could also occur. Based on the reaction products, reaction pathways of solventless palmitic acid deoxygenation reaction could be proposed. Palmitic acid is first deoxygenated through decarboxylation reaction to produce n-pentadecane and CO$_2$ and through decarbonylation reaction to produce 1-pentadecene, CO and H$_2$O. Besides that, 1-pentadecene can also be produced through dehydrogenation of n-pentadecane. Moreover, a part of palmitic acid is converted into n-hexadecane (C$_{16}$H$_{34}$) through hydrogenation reaction [5].

Kinetic of solventless deoxygenation of palmitic acid over NiMCFcatalyst

Kinetic study of palmitic acid deoxygenation over the synthesized NiMCF catalyst was investigated at three different temperatures (280, 290 and 300°C). 6 g of palmitic acid and catalyst loading of 15 wt% was used in all the experiments. The kinetic model of oleic acid under inert atmosphere is suggested to follow first order with respect oleic acid [26]. Hence, the first order kinetic model for palmitic acid deoxygenation can be expressed as follows;

$$\text{Rate} = \frac{-d[C_{PA}]}{dt} = k[C_{PA}] \tag{6}$$

where, k is the rate constant, t is reaction time and $[C_{PA}]$ is the concentration of palmitic acid. In the kinetic study, the change in the palmitic acid concentration with the time was followed from the start of the reaction, $[CP_A]_0$ at t=0 to $[C_{PA}]_t$ at time t. These are the limits between which integration is performed. Upon integration and rearrangement, the final equation obtained is;

$$\ln = \frac{\left[C_{PA}\right]_t}{\left[C_{PA}\right]_0} = -kt \tag{7}$$

The experimental results obtained in this study are shown in Figure 2. As can be seen in the figure, the increase in temperature reaction resulted in an increase in the disappearance of palmitic acid concentration. This indicated that the temperature had positive effect in accelerating the reaction rate. Since the kinetic model of the palmitic acid deoxygenation is assumed to follow a first order model with respect palmitic acid, Equation (7) can be used. Figure 2 shows the plot of Equation (7) which ln $[C_{PA}]_t/[C_{PA}]_0$ is plotted against the reaction times. All the experimental data obtained for the palmitic acid deoxygenation experiments are found to be in good agreement with the Equation 7 due to the high value of correlation coefficient (R^2 are above 0.95) for all the straight lines. This indicates that the assumption that was used in this analysis was valid.

The integrated form of the rate law allows us to find the concentration of palmitic acid at any time after the start of the reaction. The slopes of the straight lines obtained by plotting ln $[C_{PA}]_t/[C_{PA}]_0$ versus the reaction time give $-k$, as can be seen in Figure 2. The reaction rate constants (k) are summarized in Table 1. As can be seen in the table, the increase in temperature resulted in the increase in the reaction rate constant. This was because mobility of reactant molecules was higher when the temperature was increased. The higher mobility of palmitic acid molecules made it easier for the molecules to reach the metallic nickel that might be located in the cell or the window pores in the NiMCF catalyst. The metallic nickel sites were the active sites for catalyzing the palmitic acid deoxygenation to produce diesel-like hydrocarbons of n-pentadecane and pentadecene [19].

Phenomenon of molecular diffusion is generally relevant in most liquid phase reactions involving porous catalysts. Furthermore, a rather viscous liquid (palmitic acid) was used as the reactant in this reaction. However, it has been reported that catalyst having a minimum average

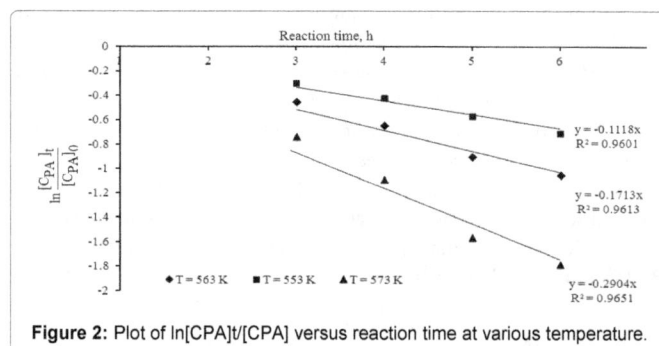

Figure 2: Plot of ln[CPA]t/[CPA] versus reaction time at various temperature.

Temperature (K)	Specific rate constant, k$_s$ (h^{-1})
553	0.1118
563	0.1713
573	0.2904

Table 1: Specific rate constants for the first order kinetic model.

pore size of 20 Å is required for liquid phase reactions to effectively avoid internal diffusion effect of the reactants and products [22]. In this study, structure of the NiMCF catalyst with large cell size (234 Å) and window pore size (90 Å) in the NiMCF catalyst should favour diffusion of the reactant and product molecules [23]. Thus, the effect of internal diffusion limitation should be negligible considering the small particle sizes of the catalyst. As such, reactant molecules reacted faster in the catalyst when mobility of reactant and product molecules was higher. Feature of molecular diffusions in three-dimensional structure of NiMCF catalyst together with its cell and window pores can be seen in Figure 3.

The rate constant, k, is related to the reaction temperature by the Arrhenius equation. The equation takes into account the rate constant of chemical reactions (k_s), reaction temperature (T), the activation energy (E_A), the pre-exponential factor (A) and the universal gas constant R. The Arrhenius equation is a relationship for the dependence of a reaction rate on temperature;

$$\ln k_s = \ln A - E_A/RT \qquad (8)$$

where; k_s=specific rate constant, h^{-1}

A=pre-exponential factor, h^{-1}

E_A=activation energy, J/mol

R=gas constant=8.314 J/mol.K

The activation energy, E_A, and pre-exponential factor, A, can be obtained by plotting ln k_s versus 1/T. The plot should produce a straight line in which a slope and intercept equal to - E_A/R and ln A, respectively. On the basis of data obtained from Table 1, a plot of ln k_s vs 1/T yields a straight line with high value of correlation coefficient (R^2>0.95). This result shows that the reaction rate constant follows the Arrhenius law, as can be seen in Figure 4. The value of activation energy (E_A) and pre-exponential factor (A) are calculated to be 111.565 kJ/mol and 38×10^8/h, respectively. The value of EA for this reaction is relatively lower than that reported in literature for Ni/MgO-Al$_2$O$_3$ (123 kJ/mol) to indicate higher activity of the catalyst used in this study [19].

With the kinetic parameters (E_A, A and k) that were obtained, mathematical model of rate expressions can be constructed as follows,

$$-r_{PA} = 38 \times 10^{8} \exp\left(-\frac{111565}{RT}\right)\left[C_{PA}\right] \, mol^{-1} \, h^{-1} \qquad (9)$$

The values of palmitic acid concentrations can be calculated using the first order kinetic model using Equation (7). The values of the palmitic acid concentration from the calculation and from the experimental results are presented in Table 2.

A parity plot between the experimental and calculated values of palmitic acid concentration is showed in Figure 5. As can be seen in the figure, the values of the palmitic acid concentration calculated at any time after the start of the reaction are in good agreement with experimental palmitic acid concentrations with a high value of correlation coefficient (R^2>0.95). It can be concluded from the results that palmitic acid deoxygenation over NiMCF catalyst satisfactorily followed first order kinetic model with respect to palmitic acid.

Reusability study of NiMCF catalyst

One of the major obstacles in the use of supported metal catalyst in fatty acid deoxygenation for production of diesel-like hydrocarbons is catalyst deactivation. Study of reusability of catalyst is important

for the economic assessment of the catalytic fatty acid deoxygenation process. Through the catalyst reusability study, stability and degree of catalyst deactivation can be examined. In this part of work, reusability of the NiMCF catalyst was studied through experiments in which the NiMCF catalyst after one cycle of use (or spent NiMCF catalyst) was re-used without regeneration for further solventless palmitic acid deoxygenation palmitic cycle. Alternatively, the catalyst was also regenerated as described in section 2.4 and used in the same reaction. Figure 6 shows results of the palmitic acid conversions obtained from various reaction times for the fresh NiMCF and the spent NiMCF catalytst.

As can be seen in Figure 6, palmitic acid conversions decreased after one cycle of use for all reaction times. This indicated the occurrence of significant deactivation of NiMCF catalyst during the palmitic acid

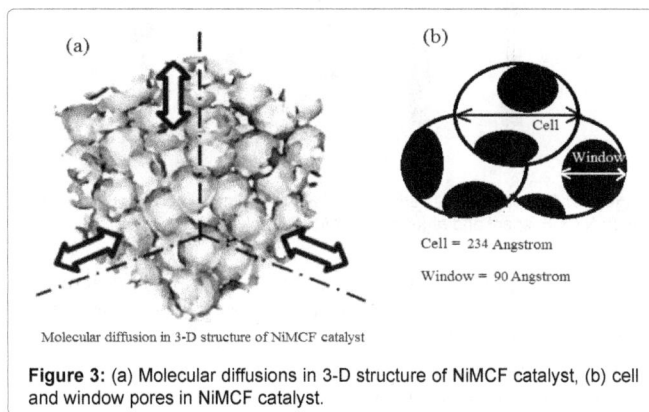

Molecular diffusion in 3-D structure of NiMCF catalyst

Figure 3: (a) Molecular diffusions in 3-D structure of NiMCF catalyst, (b) cell and window pores in NiMCF catalyst.

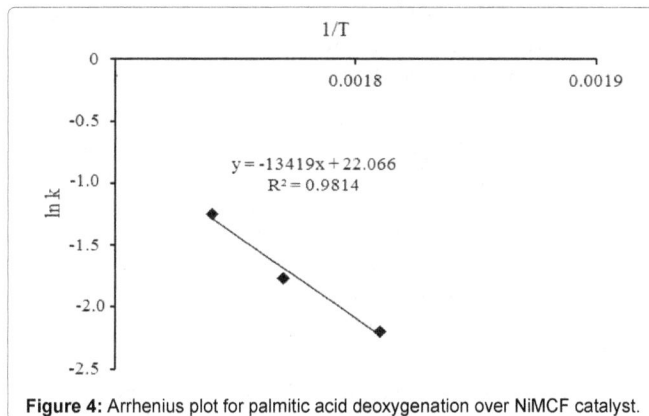

Figure 4: Arrhenius plot for palmitic acid deoxygenation over NiMCF catalyst.

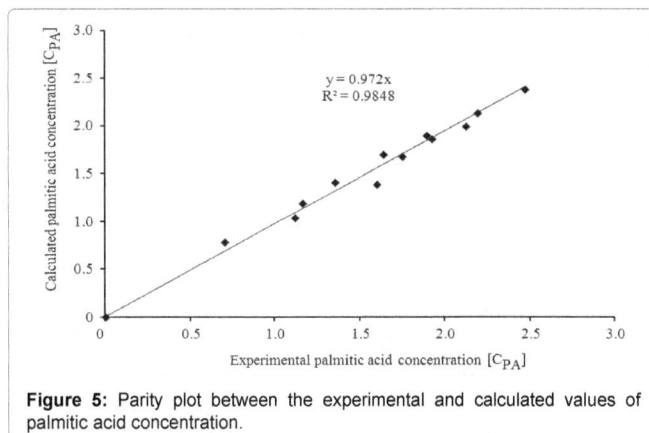

Figure 5: Parity plot between the experimental and calculated values of palmitic acid concentration.

Time (h)	[C_{PA}] at 553 K		[C_{PA}] at 563 K		[C_{PA}] at 573 K	
	Experimental	Calculated	Experimental	Calculated	Experimental	Calculated
2	-	-	-	-	1.92	1.86
3	2.47	2.38	2.12	1.99	1.6	1.39
4	2.19	2.13	1.75	1.68	1.12	1.04
5	1.89	1.9	1.35	1.41	0.7	0.78
6	1.64	1.7	1.16	1.19	0.56	0.58

Table 2: Values of experimental palmitic acid concentrations versus calculated ones.

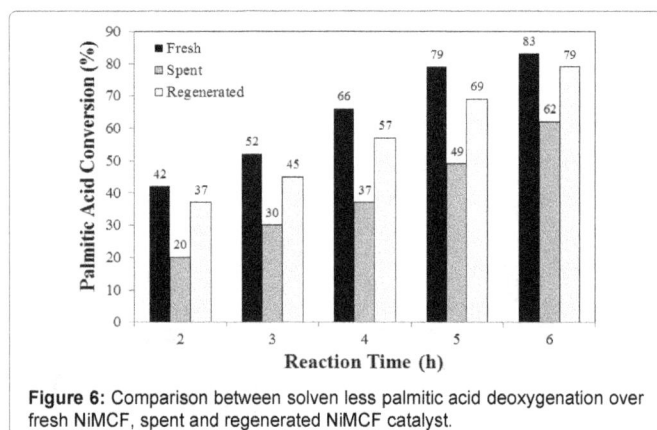

Figure 6: Comparison between solven less palmitic acid deoxygenation over fresh NiMCF, spent and regenerated NiMCF catalyst.

deoxygenation process after the first cycle. The catalyst deactivation caused reductions on the fatty acid conversion. The spent catalyst achieved palmitic acid conversion of 20% at a reaction time of 2 h. Meanwhile, fresh catalyst achieved palmitic acid conversion of 42% at a reaction time of 2 h. Thus, the reduction in palmitic acid conversion at 2 h of reaction is 52.8%. Meanwhile, reduction of palmitic acid conversion at 3, 4, 5 and 6 h of reaction time were 42.3%, 43.9%, 38.0% and 25.3%, respectively. Hence, average reduction of palmitic acid conversion using spent catalyst was 40.5%.

Figure 6 also shows palmitic acid conversions obtained by using the fresh NiMCF and regenerated NiMCF catalysts in the solventless palmitic acid deoxygenation using the same operating conditions. As can be seen, palmitic acid conversions using the regenerated catalyst were slightly lower than those using fresh catalyst for all reaction times. Reduction of palmitic acid conversion at 2, 3, 4, 5 and 6 h of reaction time were 11.9%, 13.5%, 13.6%, 12.7% and 4.8%, respectively. The average reduction of palmitic acid conversion using a semi-batch reactor was 11.3%.

Despite significant reduction in palmitic acid conversion, the values achieved by the regenerated catalyst were higher than those achieved by the spent catalyst without regeneration. By regenerating the spent catalyst, the extent on the drop in palmitic acid conversion reduced from 40.5% to 11.3%. This indicated that the regeneration of the spent catalyst using this method could improve the performance of the spent catalyst.

Deactivation of palladium supported on MCF catalyst (PdMCF) has been reported to be severe after one cycle of use for stearic acid deoxygenation in the presence of dudeen as solvent [14]. Almost no conversion of stearic acid was achieved by spent PdMCF catalyst in the deoxygenation reactions at reaction times from 0 until 6 h. PdMCF catalyst was mainly deactivated by significant organic deposition on the catalyst during the reaction. Other deactivation mechanism i.e., oxidation of metallic palladium and agglomeration of palladium particles on the catalyst was insignificant. The catalyst deactivation sources were elucidated based on results of various characterization

tests. Regeneration of the spent PdMCF catalyst was also carried out by sequential hot extractions with solvents tetrahydrofuran (THF), hexane and dichloromethane (DCM) followed by re-reduction process at 300°C. The regenerated catalyst showed a 19-fold increase in decarboxylation activity as compared to the spent PdMCF catalyst.

In order to identify mechanism of catalyst deactivation, several catalyst characterization methods such as XRD, SEM and TGA were performed. Evaluation of the deactivation mechanism involved comparison of the physical and chemical characteristics for the fresh and spent NiMCF catalysts. However, excessive discussion on the catalyst characteristics is avoided as in this liquid phase reaction, changes in the characteristics of the catalyst could significantly occur during the reaction and characteristics that are obtained ex-situ such as surface characteristics etc. could give misleading information on the reaction. In this study, the catalyst was recovered after the reaction by filtration and washed with organic solvents followed by drying. Unfortunately, complete removal of organic substances from the surface of the catalyst was not achieved. These organics would decompose during the degassing step (at 300°C) of surface analysis through nitrogen adsorption-desorption to cause errors in the results. Thus, attempts to measure the surface area after the reaction were unsuccessful. Figure 7 compares the characteristic of the fresh, spent and regenerated NiMCF catalysts by means of an XRD analysis method. For the spent catalyst sample, XRD pattern displays a peak at $2\theta=23°$ that is attributed to the presence of amorphous silica. Besides that, the XRD pattern displays peaks at $2\theta=33$ and $60°$ that are characteristic of nickel ions (Ni^{2+}) that could be in the form of nickel phylosilicates [27].

Furthermore, peak at $2\theta=44°$ attributed to metallic nickel particles is almost not detected in the spent catalyst sample. This indicated that metallic nanoparticle sintering, agglomeration and ripening that are common nanoparticles phenomena were not the main cause of deactivation in this system. In fact, higher peaks at $2\theta=33$ and $60°$ in the XRD pattern of the spent catalyst compared to those in the pattern of fresh catalyst could be due to evolvement of water vapour that oxidized the metallic nickel nanoparticles (Ni^0) during the palmitic acid deoxygenation so that they changed into nickel ions (Ni^{2+}) that could be in the form of nickel phylosilicate or nickels oxide phases.

Water vapour is a by-product in palmitic acid decarbonylation and hydrogenation. It could also be produced through water gas-shift reaction and methanation [28,29] in which molecules of H_2 from dehydrogenation reactions react with molecules of other gaseous products, i.e., CO_2 and CO (by-products from decarboxylation and decarbonylation), as given in the following reactions;

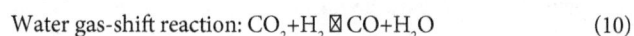

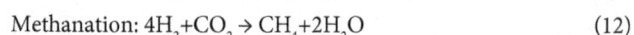

Water gas-shift reaction: $CO_2 + H_2 \rightleftharpoons CO + H_2O$ (10)

Methanation: $2H_2 + CO \rightarrow CH_4 + H_2O$ (11)

Methanation: $4H_2 + CO_2 \rightarrow CH_4 + 2H_2O$ (12)

It has been reported in literature that the presence of water vapor reduced the activity iron catalyst in Fischer-Tropsch reaction due to

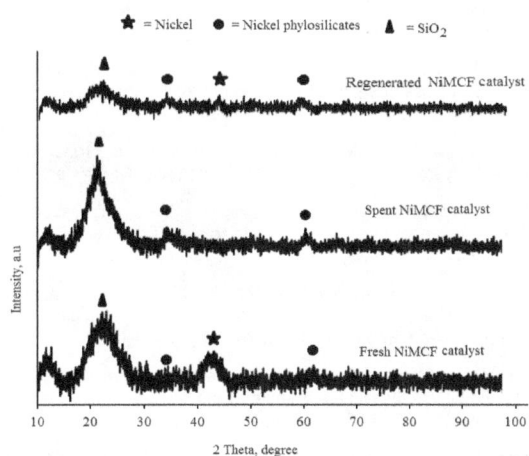

Figure 7: XRD patterns of fresh, spent and regenerated NiMCF catalysts.

the oxidation of iron catalyst by water vapor [30]. Meanwhile, nickel catalyst was found to have low activity for hydrocarbon transformation in the presence of water vapour due to oxidation metallic nickel by water vapour [31]. Through the oxidation reaction, metallic nickels in Ni/Al_2O_3 catalyst transformed into nickel ions (Ni^{2+}) in the form of nickelous oxide phases that rapidly re-crystallize [32]. This nickel catalyst deactivation mechanism can be applicable to the deactivation of NiMCF catalyst by water vapor in this study.

For the fresh catalyst sample, the presence of metallic nickel particles was clearly detected through the observation on the XRD pattern, as can be seen in Figure 7. The phase transformation from metallic nickel (Ni^0) into nickel ion (Ni^{2+}) reduced the activity of the spent catalyst in the deoxygenation reaction. This caused the decrease in palmitic acid conversion this system. It can be concluded that metallic particles (Ni^0) in NiMCF catalyst have higher active sites than nickel ion (Ni^{2+}) in NiMCF catalyst for the palmitic acid deoxygenation reaction. This result was in agreement with reports that oxide type catalysts such as MnO_3/C, MnO/C, WO_3/C, $NiOMoO_3/Al_2O_3$ and V_2O_5/C had lower activities compared to metal type catalysts such as Ni/C, Co/SiO_2, Ni/SiO_2 and Ni/Al_2O_3 for deoxygenation of bio-oil [33] and also for deoxygenation of fatty acid methyl ester.

On the basis of the aforementioned characterization results, it is hypothesized that the catalyst deactivation in this study could be attributed to organic molecules derived from reactant and products as well as small deposits of coke on the catalyst. Besides that, phase transformation from metallic nickel (Ni^0) into to nickel ion (Ni^{2+}) was also responsible for the catalyst deactivation. Therefore, regeneration of the spent catalyst was then carried out. Furthermore, the regenerated catalyst was characterized by means of XRD and SEM. Figure 7 also shows the comparison between XRD pattern of spent NiMCF catalyst and that of regenerated NiMCF catalyst.

The XRD pattern of the regenerated catalyst displayed the presence of peak at $2\theta=44°$ that is characteristics to the metallic nickel particles. However, peaks at at $2\theta=33$ and $60°$ that are characteristics of the nickel phylosilicates are still present with considerable intensities in the regenerated catalyst. This indicated that not all nickel phylosilicate was reduced into metallic nickel after the re-reduction process. The re-reduction process was carried out at 550°C for 2.5 h in this study. The unreduced nickel phylosilicate was probably due to very high interactions between nickel phylosilicates and MCF silica. As such, in order to reduce the more nickel phylosilicate, the re-reduction process should be carried out at higher temperatures and/or for a longer period.

In order to identify the possibility of catalyst deactivation due to deposition of organic molecules on the spent catalyst, weight loss, thermal behaviour and structural destruction of the spent catalyst were investigated using TGA analysis method. Figure 8 shows a TGA thermo gram of the spent catalyst together with the first derivative whether its weight change from the TGA test. A decomposition region occurred from temperatures of around 250 to 600°C. The derivative of the thermo gram clearly exposes this inflection with maximum peak at around 400°C. The decomposition of the spent catalyst in this region could be attributed to the removal of organic components that were deposited on the catalyst. Removal of organic components deposited in spent Pd functionalized MCF catalyst at the same temperature region was reported in the literature [14].

Morphologies of the fresh and spent NiMCF catalyst were examined by means of SEM. The SEM image results are shown in Figure 9. The SEM image of the spent catalyst clearly confirms layered structures in the spent catalyst. The layered structures contained nickel ions (N^{2+}) that could be in the form of nickel phylosilicate. The SEM image of the spent catalyst also confirmed that the spent catalyst had fewer porous structures compared to the fresh catalyst. Meanwhile, the SEM image of the fresh catalyst reveals high porous structures with Nano worm structures of nickel nanoparticles on the MCF silica.

Figure 9 also shows comparison between SEM images of the fresh NiMCF catalyst and regenerated NiMCF catalysts. The SEM image of the regenerated catalyst confirmed nickel nanoparticles in the form of Nano worm structures with sizes of around 19 nm. The structures in the regenerated catalyst were almost the same with those in fresh catalyst but the nickel nanoparticle sizes in the regenerated catalyst were slightly larger than those in fresh catalyst. Besides that, mesoporous structures in the regenerated catalyst were slightly lower than those in fresh catalyst. Nevertheless, the regenerated catalyst regained nearly all of lost porosity and metallic nickel particles after the sufficient removal of deposited organic molecules and followed by the re-reduction process.

Conclusion

Kinetic of solventless palmitic acid deoxygenation over nickel functionalized mesostructured cellular foam catalyst (NiMCF) to selectively synthesize n-pentadecane and 1-pentadecene (diesel-like hydrocarbons) was successfully studied in a temperature range from 280 to 300°C. The deoxygenation reaction satisfactorily followed a first order kinetic with an activation energy of 111.57 kJ/mol. From the reusability studies of the NiMCF catalyst, it was found that palmitic acid conversions significantly decreased after one cycle of use for all reaction times which suggested the occurrence of catalyst deactivation. The average reduction of palmitic acid conversion was about 40.5%. Based on characterizations of the spent catalyst, the main cause for

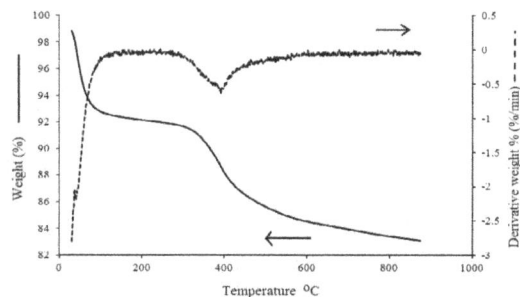

Figure 8: TGA profile (solid line) with its derivative profile (broken line) for spent NiMCF catalyst.

5. Ang GT, Tan KT, Lee KT (2015) Supercritical and Superheated Technologies: Future of Biodiesel Production. J Adv Chem Eng 5: e106.

6. Moore M, Herrell B, Counce R, Watson J (2015) A Hierarchical Procedure for Synthesis of a Base Case Solid Oxide Fuel Cell System in Aspen HYSYS. J Adv Chem Eng 5: 129.

7. Lee SP, Ramli A (2013) Methyl oleate deoxygenation for production of diesel fuel aliphatic hydrocarbons over Pd/SBA-15 catalysts. Chem Cent J 7: 149.

8. Amani H, Ahmad Z, Hameed BH (2014) Highly active alumina-supported Cs-Zr mixed oxide catalysts for low-temperature transesterification of waste cooking oil. Appl Catal A 487: 16-25.

9. Maki-Arvela P, Rozmyszowicz B, Lestari S, Simakova O, Eranen K, et al. (2011) Catalytic deoxygenation of tall oil fatty acid over palladium supported on mesoporous carbon. Energy Fuels 25: 2815-2825.

10. Amani H, Asif M, Hameed BH (2016) Transesterification of waste cooking palm oil and palm oil to fatty acid methyl ester using cesium-modified silica catalyst. J Taiwan Inst Chem Eng 58: 226-234.

11. Mansur AA, Pannirselvam M, Al-Hothaly KA, Adetutu EM, Ball AS (2015) Recovery and Characterization of Oil from Waste Crude Oil Tank Bottom Sludge from Azzawiya Oil Refinery in Libya. J Adv Chem Eng 5: 118.

12. Albrecht KO, Hallen RT (2011) A Brief Literature Overview of Various Routes to Biorenewable Fuels from Lipids for the National Alliance for Advanced Biofuels and Bio-products (NAABB) Consortium. Department of Energy, Pacific Northwest National Laboratory.

13. Sivula L, Ilander A, Väisänen A, Rintala J (2010) Weathering of gasification and grate bottom ash in anaerobic conditions. J Hazard Mater 174: 344-351.

14. Ping EW, Pierson J, Wallace R, Miller JT, Fuller TF, et al. (2011) On the nature of the deactivation of supported palladium nanoparticle catalysts in the decarboxylation of fatty acids. Appl Catal A 396: 85-90.

15. Kho ET, Scott J, Amal R (2016) Ni/TiO2 for low temperature steam reforming of methane. Chemical Engineering Science 140: 161-170.

16. Wang J, Lu AH, Li M, Zhang W, Chen YS, et al. (2013) Thin porous alumina sheets as supports for stabilizing gold nanoparticles. ACS Nano 7: 4902-4910.

17. Haitham MA, Bernd H (2015) Template-Assisted Synthesis of Metal Oxide Hollow Spheres Utilizing Glucose Derived-Carbonaceous Spheres As Sacrificial Templates. J Adv Chem Eng 5: 116.

18. Olutoye MA, Wong SW, Chin LH, Amani H, Asif M, et al. (2016) Synthesis of fatty acid methyl esters via the transesterification of waste cooking oil by methanol with a barium-modified montmorillonite K10 catalyst. Renewable Energy 86: 392-398.

19. Roh HS, Eum IH, Jeong DW, Yi BE, Na JG, et al. (2011) The effect of calcination temperature on the performance of Ni/MgO-Al$_2$O$_3$ catalysts for decarboxylation of oleic acid. Catal Today 164: 457-460.

20. Zhao M, Yang X, Church TL, Harris AT (2012) Novel CaO-SiO2 sorbent and bifunctional Ni/Co-CaO/SiO2 complex for selective H2 synthesis from cellulose. Environ Sci Technol 46: 2976-2983.

21. Lam MK, Uemura Y (2015) The Potential of Gamma-Valerolactone (GVL) Production from Oil Palm Biomass. J Adv Chem Eng 5: e105.

22. Mahmud CK, Haque MdA, Chowdhury AMS, Ahad MA, Gafur MdA (2014) Preparation and Characterization of Polyester Composites Reinforced with Bleached, Diospyros perigrina (Indian persimmon) Treated and Unbleached Jute Mat. J Adv Chem Eng 4: 114.

23. Hermida L, Abdullah AZ, Mohamed AR (2013) Nickel functionalized mesostructured cellular foam (MCF) silica as a catalyst for solventless deoxygenation of palmitic acid to produce diesel-like hydrocarbons. In: Mendez-Vilas A (Ed) Materials and Processes for Energy: Communicating Current Research and Technological Developments. Brown Walker Press, Boca Raton, USA.

24. Morales G, Melero JA, Iglesias J, Paniagua M (2014) Advanced Biofuels from Lignocellulosic Biomass. J Adv Chem Eng 4: e101.

25. Immer JG (2010) Liquid-phase deoxygenation of free fatty acids to hydrocarbons using supported palladium catalysts. North Carolina State University, Raleigh.

26. Díaz L, Brito A (2014) FFA Adsorption from Waste Oils or Non-Edible Oils onto an Anion-Exchange Resin as Alternative Method to Esterification Reaction Prior to Transesterification Reaction for Biodiesel Production. J Adv Chem Eng 4: 105.

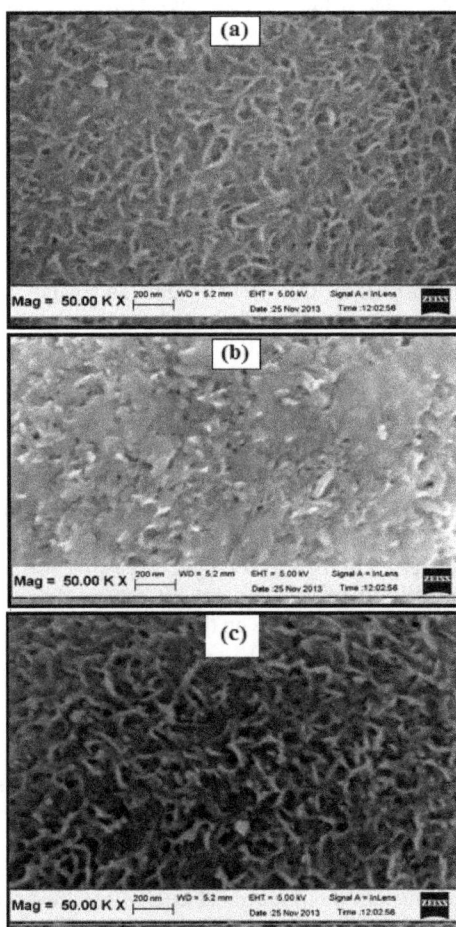

Figure 9: Comparison between SEM images of (a) fresh NiMCF, (b) spent NiMCF and (c) regenerated NiMCF catalyst obtained from catalytic deoxygenation in a semi-batch reactor.

the deactivation sources was found to be the phase transformation of metallic nickel (Ni^0) into to nickel ion (Ni^{2+}) and the deposition of organic molecules on the NiMCF catalyst during the palmitic deoxygenation. Regeneration of the spent improved the performance the spent catalyst in which the extent of drop in the palmitic acid conversion could be reduced from 40.5% to 11.3%.

Acknowledgments

The authors gratefully acknowledge a Research University grant (No. 814181) from Universiti Sains Malaysia and a Science fund grant (6013381) from MOSTI Malaysia to support this study.

References

1. Amani H, Ahmad Z, Asif M, Hameed BH (2014) Transesterification of waste cooking palm oil by MnZr with supported alumina as a potential heterogeneous catalyst. J Ind Eng Chem 20: 4437-4442.

2. Amani H, Ahmad Z, Hameed BH (2014) Synthesis of fatty acid methyl esters via the methanolysis of palm oil over Ca3.5xZr0.5yAlxO3 mixed oxide catalyst. Renewable Energy 66: 680-685.

3. Ayodelea OB, Hamisu U, Jibril MY, Uemurad WM, Dauda AW (2015) Hydrodeoxygenation of oleic acid into n- and iso-paraffin biofuel using zeolite supported fluoro-oxalate modified molybdenum catalyst: Kinetics study. J Taiwan Inst Chem Eng 50: 142-152.

4. Matzen M, Alhajji M, Yasar D (2015) Demirel Technoeconomics and Sustainability of Renewable Methanol and Ammonia Productions Using Wind Power-based Hydrogen. J Adv Chem Eng 5: 128.

27. Nares R, Ramirez J, Gutierrez-Alejandre A, Cuevas R (2009) Characterization and Hydrogenation Activity of Ni/Si(Al)- MCM-41 Catalysts Prepared by Deposition−Precipitation. Ind Eng Chem Res 48: 1154-1162.

28. Zhao X, Wei L, Cheng S, Huang Y, Yu Y, et al. (2015) Catalytic cracking of camelina oil for hydrocarbon biofuel over ZSM-5-Zn catalyst Original. Fuel Process Technol 139: 117-126.

29. Madsen AT, Rozmyszowicz B, Simakova IL, Kilpio T, Leino AR, et al. (2011) Step Changes and Deactivation Behavior in the Continuous Decarboxylation of Stearic Acid. Ind Eng Chem Res 50: 11049-11058.

30. Hermida L, Abdullah AZ, Mohamed AR (2015) Deoxygenation of fatty acid to produce diesel-like hydrocarbons. Renewable Sustainable Energy Rev 42: 1223-1233.

31. Norhasyimi R, Zuhairi A, Abdul Rahman M (2010) Recent progress on innovative and potential technologies for glycerol transformation into fuel additives: A critical review. Renewable Sustainable Energy Rev 14: 987-1000.

32. Ullah K, Ahmad M, Sultana S, Teong LK, Sharma VK, et al. (2014) Experimental analysis of di-functional magnetic oxide catalyst and its performance in the hemp plant biodiesel production. Applied Energy 113: 660-669.

33. Wolfson A, Yefet E, Alon T, Dlugy C, Tavor D (2015) Glycerolysis of Esters with Candida antarctica Lipase B in Glycerol. J Adv Chem Eng.

11

A Hierarchical Procedure for Synthesis of a Base Case Solid Oxide Fuel Cell System in Aspen HYSYS

Mark Moore*, Bonnie Herrell, Robert Counce and Jack Watson

Department of Chemical Engineering, University of Tennessee, USA

Abstract

Presented is a computer-aided process design and analysis procedure for use in creating Aspen HYSYS-based base-case design simulations for a Solid Oxide Fuel Cell. This procedure is based on the work of Douglas. This approach provides a step by step method to system design that permits preliminary assessments of the economic viability, which allows for expeditious creation of realistic and economic design simulations. Each step presents heuristics designed to aid in the specific design element addressed in that step. An example design of a Solid Oxide Fuel Cell from the Department of Energy's Fuel Cell Handbook is used to illustrate the design process. This example answers the questions presented by Douglas, and culminates in an energy balance and a capital cost table for the process. The example illustrated has an expected return on investment of 15% and serves as a guide for teaching the use of a hierarchical procedure for the synthesis of a modern chemical process.

Keywords: HYSYS; Simulation; Solid oxide fuel cell

Introduction

The rise of domestic production of natural gas is reshaping the U.S. energy economy, spurring the increased use of natural gas for electricity generation [1]. The fuel cell represents an effective process to produce electricity from natural gas. A fuel cell operates electrochemically, and is not limited by the Carnot Cycle. It therefore emits lower emissions of NO_x and CO_2 compared to systems using the combustion process [2]. A Solid Oxide Fuel Cell (SOFC) operates at a high temperature, easily reforming natural gas within the cell, promoting rapid electro catalysis with nonprecious metals, and producing high quality by product heat for cogeneration [2]. Using cogeneration a SOFC can reach efficiencies exceeding 70%, where a typical efficiency for a conventional power plant is 35% [2].

The potential of SOFC's has prompted a number of studies into the design of SOFC's. These include work at the Pacific Northwest National Laboratory on the design of low-cost modular SOFC using lower temperatures, a comparison of a 1 MW SOFC to a conventional thermal power generation plant and finding the optimal design as well as the capital and life cycle costs [3-5]. The purpose of this paper is to provide a step-by-step procedure, for use with Aspen HYSYS, for the base case design of a SOFC system in which the economics of the project are examined at every level of the design process. This procedure allows the overall economic impact to be calculated so that non-economic process/systems can be identified and rejected early in the process before expensive design or testing is carried out. Douglas' Hierarchical Decision Procedure for Process Synthesis is used for this purpose [6]. Douglas' procedure breaks down the design procedure into steps in order to determine whether a process flowsheet is viable as the process synthesis proceeds. The example process used in this paper to illustrate the process is a natural gas fueled SOFC outlined by the Fuel Cell Handbook [7]. Figure 1 shows the schematic used to represent this fuel cell process.

The hierarchy of decisions for creating a viable process synthesis expands Douglas' method to 7 levels.

Level 1: Process Classification and Input Information

Level 2: Input-Output Analysis of the Simulation

Level 3: Reactors and Compressors Analysis

Level 4: Separation Analysis

Level 5: Heat Exchanger Networking Analysis

Level 6: Energy Balance and Analysis

Level 7: Capital Cost Summary and Return on Investment Analysis

At every level, the economic potential is estimated. If the economic potential of the process proves to be undesirable, the designer may alter the decisions made to accommodate a more economical process.

Level 1

Process classification and input information

The first level of the HYSYS-based design methodology is the gathering of information about the process. The input information for the process being developed includes the following:

The components: Every chemical compound that is used or synthesized within the process must be accounted for in the HYSYS setup stage. Hypothetical components may be defined for cases which include components that are not defined within HYSYS.

The equation of state: HYSYS includes a number of methods for thermodynamic calculations, including several Equations of State.

The reactions: The reactions which occur during the simulated process may be defined in HYSYS to ensure that material and energy balance issues are handled appropriately. These reactions can be assigned to specific reactors within the simulation. It is also possible to use a HYSYS module called a Gibbs Reactor, in which HYSYS brings all of the reactants into equilibrium based on the Gibbs free energy.

Stream properties: Each component stream in the process must

---

*Corresponding author: Mark Moore, Department of Chemical Engineering, University of Tennessee, USA, E-mail: mmoore76@vols.utk.edu

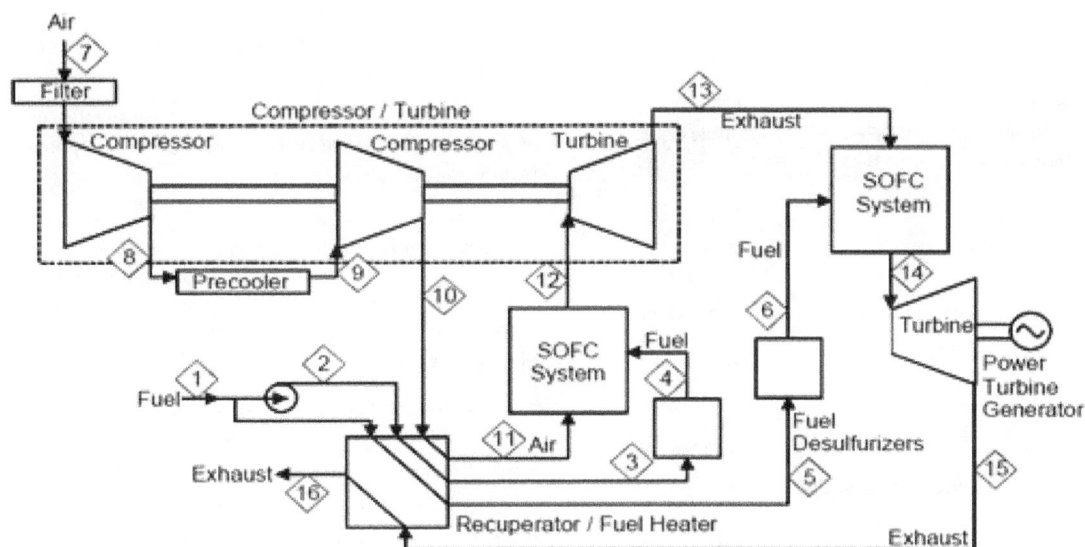

Figure 1: Schematic for a 4.49 MW natural gas-fueled SOFC Process (DOE, 2004).

have values for temperature, pressure, composition, and flow rate. HYSYS uses a degree of freedom analysis to calculate these stream properties, the user only needs to define some and HYSYS will calculate the rest. In addition, HYSYS alerts the user as to what is needed if the stream is underspecified. Downstream properties will commonly be calculated automatically in HYSYS, leading to consistency errors when the user definitions do not match the HYSYS calculations.

Level 1 example

Conditions within the simulation were chosen to match those in the DOE Fuel Cell Handbook. Some important conditions that should be met to properly simulate the fuel cell system are in Table 1. The fuel utilization value is the percentage of the fuel that goes through the fuel cell, the remaining fuel is combusted. The electrical efficiency of the fuel cell is measured with only the fuel that is used by the fuel cell. An overall efficiency measuring the amount of electricity produced from the total feed would then be the product of the electrical efficiency and the fuel utilization.

To complete the first level of the methodology, the reactions which occur within the SOFC must be taken into account, and from these reactions, the chemical compounds may be found. The following reactions are utilized within the SOFC:

- $CH_4 + H_2O \leftrightarrow CO + 3H_2$

- $CO + H_2O \leftrightarrow CO_2 + H_2$

- $CH_4 + \frac{1}{2}O_2 \leftrightarrow 2H_2O + CO$

- $2CO + O_2 \leftrightarrow 2CO_2$

- $2H_2 + O_2 \leftrightarrow 2H_2O$

- $C_2H_6 + 2H_2O \leftrightarrow 2CO + 5H_2$

- $C_2H_6 + \frac{5}{2}O_2 \leftrightarrow 3H_2O + 2CO$

These reactions all occur within the vapor phase. From these reactions it is clear that the chemical components carbon monoxide, water, carbon dioxide, methane, ethane and hydrogen gas must be

added to the component list in the basis environment. It would be impractical to use pure oxygen as the feed to the fuel cell, so argon and nitrogen gas should be added as well to simulate the air feed. With the component and reaction lists defined, an equation of state may be chosen. Because of its accuracy with hydrocarbon systems, the chosen equation of state was Peng-Robinson.

The raw material and product prices used in Level 2 for the input-output cost analysis are presented in Table 2. The main product of the system is the electricity produced by the fuel cells and the expanders, with the heat associated with the exhaust stream having value as well. The value of waste heat in the steel industry in 2004 was $0.04 per 4.18 MJ, assuming that this value would hold for the SOFC system it is used to value the waste heat in the final exhaust.

The price of natural gas used as the fuel for the system is based on the United States Energy Information Administration listed price in March of 2012, while the price of electricity per kWh sold is based on rates set by the Knoxville Utilities Board in April 2012 [8,9]. Air is supplied at atmospheric pressure and 21°C, and natural gas is supplied at 8.85 atm and 15°C.

Level 2

Input-output analysis of the simulation

The second level of the process focuses on the overall economic potential. The primary feed and effluent streams, as well as their values, must be identified. This is done with an overall mass balance of the input streams and the outlet streams of the simulation. The economic potential of the design can then be approximated by taking the difference of the value of the product produced by the system and the material costs required to create those products. Since the design process has not begun yet the values for the natural gas feed stream and the electricity produced should be the theoretical design values. It is also impossible to judge the amount of waste heat provided by the final exit stream, so calculating the value of this will be done in level 5 when heat exchangers are added [10].

Component	Value
Air to Fuel Ratio	20.8 (mole to mole)
Electrical Efficiency of Fuel Cells (LHV)	66.6%
Fuel Utilization	78%
Electrical Efficiency of Turbines	75%
Fraction of Feed of Natural Gas Going to HPFC	0.021 (mole to mole)
Fraction of Feed of Natural Gas Going to LPFC	0.021 (mole to mole)
Temperature of HPFC	860°C
Temperature of LPFC	875°C
Pressure of HPFC	8.39 atm
Pressure of LPFC	2.83 atm

Table 1: Important conditions of the fuel cell system from fuel cell handbook.

Utility	Cost
Electricity	$0.09756 per kWh
Natural Gas	$0.0466 per kg
Waste Heat	$0.04 per 4.18 MJ

Table 2: Input-Output Material and Utility Costs.

Level 2 example

The flow rates of the feed streams provided by the FCHB are in Table 3. Assuming the base cost for natural gas (listed in Level 1) in the simulation, the annual cost for the natural gas consumed would be approximately $211,619. All methane is considered to have reacted within the process. Air is not assigned a value here. The amount of electrical production in the FCHB is 4.49 MW. Table 4 indicates the annual values for the power produced by the 4.49 MW SOFC simulation, assuming continuous use throughout the year. Finally, Table 5 illustrates the summary of costs and economic potential for the Level 2 analysis.

Level 3

Reactors and compressors

Douglas provides several questions which may be adapted for consideration within this level [6].

1. How many reactors are necessary to use the design, and which material streams are associated with these reactors? Are there specifications within the design which require additional unit operation modules in HYSYS to adapt to the program's limitations?

The number of reactors in a fuel cell system will be equal to the number of fuel cells in the network. HYSYS does not have a unit operation module for the fuel cell, so a Gibbs reactor can be used as a substitution. The Gibbs reactor will combine the reactions of the anode, the cathode, the reformation reaction and, if necessary, the combustion reaction into one module. In an internally reforming SOFC the heat of the fuel cell reaction, and the heat provided by the combustion chamber, is used for the endothermic reforming reaction. A Gibbs reactor in HYSYS has no energy stream for electricity, so that all of the energy created by the fuel cell and combustion reactions that is not used by the reforming reaction goes to heat. By defining the temperature of the material stream leaving the Gibbs reactor the temperature of the Gibbs reactor is defined. HYSYS calculates the amount of heat that must be removed from the reactor to reach this temperature and assigns it to the energy stream of the Gibbs reactor. This energy stream then represents the electricity generated by the SOFC.

2. Are there recycle streams present in the design? Would the design benefit from additional recycle streams?

Possible benefits to a recycle stream are to recycle any unused fuel into the anode to provide a higher utilization rate of the fuel source, to recycle any carbon dioxide into the cathode to act as an oxidant, or to recycle any water formed as a result of the fuel cell reaction for use in the reforming reaction.

3. Does using an excess feed stream have an effect on the production of the desired product(s)? Is there an economic potential gain or loss from using an excess?

An excess feed stream of air makes the fuel the limiting reactant in the reactions, ensuring that the fuel is fully utilized in the system. Excess air is also a method for controlling the temperature of the fuel cells and the material streams by removing heat from the reactor. The temperature can be controlled in this manner without the utilization of cooling water.

4. How many compressors and expanders are required by the design, and how many may be excluded? What is the effect of these units on the economic potential of the design?

Compressors will be necessary to bring the air feed stream to the pressure of the fuel cell. Caution must be taken when compressing the air stream so that the stream does not become too hot and damage the compressor. It may be necessary to compress the stream in 2 or more steps while cooling the stream between the compressors. The general rule is that if the desired ending pressure is more than 4 times the initial pressure another step should be added [11]. Providing expanders associated with the effluent stream from the SOFC systems recovers some of the energy associated with the SOFC process.

Once the fuel cells, expanders and compressors are added to the HYSYS simulation, the next step in Level 3 is to estimate the cost of the compressors, expanders, and SOFCs to determine their impact on the economic potential. Compressors and expanders are valued similarly by their respective power in Watts. The value of the power of a compressor/expander in HYSYS is provided by the energy stream of the unit operation module. HYSYS assigns the energy streams on compressors and expanders as positive, regardless of whether electricity is being produced or consumed. 75% adiabatic efficiency is assumed for all compressors and expanders. These values for the efficiency, and the costs of the equipment, are from "Chemical Engineering Process Design and Economics: A Practical Guide" by Ulrich [11].

The fuel cells have their capital costs approximated at $175 per kW produced [12]. To estimate the economic potential for level 3 and beyond it is necessary to annualize the capital costs. The annual expenses including equipment depreciation are used in the annualization calculation. Annual expenses are those that are directly proportional to the capital costs and are listed in Table 6. Straight line depreciation over a lifespan of 10 years is used to calculate depreciation. Cost of capital is not included in the annualization of the capital costs.

The economic potential for level 3 is then:

$$EP\,3 = EP\,2 - Capital\ costs\ for\ Level\,3 \times 0.24$$

Level 3 example

The first SOFC system operates at a high pressure (~9 atm) and the feed air will require compression prior to entering the fuel cell. The effluent from the first SOFC then becomes the air feed for the second lower pressure SOFC. An expander is added to the effluent stream of the first SOFC to lower the pressure of the material stream before it is fed into the lower pressure SOFC. A final expander is used on the LP Exhaust recovering energy from the second SOFC effluent as electricity.

Material component	Hourly projections	Annual projection
Consumed Air	18,540 kg	1.62×10^8 kg
Consumed Natural Gas	518.4 kg	4.54×10^6 kg

Table 3: Projected annual consumption and production of material components in a 4.49 MWh SOFC.

Component	Approximate value in 2012 US dollars
Annual Electrical Production from SOFC	3.94×10^7 kWh
Annual Electrical Value from SOFC	$3,837,268

Table 4: Electrical cost analysis for a 4.49 MW SOFC located in the Southeastern United States.

Annual component consumption/ production	Approximate annual cost/value
Natural Gas Costs	$211,619
Economic Potential for Level 2	$3,625,650

Table 5: Summary of Level 2 Costs and Profits for a 4.49 MW SOFC system.

Capital-related cost item	Fractions of fixed capital
Maintenance and repairs	0.06
Operating supplies	0.01
Overhead, etc.	0.03
Taxes and insurance	0.03
General Depreciation	0.01 0.10
Total	0.24

Table 6: Annual expenses proportional to fixed capital.

Compressor	Adiabatic efficiency	Shaft work (kW)	Capital cost	Annualized cost
Compressor1	75.00%	634.3 kW	$1,757,325	$421,758
Compressor2	75.00%	663.2 kW	$1,820,750	$436,980
Expander1	75.00%	1408 kW	$3,446,813	$827,235
Expander2	75.00%	1292 kW	$2,409,063	$578,175
HPFC			$303,800	$72,912
LPFC			$239,050	$57,372

Table 7: Specifications and approximated costs for compressors, expanders, and fuel cells.

Table 7 shows the capital costs for the compressors, expanders, and the fuel cells. A lifespan of 10 years was used for the annualization costs for depreciation and the return of investment factor. The economic potential for level 3 is then: $1,231,218.

Level 4

Structuring a separations system

Douglas devotes Level 4 of his design method to the creation of a separations system within the flowsheet to recover products and byproducts at marketable concentrations. While the design of a separations system applied to SOFCs is unnecessary due to the nature of the example, it is a notable step in the design process and therefore is mentioned here. Many HYSYS design simulations do include separations systems, and as such, principles which Douglas takes into consideration should be referred to in his 1985 work [6].

Level 5

The heat exchanger network

The use of heat exchangers in a simulation serves to reduce overall energy costs by recycling heat produced by the system to heat the feed streams. Heat exchangers which utilize the exhaust streams in the process are a cost-effective way of optimizing operations. To add heat exchanger networks, a number of questions must be asked and answered:

1. What type of heat exchanger should be used?

The academic version of HYSYS gives a few options when selecting a heat exchanger design, with shell and tube heat exchangers being standard. TEMA type E (one shell pass) or type F (two shell passes with a longitudinal baffle) can be selected, co-current or countercurrent flow can be selected, and the number of tube passes can be defined.

Solid Oxide Fuel Cells operate at high temperatures, so it can be anticipated that the effluent streams will also be at high temperatures while the air and natural gas feed streams are at room temperature. Because of the large temperature difference between the streams, countercurrent exchangers are recommended for these heat exchangers [13]. Since the pressure of the feed streams is higher than the exhaust streams, the feed streams should be on the tube side while the exhaust stream can be on the shell side. The high temperatures of the effluent streams can also cause expansion of the tubes during start up, so that a U-Tube type arrangement (2 tube passes) of the tubes is also recommended [11]. If the heat flow is such that two Shell and Tube heat exchangers are necessary, a second shell pass can be added to save the expense of a second heat exchanger [13]. One shell pass exchangers are the most common, however [13].

The heat exchanger used to remove heat from the air feed between compressors needs other considerations than the exchangers between the effluent and the feed streams. Countercurrent flow allows the hot stream to approach the temperature of the entering cold stream, and generates less entropy than co-current flow [13]. The use of cooling water means the possibility of mineral deposit build up in the exchanger, so the ease of cleaning the exchanger is an issue. Running the water through the tubes, and designing the exchanger for one tube pass, allows for easy cleaning of these mineral deposits. Expansion of the tubes should not be an issue because of the relatively low temperature of the air feed stream.

1. Does a product stream within the simulation have a high enough temperature to provide heat for other streams?

Setting the tube side exit temperature and pressure will allow HYSYS to calculate the exit temperature of the shell side. If the temperature difference is inadequate HYSYS will give an error message and the tube side exit temperature can be adjusted.

2. Is it possible to link several heat exchangers within the simulation with the same exhaust stream feed, thus maximizing the exhaust's potential within the plant?

Once the exhaust stream being used for the shell-side feed of the heat exchanger has passed through, it may be possible to use the exhaust from the heat exchanger shell to feed another heat exchanger. This can continue on until all heating needs have been satisfied or the heat of the effluent stream is exhausted. If all of the feed streams have been heated to the desired temperature the final exhaust stream may still have value. Exhaust streams from natural gas can be cooled to a temperature of 120°C with the sensible heat taken from the stream sold at the rate of $0.04 per 4.18 MJ [13,14]. A unit operation module called a cooler can be used to cool the final exit stream to 120°C with the energy stream giving the value of heat removed from the stream. This value is assuming 100% efficiency of heat exchange, which is optimistic, but for simplicity reasons is used in the example calculations.

The level 5 EP is then:

EP5 = EP3 + Value of sensible heat sold − Annualized cost of heat exchangers

Level 5 example

The exhaust from the final expander in the simulation has a high temperature and is utilized in a heat exchanger network to provide heat for the feed streams. This may be simulated through placement of a shell-and-tube heat exchanger on each feed stream so that the feed streams may be heated in series by the hot effluent stream. A cooler can then be place on the final exhaust stream to cool the stream to 120°C so that the value of the stream as a heat source can be estimated.

The next step of Level 5 is the calculation of the effect on the economic potential of the heat exchangers added. The relevant figures for determination of heat exchanger costs are found in the book by Ulrich [11]. The method for calculating the capital costs for heat exchangers in Ulrich is by the heat transfer area. The academic version of HYSYS uses a heat transfer area of 60.32 m^2 regardless of the use or function of the heat exchanger. This means that it is necessary to calculate the size of the heat exchangers by hand. HYSYS provides the stream temperatures, pressures and enthalpies of the material streams so that a short cut calculation is relatively easy. The capital costs of the heat exchangers and their annualized costs are in Table 8. The value of the heat in the final exhaust stream was found to be $315,058 per year with a rate of sensible heat removal of 1,044 kW. The Economic Potential for Level 5 is then: $1,516,069.

Level 6

Energy analysis

One of the most important considerations of any process design is that of the energy balance of the process. In Level 6, one must not only perform an energy balance on the process as a whole, but also on the individual energy-producing elements of the process (such as fuel cell stacks). An advantage of using computer design software, such as HYSYS, is that the enthalpy of the streams and energy required by the unit operation modules is readily available. There are several questions which must be asked by the process designer:

1. Is the process as a whole balanced? If not, how must it be changed?

Aspen HYSYS provides indicators if part or all of a simulation is unbalanced. Unit operation modules and material streams will appear yellow if they are under-defined or if there is a conflict between a user input value and a calculation performed by HYSYS. Many times HYSYS will provide a notification about where the error is being detected. The energy streams of the unit operation modules indicate the amount of energy (heat or electricity) needed to operate the module at the user defined quantities. Care must be taken in examination of these values as it is often difficult to distinguish if the energy is being produced by the module or is being supplied to the module.

The energy associated with the material streams is represented by the "heat flow" of the stream in the dialog window of the stream. This heat flow is based on the enthalpy calculation of the material stream and is relatively meaningless on its own. The heat flow becomes meaningful when used in an energy balance equation since it represents the change in the heat flow of the process. HYSYS provides a spreadsheet applet that can be used for the energy balance. Values associated with a material stream, energy stream, or unit operation module can be input into the spreadsheet by dragging the value into the spreadsheet while holding down the right mouse button. When inputting the values in

this manner any change in the HYSYS simulation will be represented automatically in the spreadsheet. This dynamic property of the applet makes it a convenient tool for energy and material balances.

The balance equation for the entire process is then: heat flow of input streams + energy in = heat flow of output streams + energy out. The energy in and energy out terms represent the energy streams of the unit operations, in this case the expanders, compressors, pumps and Gibbs reactors. To calculate the total electrical energy produced by the system subtract the energy in from the energy out. To calculate the energy produced by just the fuel cells take the sum of the energy streams of the Gibbs reactors.

1. Are the energy-producing elements of the process balanced? If not, how must they be changed?

If the energy producing elements are outlined in yellow then they are unbalanced, if the dialog box of the unit operation module has a green bar at the bottom then it is balanced. Another indication is the color of the material streams; a yellow stream represents an unbalanced system while a blue stream represents a balanced system.

2. Is the energy production by the process the desired amount? If not, how can the process be scaled to fit the needs of the designer?

The system takes into account energy production and heat production. Calculations are required to determine how much energy is being produced by the simulation. This information may be used to scale the simulation to the desired result. Achieving the exact desired number may be difficult within HYSYS.

Level 6 example

The task for Level 6 is to perform required energy balances for the process. This includes not only the system as a whole, but each individual SOFC system within the simulation.

Level 7

The capital cost table is shown in Table 9. The actual costs are the installed costs of the equipment. The costs of contingencies and auxiliary facilities have been neglected.

Results and Conclusion

The goal has been to use the procedure and heuristics outlined by Douglas to design a system for a Solid Oxide Fuel Cell [6]. The Douglas procedure proceeds as a series of levels, 1 to 7, in which design decisions must be made and questions presented by Douglas must be answered. The process is illustrated at every level using an example SOFC system outlined in the Department Of Energy's Fuel Cell Handbook. With this example process the questions and design decisions in each level have been answered and addressed (Table 10).

In level 1 the components, chemical reactions, and stream properties of the system are identified, as well as the equation of state to be used by HYSYS for the calculations. In level 2 the total economic potential of the system is evaluated by calculating value of the natural gas needed as a raw material for one year of operation and subtracting that from the value of the electricity to be generated. Level 3 proceeds by determining the number of reactors (fuel cells) necessary, whether recycle streams are needed, if using an excess of the feed stream benefits the process, and the number of compressors and expanders required for the design. Level 4 of the Douglas method is dedicated to the structure of a separations system, which is not required in a fuel cell process. The heat exchanger network is the focus of level 5. The

Exchanger	Capital cost	Annualized cost
E-101	$44,475	$10,674
E-102	$8,006	$1,921
E-103	$6,671	$1,601
E-104	$66,713	$16,011

Table 8: Heat exchanger capital and annualized costs.

Equipment ID	Number	Capacity	Cost	Bare Module Factor	Actual Cost
Compressor1	K-101	643.3 kW	$667,350	2.5	$1,668,375
Electric Motor - Compressor1	S-101	643.3 kW	$59,300	1.5	$88,950
Compressor2	K-102	663.2 kW	$685,700	2.5	$1,714,250
Electric Motor - Compressor2	S-102	663.2 kW	$71,000	1.5	$106,500
Expander1	K-103	1406 kW	$1,334,250	2.5	$3,335,625
Electric Motor - Expander1	S-103	1406 kW	$74,125	1.5	$111,188
Expander2	K-104	1289 kW	$924,487	2.5	$2,311,218
Electric Motor - Expander2	S-104	1289 kW	$65,230	1.5	$97,845
HPFC	G-101	1736 kW	$303,800	1	$303,800
LPFC	G-102	1366 kW	$239,050	1	$239,050
Air Feed Cooler Heat Exchanger	E-101	91.37 m^2	$14,825	3	$44,475
HP Fuel Cell Fuel Feed Heat Exchanger	E-102	0.69 m^2	$2,669	3	$8,006
LP Fuel Cell Fuel Feed Heat Exchange	E-103	0.55 m^2	$2,224	3	$6,671
Air Feed Heater Heat Exchanger	E-104	159.3 m^2	$22,238	3	$66,713
Total Cost					$10,101,666

Table 9: Capital cost summary.

	FCHB	Simulation
Air to Fuel Ratio	20.8 (mole to mole)	21.05 (mole to mole)
Electrical Efficiency of Fuel Cells (LHV)	66.6%	NA
Fuel Utilization	78%	NA
Combined Electrical Efficiency and Fuel Utilization	51.9%	51.2%
Electrical Efficiency of Turbines	75%	75%
Fraction of Feed of Natural Gas Going to HPFC	0.021 (mole to mole)	0.025 (mole to mole)
Fraction of Feed of Natural Gas Going to LPFC	0.021 (mole to mole)	0.020 (mole to mole)
Temperature of HPFC	860°C	860°C
Temperature of LPFC	875°C	874°C
Pressure of HPFC	8.39 atm	8.37 atm
Pressure of LPFC	2.83 atm	2.96 atm

Table 10: Important conditions of the fuel cell system from fuel cell handbook and of the simulation.

type of heat exchangers needed and the possibility of linking several heat exchangers together are addressed as well as whether the product stream has the heat flow necessary to heat the feed stream. An energy balance is performed in level 6, and the procedure culminates in level 7 with a capital cost table.

Level 2 requires an economic analysis of the costs of the raw materials vs the potential income of the finished product. Levels 3 through 5 each continue this analysis by deducting the annualized capital costs of the equipment added to the design at that level from the profit calculated in the previous level. An approximate rate of return of the example design is about 15%. This differs from a standard rate of return in that the annual profit approximated by the economic potential calculated in level 5 only includes capital related expenses with the working capital being neglected.

References

1. Annual Energy Outlook 2014 with projections to 2040.

2. Chouldhury A, Chandra H, Arora A (2013) Application of solid oxide fuel cell technology for power generation-A review. Renewable and Sustainable Energy Reviews 20: 430-442.

3. Singhal SC (2002) Solid oxide fuel cells for stationary, mobile, and military applications. Solid State Ionics 152: 405-410.

4. Tanaka K, Wen C, Yamada K (2000) Design and evaluation of combined cycle system with solid oxide fuel cell and gas turbine. Fuel 79: 1493-1507.

5. Braun RJ (2002) Optimal design and operation of solid oxide fuel cell systems for Small-scale Stationary Applications. (Doctoral dissertation). University of Wisconsin-Madison, USA.

6. Douglas JM (1985) A hierarchical decision procedure for process syntheses. AIChE J 21: 353-362.

7. EG&G Technical Services Inc. (2004) Fuel Cell Handbook. U.S. Department of Energy. Office of Fossil Energy, National Energy Technology Laboratory.

8. http://www.eia.gov/dnav/ng/ng_pri_sum_dcu_nus_m.html

9. Knoxville Utilities Board (2015) Electric Rate Schedules.

10. http://www.eia.gov/dnav/ng/ng_pri_sum_dcu_nus_m.html

11. Urich GD, Vasudevan PT (2004) Chemical engineering process design and economics: A practical guide (2ndedn), Process Publishing, Durham NH, USA.

12. http://energy.gov/fe/articles/seca-fuel-cell-program-moves-two-key-projects-next-phase

13. Perry RH, Green DW (1997) Perry's chemical engineer's handbook (7thedn), McGraw-Hill, New York, USA.

14. Maruoka N (2003) Feasibility Study for Recovering Waste Heat in the Steelmaking Industry Using a Chemical Recuperator. ISIJ INT 44: 257-262.

Recovery and Characterization of Oil from Waste Crude Oil Tank Bottom Sludge from Azzawiya Oil Refinery in Libya

Abdulatif A Mansur[1,2]*, Muthu Pannirselvam[3], Khalid A Al-Hothaly[1,4], Eric M Adetutu[1] and Andrew S Ball[1]

[1]*School of Applied Sciences, RMIT University, Bundoora 3083, Australia*

[2]*Environmental and Natural Resources Engineering, Faculty of Engineering, Azawia University, Libya*

[3]*School of Civil, Environmental and Chemical Engineering, RMIT University, Melbourne 3000, Australia*

[4]*Department of Biotechnology, Faculty of Science, Taif University, Kingdom of Saudi Arabia*

Abstract

In this work we present the results of quantitative and qualitative analyses of oil obtained from crude oil tank bottom sludge (COTBS) generated from Azzawiya oil refinery in Libya. The aim of the study was to recover and evaluate oil from waste oily sludge and to compare it with parent oil (Hamada crude oil) in order to assess the commercial potential of recycling the oil. The benefits would be two-fold, firstly to improve oil utilisation efficiency and secondly in reducing the environmental contamination associated with the petrogenic hydrocarbon industry. Oily COTBS and extracted oil were characterised and key properties were measured including water and oil content, light and heavy hydrocarbon content, solid content and organic matter content for COTBS and water content, density, specific gravity, API (American Petroleum Institute) gravity, viscosity, salt and ash content for the extracted oil. Solvent (hexane) extraction confirmed that the oily sludge contained 42.08% (± 1.1%) oil composed of light hydrocarbons (30.7 ± 0.07%) and heavy hydrocarbon (69.3 ± 0.4%) fractions. The water and solid contents were 2.9% (± 0.2%) and 55.02% (± 0.6%) respectively. The properties of the recovered oil were assessed; gas chromatograph spectrophotometer (GC-MS) results indicated that the oil contained 139 different hydrocarbon fractions with a total petroleum hydrocarbon (TPH) concentration of 29,367 mgkg^{-1} and a polycyclic aromatic hydrocarbons (PAH) concentration of 11,752 mgkg^{-1}. Several parameters of the oil were measured and compared to the parent oil (Hamada crude oil) including density, specific gravity, viscosity, salt and ash content. The API of the extracted oil (33.03) was lower than the parent oil (38.8) due to a reduced light hydrocarbon (LHC) content. TGA-FTIR hyphenation shows both mass loss of hydrocarbons— low, medium and high molecular mass over a range of temperatures between 60°C and 450°C. crude oil extract exhibited a non-Newtonian behaviour (shear thinning) for the shear rate sweep between 10 and 500/s. dynamic shear rheology data showed that the extracted oil exhibit more like a solid than liquid. Overall the findings of the study confirmed that COTBD has a significant amount of oil similar in properties to Hamada crude oil. This large amount can be reclaimed and recycles. Depending on this essay, a commercial process could be performed which in parallel will reduce the environmental contamination with hydrocarbons.

Keywords: Solvent extraction; Oil recovery; Total petroleum hydrocarbon; Hyphenation; Dynamic shear rheology; Shear thinning; Storage modulus; Loss modulus; Thermogravimetry

Introduction

Petroleum crude oil represents one of the main current sources of energy. With the continuous increase in world population and industrialization, there is an increase in the global demand for petroleum crude oil and downstream products. In July 2012, Endurance International Group, Inc. (EIG) reported total global crude oil stocks of 7148 million barrels, with an estimation daily flow of oil production of around 75 million barrels [1]. In the processing of this crude oil the oil industry annually generates massive quantities of oily sludge during the different crude oil operations from exploration to refining [2]. The largest amount of the oily sludge is generated in oil refineries during oil storing operations. Most of the crude oil storage tanks contain bottom settling sediments accumulated over the time which are called crude oil tank bottom sludge (COTBS). During the cleaning processes, all the waste (COTBS) is removed and dumped in designated ponds. The continuous generation of COTBS during the bulk storage of crude oil is an unavoidable phenomenon [3]. COTBS usually contains a significant amount (30-50%) of oil (heavy hydrocarbons) [4], in addition to water (30-50%) and solids (10-12% (w/v) [5]. However, the global composition of COTBS is highly variable varying from one facility to another and from tank to tank within the same facility. COTBS composition is dependent on the composition of the stored oil, storage

conditions, storage period and the design and mechanical conditions of the storage tank [6].

Due to the accumulation of large quantities of COTBS together with its hazardous nature and associated waste management difficulties, COTBS has become a critical problem in most oil refineries [7]. In 2001, the USA petroleum industry was estimated to generate about 1.5×10^6 barrels of COTBS per annum [8] while large oil refineries (processing $2\text{-}5 \times 10^5$) barrels/day were estimated to produce 10×10^3 m^3 per year [9] . In India, petroleum oil refineries generate around 50×10^3 tonnes (t) per annum of oil rich COTBS (30-40% oil) [7]. In China, the petrochemical industry discharges nearly 3×10^6 t of COTBS per annum. One third of this amount (1×10^6 t) was derived from cleaning

*Corresponding author: Abdulatif A Mansur, School of Applied Sciences, RMIT University, Bundoora3083, Australia
Email: S3370890@student.rmit.edu.au, mansour2001uk@yahoo.com

operations associated with crude oil storage tanks [10]. In parallel with oil refineries, oilfields also generate significant amounts of COTBS. In 2010, Shengli Oilfield alone discharged more than 10×10^4 t of COTBS [11].

The accumulation of COTBS inside the oil tanks reduces their oil storing capacities and introducing the oily sludge into the refinery can ultimately disturb the refining processes [12]. In contrast, spilling of COTBS in the environment without treatment poses a significant risk to the surrounding environment and population. Importantly, prolonged storage of COTBS in accumulation ponds leads to seepage and contamination of ground water as well as the reduction in the light (volatile) fractions. Many of these volatile compounds are known of suspected carcinogens and mutagens and their release into the air poses a significant threat to the ecosystem and human population. Moreover, If COTBS is disposed of inappropriately, the oily sludge will splash into the soil where hundreds of individual compounds will contaminate the soil [11,13,14]. As many of COTBS hydrocarbon components are considered as toxic, mutagenic or carcinogenic [15], in 1992 the United States Environmental Protection Agency (US EPA) announced a final rule (57 FR 37194, 37252) stating treatment regulations and standards under the land disposal restrictions program for several hazardous wastes including hydrocarbon materials (COTBS) [16].

COTBS is continuously generated and disposed of in large quantities [17]. Recently, development of treatment strategies for COTBS to reduce their environmental burden has received increased global attention [4] and different effective remediation techniques have been proposed [17] including physical, chemical and biological methods. Among the techniques described, landfilling, incineration, microwave liquefaction, centrifugation, encapsulation, biodegradation in landfarming, biopiles and bioreactors have all featured [18]. However, some methods (e.g. incineration) have become restricted in some countries through the implementation of rigorous environmental standards because of their potential environmental impact [19].

Given the high hydrocarbon content of the oily COTBS, the conventional treatment methods such as land farming, landfilling and incineration are time-consuming, ineffective, expensive and may potentially release more unwanted environmental pollutants [20,21]. Also as oily COTBS is recognized as a potentially valuable energy resource, decomposition (bioremediation) techniques using microorganisms are also inadvisable [21]. The current driving force for increased interest in studying and characterising the COTBS are to recover oil from waste oily sludge in order to assess the commercial potential of recycling the oil. The benefits would be two-fold, firstly to improve oil utilisation efficiency and secondly in reducing the environmental contamination associated with the petrogenic hydrocarbon industry [22,23].

To characterise and classify recovered oil for commercial use, its physiochemical properties should be known. To classify the oil, the API gravity is the most important property. It is the relative density of the petroleum liquids and the density of water, and used to compare the relative densities of petroleum products. API is a scale for denoting the 'lightness' or 'heaviness' of petroleum crude oils and products. The lighter hydrocarbon the higher API gravity and the lighter hydrocarbon the higher market value. Oils with API more than 30° are known as light while oils in the range between 22° and 30° are medium, but API less than 22° are heavy and below 10° are extra heavy. It is preferable to between 25° and 30° [24]. In addition to the density, the viscosity which is the resistance to flow is another important factor that affects the pumping and transportation abilities

through the pipelines. Dealing with high viscosity oil is one of the main difficulties in transportation through the piping network [25]. Usually the viscosity of hydrocarbon oils ranges from 100 mPa to 10^5 mPa and the maximum desired viscosity is 400 mPa [26], but the high viscosity can be reduced to the desired value by reducing the liquid temperature by adding gaseous or liquid diluents [25]. In addition, flash point which is the minimum temperature at which the vapours of the material can ignite is the indicator of the flammability of the hydrocarbon oils. Safe handling of oils including processing, storage and transportation needs knowing the accurate values of flash point [27]. Moreover; presence of ash in the oil can affect the quality of oil. The ash content provides knowledge of metallic constituent left after complete combustion of the oil under specific condition. High ash content lowers the heating values and it is undesirable for direct combustion due to fouling and slagging [28]. Usually the petroleum crude oil contains a small amount of salts expressed as the presence of NaCl. If the salt content is higher the 1000bbl, the salt need to be minimized to reduce the fouling and corrosion in addition to the formation of acids by salts chlorides [29].

Therefore this work aims to assess the quality of recovered oil from waste oily sludge and to compare it with parent oil (Hamada crude oil) for recycling purposes and (ii) to reduce the environmental impact of COTBS by reducing the oil content of the soil to the minimum possible levels.

Materials and Methods

Characterization of sludge

Petroleum-based COTBS samples used for this study were obtained from collection lagoons at Azzawiya oil refinery in Libya. After collection, the sludge samples were kept at room temperature for the duration of the study. The sludge was mixed well manually before each sample was taken.

Water content (wt%): The water content of the COTBS samples was measured as indicated by the American Society for Testing and materials (ASTM) standard method (D95). COTBS samples (25 g, in triplicate) were taken and placed in an extraction thimble and 75 ml of dichloromethane (DCM) (solvent) were added (1:3 soil: solvent ratio) [30].The oil, solvent and water were then distilled and the condensate (water and solvent) continuously separated in a trap and transferred to a graduated cylinder. The triplicate condensates were pooled together [31]. It should be noted that due to the density of water ($1g/cm^3$) being less than the density of solvent, the solvent layer settled at the bottom of the separation funnel and was measured.

Volatile hydrocarbons and moisture content (wt%): Volatile hydrocarbons (VH) and moisture content of the COTBS samples were determined in triplicate by weighing (8 g) in ceramic crucibles and heating to 105°C in a ventilated incubator for 24 h. The lost mass was attributed to light volatile hydrocarbons and moisture content [32]. The light volatile hydrocarbons were calculated using the following equation:

$$\text{Light hydrocarbons} = \frac{\text{reduced mass (g)}}{\text{mass of tested sample (g)}} \times 100\% - \text{water content (\%)} \quad (1)$$

Solid content (wt%): Solid materials (sediment, ash and organic) content were measured according to the method described by [2] with some adjustments. After measuring the light hydrocarbons and moisture content, the dried COTBS samples (at 105°C) were heated in a muffle furnace (LABEC, Laboratory Equipment. Pty Ltd, Australia) to 550°C for 30 min [32] and the remaining samples re-weighed. The solid (sediment and ash) content of the COTBS was calculated using the following equation:

$$\text{Solid content} = \frac{\text{reduced mass (g)}}{\text{mass of tested sample (g)}} \times 100\% \qquad (2)$$

Organic matter content: The organic matter concentration was measured by loss on ignition of dry solid material in the muffle furnace (550°C for 30 min). The mass that was lost by the sample was attributed to organic material.

Non-volatile hydrocarbons content (wt%): Non-volatile hydrocarbons (NVH) was calculated according to [32] using the following equations:

$$\text{NVH} = 100\% - \text{VH} + \text{SC} + \text{WC} \qquad (3)$$

Where: NVH is non-volatile hydrocarbons (wt%)

VH is volatile hydrocarbons (wt%)

SC is solid content (wt%)

WC is water content (wt%)

Oil recovery: Oil was recovered from 6 sludge samples by solvent extraction (dichloromethane (DCM)) using a previously described protocol [30]. Briefly, a known amount (74.3 g, 76.1 g, 71.00 g, 67.5 g, 70.6 g and 67.1 g) of COTBS samples were weighed at room temperature and placed in Teflon coated (250 ml) centrifuge tubes and (DCM) was added (1:1 soil: solvent ratio). The oil in this mixture was extracted by agitation (130 rev min^{-1}) for 30 min, followed by centrifugation at 5000 rev min^{-1} for 5 min. The supernatants containing oil and solvent were removed and collected into glass bottles (500 ml).

Solvent recovery: To concentrate the reclaimed oil, the solvent (DCM) soluble fraction was rotary evaporated (250 ml a time) in a Buchi 461 water bath (Buchi RE111 Rotavapor, Buchi, Switzerland) at 40°C. The remaining oil was measured and reported as the volume of oil content in the COTBS and prepared for further analysis [33,34].

Characterization of recovered oil

Determination of hydrocarbon fractions concentration: For the analysis of hydrocarbons fractions within the recovered oil, including aromatic and aliphatic compounds, a combined paraffins, isoparaffins, aromatic, naphthalenes and oleffins (PIANO) PIANO-5-Piano (DHA) standard combined set (2 ml ampule) (Spectrum Quality Standards, Ltd. Sugarland, TX, USA) was used. The hydrocarbon content of the extracted oil was analysed as described in ASTM D513 [30] using gas chromatograph mass spectrometer (GC-MS) equipped with autosampler (Aglient 6890 GC and Leco Pegasus III TOF-MS). Samples were injected and separated on a capillary column Agilent DB-5MS (60 m by 0.25 mm with 0.25 μm film thickness). The injection temperature and volume was 225°C and 0.2 μl respectively. Helium (1.8 ml min^{-1}) was used as a carrier gas at a constant flow rate. The concentration of each hydrocarbon fraction was analyzed and the total peak area of each fraction was compared to the peak area of each fraction in the PIANO standard curve [35].

Density: The density (ρ) of extracted oil was estimated by dividing a known mass of the oil to its volume. Briefly, 10 ml of the recovered oil was measured and weighed (Mettler AE 260, Mettler-Toledo, Switzerland). The density of the oil was derived from the following equation:

$$\rho = \text{mass / volume} \qquad (4)$$

Specific gravity: Specific gravity (SG$_{true}$) is the ratio of the density of a liquid to the density of water (g/l). The specific gravity of the claimed oil was measured according to [36] and can be expressed mathematically from the following equation:

$$\text{SG}_{true} = (\tilde{n} \text{ sample}) / (\tilde{n} \text{ H2O}) \qquad (5)$$

API gravity: API gravity was calculated using the specific gravity of the oil extract, a unit-less property and determined at 60°F. API gravity was calculated according to [37] using the following equation:

$$\text{API gravity} = (141.5/\text{Specific Gravity}) - 131.5 \qquad (6)$$

Viscosity: Viscosity (υ) was measured using a Cannon-Fenske (Fisher Scientific, Pittsburgh, PA) glass capillary kinematic viscometer in a constant temperature bath in accordance with ASTM D445. Kinematic viscosity is determined by measuring the time (t) for a known volume of liquid flowing under gravity to pass through a calibrated glass capillary viscometer tube. The manufacturer of the Cannon- Fenske type viscometer tubes supplied calibration constants (c) at a range of temperature 40°F and 100°F [38,39].

Kinematic viscosity (υ) in centistoke (cSt) was calculated from the following equation.

$$\upsilon = c.t \qquad (7)$$

Ash content: Ash content of the extracted oil was determined using a loss-on-ignition procedure according to [40] with some adjustments. Triplicate samples (5 g) were heated overnight at 105°C and then transferred to a muffle furnace held at 550°C for 30 min to burn the organic matter. Ash content was calculated from the ratio of pre- and post-ignition sample mass.

Salt content: The salinity of the extracted oil was determined using the electrometric method according to ASTM D 3230 procedures using a Pro 2030 multimeter (YSI Incorporation, Yellow Springs. OH 45387. USA). In this method, the sample was dissolved in a mixed solvent and placed in a test cell consisting of a beaker and two parallel stainless steel plates. An alternating voltage was passed through the plates, and the salt content was obtained by reference to a calibration curve of the relationship of salt content of known mixtures to the current [41,42].

Thermogravimetric analysis (hyphenation with FTIR spectroscopy): The sample (COTBS extract) was used to measure mass loss and to determine functional group of the sample at a given time and temperature. Wilkie previously described a TGA/FTIR hyphenation technique that could be applied to investigate the degradation of crude oil extract [43]. Thermogravimetric analysis was performed on STA6000 operating under nitrogen with a flow rate of 20 ml/min through the furnace in the following conditions. A sample mass of ~40 mg was heated in the crucible with heating rate of 20°C/min from 50 to 950°C in an inert atmosphere (nitrogen). The gas evolving from STA6000 was transferred via gas transfer line. This transfer line allows the transfer of combustion of pyrolysis products from thermal analyser to FTIR 100 through the gas flow cell. Spectrum time base was used to analyse the sp files collected during the testing to analyse the spectrum (collected continuously for over 3600 s).

Shear rheology testing to measure viscosity: HR3 (Hybrid Discovery) rheometer was used to measure the rheological properties of sample (COTBS extract). The instrument (HR3 rheometer) was calibrated with viscosity standard including Polydimethyl Siloxane (PDMS). Crossover frequency of G' and G" of calibration results matched with the value recommended by instrument suppliers–TA instruments. Shear rheology was conducted on the sample at a constant temperature of 50°C and a constant strain of 1%, angular frequency of 1 to 100 rad/s. Parallel plate geometry (40 mm smart swap, stainless steel) was used in this research. It should be noted that the sample was tested as received.

Results

Composition of the oily sludge

Several key properties of the sludge were analysed and indicated in Table 1. The water content of the sludge was 2.9 (± 0.2%) and the solid content was 55.02 (± 0.6%). The amount of organic material in the solid content was found to be 70 (± 0.6%) of the original dry mass of the sludge. Solvent extraction of the hydrocarbon oil from COTBS using dichloromethane (DCM) showed that the sludge contained a significant amount of oil (oil content 42.08 ± 1.1%) compsed of light (volatile) hydrocarbon (VH, 30.7 ± 0.07%) and non-volatile hydrocarbons (NVH, 69.3 ± 0.4%).

Composition of recovered oil

In addition to the amount of oil recovered, the quality of the oil has a major influence on determining the commercial viability of the recycling process. Selected physiochemical properties of the recovered oil from COTBS were tested including water content, organic material, density, specific gravity, API gravity, viscosity, salt content and ash content. The selected properties were compared with parent oil

Property %	Sludge
Water content	2.9 ± 0.2
Oil content	42.08 ± 1.1
Light hydrocarbons and moisture content in the recovered oil	30.7 ± 0.07
Non-volatile hydrocarbons in the recovered oil	69.3 ± 0.4
Solid content	55.02 ± 0.6
Organic matter content in solids	70 ± 0.6

Table 1: Properties of studied sludge.

Property	Extracted oil from sludge	Hamada petroleum crude oil
Water	–	0.1
Density @ 15°C, g/ml	0.86	0.8304
Specific gravity@60/60°F	0.86	0.8311
API gravity	33.03	38.8
Viscosity@70°F, cSt	7.01	6.8431
@100°F, cSt	3.655	3.5742
Salt content (as NaCl) mg/l	2.30	2.14
Ash content g/g oil	0.007	0.004

Table 2: Properties of extracted oil from COTBS and Hamada petroleum crude oil.

Parameter	Importance	Reference
Water content	Increases the process pressure due to steam formation	[57]
High viscosity	Resist to flow, required more energy for pumping and decreases the heat transfer efficiency	[59]
High density	Resist to flow and requires more energy for pumping	[58]
API gravity	Indication of oil grade and quality	[58]
Salt content	High salt content causes corrosion and fouling of process equipment and hydrolysed to hydrochloric acid	[70]
Ash content	Decreases the heat value	[28]
Solid content	Increase the viscosity	[2]

Table 3: Importance of oil parameters.

Temperature	Major compound evolving from the sample at that temperature
170°C	Methylene Chloride
410°C	trans-1,4-dimethylcyclohexane
450°C	Cyclohexane 1-hexyl 4-tetradecyl
570°C	Isobutyl cyclo-hexane
770°C	1-ethyl 2-methylcyclohexane

Table 4: Gases evolving at various temperatures during hyphenation studies.

(Hamada crude oil) properties are summarized in Tables 2 and 3. GC-MS based analysis of the recovered oil resulted in up to 136 different hydrocarbon fractions being detected including aromatic compounds (45.6%) and aliphatic compounds (34.6%) with some of these fractions (19.8%) being undefined (Figure 1). The concentrations of TPH and PAHs were 29,367 mgkg^{-1} and 11,752 mgkg^{-1} respectively. API gravity was calculated mathematically after determining the density of the oil (0.68 gl^{-1}) and was 33.03, confirming that this oil was as light as the parent oil. In addition, the kinematic viscosity of the recovered oil measured at 70°F and 100°F was (7.01 and 3.655 cSt) respectively. According to the loss in ignition results, the ash content of this oil was 0.007 g/g oil. Finally, the salinity test (as NaCl content) of the examined oil indicated that the salt content was 2.30 mgl^{-1}.

Thermogravimetric analysis

Mothe et al. [44] concluded that the study of thermogravimetry of crude oil based materials is very complicated due to the presence of many complex constituents [44]. In this research, we applied a hyphenation technique to analyse the gas evolving from the sample during thermogravimetric analysis (Table 4). The crude oil was exposed to a nitrogen atmosphere at heating rate of 20°C/min for 45 min (2700 s) from 50 to 950°C.

Rheology section

The International Energy Agency reported that heavy crude oil represents over 50% of the world's recoverable oil resources. Crude oil is a composition of large amount of hydrocarbons and varying amount of waxes. Ghannam stated the dynamic shear rheology test is a rheological investigation to study the viscoelastic behaviour of crude oil [45]. Rheological properties for petroleum oils are very useful for all processes in which fluids are transferred from one location to another. Evdokimov et al concluded that limited numbers of crude oil rheological properties are currently available, in particular for heavy crude oil [46].

The dynamic frequency sweep test shows the effect of oscillating stresses on this extracted crude oil. The storage modulus exhibits solid behaviour of sample and loss modulus exhibits liquid behaviour of samples. Storage modulus shows the contribution of stress energy that is stored during the test and can be recovered. As a standard procedure, linear viscoelastic region of sample was studied of the sample using strain sweep test (in dynamic mode). Three tests of the above mentioned methodology was conducted to check the reliability of the measured data. The rheograms matched well within the tolerance limits of about ± 3%.

Discussion

Due to the undesirable environmental impact of dumping COTBS with high hydrocarbon content and the economic benefits of the COTBS as a source of petroleum oil, there is current interest in studying the quantitative and qualitative characteristics of COTBS and the recovered oil. The management of oily wastes involves the analysis and characterization of both recovered oil and sludge. Knowing the

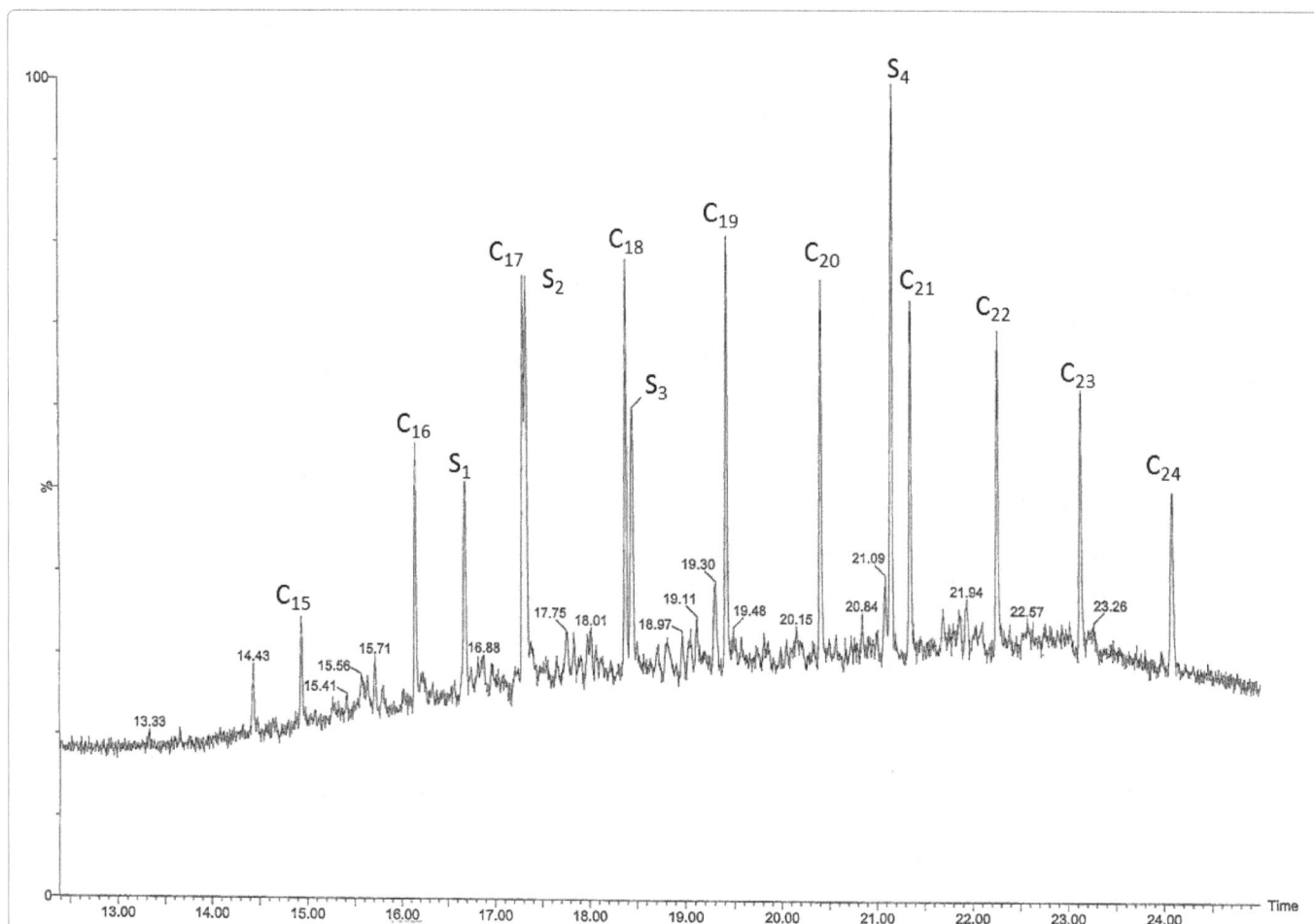

Figure 1: GC-MS chromatogram of oil extracted from COTBS with some selected peaks identified. S1 2,6,10-trimethyldodecane, S2 2,6,10,14-tetramethylpentadecane, S3 2,6,10,14-tetramethylhexadecane, S4 cyclic octa-atomic sulphur.

amounts produced, physical and chemical properties of sludge are important parameters in defining the applied conventional treatment strategies [18]. For the recovered oil, quantitative and qualitative assessment represents the most important parameters in determining the applicability of the oil for use as crude or fuel oil. In this research, the results of the studied COTBS showed that the average water content was low (2.9%). This low water content percentage was expected because the sludge studied in this research has been accumulated for several years and the water was separated into the bottom of the collection ponds due to the density difference between water and sludge. Comparing with other sludges, this percentage was very low. Heidarzadeh et al. [47] showed that the water content of a studied COTBS obtained from Iran refinery was 28.3%. Similarly, other research [48] determined the water content of 2 different oily sludges in China and found a significant amount of water (16.2 and 27.6%). On the basis of dry mass, the solid content of the oily sludge was 55.02%, of which 70% was organic matter. This relatively high solid content was expected since the stored COTBS was not exposed to a solid removal process. The presence of solid materials including sand and rust significantly increases the sludge viscosity [2,48]. Extraction of the oil from COTBS using a 1:1 ratio of COTBS: solvent revealed a high oil content (42.08%) of which with 29.7% were VH and 70.3% NVH. This oil could be potentially extracted and recycled. Recovering and recycling of valuable hydrocarbon oils aids the conservation of environment and energy resources [2,49] and decreases the consumption of non-renewable energy resources [50].

According to the American Petroleum Institute (USAPI), the primary environmental consideration in handling oily sludge is the maximum hydrocarbon recovery [51]. In the oil refining industry in the USA, more than 80% of the generated waste hydrocarbons were recycled while the remaining (20%) were disposal according to the (EPA) standards [52]. Generally, some studies suggested that high oil concentrations in COTBS (>50%) and a relatively low concentration of solids (<30%) are preferable for recycling [53]. Others suggested that even at low COTBS oil content (>10%), oil recovery is still accepted [7]. In this study, within the extracted oil (42.08%), the light hydrocarbon fractions (LHF) were not very high because of evaporation since the collection ponds were exposed to the environmental elements (wind, sun, etc) for many years. A similar study on exposed COTBS [47] indicated that the LHF was also very low (<10%). Storing COTBS in closed facilities results in increased volatile fractions [32] studied COTBS obtained from the Niger delta, Nigeria and determined the oil content to be 73.24%, of which 45.84% were light hydrocarbons. Light hydrocarbon fractions are important in qualifying the oil grade. The presence of (LHF) decreases the density and viscosity and increases the API of the oil. Hu et al. [4] reported that the oil content in COTBS range from 5% to 86.2% (w/v), although oil content in the range of 15-50% (w/w) was more frequent while the solids and water contents were in the range of 5-46% (w/w) and 30-85% (w/w) respectively. Increasing the solvent:sludge ratio could increase the amount of recovered oil. Zubaidy and Abouelnasr [2] studied various ratios and found that (4:1) solvent:sludge ratio facilitated the extraction of the highest amount of oil. Zubaidy and Abouelnasr [2] studied the solvent extraction of hydrocarbon compounds and found that heavy molecular mass hydrocarbons were extracted more when less solvent was used. However, increasing the solvent:sludge ratio is not economical as most of the solvents used are expensive; in addition, in low condensation efficiencies, large amounts of solvent could not be reclaimed. Moreover, since the solid content was high (55.02%), a large amount of solvent will be adsorbed within the porous particles. The solvent path within a porous media depends largely on pore size; the more porous media the less solvent reclaimation back [54]. Consequently, 1:1 solvent: sludge ratio is recommended for reclamation and economic purposes.

In addition to the amount of oil recovered, the quality of the oil is also a major concern. A set of oil grading physiochemical properties were conducted. Comparing with properties of Hamada crude oil, some properties of the recovered oil were higher including density, specific gravity (*SG*), viscosity, salt content and ash content; in contrast, water content and API gravity were less than those of Hamada crude oil (Table 2). Gas chromatographic mass spectrophotometer (GC-MS) analysis of the extracted oil showed the presence of a range of hydrocarbon fractions composed of 136 different compounds ranging from C_{14}-C_{24} (Figure 1) including both aromatic (45.6% v/w) and aliphatic compounds (34.6% v/w) with 19.8% being undefined. The total petroleum hydrocarbon (TPH) concentration of the recovered oil was 29,367 mgkg^{-1} and polycyclic aromatic hydrocarbon (PAH) concentration was 11,752 mgkg^{-1}. For comparison, [55] conducted a similar study on COTBS and found that the TPH of extracted oil was higher than 29,367 mgkg^{-1} (500,000 mgkg^{-1}), while another study found the TPH of an extracted oil to be very low (850 ± 150 mgkg^{-1}) [31]. Water content analysis indicated that the extracted oil was free of water. The presence of water lowers the heating value of the oil [56]; in addition, it generates steam and builds pressure in the refining processes [57]. The density of the oil was measured and found to be 0.86 which was higher than the density of the parent oil. Also, the *SG* has also been calculated at 60°F and found to be higher (0.86) than the original crude oil. Depending on the *SG* of the extracted oil, the mathematical calculation results show the API density to be 33.03. The API density is one of the main parameters used to grade the crude oil. Martínez-Palou et al. [58] indicated that oil with API<10 is classified as extra heavy while API<22.3 is heavy, API 22.3 to 31.1 is medium and API>31.1 is light. The lighter the oil is the higher content of light hydrocarbon compounds, and the less wax and asphaltenes are present. Viscosity is the resistance to flow and a measure of the internal molecular fraction of the fluid and is an important parameter affecting the pumping of oil and atomization of fuel [59]. The kinematic viscosity of the extracted oil was measured at 2 different temperatures and compared to that of Hamada crude oil. At 60°F viscosity was 7.01 cSt and at 100°F was 3.655 cSt while the viscosity of the Hamada oil was lower (6.8431 cSt and 3.5742 cSt) respectively. Lower LHF content in the oil increases the density and viscosity of the oil. Consequently more power is needed for pumping [59].

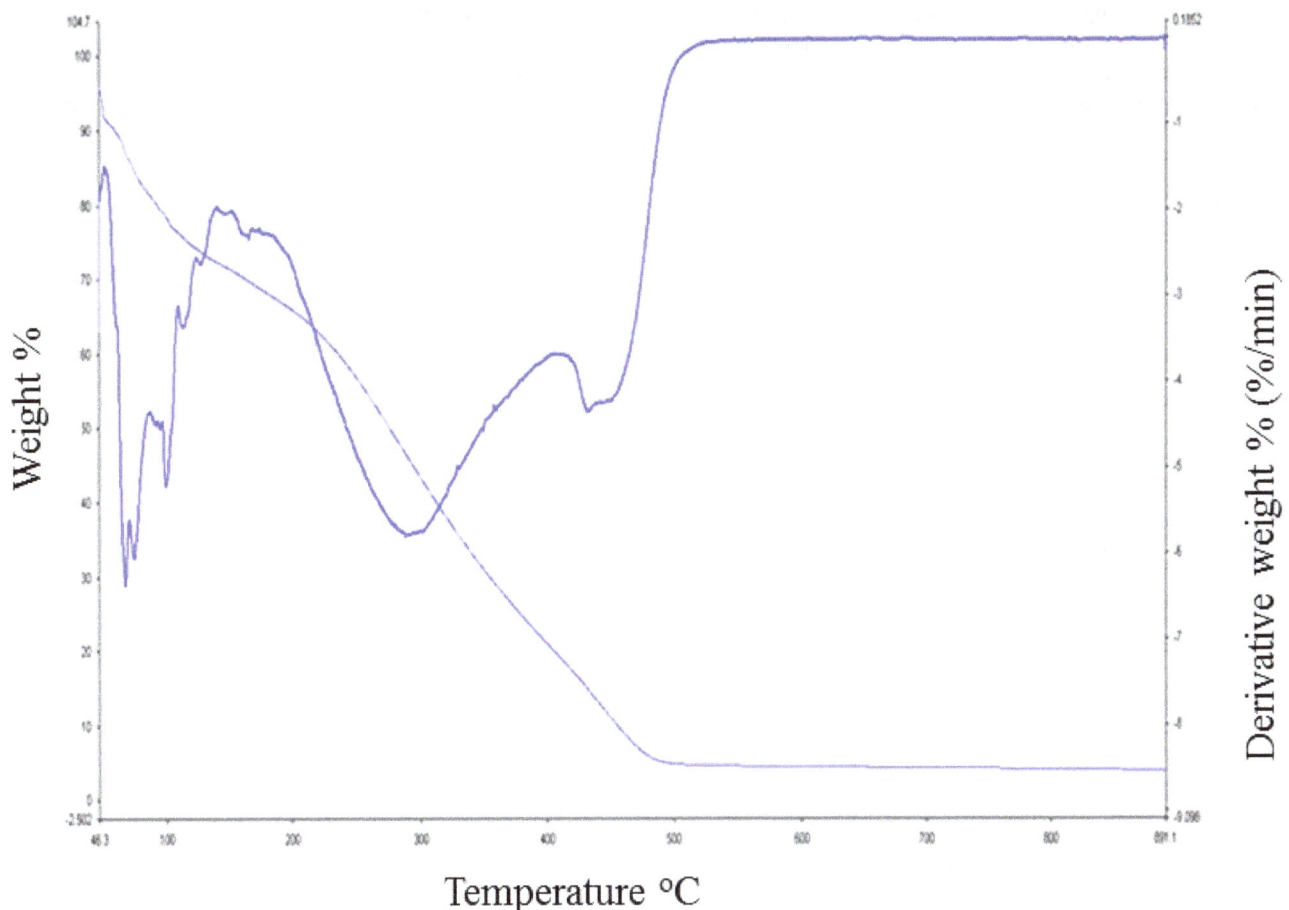

Figure 2: Mass loss curve of the extracted oil.

In addition to magnesium chloride ($MgCl_2$), salts in crude oils consist of up to 75% sodium chloride (NaCl) [60]; typically the salinity of oil is measured as NaCl content. The salinity of the recovered oil was 2.30 mgl^{-1}. This value was slightly higher than the salt content of Hamada oil (2.14 mgl^{-1}). Comparing with other studies, this value was very low. Zubaidy and Abouelnasr [2] investigated the properties of 3 extracted oils and found the salt content to be 5, 7 and 196.4 mgl^{-1}). The presence of salts in oil is not favourable; even small concentrations of salt will accumulate in process equipment leading to fouling. In addition and more importantly, NaCl and $MgCl_2$ can be hydrolysed to hydrochloric acid as indicated in the following equations:

$$2NaCl + H_2O \rightarrow 2HCl + Na_2O \qquad (8)$$

$$MgCl_2 + H_2O \rightarrow 2HCl + MgO \qquad (9)$$

The produced hydrochloric acid is known to be extremely corrosive [42]. Ash content represents organic materials [61] and is another property used to assess the heating and calorific values of the oil [28,62]. The ash content in the recovered oil contains a higher ash (0.007 g/g oil) than Hamada oil (0.004 g/g oil). Previous studies confirm that lower ash content is indicative of high quality oil [63]; [28] studied the effects of ash content and found that presence of ash in oil reduces the heating value.

Thermogravimetric analysis

Thermogravimetric curve did not show decomposition stage at all temperatures as clearly shown in Differential Thermogravimetric (DTG) curve) (Figure 2), however there was a sharp mass loss at the following temperatures: 66.82, 74.03, 89.40, 101.96, 114.73, 129.37, 146.07, 161.39, 289.28, 431.48, and 456.51°C. The decomposition of crude oil at various temperatures shows the mass loss of hydrocarbons (low, medium and high molecular weight) respectively at 161.39, 289.28, 431.48, 456.51°C.

Figures 3 and 4 show the spectra obtained at various time intervals. EPA and NIST gas phase libraries were used to search/match the spectra obtained in this hyphenation experimentation and to compare the best possible match spectrum (Figures 5 and 6). The following are the gases evolved at various temperatures: methylene chloride, trans-1,4-dimethylcyclohexane, cyclohexane1-hexyl4-tetradecyl, isobutyl cyclo-hexane and 1-ethyl 2-methylcyclohexane (Table 4). The spectra were obtained via search/match technique using EPA and NIST gas phase library. These values are matching similar to results published in the article Mothe et al in the article titled "Thermal Evaluation of Heavy Oils by Simultaneous TG/DTG/FTIR" [44].

Dynamic shear rheology

It has been observed that 1% is within the linear viscoelastic region of crude oil (internal bonds of the sample tested are intact at that strain value). Figure 7 shows the storage and loss moduli versus frequency: storage modulus G' values are higher than the values of loss modulus G''. It can be inferred that crude oil sludge exhibits solid like behaviour (sludge) rather than the viscous liquid behaviour (oil). This behaviour could be due to the large solid particles present in the sludge. In general, it has been stated that crude oil exhibits more viscous liquid like behaviour than solid-like material. This change in rheological behaviour could be due to the presence of solid waste present in this oil. The viscosity of the crude oil decreases with increasing temperature as expected. It can be inferred from Figure 8 that the crude oil clearly exhibited a non-Newtonian behaviour (shear thinning) for the shear rate sweep between 10 and 500/s.

Figure 7 shows the storage and loss moduli of crude oil for a frequency sweep (1 to 100 rad/s). The storage modulus and loss moduli increased with increase in angular frequency (1 to 100 rad/s). Viscosity and other rheological parameters are very important for any

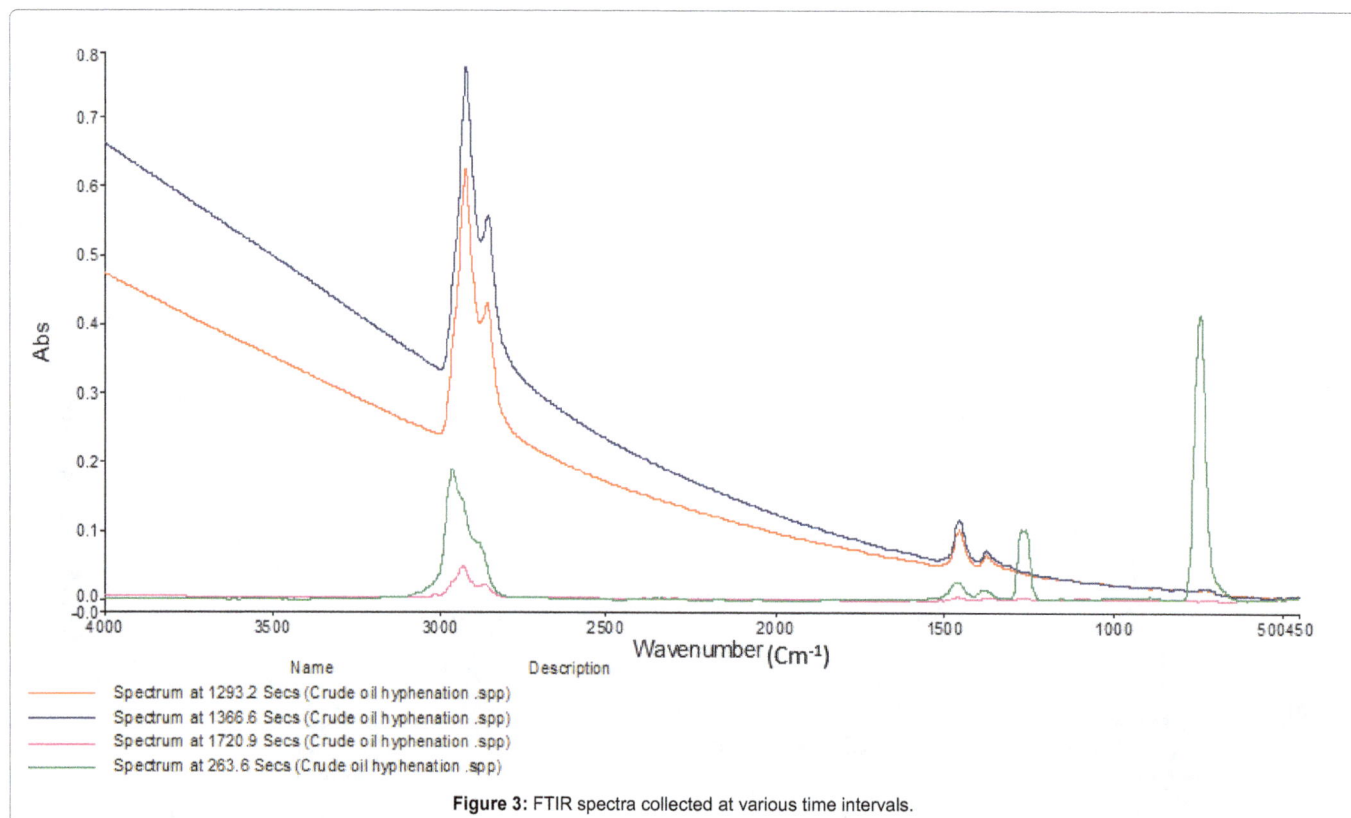

Name   Description
—— Spectrum at 1293.2 Secs (Crude oil hyphenation .spp)
—— Spectrum at 1366.6 Secs (Crude oil hyphenation .spp)
—— Spectrum at 1720.9 Secs (Crude oil hyphenation .spp)
—— Spectrum at 263.6 Secs (Crude oil hyphenation .spp)

Figure 3: FTIR spectra collected at various time intervals.

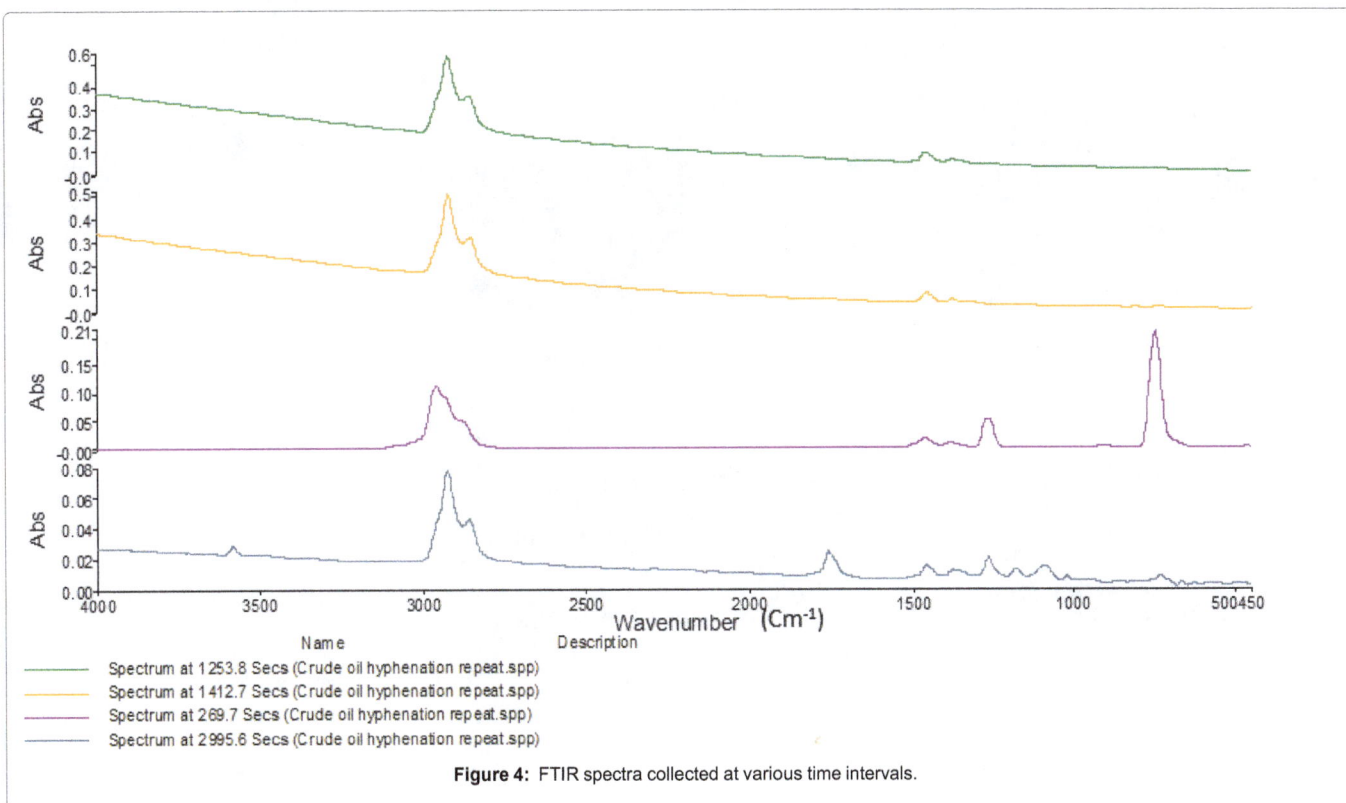

Figure 4: FTIR spectra collected at various time intervals.

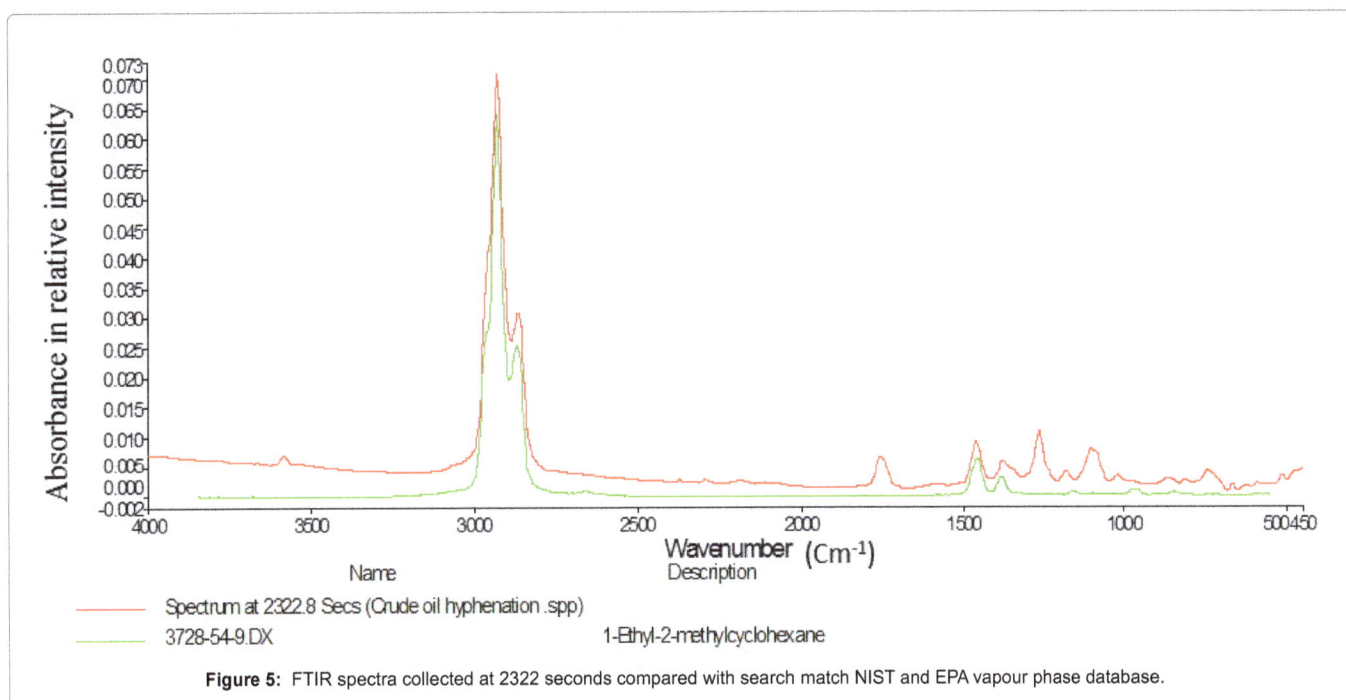

Figure 5: FTIR spectra collected at 2322 seconds compared with search match NIST and EPA vapour phase database.

liquids including materials like crude oil that are often transported via pipelines [45,46,64]. The transportation of crude oil is very challenging due to the high viscosity (medium and high molecular weight and low flowability). Figure 8 shows the viscosity over a temperature range (35 to 50°C). The viscosity decreases with increasing temperature. The low and medium molar mass compounds tend to affect the bonds between the solid particles, eventually reducing the oil viscosity. Ghannam et al

concluded that the decrease in viscosity with increase in temperature could be due to the effect of temperature on the chemical structure of the ingredients of the crude oil [45]. The change in viscosity over temperature could be attributed to changes in the chemical structure of high molecular mass components of the crude oil, in particular wax and asphalt [65].

For future work, the concentrations of heavy metals such as P, Cd,

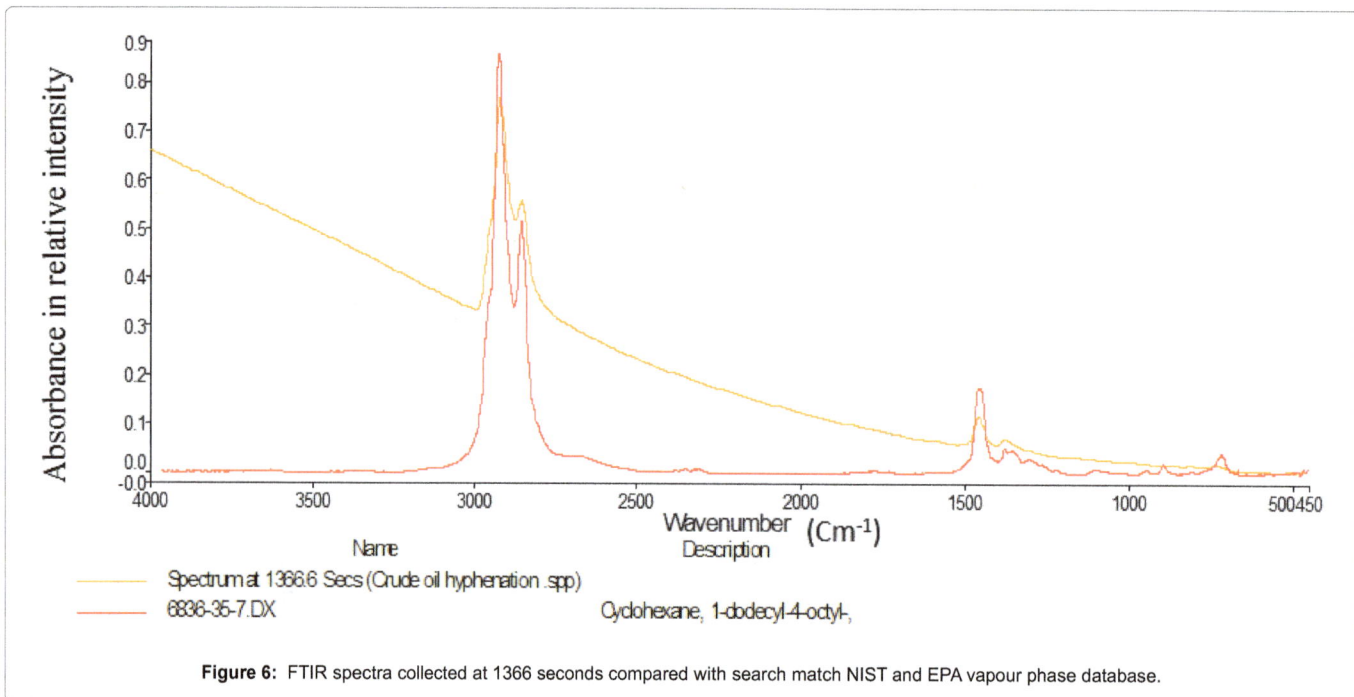

Figure 6: FTIR spectra collected at 1366 seconds compared with search match NIST and EPA vapour phase database.

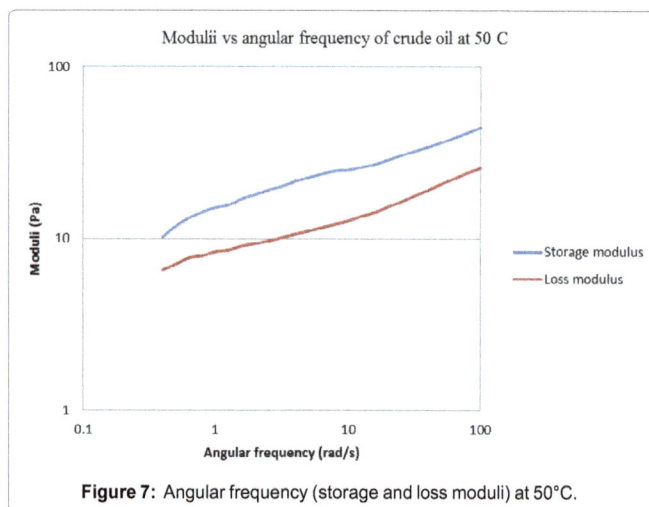

Figure 7: Angular frequency (storage and loss moduli) at 50°C.

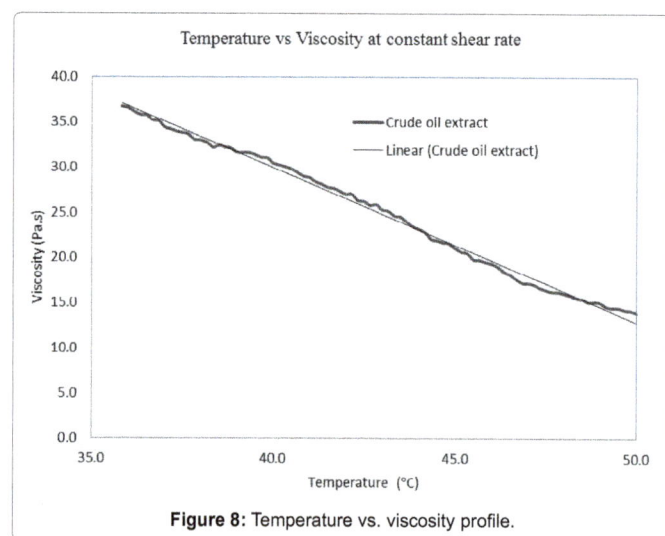

Figure 8: Temperature vs. viscosity profile.

Cr, Cu, Ni, Pb, Zn and Fe can be carried out. This can be determined using inductively coupled plasma optical emission spectrometry (ICP-OES) (Optima 2000DV; Perkin Elmer, Waltham, MA, USA) (detection limit 0.001-0.030 mgl^{-1}) and inductively coupled plasma-mass spectrometry (ICP-MS) (7500a; Agilent Technologies, Santa Clara, CA, USA) (detection limit 0.015-0.120 mgl^{-1}) [66]. In addition, luminescent iridium (III) complex-base chemosensor and probes can be used to detect the heavy metals ions [67,68]. Moreover, logic gates technique is one of the useful tool that respond to the ions of metal [69,70].

Conclusion

The purpose of this study was (i) to extract oil from Libyan COTBS and present the most important properties through a qualitative and quantitative assessment and to compare it with the parent oil (Hamada crude oil), and (ii) to reduce the oil content of the sludge to the minimum level to reduce the environmental impact of COTBS. Some important properties of the COTBS were studied (Table 1). To determine the applicability of recovered oil for use as feedstock for

recycling, the physiochemical properties were investigated (Table 2). The extraction yield confirmed that the oil content of COTBS was significant (42.08%). GC-MS analysis of the recovered oil indicated that the oil contained hydrocarbon fractions in the range of C_{14}-C_{24} with TPH and PAHs concentrations of 29,367 mgkg^{-1} and 11,752 mgkg^{-1} respectively. Comparing the properties of the recovered oil with the parent oil which was the original source of this sludge and classified as light oil, the reclaimed oil was heavier. Lower LHF content decreased the API gravity (33.03), but the API still higher than (31.1) meaning that the reclaimed oil is light oil. Consequently, the recovered oil can be classified as light oil and potentially used as good feedstock oil. The candidate (crude oil) exhibits non-Newtonian shear thinning behaviour over the shear rates sweep between 10 and 500/s. Viscosity of crude oil decreases significantly with increase in temperature due to the effect of temperature on chemical structure of the ingredient of the crude oil. This is the first report of the oil recovery from COTBS from Libya having significant potential of oil for use in the oil industry.

Acknowledgement

This work was supported by the Libyan Ministry of Higher Education and Science Research. The authors thank the Environmental and Natural Resources Engineering, faculty of engineering. Sabratah. Azawia University- Libya, the management of Azzawiya oil refinery- Libya and brother Ali Mansour for his unlimited support and supplying soil and COTBS samples.

References

1. Kilian L, Lee TK (2014) Quantifying the speculative component in the real price of oil: The role of global oil inventories. Journal of International Money and Finance 42: 71-87.

2. Zubaidy EA, Abouelnasr DM (2010) Fuel recovery from waste oily sludge using solvent extraction. Process Safety and Environmental Protection 88: 318-326.

3. Giles HN, Koenig JJ, Neihof RA, Shay JY, Woodward PW (1991) Stability of refined products and crude oil stored in large cavities in salt deposits: Biogeochemical aspects. Energy & fuels 5: 602-608.

4. Hu G, Li J, Zeng G (2013) Recent development in the treatment of oily sludge from petroleum industry: a review. J Hazard Mater 261: 470-490.

5. Saikia M, Bora M, Dutta N (2003) Oil recovery from refinery sludge-a case study. Chemcon CHM 27.

6. Sanders DA, Veenstra JN (2001) Pollution prevention and reuse alternatives for crude oil tank bottom sludges. Proceedings of the 8th international petroleum environmental conference, Houston, USA.

7. Ramaswamy B, Kar DD, De S (2007) A study on recovery of oil from sludge containing oil using froth flotation. J Environ Manage 85: 150-154.

8. Striegel J, Sanders DA, Veenstra JN (2001) Treatment of contaminated groundwater using permeable reactive barriers. Environmental Geosciences 8: 258-265.

9. Ward O, Singh A, Van Hamme J (2003) Accelerated biodegradation of petroleum hydrocarbon waste. J Ind Microbiol Biotechnol 30: 260-270.

10. Liu J, Jiang X, Han X (2011) Devolatilization of oil sludge in a lab-scale bubbling fluidized bed. J Hazard Mater 185: 1205-1213.

11. Du W, Wan Y, Zhong N, Fei J, Zhang Z, et al. (2011) Status quo of soil petroleum contamination and evolution of bioremediation. Petroleum Science 8: 502-514.

12. Kam EK (2001) Assessment of sludges and tank bottoms treatment processes. Proceedings of the 8th international petroleum environmental conference, Houston, USA.

13. Jing G, Luan M, Du W, Han C (2012) Treatment of oily sludge by advanced oxidation process. Environmental Earth Sciences 67: 2217-2221.

14. Avilachavez MA, Eustaquio-Rincon R, Reza J, Trejo A (2007) Extraction of hydrocarbons from crude oil tank bottom sludges using supercritical ethane. Separation Science and Technology 42: 2327-2345.

15. Liu W, Luo Y, Teng Y, Li Z, Christie P (2009) Prepared bed bioremediation of oily sludge in an oilfield in northern China. J Hazard Mater 161: 479-484.

16. Lin GH, Sauer NE, Cutright TJ (1996) Environmental regulations: a brief overview of their applications to bioremediation. International biodeterioration & biodegradation 38: 1-8.

17. Zhang J, Li J, Thring RW, Hu X, Song X (2012) Oil recovery from refinery oily sludge via ultrasound and freeze/thaw. J Hazard Mater 203-204: 195-203.

18. da Silva LJ, Alves FC, de França FP (2012) A review of the technological solutions for the treatment of oily sludges from petroleum refineries. Waste Manag Res 30: 1016-1030.

19. Jin Y, Zheng X, Chu X, Chi Y, Yan J, et al. (2012) Oil recovery from oil sludge through combined ultrasound and thermochemical cleaning treatment. Industrial & Engineering Chemistry Research 51: 9213-9217.

20. Buyukkamaci N, Kucukselek E (2007) Improvement of dewatering capacity of a petrochemical sludge. J Hazard Mater 144: 323-327.

21. Shie JL, Chang CY, Lin JP, Wu CH, Lee DJ (2000) Resources recovery of oil sludge by pyrolysis: kinetics study. Journal of Chemical Technology and Biotechnology 75: 443-450.

22. Mishra S, Jyot J, Kuhad RC, Lal B (2001) Evaluation of inoculum addition to stimulate in situ bioremediation of oily-sludge-contaminated soil. Appl Environ Microbiol 67: 1675-1681.

23. Liu W, Luo Y, Teng Y, Li Z, Ma LQ (2010) Bioremediation of oily sludge-contaminated soil by stimulating indigenous microbes. Environ Geochem Health 32: 23-29.

24. Pusch G, Gaida KH (1981) Method of recovering petroleum and bitumen from subterranean reservoirs. Google Patents US4252191 A.

25. Centeno G, Sanchez-reyna G, Ancheyta J, Munoz JA, Cardona N (2011) Testing various mixing rules for calculation of viscosity of petroleum blends. Fuel 90: 3561-3570.

26. Hasan SW, Ghannam MT, Esmail N (2010) Heavy crude oil viscosity reduction and rheology for pipeline transportation. Fuel 89: 1095-1100.

27. Gharagheizi F, Tirandazi B, Barzin R (2008) Estimation of aniline point temperature of pure hydrocarbons: A quantitative structure property relationship approach. Industrial & Engineering Chemistry Research 48: 1678-1682.

28. Biller P, Ross AB (2011) Potential yields and properties of oil from the hydrothermal liquefaction of microalgae with different biochemical content. Bioresour Technol 102: 215-225.

29. Gary JH, Handwerk GE, Kaiser MJ (2010) Petroleum refining: technology and economics, CRC press.

30. Mansur A, Adetutu E, Kadali K, Morrison P, Nurulita Y, et al. (2014) Assessing the hydrocarbon degrading potential of indigenous bacteria isolated from crude oil tank bottom sludge and hydrocarbon-contaminated soil of Azzawiya oil refinery, Libya. Environ Sci Pollut Res Int 21: 10725-10735.

31. Joseph PJ, Joseph A (2009) Microbial enhanced separation of oil from petroleum refinery sludge. J Hazard Mater 161: 522-525.

32. Taiwo E, Otolorin J (2009) Oil recovery from petroleum sludge by solvent extraction. Petroleum Science and Technology 27: 836-844.

33. Pradhan P, Giri J, Rieken F, Koch C, Mykhaylyk O, et al. (2010) Targeted temperature sensitive magnetic liposomes for thermo-chemotherapy. J Control Release 142: 108-121.

34. Garcia-Perez M, Shen J, Wang XS, Li CZ (2010) Production and fuel properties of fast pyrolysis oil/bio-diesel blends. Fuel Processing Technology 91: 296-305.

35. Sabate J, Vinas M, Solanas A (2004) Laboratory-scale bioremediation experiments on hydrocarbon-contaminated soils. International Biodeterioration & Biodegradation 54: 19-25.

36. Demirbas A (2008) Relationships derived from physical properties of vegetable oil and biodiesel fuels. Fuel 87: 1743-1748.

37. Yuying L, Bing L (2011) The Changes of Crude Oil during Evaporation Process in Environment. Bioinformatics and Biomedical Engineering, 5th International Conference on IEEE, 1-4.

38. Tat ME, Van Gerpen JH (1999) The kinematic viscosity of biodiesel and its blends with diesel fuel. Journal of the American Oil Chemists' Society 76: 1511-1513.

39. Noureddini H, Teoh B, Clements LD (1992) Viscosities of vegetable oils and fatty acids. Journal of the American Oil Chemists Society 69: 1189-1191.

40. Xu Y, Lu M (2010) Bioremediation of crude oil-contaminated soil: comparison of different biostimulation and bioaugmentation treatments. J Hazard Mater 183: 395-401.

41. Fortuny M, Oliveira CB, Melo RL, Nele M, Coutinho RC, et al. (2007) Effect of salinity, temperature, water content, and pH on the microwave demulsification of crude oil emulsions. Energy & fuels 21: 1358-1364.

42. Speight JG, Speight J (2002) Handbook of petroleum product analysis, Wiley-Interscience, New Jersey, USA.

43. Wilkie CA (1999) TGA/FTIR: an extremely useful technique for studying polymer degradation. Polymer Degradation and Stability 66: 301-306.

44. Michelle Mothe CM, Carvalho CH (2013) Thermal Evaluation of Heavy Oils by Simultaneous TG/DTG/FTIR. 41st Annual Conference of the North American Thermal Analysis Society.

45. Ghannam MT, Hasan SW, Abu-Jdayil B, Esmail N (2012) Rheological properties of heavy & light crude oil mixtures for improving flowability. Journal of Petroleum Science and Engineering 81: 122-128.

46. Evdokimov IN, Eliseev DY (2001) Rheological evidence of structural phase transitions in asphaltene-containing petroleum fluids. J Pet Sci Eng 199-211.

47. Heidarzadeh N, Gitipour S, Abdoli MA (2010) Characterization of oily sludge from a Tehran oil refinery. Waste Manag Res 28: 921-927.

48. Wang J, Yin J, Ge L, Shao J, Zheng J (2010) Characterization of oil sludges from two oil fields in China. Energy & Fuels 24: 973-978.

49. Elektorowicz M, Habibi S (2005) Sustainable waste management: recovery of fuels from petroleum sludge. Canadian Journal of Civil Engineering 32: 164-169.

50. Xiaoxue HSLZL (2002) The Current Status of Oily Sludge and its Treatment Technique in Gudong Oil field. J Environmental Protection of Oil & Gas Fields 1: 007.

51. American Petroleum Institute (1989) Category Assessment Document for Reclaimed Petroleum Hydrocarbon: Residual Hydrocarbon Waste from Petroleum Refining. In: Program UEHC (ed.) Washington DC, USA.

52. American Petroleum Institute (1992) API Environmental Guidance Document: Onshore Solid Waste Msanagment in Exploration and Petroleum Operations. Washing DC, USA.

53. Hahn W (1994) High-temperature reprocessing of petroleum oily sludges. SPE Production & Facilities 9: 179-182.

54. Poulsen MM, Kueper BH (1992) A field experiment to study the behavior of tetrachloroethylene in unsaturated porous media. Environmental science & technology 26: 889-895.

55. Al-Futaisi A, Jamrah A, Yaghi B, Taha R (2007) Assessment of alternative management techniques of tank bottom petroleum sludge in Oman. J Hazard Mater 141: 557-564.

56. Zhang Q, Chang J, Wang T, Xu Y (2007) Review of biomass pyrolysis oil properties and upgrading research. Energy conversion and management 48: 87-92.

57. Gregoli AA, Rimmer DP (2000) Production of synthetic crude oil from heavy hydrocarbons recovered by in situ hydrovisbreaking. Google Patents, US 6016867 A.

58. Martínez-Palou R, Mosqueira MDL, Zapata-Rendón B, Mar-Juárez, E, Bernal-Huicochea C, et al. (2011) Transportation of heavy and extra-heavy crude oil by pipeline: A review. Journal of Petroleum Science and Engineering 75: 274-282.

59. Johnson LA, Lusas E (1983) Comparison of alternative solvents for oils extraction. Journal of the American Oil Chemists' Society 60: 229-242.

60. Guedes Soares C, Garbatov Y, Zayed A, Wang G (2008) Corrosion wastage model for ship crude oil tanks. Corrosion Science 50: 3095-3106.

61. Vardon DR, Sharma BK, Scott J, Yu G, Wang Z, et al. (2011) Chemical properties of biocrude oil from the hydrothermal liquefaction of Spirulina algae, swine manure, and digested anaerobic sludge. Bioresour Technol 102: 8295-8303.

62. Anastasakis K, Ross AB (2011) Hydrothermal liquefaction of the brown macro-alga Laminaria saccharina: effect of reaction conditions on product distribution and composition. Bioresour Technol 102: 4876-4883.

63. Sarma AK, Konwer D, Bordoloi P (2005) A comprehensive analysis of fuel properties of biodiesel from Koroch seed oil. Energy & fuels 19: 656-657.

64. Marchesini FVH, Alicke AA, De Souza Mendes PR, Ziglio CUM (2012) Rheological characterization of waxy crude oils: sample preparation. Energy & Fuels 26: 2566-2577.

65. Khan MR (2007) Rheological Properties of Heavy Oils and Heavy Oil Emulsions. 385-391.

66. Tang W, Shan B, Zhang H, Zhang W, Zhao Y, et al. (2014) Heavy metal contamination in the surface sediments of representative limnetic ecosystems in eastern China. Sci Rep 4: 7152.

67. Ma DL, He HZ, Zhong HJ, Lin S, Chan DS, et al. (2014) Visualization of Zn^{2}□ ions in live zebrafish using a luminescent iridium(III) chemosensor. ACS Appl Mater Interfaces 6: 14008-14015.

68. He HZ, Leung KH, Yang H, Chan DS, Leung CH, et al. (2013) Label-free detection of sub-nanomolar lead(II) ions in aqueous solution using a metal-based luminescent switch-on probe. Biosens Bioelectron 41: 871-874.

69. Ma DL, He HZ, Chan DSH, Leung CH (2013) Simple DNA-based logic gates responding to biomolecules and metal ions. Chemical Science 4: 3366-3380.

70. Ramachandran SD, Sweezey MJ, Hodson PV, Boudreau M, Courtenay SC, et al. (2006) Influence of salinity and fish species on PAH uptake from dispersed crude oil. Mar Pollut Bull 52: 1182-1189.

13

Evaluation and Characterisation of Composite Mesoporous Membrane for Lactic Acid and Ethanol Esterification

Edidiong Okon, Habiba Shehu and Edward Gobina*

Center for Process Integration and Membrane Technology (CPIMT), School of Engineering, The Robert Gordon University, Aberdeen, AB10 7GJ, United Kingdom

Abstract

Recently, the use of inorganic composite mesoporous membranes in chemical industries has received a lot of attention due to a number of exceptional advantages including thermal stability and robustness. Inorganic mesoporous membranes can selectively remove water from the reaction product during lactic esterification reactions in order to enhance product yield. In this work, the characterization and evaluation of a catalytic mesoporous membrane with 15 nm pore size was tested with different carrier gases before employing the gases for lactic acid and ethanol esterification product analysis with Gas Chromatography coupled with mass spectrometry (GC-MS). The membrane was coated once with silica solution before the permeation test with carrier gases. Helium (He), nitrogen (N_2), argon (Ar) and carbon dioxide (CO_2) were used for the permeation tests conducted at the feed pressure of 0.10–1.00 bar and at the temperature of 413 K. The gas flow rate showed an increase with respect to feed pressure indicating Knudsen flow as the dominant transport mechanism. The order of the gas flow rate with respect to the feed pressure drop was Ar>CO_2>He>N_2. The morphological characteristic of the membrane was determined using scanning electron microscopy coupled with energy dispersive x-ray analyzer (SEM/EDXA). The SEM result of the membrane showed the distribution of the silica on the surface of the membrane. The surface area and pore size distribution of the silica membrane was analyze using liquid nitrogen adsorption/desorption method. The surface area results obtained from the Brunauer-Emmett-Teller (BET) isotherm for the 1st and 2nd dip-coated membranes were 1.497 and 0.253 m^2/g whereas the Barrette-Joyner-Halenda (BJH) curves for the pore size of the 1st and 2nd dip-coated membranes were 4.184 and 4.180 nm respectively, corresponding to a mesoporous structure in the range of 2-50 nm. The BET isotherms of the silica membranes showed a type IV isotherm with hysteresis. The BJH curve for the 2nd dip-coated membrane showed a 4% reduction in pore size after the modification process.

Keywords: Characterisation; Permeation; Inorganic membrane; Esterification; Gas transport; Isotherm

Introduction

Solvents play a major role in all stages of industrial manufacturing sector. The environmental and toxicological effects of solvents have become important in chemical processes. Because environmental problems have threatened the natural order including climate change and global warming, a lot of research is being carried out to find environmentally safe chemicals and processes [1]. Lactic acid is the simplest hydroxycarboxylic acid with an asymmetric carbon atom. It can be obtained from biomass, petroleum and coal. Copolymers and polymers of lactic acid are known to be eco-friendly and are compatible due to their degradability mild products, which makes them desirable as an alternative petrochemical polymer [2]. Lactic acid can react with ethanol during esterification process to produce ethyl lactate which is used as flavour chemicals, food additive, perfumery and as solvent. Ethyl lactate can replace environmentally damaging solvents including toluene, acetone, N-methyl pyrrolidone and xylene [3]. The use of inorganic ceramic membrane to selectively eliminate water from the reaction product during esterification of lactic acid is yet another important application that has attracted a lot of attention [4]. Membrane-based separation technologies have shown a wide range of application in food, biotechnology, pharmaceutical and in the treatment of other industrial effluents [5].

Membrane may be classified as heterogeneous or homogeneous, asymmetric or symmetric, solid or liquid, charged or uncharged and organic or inorganic membranes. The types of membranes include: dense, porous and composite membranes. They can also be inorganic and organic membranes. Table 1 shows the classification of inorganic membrane base on their nature and their most essential characteristics, permeability and selectivity [6]. Inorganic membranes have shown an increasing interest in the separation of gas mixtures

at high temperatures. However, one of the most promising uses of inorganic membrane is in the reactors where product purification by separation and chemical conversion occurs in the same device resulting in process intensification. Moreover, it is possible to obtain important enhancement over the equilibrium conversion of the reactor feed stream by selectively separating one or more reaction products across the membrane wall [7]. The function of the porous membrane during esterification reaction is to selectively remove water from the reaction mixture which will result in an equilibrium shift thus, driving the reaction towards completion [8,9]. Membranes different geometry and characteristics are generally made from a wide variety of chemically and thermally stable polymers. Other materials including alumina, titania, zirconia and silica are also being used [6].

Inorganic membrane can be prepared using different methods including sol-gel, chemical vapour deposition and sintering methods. Sol-gel method of preparation has been found to be the most suitable method for porous membrane preparation in contrast to chemical vapour deposition and sintering methods [10]. The mechanism of gas transport through membranes is generally divided into 5 groups: surface

***Corresponding author:** Edward Gobina, Center for Process Integration and Membrane Technology (CPIMT), School of Engineering, The Robert Gordon University, Aberdeen, AB10 7GJ, United Kingdom' E-mail: e.gobina@rgu.ac.uk

Types of membrane	Material	Permeability	Selectivity
Composite	Metal-metal	Moderate	Very selective
	Ceramic-metal		
Dense	Metallic	Low/moderate	High
	Solid-electrolyte		
Porous	Microporous	Moderate	Very selective
	Mesoporous	Moderate/high	Low/moderate
	Macroporous	high	Non-selective

Table 1: Summary of the classification of inorganic membrane.

diffusion, capillary condensation, Knudsen diffusion, viscous flow and molecular sieving mechanisms [11]. In Knudsen diffusion mechanism, gas molecules diffuse through the pores of the membrane and then get transported by colliding more frequently with the pore walls. Viscous flow mechanism takes place if the pore radius of the membrane is larger than the mean free path of the permeating gas molecule. Gas separation by molecular sieving mechanism takes place when the pore dimensions of the inorganic ceramic membrane approach those of the permeating gas molecules [12]. However, in capillary condensation mechanism, separation can takes place in the pores of the membrane with mesoporous layer in the presence of condensable gas specie [12].

Surface diffusion mechanism occurs when the adsorption of the permeating gas molecule occurs on the pore surface of the membrane material there by increasing the gas transport performance [12]. Physisorption is known as one of the most important method to characterised nanosized porous material with respect to the pore volume, pore size distribution and specific surface area [13]. According to Lee et al. [14], when the gas molecule interacts with the solid surface, the amount of the gas molecule adsorbs on the surface equals the total partial pressure of the gas molecule. However, the measurement of the adsorbed amount of gas over a range of partial pressure at a single temperature produces an adsorption isotherm. Hence, the resulting adsorption isotherm shows the various types based on the pore structure of a porous media and intermolecular interaction between the gas molecule and the surface [14].

According to IUPAC (International Union of Pure and Applied Chemistry), the physisorption isotherm can be classified into six different types as explained below:

• Type I isotherm is characterised by the adsorption in the non-porous microporous region at a low relative pressure.

• Type II is characteristic of non-porous or macroporous adsorbents with the formation of a multilayer of adsorbate (gas molecule) on the surface of the adsorbent (membrane material).

• Type III is characteristic by a non-porous or macroporous layer with weak interaction between the gas molecule and the membrane material.

• Type IV isotherm reflects a macroporous material which involves the coverage of the monolayer-multilayer on the external and mesoporous surface which is followed by capillary condensation in the mesoporous region with the formation of several hysteresis loop based on the shape of the pores.

• Type V isotherm is characteristic of a mesoporous material and involves the weak interaction between the permeating gas molecule and the membrane material.

• Type VI isotherm takes place in a highly uniform surface [13]. During the specific surface area analysis of the internal and external surface of porous material using physical gas adsorption, the amount of gas adsorbed depends on the relative vapour pressure [13].

In this study, the characterisation and evaluation of a mesoporous composite membrane with 15 nm pore size was investigated to determine the permeation properties of carrier gases (Ar, He, N_2 and CO_2) with the composite membrane before lactic acid and ethanol esterification reactions process.

Experimental

Permeation tests

The four carrier gases used for the permeation analysis include; nitrogen (N_2), argon (Ar), helium (He) and carbon dioxide (CO_2). The gases were supplied by BOC, UK. The permeation analysis was carried out at the feed pressure drop of 0.10–1.00 bar and at 413 K. The porous tube support was modified once before the permeation analysis. The total length of the membrane was 36.6 cm, while the inner and outer radius of the membrane was 7 and 10 mm respectively. The support modification process was carried out based on the procedure developed by Gobina [15]. Figure 1 shows the single gas permeation setup [16].

Liquid nitrogen adsorption

The surface area analysis of the support and coated support was examined using liquid nitrogen adsorption instrument, prior to the analysis, a small fragment of the silica membrane was crushed and used for the liquid nitrogen analysis. The actual weight of the 1st and 2nd dipping membrane samples was 4.3 g and 4.2 g respectively. The sample cells weight for the 1st and 2nd dipping was 16.8 g and 27.7 g respectively. The degassing of the system was carried out using a flow of a dry helium gas with some heat in order to remove moisture from the sample. A similar procedure to that of Vospernik et al. [17] was use for the liquid nitrogen analysis with some modification in the temperature. The liquid nitrogen temperature was 77 K.

The actual weights of the silica membranes, the cell weight and the weight of sample + cell before and after the degasing process are presented in Table 2. The pictorial diagram of the liquid nitrogen adsorption instrument is shown in Figure 2.

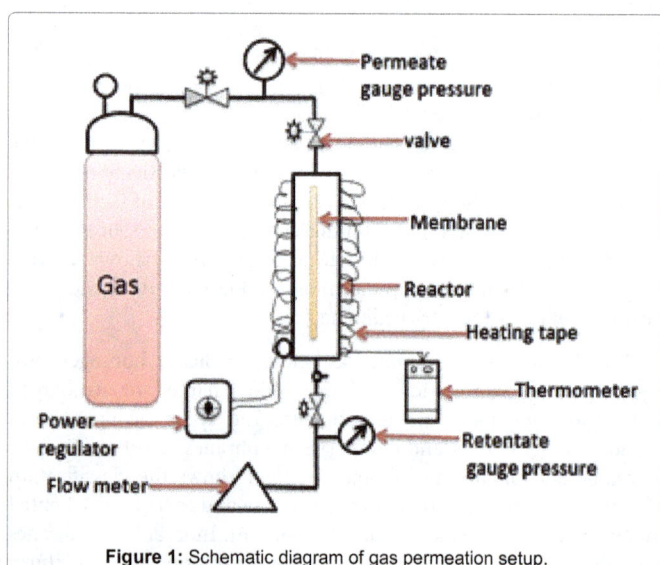

Figure 1: Schematic diagram of gas permeation setup.

Sample	Cell weight (g)	Weight of sample+cell before degasing(g)	Actual weight of sample (g)	Weight of sample after degasing (g)
1st dipping	16.8	21.1	4.3	21.0
2nd dipping	27.7	31.9	4.2	31.8
Support	16.8	21.1	4.2	21.0

Table 2: Sample and cell weights before and after degasing process.

Results and Discussion

Figure 3 shows the relationship between the gas flow rate (mol/sec) and the feed pressure drop (bar) at 413 K. From the result obtained, it was found that the gas flow rate increase with respect to feed pressure (bar). The order of the gas flow rate through the silica membrane as shown in Figure 3 was: Ar (40 mol/g)>CO_2 (44 mol/g)>He (4 mol/g)>N_2 (28 mol/g). It can be seen that Ar gas showed a higher flow than CO_2 with the highest molecular weight. However, N_2 gas with the higher molecular weight showed a higher flow in contrast to He gas with a least molecular weight. According to Knudsen flow mechanism, gases with the least molecular weight can permeate faster than gases with the heavier molecular weight. It was found that the order of gas flow rate was not based on the Knudsen flow mechanism of transport which as a relationship with the gas molecular weight. From the gas flow rate relation with the feed pressure drop, Ar gas was suggested as the suitable carrier and detector gas to be coupled with GC-MS for the analysis of the esterification lactic acid and ethanol esterification product.

Figure 4 shows the relationship between the gas flow rate (mols^{-1}) and the inverse square root of the gas molecular weight. According to Markovic et al. [18], Knudsen diffusion is the dominant flow mechanism of transport in silica membrane if gas flow rate is proportional to the inverse square root of the gas molecular weight. From the result obtained in Figure 4, it was observed that there was no linear proportionality correlation of the gases with the inverse square root of the molecular weight indicating that the Knudsen flow mechanism was non-operative.

The relationship between the gas permeance and inverse of the gas viscosity was also plotted at the gauge pressure drop of 0.9 bar and 413 K. From Figure 5, it was observed that gases with highest viscosity value showed the least permeance which suggests that the gas flow through the silica membrane was based on viscous flow mechanism of gas transport.

From the result obtained in Table 2, it was observed that the sample weight reduced in size by 0.1 g in both silica membrane samples after the degasing process. The BET and BJH of the two membrane samples were obtained from the liquid nitrogen adsorption. The amount of the adsorbed gas molecule depends on the relative vapour pressure (P/P_O). Figures 6 and 7 shows the plot of the amount of gas adsorbed (volume at STP (cc/g)) against the relative pressure P/P_O for the 1st and 2nd dip-coated fragments. The BET isotherm of the both membranes possess hysteresis loop in their curves indicating the pore size of 4.184 nm for the 1st dip-coated fragment and 4.180 for the 2nd dip- coated fragment respectively as shown in Table 3. The pore size of the dip-coated silica membranes was characteristic of mesoporous structure with hysteresis which was indicative of a capillary condensation in the mesoporous region.

From Figure 6, the nitrogen adsorption isotherm for the 1st dip-coated membrane was observed to be similar to that of a type IV physisorption isotherm with the presence of hysteresis based on the six classification of the BET isotherm. A similar result was obtained

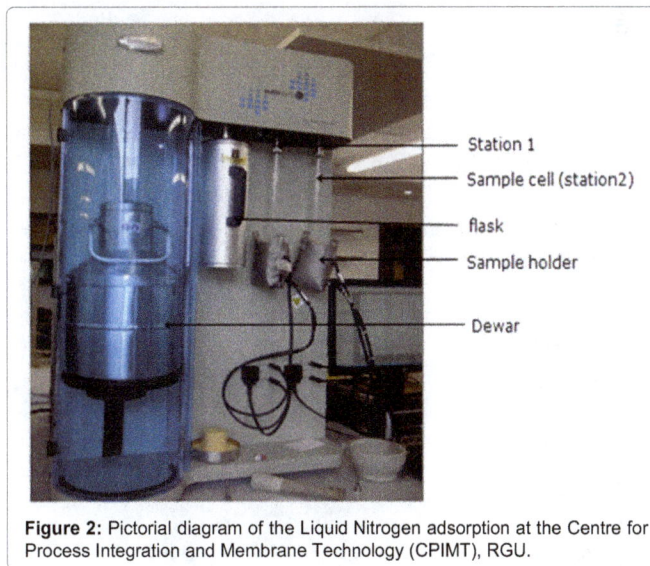

Figure 2: Pictorial diagram of the Liquid Nitrogen adsorption at the Centre for Process Integration and Membrane Technology (CPIMT), RGU.

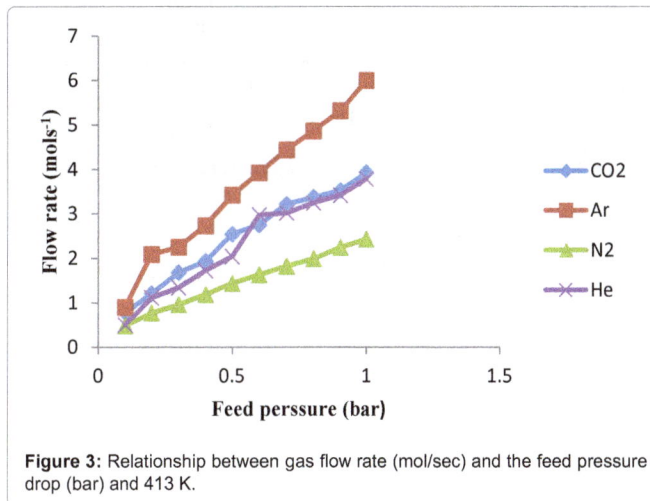

Figure 3: Relationship between gas flow rate (mol/sec) and the feed pressure drop (bar) and 413 K.

Sample	BET Surface area (m²/g)	BJH Pore diameter(nm)	Pore Volume (cc/g)
1st dipping	0.253	4.184	0.006
2nd dipping	1.497	4.180	0.004

Table 3: BET and BJH values for 1st and 2nd dip-coated silica membrane.

by Lee et al. [7]. Comparing the BET isotherms of the 1st and 2nd dip-coated membranes, it was observed that the hysteresis in the BET curve of the 1st dip-coated fragment was higher compared to the 2nd dip-coated fragment. This was due to the fact that the 1st dip-coated membrane exhibited a higher surface area and pore size before the dip-coating process however, hysteresis effect decreases after each dipping process which was also attributed to the reduction in the pore size of the membrane.

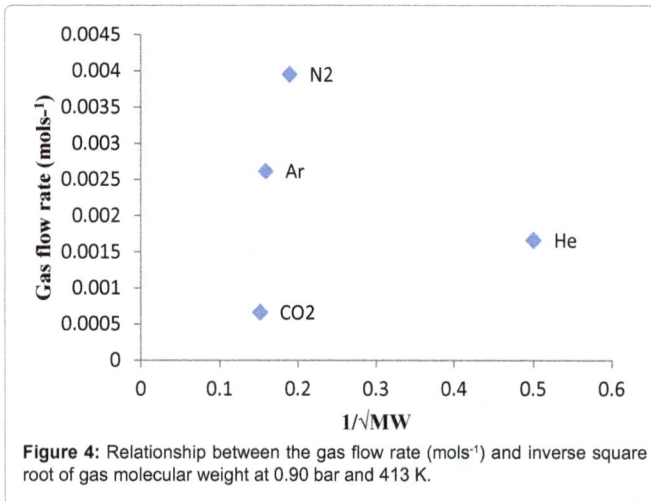

Figure 4: Relationship between the gas flow rate (mols⁻¹) and inverse square root of gas molecular weight at 0.90 bar and 413 K.

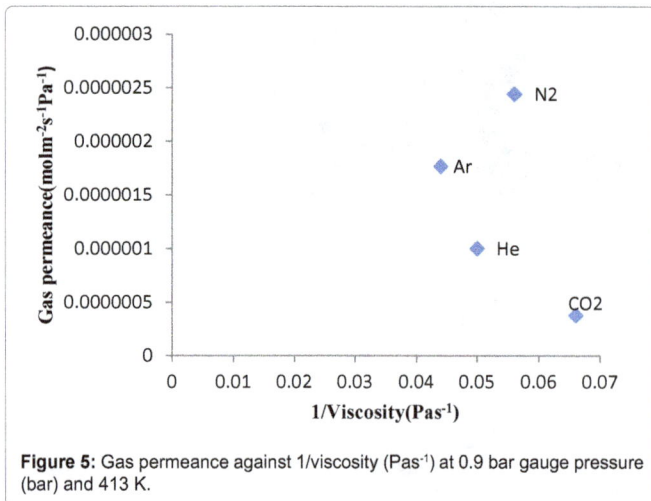

Figure 5: Gas permeance against 1/viscosity (Pas⁻¹) at 0.9 bar gauge pressure (bar) and 413 K.

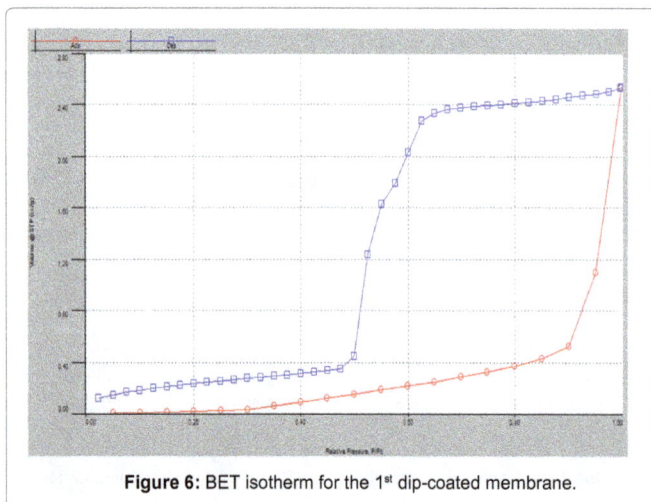

Figure 6: BET isotherm for the 1st dip-coated membrane.

The pore size distribution for the silica material was estimated based on the BJH results. From Figure 8, it was observed that the silica membrane showed a mesoporous structure with the pore diameter and pore volume of 4.184 nm and 0.006 cc/g respectively for the 1st dip-coated membrane whereas the pore diameter and the pore volume of the 2nd dipping from Figure 9 were 4.180 nm and 0.004 cc/g respectively.

However, the surface area of the 1st dip-coated membrane was found to be lower in contrast to the 2nd dip-coated membrane and shown in Table 2. It was observed that there was a 4% reduction in the pore diameter of the membrane after the dipping process, which confirmed the fact that the membrane was characteristic of a mesoporous structure and the silica membrane was suitable for gas molecules to penetrate through the pore walls. A similar result was obtained by Markovic et al. [18].

The surface of the silica membrane was examined using Scanning electron microscopy and energy dispersive x-ray analyser (SEM-EDXA) to determine the structural morphology of the membrane sample. From Figure 10, it was found that a defect-free surface with no evidence of crack on the surface of the silica membrane. It was also found that the silica solution was distributed evenly on the surface of the membrane.

Conclusion

The gas transport across silica membrane was not entirely based by Knudsen mechanism of transport but included other mechanism such as viscous flow. The gas flow rate showed an increase with respect to feed pressure drop. The SEM image of the silica membrane showed a clear surface without any crack on the surface. The BET of the 1st and 2nd coated silica membrane showed a type IV isotherm with the presence of hysteresis in both curves, indicating that the membrane

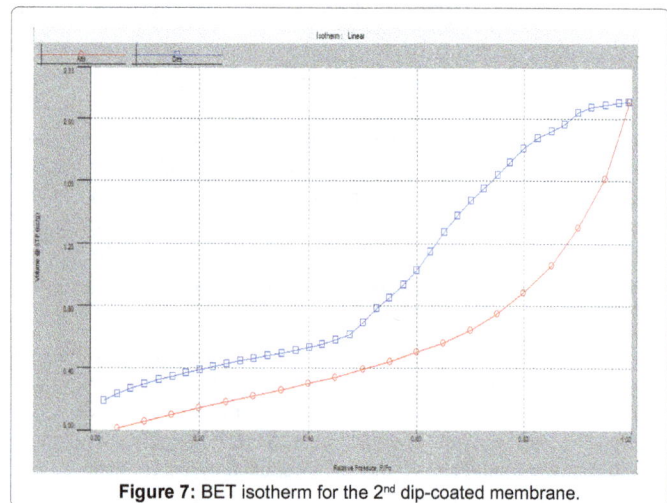

Figure 7: BET isotherm for the 2nd dip-coated membrane.

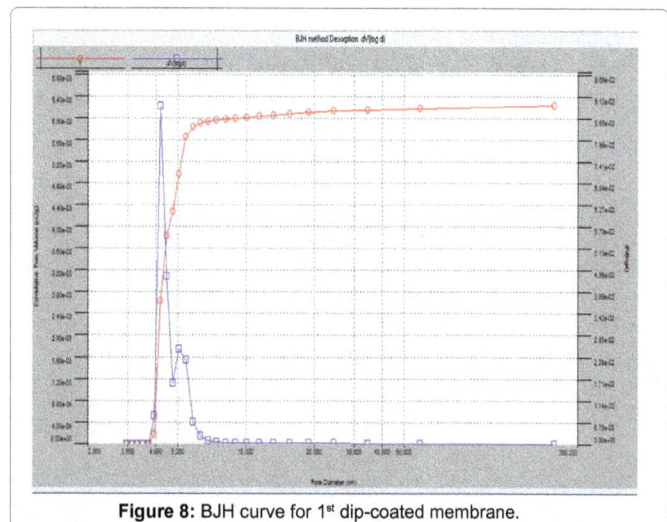

Figure 8: BJH curve for 1st dip-coated membrane.

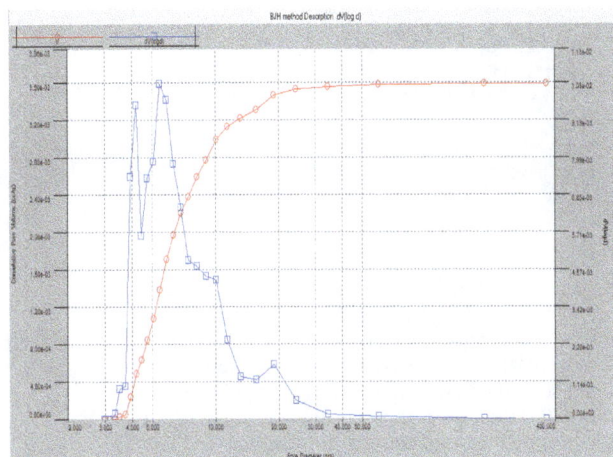

Figure 9: BJH curve for 2nd dip-coated membrane.

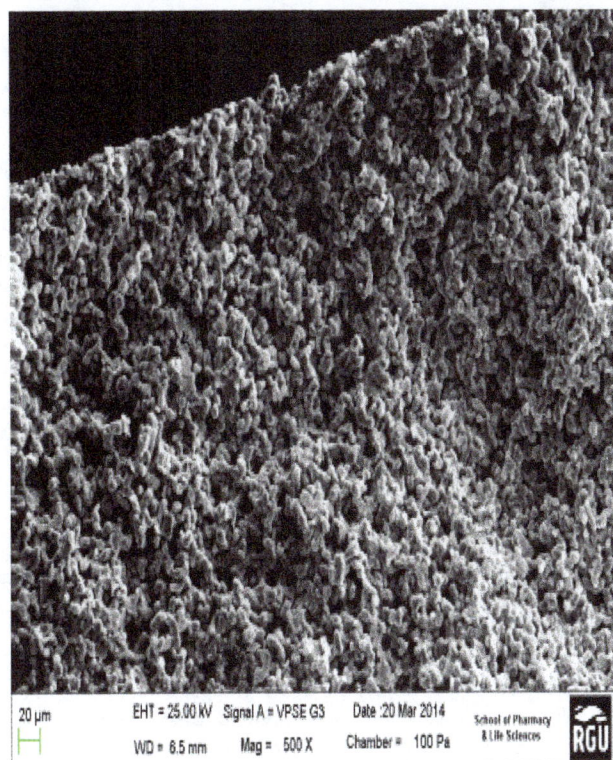

Figure 10: SEM surface image of cross-section of the silica membrane.

was characteristic of a mesoporous classification in the range of 2-50 nm. The hysteresis in the BET isotherm of the 1st dip-coated membrane was found to be larger than that of the 2nd dip-coated membrane indicating the effect of the dip-coating process. The BJH of 2nd coated silica membranes showed an average pore reduction of 4% after the modification process.

Acknowledgments

The authors wishes to express sincere thanks to and CPIMT for supplying the membrane and other infrastructures used in the study, School of Pharmacy and Life Science, RGU for the SEM images. Additionally, CCEMC is great fully acknowledged for their financial contribution towards the research work.

References

1. Nigiz FU, Hilmioglu ND (2015) Green solvent synthesis from biomass based source by biocatalytic membrane reactor. International Journal of Energy Research 40: 71-80.

2. Sanz MT, Murga R, Beltrán S, Cabezas JL, Coca J (2004) Kinetic study for the reactive system of lactic acid esterification with methanol: Methyl lactate hydrolysis reaction. Industrial & Engineering Chemistry Research 43: 2049-2053.

3. Engin A, Haluk H, Gurkan K (2003) Production of lactic acid esters catalyzed by heteropoly acid supported over ion-exchange resins. Green Chemistry 5: 460-466.

4. Pereira CS, Silva VM, Rodrigues AE (2011) Ethyl lactate as a solvent: Properties, applications and production processes-a review. Green Chemistry 13: 2658-2671.

5. Calvo JI, Bottino A, Capannelli G, Hernández A (2008) Pore size distribution of ceramic UF membranes by liquid–liquid displacement porosimetry. J Memb Sci 310: 531-538.

6. Mulder M (1996) Basic Principles of Membrane Technology. Kluwer Academic Publishers, The Netherlands.

7. Collins JP, Way JD (1993) Preparation and characterization of a composite palladium-ceramic membrane. Industrial & Engineering Chemistry Research 32: 3006-3013.

8. Zhang Y, Ma L, Yang J (2004) Kinetics of esterification of lactic acid with ethanol catalyzed by cation-exchange resins. React Funct Polym 61: 101-114.

9. Dassy S, Wiame H, Thyrion FC (1994) Kinetics of the liquid phase synthesis and hydrolysis of butyl lactate catalysed by cation exchange resin. J Chem Technol Biotechnol 59: 149-156.

10. Ju X, Huang P, Xu N, Shi J (2000) Influences of sol and phase stability on the structure and performance of mesoporous zirconia membranes. J Memb Sci 166: 41-50.

11. Abedini R, Nezhadmoghadam A (2010) Application of Membrane in Gas Separation Processes: Its Suitability and Mechanisms. Petroleum & Coal 52: 69-80.

12. Lee H, Suda H, Haraya K (2005) Gas permeation properties in a composite mesoporous alumina ceramic membrane. Korean J Chem Eng 22: 721-728.

13. Weidenthaler C (2011) Pitfalls in the characterization of nanoporous and nanosized materials. Nanoscale 3: 792-810.

14. Lee H, Yamauchi H, Suda H, Haraya K (2006) Influence of adsorption on the gas permeation performances in the mesoporous alumina ceramic membrane. Sep Purif Technol 49: 49-55.

15. Gobina E (2006) Apparatus and Method for separating gases. United States patent. Patent No.7048778 B2, Robert Gordon University, Aberdeen, UK.

16. Okon E, Shehu H, Gobina E (2014) Synthesis of Gas Transport through Nano Composite Ceramic Membrane for Esterification and Volatile Organic Compound Separations. Journal of Mechanics Energy and Automation 4: 905-913.

17. Vospernik M, Pintar A, Bercic G, Batista J, Levec J (2004) Potentials of Ceramic Membranes as Catalytic Three-Phase Reactors. Chemical Engineering Research and Design 82: 659-666.

18. Markovic A, Stoltenberg D, Enke D, Schlünder E, Seidel-Morgenstern A (2009) Gas permeation through porous glass membranes: Part I. Mesoporous glasses-effect of pore diameter and surface properties. J Memb Sci 336: 17-31.

14

Banded Crystalline Spherulites in Polymers and Organic Compounds: Interior Lamellar Structures Correlating with Top-Surface Topology

Eamor M Woo[1]*, **Graecia Lugito[1]**, **Cheng-En Yang[1]**, **Shi-Ming Chang[1]** and **Li-Ting Lee[2]**

[1]*Department of Chemical Engineering, National Cheng Kung University, Taiwan*
[2]*Department of Materials Science and Engineering, Feng Chia University, Taiwan*

Abstract

By using several different polymers and compounds, organics or inorganic, low-molecular-weight or high-molecular-weight, this study has proven that they can be packed into concentric ring bands, usually circular, but other shapes like hexagons or flower-like petals can also be possible. With these traits taken into consideration, it is difficult to generalize the ring bands in spherulites to a single cause of lamellae twist/spiral. Instead, exposure of lamellae underneath the top layers beyond thin films becomes essential for shedding new light on all these intricately complex issues. The novel approaches in this study circumvent such limitation by interior dissection of PEA, which clearly reveals that no continuous spiraling twist, as the cross sections show a corrugated-board structure with layers resembling a peel-able onion, where each radially oriented layer is sandwiched with a tangential layer of lamellae.

Keywords: Ring-banded spherulite; Interior lamellae; Poly(L-lactic acid); Poly(ethylene adipate); Phthalic acid

Introduction

Compounds, organics or inorganic, low-molecular-weight or high-molecular-weight, can crystallize if their chemical structures are relatively ordered. In addition to many regular shapes that are widely known, crystals can be packed into concentric ring bands, usually in circular shape, but other shapes like hexagons or flower-like petals can also be possible. Banded patterns in spherulites mean that concentric rings (Note: sometimes not necessarily concentric rings with a common nucleus, but the crystal plates are aligned as spirals [clockwise or counterclockwise] from center to outer rims on the same substrate plane) are packed as a result of crystallization, usually under some specific conditions, such as temperature of crystallization, confinement, diluents, and/or solvent evaporation rates. In banded spherulites, the rings possess different optical properties, surface topology, and sometimes lamellar orientations. These banded patterns in spherulites are sporadically seen in crystallization of many materials; yet the mechanisms remain elusive and have been highly debated. Among the small-molecule compounds known to display rings in crystallized spherulites are: hippuric acid [1], phthalic acid [2], testosterone propionate [3], aspirin [4], D-mannitol [5], as reported by Kahr et al., and inorganic salts such as ($K_2Cr_2O_7$) [6-8]. In the fields of crystallography for small molecules, there are two proposals for accounting the ring bands: helical crystals, and rhythmic precipitation. There are also occasions that both mechanisms are simultaneously responsible for the banded patterns.

In addition to many literature reported on ring-banded spherulites in many other polymers, aromatic as well as some aliphatic polyesters, are among the most investigated materials for crystalline behavior including banded spherulites, such as poly(trimethylene terephthalate) (PTT) [9,10], poly(pentamethylene terephthalate) (PPT) [11,12], poly(octamethylene terephthalate) (POT) [13], poly(nonamethylene terephthalate) (PNT) [14,15], and poly(ethylene adipate) (PEA) [16-18]. The reasons these polyesters have been the focus of studies in crystal structures were due to that they always could be packed into the ring-banded spherulites within a certain temperature range. Most investigations approach the issue from surface views on thin-film samples that have been crystallized in controlled conditions. For different materials, the banding patterns may look similar, but the top-surface topological patterns differ subtly, which suggest that the internal lamellae assemblies also vary correspondingly.

Recently, biodegradable polymers, such as poly(L-lactic acid) (PLLA), poly(3-hydroxybutyrate-co-3-hydroxyvalerate) (PHB-co-HV), and poly(3-hydroxybutyrate) (PHB), etc. [19-23], have been shown to exhibit ring-banded spherulites just like many of other aryl-polyesters (PPT, POT, PNT, etc.) [9-15] or aliphatic polyesters (PEA, PBA, etc.) [16-18]. However, most investigators only focused on analyses based on top-surfaces of thin films, and very rarely the inner structures of banded materials were examined. Polymers are known to have chain folding from classical literature and some investigators including Lotz and Cheng [24] argued that chain folding induced surface stresses, which in turn were responsible for crystal twisting. By contrast, small-molecule compounds, including hippuric acid [1], phthalic acid [2], testosterone propionate [3], and aspirin [4], etc., do not have chain folding in crystals at all, but the crystals of some of the compounds also twist. Kahr et al [1-5], following the initiating arguments of a classical work by Bernauer [25] on many organic compounds using optical microscopy in 1929, have argued that temperature-gradient induced stresses, and other stresses, etc., may cause helice shapes in these small-molecule crystals. Apparently, the above comparisons between long-chain polymers vs. small-molecule compounds clearly hint that crystal twist is a habit that not necessarily is related to or induced by chain folding. A small-molecule compound, phthalic acid (PA), has been extensively investigated by Kahr et al. [2], in revealing ring-banded spherulites by evaporating PA from 20/80 water/ethanol solution at ambient temperature (25°C). This PA compound and its spherulites serve as a useful low-molecule model in comparison to the ring-banded long-chain polymer (PLLA) or aliphatic polyester (PEA), aryl polyester (PDoT) to be dicussed in this work.

*****Corresponding author:** Eamor M Woo, Department of Chemical Engineering, National Cheng Kung University, Tainan, 701-01, Taiwan
E-mail: emwoo@mail.ncku.edu.tw

An even more critical issue is that albeit the facts of stress-incduced twisting in single-crystals of either small-molecule compounds or polymers, are there correlations between the ring-banded patterns and the twisting of single crystals. Key point may be in providing direct evidence of whether or not the observed pitch of the twist single crystals is in agreement with the pitch of optical birefringent patterns in ring-banded spherulites. This single most critical evidence (twist pitch of single crystals being in agreement with optical ring interspacing in banded spherulites) appears to be missing in numerous works reported in the literature. Many investigators only focused on analyzing the twisting of single crystals and mechanisms therein, and argued that these twist crystals could be retrived from the ring-banded spherulites. However, these proposed correlations may be an Achilles-knee problem that should be tackled with more careful and deeper analyses, as they may run riskes of the facts that twisted crystals may also be retrived from ringless and non-banded spherulites.

To tackle the complex issues, novel approaches were taken in the work to examine the interiors of crystallized samples that have been prepared into bulk forms. In addition, thickness was varied from very thin to thick to observed changes in crystallized patterns. Top morphology and crystal patterns on the top surface in correlation with the interior crystal lamellae was established. As the surface topology differs among different materials with banding patterns, the inner structures may also differ. Several polymers, mainly synthetic biodegradable polyesters, as well as small-molecules were used as models for comparison and testing for universal behavior in crystals that form ring bands. In addition, different types of ring bands were analyzed and mechanisms exemplified.

Experimental

Materials and preparation

Poly(L-lactic acid) (PLLA) of two grades, M_w=11,000 g/mol and M_w=119,400 g/mol, were purchased from Polyscience, Inc. (USA) and from NatureWorks, respectively. Poly(ethylene adipate) (PEA), with M_w=10,000 g/mol, T_g=-52°C, and T_m=43°C, was purchased as research-grade material from Aldrich Co. (USA). Poly(dodecamethylene terephthalate), (PDoT) was synthesized with a two-step polymerization from 1,10-dodecanediol and dimethyl terephthalate (DMT) with 0.1% butyl titanate as a catalyst. PDoT is an aryl polyester and has a structure similar to commercial poly(ethylene terephalate) (PET), except that PDoT has a much longer methylene segment (12 CH_2 units) between the terephthalate groups in its main-chain repeat units. The weight-average molecular weight (M_w) and the polydispersity index (PDI) as determined by gel permeation chromatography (GPC, Waters) were 25,300 g/mol and 2.07, respectively. Phthalic acid (PA), with M_w=166.13 g/mol and T_m=205°C, was purchased from Alfa Aesar, a Johnson Matthey Company. To ensure there was no impurity, those polymer materials were firstly dissolved into chloroform ($CHCl_3$) and filtered by PTFE syringe filter 0.45 μm and dried in vacuum oven for 7 days at room temperature. Samples of polymers (PEA, PLLA, PDoT) were prepared by dissolving each polymer into common solvent of chloroform, or other suitable solvents in various concentrations depending on film thickness required. Polymers were either drip- or spin-cast into thin films on glass substrates. For PEA, a few drops were repeatedly deposited on a glass substrate to stacked into a thick bulk sample (~100 μm). Degassing was performed at 45°C in a vacuum oven for 24 h to remove the residual solvent. The crystallization of polymers was conducted at a suitable temperature of crystallization directly from their melt state for displaying ring-banded spherulites.

Apparatus and procedures

A polarized optical microscope (POM, Nikon Optiphot-2), equipped with a Nikon Digital Sight (DS)-U1 camera control system and a microscopic hot stage (Linkam THMS-600 with T95 temperature programmer), was used to confirm the crystal morphology prior to the AFM and SEM observation. A sensitive tint plate with 530 nm optical path difference was inserted in between the polarizer and analyzer for enhancing the contrast of birefringence.

Scanning electron microscope (SEM, FEI Quanta-400F) was used to characterize both fractured and top free surfaces of bulk form PEA samples. PEA in thick bulk forms was heated to T_{max}=90°C, rapidly dipped in a silicone oil bath set at controlled temperature of 28°C for crystallization. The fully crystallized bulk PEA samples were washed with petroleum ether before observation. The samples were fractured into halves, then coated with gold vapor deposition using vacuum sputtering prior to SEM characterization.

Atomic-force microscopy (AFM, diCaliber, Veeco Corp., Santa Barbara, USA) investigations were made in the intermittent tapping mode with a silicon-tip (f_o=70 kHz, r=10 nm) installed. AFM measurements were carried out to determine the phase images and height profiles in the morphology topology of thin-film samples.

Results and Discussion

Poly(ethylene adipate) (PEA)

Poly(ethylene adipate) has long been known to exhibit banded spherulite when melt-crystallized at T_c=27-30°C. Many investigators seemed to be quick to jump into a presumption by following Keith and Padden who proposed earlier that thousands or millions of lamellae spiral and twist in coordinated fashion to generate such banding patterns seen in PEA. However, since the earlier proposal by Keith and Padden more than 60 years ago, no one ever attempted to answer why millions of lamellae were able to spiral coordinately to generate such smooth ring bands; or faced deeper questions why banding is displayed by PEA spherulites only in this narrow temperature range (27-28°C), but not in any other temperatures? Furthermore, most polymer thin films in many investigations were not so thin enough to accommodate only one single layer of lamellar plates to do spiraling from nucleation center; there were actually tens or hundreds of lamellae (each averaging for 15 μm), i.e., many more than just one single layers that have to be co-spiral without any slips or mismatches to generate such smooth, or almost perfect, ring bands in spherulites. All past investigations on ring bands had been mostly based on analyses on top surfaces of thin films; then a deeper question for thoughts is that what happen to the lamellae underneath the top surfaces showing ring bands? The spiral of lamellae plates in accounting for the observed ring bands in spherulites has been conceived to be somewhat like that in a-helix in proteins. However, in proteins, it is known and well established that it takes strong intra-molecular hydrogen bonds for stabilizing an alpha-helix in maintaining the spiral conformation; many investigators seem to have overlooked a very essential point that in crystal lamellae of most polymers, there are no such H-bonding for holding the lamellae in long spirals from the nucleation center all the way to the edge of the spherulites. Surface stress may cause occasional and irregular re-orientation of the lamellae plates, but consistent and coordinated spiral of the lamellae plates obviously need more convincing mechanisms than just stresses.

Observation as an example is placed on ring bands in poly(ethylene adipate) (PEA), whose top surface banding patterns and interior lamellar assembly would be compared. Figure 1 shows double ring-

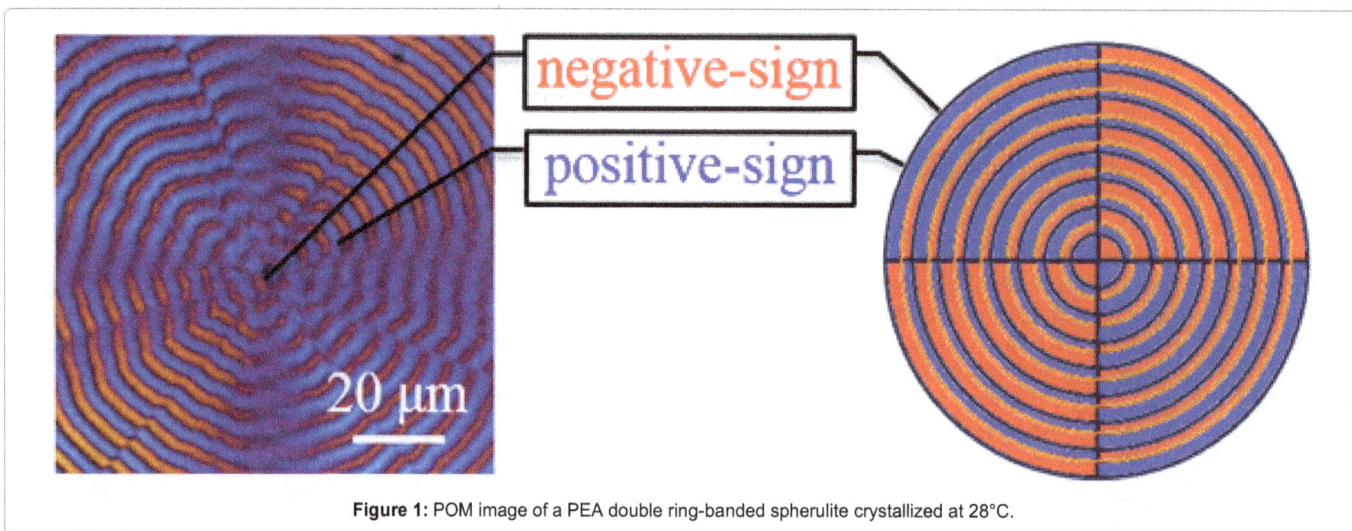

Figure 1: POM image of a PEA double ring-banded spherulite crystallized at 28°C.

banded spherulite of PEA crystallized at 28°C. At the beginning (center part), PEA spherulite shows the sign of negative-type spherulite, after about 4 μm in radius, the sign changes to positive-type for another 2.5 μm, and so on, composing the concentric pattern of alternating interference colors. One may sometimes misinterpret this kind of double ring-banded spherulite as negative single ring-banded spherulite due to its wider band of negative-type compared to the positive one.

The previous study clearly reveals the formation of mutually perpendicular orientations of crystal lamellae and its relation to the alternating birefringence in PEA double ring-banded spherulites [16-18]. Moreover, in this study, the interior architecture of bulk samples has been studied in order to give a manifest about the mechanisms and behavior of the lamellar assembly in case of PEA double ring-banded spherulite.

Poly(L-lactic acid) (PLLA)

In contrast to the ring band patterns composed o obviously blue/orange birefringence bands in PEA, poly(L-lactic acid) (PLLA), spin-cast to ~300-500 nm film thickness, can display concentric ring-banded spherulite with no birefringence colors. POM and AFM characterizations were performed on PLLA crystallized at T_c=115°C, and results are shown in Figure 2. The POM graph, which demonstrates no birefringence colors in the banding patterns, is partially owing to the film thickness being smaller than (or roughly equal to) the visible optical light wavelength. The AFM analyses revealed concentric rings (of roughly hexagonal shape) in PLLA spherulites with inter-ring spacing ca. 7-10 μm, depending on the film thickness. An up-and-down topology for the ring-banded spherulite is clearly demonstrated by AFM analyses with height drop of 300 nm. The dramatic vertical and sudden drop from one band to the next one would rule out the possibility that lamellae are continuously spiral (twist) gradually from nucleous center to edge of the spherulites. Instead, one clearly sees that one band is almost detached from the next one, as the drop of height between the neighboring band almost equals to the film thickness, which means that the band drops to the substrate surface before it rises again to form the next bands in spin-cast PLLA (T_c=115°C). Spiral or twist of lamellae plates would not cause a sudden drop to the substrate surface at all. The significant differences between the ring-band characteristics in PLLA spherulites from those in PEA spherulites should yield a hint that ring bands and their formation may not be from a same mechanism.

Ring bands in spherulites could be influenced with samples being covered or not by another top-glass. The fact suggests that the top topology responsible for the ring-banded patterns in the spherulites could be influenced or altered by physical contact with a top surface that makes contact with the lamellae (or crystal plates). Figure 3 shows POM graphs for high-M_w poly(L-lactic acid) (M_w=119,400, NatureWorks) crystallized at T_c=130°C: (a) uncovered, (b) covered samples. If samples were crystallized with top surface free (no top glass cover on samples), the PLLA polymer was melt-crystallized at 130°C into ringless (no ring bands) spherulites; by contrast, with top glass on samples being crystallized, PLLA was melt-crystallized at 130°C into apparently ring-banded spherulites of large banding space (inter-ring spacing=100 μm). If chain folding and its effect on surface stresses between the lamellae were the sole causes responsible for the ring bands in spherulites, then top glass in contact with the lamellae should not be an influential factor for the ring-band patterns. The lamellae only make contact with the top glass in the interfaces between polymer and glass on top, and only the very top-layer lamellae are in contact with the glass, and the other lamellae inside should not be influenced at all. Furthermore, if making contact with the top glass would erase or diminish the surfaces between lamellae, then polymers on bottom glass substrates should also lead to ringless spherulites. However, these numerous results are opposite to such arguments, as all polymers showing ring banded spherulites have always been observed on bottom glass substrates.

Poly(dodecamethylene terephthalate) (PDoT)

Ring bands in a same polymer can exhibit many different types of patterns, which is difficult or almost impossible to justify all these ring bands to be originated from a single mechanism, e.g., lamellar spiral, that is responsible and workable for formation of all these wildly different patterns of ring bands. Examples are given in Figure 4 that shows ring bands POM patterns for poly(dodecamethylene terephthalate) (PDoT) crystallized at 90°C and 120°C, respectively. The pattern of ring bands in PDoT crystallized at T_c=90°C (Figure 4a) may be similar to those in PEA crystallized at T_c=28°C (Figure 1). However, the pattern of the ring bands in PDoT spherulites crystallized at higher T_c=120°C (Figure 4b) is completely different. Obviously, lamellar spiral/twisting does not seem plausible for the ring bands in PDoT crystallized at T_c=120°C at all (Figure 4b), as across the large pieces of bands (~50 μm), it is easy to tell from the POM graphs that there is no gradual spiral from the beginning to end of one large band. If there were spiral of the lamellae across the

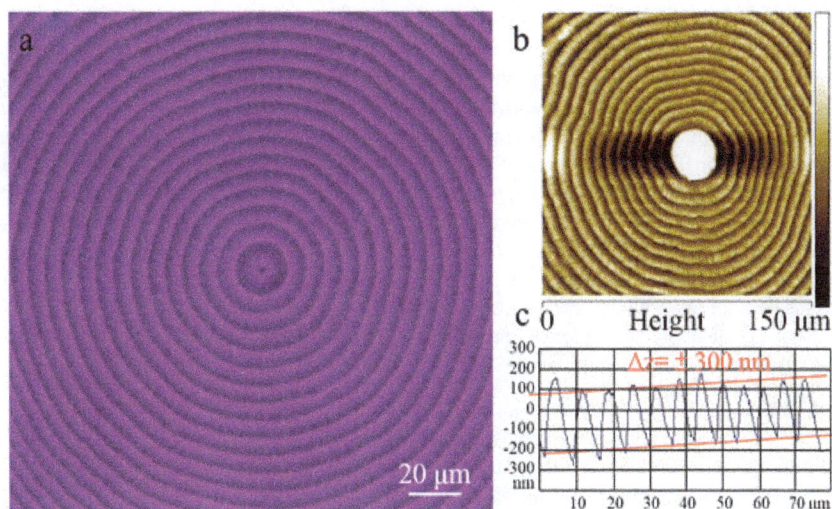

Figure 2: POM and AFM characterizations performed on low- M_w PLLA [M_w=11,000 g/mol, Polyscience, Inc. (USA)] crystallized at T_c=115°C: (a) POM graph with 530 tint plate, (b) AFM height image, and (c) AFM height profile [film thickness=0.5 µm].

Figure 3: POM graphs for high-M_w poly(L-lactic acid) [M_w=119,400 g/mol, NatureWorks] crystallized at T_c=130°C: (a) uncovered and (b) covered samples.

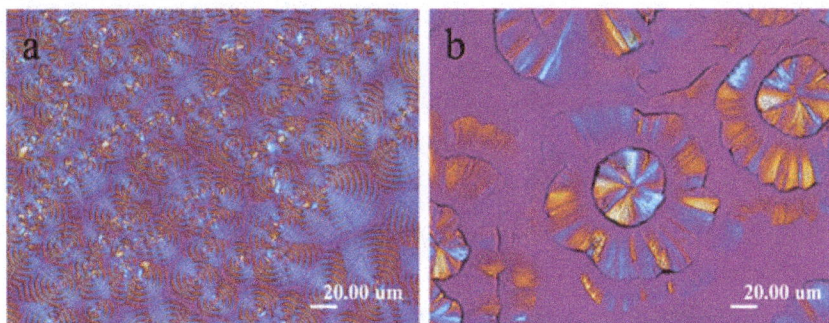

Figure 4: POM graphs for poly(dodecamethylene terephthalate) (PDoT) melt-crystallized at (a) T_c=90°C, (b) T_c=120°C, respectively.

band (spiral in radial direction), the POM birefringence would display gradual birefringence color transition. Obviously, the birefringence color remains the same across the band in the radial direction. In Figure 4a, the width of the band is too small (5 µm) to tell whether or not there is gradual spiral. But obviously, the two neighboring bands are of two opposite birefringence colors (Blue/Orange). It remains to be quested whether these two opposite colors in two neighboring bands were a result of gradual "spiral/twist"? Or simply, there were two lamellae intersecting at perpendicular orientations? Again, dissecting into the interior of the lamellae underneath these banding patterns was the only way to answer for clarification.

Ring-banded spherulites in phthalic acid

Phthalic acid (PA), a low-M_w organic compound, is known to show concentric ring bands in spherulites. Figure 5 shows POM graphs for PA (dissolved in 80% water+20 wt% ethanol), which was crystallized at ambient (28°C), 80°C, and 110°C, respectively. Note that ring bands in PA were formed in solvent-induced crystallization, but not in melt-crystallization. The ring banded patterns in low-M_w PA spherulites are in every aspects very similar to those seen in the ring-banded spherulites of many polymers, such as poly(ethylene adipate) PEA or poly(butylene adipate) PBA. Contrary to the popular interpretations taken by many polymer scientists that ring banded spherulites in polymers are a result of chain folding in lamellae, which leads to surface stresses, which in turn cause lamellar spiral twists to form ring banded patterns in spherulites, it should be emphatically pointed out that PA, being a low-M_w organic compounds, there is obviously no chain folding in the crystals it packs into. With no chain folding or surface stresses in PA crystals, the ring banded spherulites in PA obviously cannot be attributed to chain folding induced surface stresses and lamellar spiral twists, etc. This fact yields strong evidence that ring bands in spherulites cannot be attributed to only a single cause.

Apparently, analyses of these banding phenomena have traditionally been limited to crystals in thin films and microscopic views on top surfaces of these thin films. How the crystal plates in interiors are connected to the top-surface topology in ring-banded spherulites has not been addressed. Poly(ethylene adipate) (PEA) was used as a model system for interior investigation. The results in this model system might be equally applicable to other polymers or even low-M_w organic compounds.

In Figure 6a, it can be seen that the tangential layers (in skin layers) occasionally bend, curve, or twist when traversing from tangential layer to the radial layer. However, the tangential lamellae are not physically extending between the two sequential layers, as what would be proposed in a continuous twist (spiral) going only in the radial direction. Internal views are apparent in revealing two main points of observations: (1) radial lamellae taper gradually and form thinner cilia-like lamellae before being attached by the next crystalline polymer chains tangentially in next layer; and (2) when there is no more cilia-like lamellae could be attached, yet the lamellae start to bend, curve, or twist 90° to continue the growth in the radial direction between the perpendicular layers located in the neighboring layers. But no matter they bend or curve, the radial lamellae are always positioned 90° to the tangential skin layers. It could be attributed to that the attachment of tangential lamellae to the cilia-like lamellae was detected as continuous bends, curves, or twists. Previous analysis on thin-film and thickness confined samples might have taken these traits as spiral or twisting of lamellae; however, as pointed out earlier, lamellae of radial and tangential directions do intersect at 90° (or nearly), but they do not spiral gradually from one to another at all. Opposite to such views generated from classical observations on thin-films samples, the internal dissection views on PEA bulks have clearly led to that there are always perpendicular radial lamellae plates between two tangential layers. Figure 6b shows that SEM characterization on PEA bulk

crystallized at T_c=28°C, whose fractured interior shows alternating layers of radial and tangential lamellae intersecting at 90° to each other (not spiraling from one to another). The cilia fibrous layers are cleanly cleaved from tangential skins (composed by cilia-like lamellae), and radial lamellae are 90° perpendicular to the tangential skins. Each layer is composed of a tangential and a radial lamellae plate, together, these two perpendicularly intersected plates assume a dimension of 7-8 microns thick. Such internal analyses have also been discussed in a recent relevant study on PEA [17], and similar results have been shown. Here, further details of inter-lamellar intersection, yield views from a wider perspective.

Conclusion

By using several different polymers and compounds, organics or inorganic, low-molecular-weight or high-molecular-weight, this study has proven that they can be packed into concentric ring bands, usually circular, but other shapes, like hexagons or flower-like petals, can also be possible. Ring-banded spherulites in polymers can assume many types and inter-band spacing varies significantly, depending on T_c, polymer structures, film thickness, etc. Ring-banded spherulites of large spacing (20-50 μm) usually occur at higher T_c's (130°C for PLLA, 120°C for PDoT). These large-spacing ring-banded spherulites are influenced by the top glass cover; for high-M_w PLLA, the spherulites are ringless if samples are uncovered with glass plates, but ring-banded if samples are covered with glass plates, when both are crystallized at same T_c=130°C. On the other hand, the narrow-spacing ring-banded spherulites (spacing=5-10 μm) in PEA, PDoT, PLLA, PA etc, at lower T_c's (28°C for PEA, 80°C for PDoT, 115°C for low-M_w PLLA, 28°C for PA) are not influenced by the top glass cover on samples; they all develop ring-banded spherulites regardless of top glass cover or not when crystallized at the same designated T_c's for respectively compounds or polymers.

Obviously, there are three main traits for ring-banded pattern in compounds or polymers. (I) there are many types of ring bands (birefringent or not, large vs. small inter-ring spacing, etc.) in crystallized polymers; (II) ring banded spherulites can occur not only in polymers with chain-folded lamellae but also in low-M_w compounds with not chain folding at all; (III) top-glass cover on crystallizing samples can influence ring-banded spherulites of large spacing crystallized higher T_c's but no effects at all on narrow-spacing ring banded spherulites crystallized at lower T_c's of the same polymers. With these three traits, it is difficult to generalize the ring bands in spherulites to a single cause of lamellae twist/spiral; instead, exposure of lamellae underneath the top layers becomes essential for shedding new light on these intricately complex issues. The dissected cross sections show a corrugated-board structure with layers resembling a peel-able onion, where each radially oriented layer is sandwiched with a tangential layer of lamellae.

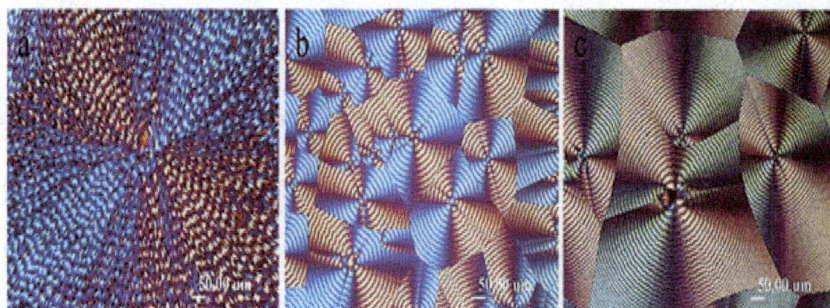

Figure 5: POM graphs for phthalic acid crystallized at (a) ambient, (b) 80°C, (c) 110°C, respectively.

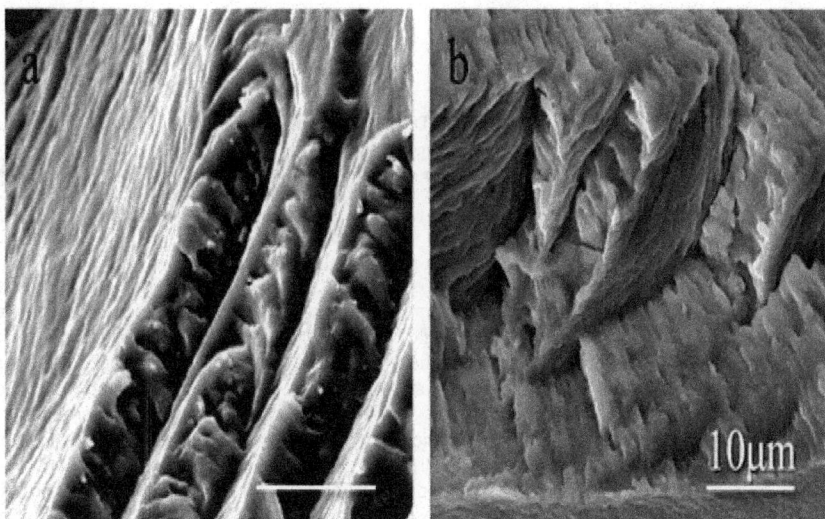

Figure 6: SEM graphs of the PEA bulk crystallized at T_c=28°C: (a) radial and tangential lamellae perpendicularly interacting at 90°; (b) tangential lamellae being cleaved cleanly from neighboring radial lamellae plates interacting at 90°.

Acknowledgment

This work has been financially supported by basic research grants (NSC 99-2221-E-006-014-MY3) in three consecutive years from Taiwan's National Science Council (NSC) – now Ministry of Science and Technology (MOST), to which the authors express their gratitude. This research was also partially supported by the Ministry of Education, Taiwan, R.O.C., in aim for the Top University Project to National Cheng Kung University (NCKU).

References

1. Shtukenberg AG, Freudenthal J, Kahr B (2010) Reversible twisting during helical hippuric acid crystal growth. J Am Chem Soc 132: 9341-9349.

2. Gunn E, Sours R, Benedict JB, Kaminsky W, Kahr B (2006) Mesoscale chiroptics of rhythmic precipitates. J Am Chem Soc 128: 14234-14235.

3. Shtukenberg A, Freundenthal J, Gunn E, Yu L, Kahr B (2011) Glass-crystal growth mode for testosterone propionate. Cryst Growth Des 11: 4458-4462.

4. Shtukenberg AG, Cui X, Freudenthal J, Gunn E, Camp E, et al. (2012) Twisted mannitol crystals establish homologous growth mechanisms for high-polymer and small-molecule ring-banded spherulites. J Am Chem Soc 134: 6354-6364.

5. Cui X, Rohl AL, Shtukenberg A, Kahr B (2013) Twisted aspirin crystals. J Am Chem Soc 135: 3395-3398.

6. Shtukenberg A, Gunn E, Gazzano M, Freudenthal J, Camp E, et al. (2011) Bernauer's bands. Chemphyschem 12: 1558-1571.

7. Imai H, Oaki Y (2010) Emergence of helical morphologies with crystals: twisted growth under diffusion-limited conditions and chirality control with molecular recognition. Cryst Eng Comm 12: 1679-1687.

8. Oaki Y, Imai H (2004) Amplification of chirality from molecules into morphology of crystals through molecular recognition. J Am Chem Soc 126: 9271-9275.

9. Chen HB, Chen L, Zhang Y, Zhang JJ, Wang YZ (2011) Morphology and interference color in spherulite of poly(trimethylene terephthalate) copolyester with bulky linking pendent group. Phys Chem Chem Phys 13: 11067-11075.

10. Wu PL, Woo EM (2003) Correlation between melting behavior and ringed spherulites in poly (trimethylene terephthalate). J Polym Sci Part B: Polym Phys 41: 80-93.

11. Wu PL, Woo EM, Liu HL (2004) Ring-banded spherulites in poly (pentamethylene terephthalate): a model of waving and spiraling lamellae. J Polym Sci Part B: Polym Phys 42: 4421-4432.

12. Wu PL, Woo EM (2004) Crystallization regime behavior of poly (pentamethylene terephthalate). J Polym Sci Part B: Polym Phys 42: 1265-1274.

13. Chen YF, Woo EM, Li SH (2008) Dual types of spherulites in poly(octamethylene terephthalate) confined in thin-film growth. Langmuir 24: 11880-11888.

14. Chen YF, Woo EM (2009) Annular Multi-Shelled Spherulites in Interiors of Bulk-Form Poly(nonamethylene terephthalate). Macromol Rapid Commun 30: 1911-1916.

15. Woo EM, Nurkhamidah S (2012) Surface nanopatterns of two types of banded spherulites in poly(nonamethylene terephthalate) thin films. J Phys Chem B 116: 5071-5079.

16. Meyer A, Yen KC, Li SH, Förster S, Woo EM (2010) Atomic-force and optical microscopy investigations on thin-film morphology of spherulites in melt-crystallized poly(ethylene adipate). Ind Eng Chem Res 49: 12084-12092.

17. Woo EM, Wang LY, Nurkhamidah S (2012) Crystal lamellae of mutually perpendicular orientations by dissecting onto interiors of poly (ethylene adipate) spherulites crystallized in bulk form. Macromolecules 45: 1375-1383.

18. Lugito G, Woo EM (2013) Lamellar assembly corresponding to transitions of positively to negatively birefringent spherulites in poly (ethylene adipate) with phenoxy. Colloid Polym Sci 291: 817-826.

19. Hsieh YT, Woo EM (2013) Microscopic lamellar assembly and birefringence patterns in poly (1,6-hexamethylene adipate) packed with or without amorphous poly(vinyl methyl ether). Ind Eng Chem Res 52: 3779-3786.

20. Hsieh YT, Ishige R, Higaki Y, Woo EM, Takahara A (2014) Microscopy and microbeam X-ray analyses in poly (3-hydroxybutyrate-co- 3-hydroxyvalerate) with amorphous poly (vinyl acetate). Polymer 55: 6906-6914.

21. Nurkhamidah S, Woo EM (2013) Unconventional nonbirefringent or birefringent concentric ring-banded spherulites in poly (L-lactic acid) thin films. Macromol Chem Phys 214: 673-680.

22. Nurkhamidah S, Woo EM, Tashiro K (2012) Optical birefringence patterns and corresponding lamellar alteration induced by solvent vapor on poly (L-lactic acid) diluted with poly (1,4-butylene adipate). Macromolecules 45: 7313-7316.

23. Lee LT, Woo EM, Hsieh YT (2012) Macro- and micro-lamellar assembly and mechanisms for unusual large-pitch banding in poly (L-lactic acid). Polymer 53: 5313-5319.

24. Lotz B, Cheng SZD (2005) A critical assessment of unbalanced surface stresses as the mechanical origin of twisting and scrolling of polymer crystals. Polymer 46: 577-610.

25. Bernauer F (1929) Gedrillte Kristalle. Gebroeder Borntraeger, Berlin, Germany.

15

Physical, Atomic and Thermal Properties of Biofield Treated Lithium Powder

Mahendra Kumar Trivedi[1], Rama Mohan Tallapragada[1], Alice Branton[1], Dahryn Trivedi[1], Gopal Nayak[1], Omprakash Latiyal[2] and Snehasis Jana[2]*

[1]*Trivedi Global Inc., 10624 S Eastern Avenue Suite A-969, Henderson, NV 89052, USA*
[2]*Trivedi Science Research Laboratory Pvt Ltd, Hall-A, Chinar Mega Mall, Chinar Fortune City, Hoshangabad Rd, Bhopal, Madhya Pradesh, India*

Abstract

Lithium has gained extensive attention in medical science due to mood stabilizing activity. The objective of the present study was to evaluate the impact of biofield treatment on physical, atomic, and thermal properties of lithium powder. The lithium powder was divided into two parts i.e., control and treatment. Control part was remained as untreated and treatment part received Mr. Trivedi's biofield treatment. Subsequently, control and treated lithium powder samples were characterized using X-ray diffraction (XRD), Differential scanning calorimetry (DSC), Thermogravimetric analysis-differential thermal analysis (TGA-DTA), Scanning electron microscopy (SEM) and Fourier transform infrared spectroscopy (FT-IR). XRD data showed that lattice parameter, unit cell volume, density, atomic weight, and nuclear charge per unit volume of lithium were altered after biofield treatment. The crystallite size of treated lithium was increased by 75% as compared to control. DSC analysis exhibited an increase in melting temperature of treated lithium powder upto 11.2% as compared to control. TGA-DTA analysis result showed that oxidation temperature, which found after melting point, was reduced upto 285.21°C in treated lithium as compared to control (358.96°C). Besides, SEM images of control and treated lithium samples showed the agglomerated micro particles. Moreover, FT-IR analysis data showed an alteration in absorption band (416→449 cm^{-1}) in treated lithium sample after biofield treatment as compared to control. Overall, data suggested that biofield treatment has significantly altered the physical, atomic, and thermal properties of lithium powder.

Keywords: Biofield treatment; Lithium; X-ray diffraction; Differential scanning calorimetry; Thermogravimetric analysis-differential thermal analysis; Scanning electron microscopy; Fourier transform infrared spectroscopy

Introduction

Lithium is highly reactive, light metal, which is commonly found in various foods such as grains, vegetables, mustard, kelp, and fish blue corn etc. Several lithium salts are used as mood stabilizing drugs, mainly in the treatment of bipolar disorder [1]. Lithium is primarily responsible to prevent mania and reduces the risk of suicide tendency in humans [2]. Overall, in placebo-controlled trials, lithium has been found useful as an adjunct medication for 45% of patients [3]. In addition, it is widely spread in central nervous system and interacts with many neurotransmitters and receptors, thus increasing serotonin synthesis [4]. Further, it is also reported that lithium ions (Li$^+$) can increase the release of serotonin or 5-hydroxy tryptamine by neurons in the brain [5]. Furthermore, the most commonly prescribed lithium salts include lithium carbonate (Li$_2$CO$_3$), lithium orotate (C$_5$H$_3$LiN$_2$O$_4$), and lithium citrate (Li$_3$C$_6$H$_5$O$_7$) for pharmacological treatment in mentally disordered patients [6,7]. Thus, by conceiving the usefulness of lithium in pharmaceutical industry, the present study was attempted to investigate an alternative way, which can modify the physical, atomic and thermal properties of lithium powder.

Harold Saton Burr had performed the detailed studies on the correlation of electric current with physiological process and concluded that every single process in the human body had an electrical significance [8]. Recently, it was discovered that all electrical process happening in body have strong relationship with magnetic field as mentioned by Ampere's law ($\oint B.dl = \mu_o I$) which states that the moving charge produces magnetic fields in surrounding space [9,10]. Thus, the human body emits the electromagnetic waves in form of bio-photons, which surrounds the body and it is commonly known as biofield. Therefore, the biofield consists of electromagnetic field, being generated by moving electrically charged particles (ions, cell, molecule etc.) inside the human body. Further, electrocardiography has been extensively

used to measure the biofield of human body [11]. Thus, human has the ability to harness the energy from environment or universe and can transmit into any living or non-living object(s) around the Globe. The objects always receive the energy and responding into useful way that is called biofield energy and the process is known as biofield treatment. Mr. Trivedi's unique biofield treatment (The Trivedi effect') has been known to transform the structural, physical and thermal properties of metals [12,13] and ceramics [14] in material science. In addition biofield treatment had improved the growth and production of agriculture crops [15-17], significantly altered the phenotypic characteristics of various pathogenic microbes [18,19], and altered the medicinal, growth and anatomical properties of ashwagandha [20].

Based on the excellent outcomes of biofield treatment, authors were interested to investigate the effect of biofield treatment on physical, atomic and thermal characteristics of lithium powder using X-ray diffraction (XRD), Differential Scanning Calorimetry (DSC), Thermogravimetric analysis-differential thermal analysis (TG-DTA), Scanning electron microscopy (SEM), and Fourier transform infrared spectroscopy (FT-IR).

Materials and Methods

The lithium powder was purchased from Alfa Aesar, USA. The sample was equally divided into two parts, considered as control and

*Corresponding author: Snehasis Jana, Trivedi Science Research Laboratory Pvt Ltd, Hall-A, Chinar Mega Mall, Chinar Fortune City, Hoshangabad Rd, Bhopal-462 026, Madhya Pradesh, India, E-mail: publication@trivedisrl.com

treatment. Control part was remained untreated and treatment group was subjected to Mr. Trivedi's biofield energy treatment.

Biofield energy treatment

The treatment sample was in sealed pack, handed over to Mr. Trivedi for biofield treatment under laboratory conditions. Mr. Trivedi provided the biofield treatment through his energy transmission process to the treated group without touching the sample. The control and treated samples were characterized using XRD, DSC, TGA-DTA, SEM, and FT-IR.

X-ray diffraction (XRD) study

XRD analysis of control and treated lithium powder was carried out on Phillips, Holland PW 1710 X-ray diffractometer system, which had a copper anode with nickel filter. The radiation of wavelength used by the XRD system was 1.54056 Å. The Kapton tapes were used to prevent the oxidation of the samples from air. The data obtained from this XRD were in the form of a chart of 2θ vs. intensity and a detailed table containing peak intensity counts, d value (Å), peak width (θ°), relative intensity (%) etc.

Additionally, PowderX software was used to calculate lattice parameter and unit cell volume of control and treated lithium powder samples. The crystallite size (G) was calculated by using Scherrer formula:

$G=k\lambda/(bCos\theta)$,

Here, λ is the wavelength of radiation used, b is full width half maximum (FWHM) and k is the equipment constant (0.94). Furthermore, the percent change in the lattice parameter was calculated using following equation:

$$\% \, change \, in \, lattice \, parameter = \frac{\left[A_{Treated} - A_{Control}\right]}{A_{Control}} \times 100$$

Where A $_{Control}$ and A $_{Treated}$ are the lattice parameter of treated and control samples respectively. Similarly, the percent change in all other parameters such as unit cell volume, density, atomic weight, and crystallite size were calculated.

Differential scanning calorimetry (DSC)

Differential Scanning Calorimeter (DSC) of Perkin Elmer/ Pyris-1, USA, with a heating rate of 10°C/min and nitrogen flow of 5 mL/min was used. The melting point and latent heat of fusion of control and treated lithium were recorded from their respective DSC curves. This system had accuracy of ± 0.2 K in the measurement of melting point.

The percent change in melting point was computed using following equations:

$$\% \, change \, in \, melting \, po \text{int} = \frac{\left[T_{Treated} - T_{Control}\right]}{T_{Control}} \times 100$$

Where, $T_{Control}$ and $T_{Treated}$ are the melting point of control and treated samples, respectively. Similarly, the percent change in the latent heat of fusion was computed.

Thermogravimetric analysis-differential thermal analysis (TG-DTA)

For TG-DTA analysis, Mettler Toledo simultaneous TG and Differential thermal analyser (DTA) was used. The samples were heated from room temperature to 400°C with a heating rate of 5°C/min under air atmosphere.

Scanning electron microscopy (SEM)

Surface morphology is the unique properties of lithium powder. Control and treated lithium samples were observed using JEOL JSM-6360 SEM instrument at 2000X magnification. In order to prevent the sample from oxidising, the environment holder and airlock system were used. With the help of these systems, the sample were prepared and mounted on environmental holder in a sealed glove box and kept in SEM for analysis. The differences in the tendency of the particles to clump were easily seen at the lower magnifications, while variations in size and morphology become clearer at higher magnification [21].

Fourier transform infrared spectroscopy (FT-IR)

FT-IR spectroscopic analysis was carried out to evaluate the impact of biofield treatment at atomic and molecular level like bond strength, stability, and rigidity of structure etc. FT-IR analysis of control and treated Lithium samples were performed on Shimadzu, Fourier transform infrared (FT-IR) spectrometer with frequency range of 300-4000 cm^{-1}.

Results and Discussion

X-ray diffraction (XRD) study

XRD diffractograms of control and treated lithium powders are shown in Figure 1. XRD patterns of control sample showed intense peaks at 2θ equal to 32.58°, 35.56°, 35.74°, 36.13°, 51.48°, 51.87°, 64.57° and 76.73°. However, crystalline peaks in treated lithium sample were observed at 2θ equal to 32.67°, 36.15°, 52.16°, 64.56°, 64.84° and 65.02°. The intense peaks were found in both control and treated samples indicated the crystalline nature of lithium powder. Furthermore, the peaks intensity at 2θ equal to 36.15° and 52.16° in treated samples were significantly reduced as compared to control. Whereas, the intensity of peak at 64.57° (control), which shifted to 64.84° (treated), was increased after biofield treatment. The intensity of the diffraction peaks are determined by the arrangement of atoms in the entire crystal and it sums the result of scattering from all atoms in the unit cell to form a diffraction peak (2θ) from the particular planes of atoms [22,23]. In addition, long range order of atoms along a plane shows higher intensity in XRD as compared to atoms with short range order. Thus, the alteration in intensity of XRD peaks in treated lithium powder as compared to control indicated that arrangement of atoms probably changed after biofield treatment. It is possible that atoms situated along the plane corresponding to 2θ equal to 36.15° and 52.16° may reorient themselves in another direction i.e., along plane attributed to 2θ equal to 64.57°, after biofield treatment. For further analysis, the XRD peaks were indexed with body centred cubic (BCC) crystal structure [24] and crystal structure parameters such as lattice constant, unit cell volume etc. were computed using PowderX software and results are presented in Table 1.

Data exhibited that lattice parameter and unit cell volume of treated lithium powder were reduced by 0.15 and 0.46%, respectively as compared to control. The reduction of lattice constant and unit cell volume indicated that a compressive strain might present in unit cell of treated lithium. It is assumed that biofield energy, which probably transferred through biofield treatment, might induce a compressive stress in treated sample. It is reported that high stress on lithium unit cell can change the crystal structure from BCC to face centred cubic (FCC) [25]. Previously, our group reported that biofield treatment had altered the unit cell volume in carbon allotropes [26]. Furthermore, the density and nuclear charge per unit volume of treated lithium powder were increased by 0.45 and 0.46%, respectively; however atomic weight was

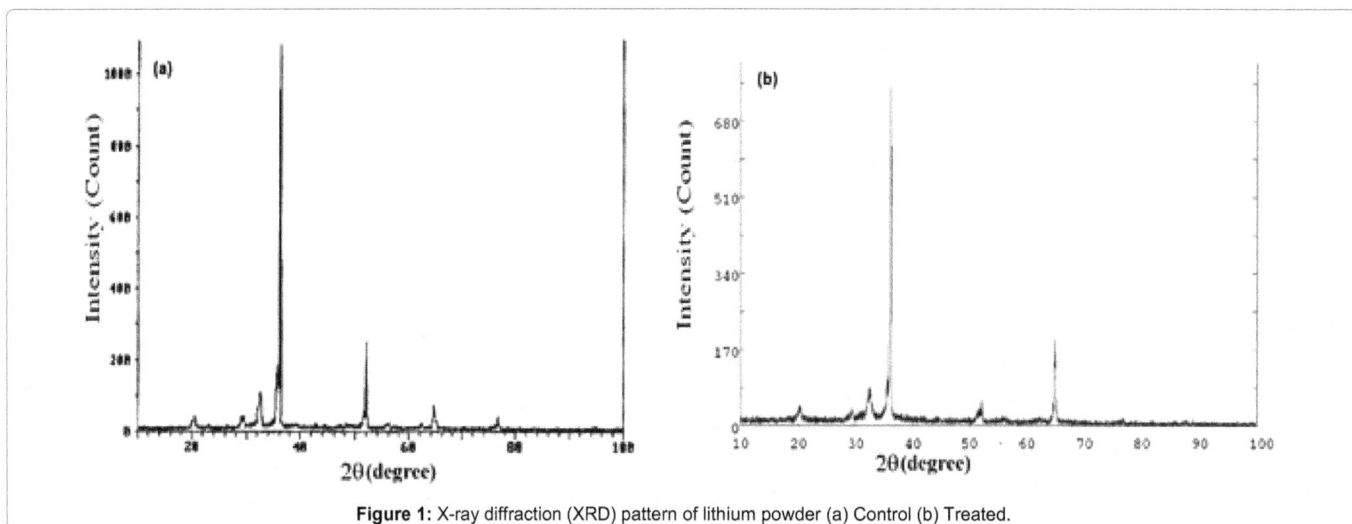

Figure 1: X-ray diffraction (XRD) pattern of lithium powder (a) Control (b) Treated.

Group	Lattice parameter (A)	Unit Cell volume (× 10⁻²³ cm³)	Density (g/cc)	Molecular weight (g/mol)	Nuclear charge per unit volume (C/m³)	Crystallite size (nm)
Control	3.52	4.36	0.537	7.06	21005	62.17
Treated	3.51	4.34	0.539	7.03	21102	108.80
Percent Change	-0.14	-0.46	0.46	-0.45	0.46	75.0

Table 1: X-ray diffraction (XRD) analysis result of control and treated lithium powder samples.

reduced (7.060→7.028) by 0.46% as compared to control. The increase in nuclear charge per unit volume indicated that nuclear strength of Li⁺ ions in treated lithium powder probably increased after biofield treatment. It is reported that Li⁺ plays an important role in central nervous system in releasing the serotonin from neurons [5,27,28]. Thus, it is assumed that serotonin releasing activity of Li⁺ in treated sample may be higher as compared to control. Besides, crystallite size (G), computed using Scherrer formula (G=kλ/bcosθ), are presented in Table 1. The crystallite size was increased from 62.17 nm (control) to 108.8 nm in treated lithium powder after biofield treatment. It indicated that crystallite size of lithium powder was significantly increased by 75% as compared to control, after biofield treatment. It is reported that crystallite size of metals and ceramics can be increased by increasing the temperature [29,30]. Recently, the increase in crystallite size in nickel and copper through biofield treatment had been reported by our group [31]. Thus, it is assumed that the energy transferred through biofield treatment probably initiated the movement of crystallite boundaries, which might lead to increase the crystallite size. Hence, XRD data revealed that biofield treatment has altered the physical and structural properties of lithium powder.

Differential scanning calorimetry (DSC)

Melting point and latent heat of fusion are the two key parameters for thermal analysis of metal powder. Fundamentally, melting point is related to the kinetic energy (thermal vibration) of atoms, whereas the potential energy is the energy required to overcome the interatomic interaction for phase change, which is related to latent heat of fusion (ΔH) [32]. The melting temperature and latent heat of fusion of control and treated lithium powder are presented in Table 2. The melting temperature of control lithium sample was found at 181.86°C which changed to 181.20°C, 202.21°C, and 200.34°C in treated samples i.e., T1, T2, and T3, respectively. It showed that melting temperature of treated lithium powder was increased by 11.2 and 10.2% in T2 and T3, respectively, though it was slightly decreased (0.36%) in T1, as

compared to control. Thus, the alteration of melting point was found in treated lithium powder indicated that the thermal vibrations of atoms probably changed after biofield treatment. The latent heat of fusion in control sample was 309.15 J/g, which changed to 42.41, 234.48, and 404.38 J/g in treated lithium T1, T2, and T3, respectively as compared to control. Recently, our group reported that biofield treatment had altered the melting point and ΔH in lead and tin powder [33]. In addition, the change in ΔH suggests that potential energy of treated lithium atoms possibly changed after biofield treatment. Thus, it is assumed that the biofield treatment probably transferred the energy to lithium powder and that might be responsible for alteration in kinetic and potential energy of treated atoms. Additionally, the increase in melting temperature in treated sample also suggests that interatomic interaction of treated lithium probably enhanced after biofield treatment. Furthermore, it is reported that Li⁺ interact with nitric oxide (NO) in CNS of human, which plays a crucial role in the neural plasticity [34,35]. The interaction of two atoms directly depends on their mobility and interatomic interaction of respective atoms [36]. Hence it is assumed that the alteration in interatomic interaction of treated lithium atoms may change the interaction of Li⁺ with NO and that can ultimately influence the mood stabilizing activity of lithium.

Thermogravimetric analysis-differential thermal analysis (TG-DTA)

Analysis result of TG-DTA is presented in Table 3. Data showed the exothermic peak at 358.96°C (control), which reduced to 305.42°C, 349.56°C, 285.21°C and 328.06°C in treated lithium samples i.e., T1, T2, T3, and T4, respectively. It could be due to oxidation of control and treated lithium powder samples. It indicated that oxidation temperature was reduced by 14.9, 2.61, 20.5, and 8.60% in treated lithium powder T1, T2, T3, and T4, respectively as compared to control. The reduction of oxidation temperature of treated samples as compared to control indicated that thermal stability of lithium powder probably decreased after biofield treatment. Therefore, based on DSC and TG-DTA data, it

Parameter	Control	T1	T2	T3
Melting Temperature (°C)	181.86	181.20	202.21	200.34
Percent change	-	-0.36	11.2	10.2
Latent heat of fusion, ΔH (J/g)	309.15	42.41	234.48	404.38
Percent change in ΔH	-	-86.3	-24.1	30.8

Table 2: Differential scanning calorimetry (DSC) analysis of control and treated of lithium powder samples.

Parameter	Control	T1	T2	T3	T4
Oxidation Temperature (°C)	358.96	305.42	349.56	285.21	328.06
Percent increase/ decrease		-14.9	-2.61	-20.5	-8.60

Table 3: Thermogravimetric analysis-differential thermal analysis (TG-DTA) of control and treated lithium powder samples.

is concluded that biofield treatment has altered the thermal behaviour of lithium powder.

Scanning electron microscopy (SEM)

SEM images of control and treated lithium powders are shown in Figure 2. It showed that powder particles were irregular and highly agglomerated in control and treated lithium powders. The SEM micrograph of control showed inter-particles and inter-agglomerated boundaries whereas treated sample showed the possible fracture and welding at the surface on the particles. Recently, our group had studied the effect of biofield treatment on antimony and bismuth powders using SEM, in which fractured surfaces were observed after treatment [37]. Thus, it is assumed that biofield treatment may induce the fracture in treated powder particles, which led to generate fresh surfaces. Further, these fresh surfaces welded together to form agglomerated powders. Therefore, SEM images revealed that biofield treatment has altered the surface morphology of lithium powder.

Fourier transform infrared spectroscopy (FT-IR)

The FT-IR spectrum serves as compound's fingerprint and provides specific information about chemical bonding and molecular structure. Thus FT-IR is more advanced and powerful analytical tool for characterization and identification of molecules. The FT-IR spectra of control and treated lithium powders are presented in Figure 3. In these spectra, the absorption band was observed at 3566 and 3674 cm^{-1} in control and treated lithium samples respectively, which were attributed to O-H stretching vibrations. Brooker et al. reported that the lithium compounds are highly air-sensitive so it can absorb the air and water easily [38]. Thus, it is possible that the lithium metal powder used in this experiment may absorb moistures from the environment. Due to which, the O-H bands were emerged in FT-IR spectra of control and treated samples. Furthermore, the absorption band found at 862, 1001, and 1446 cm^{-1} in control and 867, 1085, and 1446 cm^{-1} in treated sample were corresponding to bending, symmetric stretching, and asymmetric stretching vibrations of -CO$_3$ group. The emergence of -CO$_3$ band could be due to CO$_2$ absorption by samples. In addition, the absorption band corresponding to Li-O bond vibrations was observed at 416 cm^{-1} in control and it was shifted to 449 cm^{-1} in treated lithium sample. Simonov et al. reported the Li-O bond vibration at around 428 cm^{-1} in lithium containing compound [39]. Recently, our group reported that the alteration of absorption band in FT-IR spectra of zinc oxide, iron oxide, and copper oxide powders after biofield treatment [40]. Thus, based on this, it is assumed that biofield energy treatment might alter the bonding properties in lithium powder.

Conclusion

XRD data showed that biofield treatment results in reduction of unit cell volume and atomic weight by 0.46% as compared to control; however density and nuclear charge per unit volume were increased by 0.45 and 0.46%, respectively as compared to control. Based on the increase in nuclear charge per unit volume in treated lithium sample, it is assumed that nuclear strength of Li$^+$ ions might enhanced after biofield treatment. It may lead to increase the efficacy of Li$^+$ ions in human brain as mood stabilizer. Besides, the crystallite size was increased from 62.17 nm (control) to 108.8 nm in treated lithium powder. The melting point of treated lithium was increased upto 202.21°C as compared to control (181.86°C). Further, the change in melting point can be correlated with the change in interatomic interaction of treated lithium atoms after biofield treatment. It is assumed that the change in interatomic interaction may lead to alter the interaction of Li$^+$ ions with NO in CNS of human. In addition, TG-DTA study revealed that oxidation temperature of lithium was reduced upto 285.21°C as compared to control (358.96°C). SEM image of treated lithium sample showed the fractured and welded surface as compared to inter-particle and agglomerated boundaries in control. FT-IR result showed that, Li-O bond in treated sample (449 cm^{-1}) was altered as compared to control (416 cm^{-1}). Overall, data suggested that biofield treatment has altered the physical, atomic, and thermal properties of lithium powder. Therefore, it is assumed that biofield treated lithium powder could be more useful in mood stabilizer drug as compared to control.

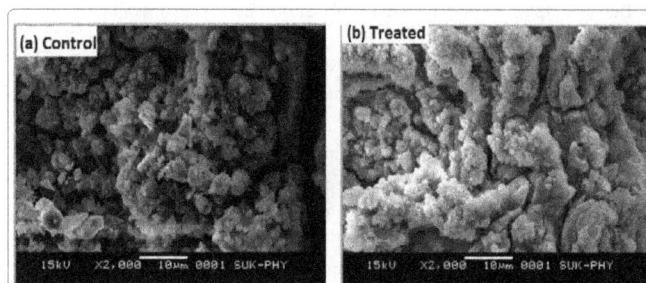

Figure 2: Scanning electron microscope (SEM) images of lithium powder.

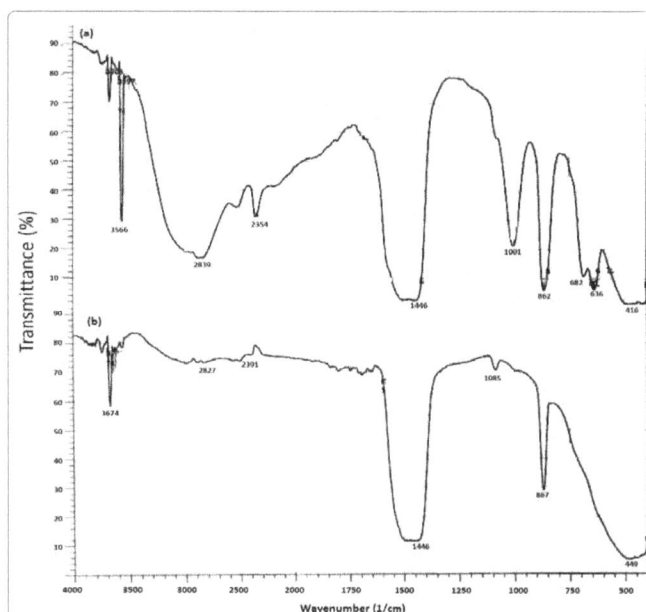

Figure 3: FT-IR spectrum of lithium powder (a) Control and (b) Treated.

Acknowledgements

Authors gratefully acknowledged to Dr. Cheng Dong of NLSC, Institute of Physics, and Chinese academy of Sciences for providing the facilities to use PowderX software for analyzing XRD data. Authors also would like to thank Trivedi Science, Trivedi master wellness and Trivedi testimonials for their support during the work.

References

1. Gelenberg AJ, Kane JM, Keller MB, Lavori P, Rosenbaum JF, et al. (1989) Comparison of standard and low serum levels of lithium for maintenance treatment of bipolar disorder. N Engl J Med 321: 1489-1493.

2. Baldessarini RJ, Tondo L, Davis P, Pompili M, Goodwin FK, et al. (2006) Decreased risk of suicides and attempts during long-term lithium treatment: a meta-analytic review. Bipolar Disord 8: 625-639.

3. Bauer M, Forsthoff A, Baethge C, Adli M, Berghöfer A, et al. (2003) Lithium augmentation therapy in refractory depression-update 2002. Eur Arch Psychiatry Clin Neurosci 253: 132-139.

4. Brunton L, Chabner B, Knollman B (2011) Goodman and Gilman's the pharmacological basis of therapeutics. 12th edn, McGraw-Hill Professional, New York, USA.

5. Massot O, Rousselle JC, Fillion MP, Januel D, Plantefol M, et al. (1999) 5-HT1B receptors: a novel target for lithium. Possible involvement in mood disorders. Neuropsychopharmacology 21: 530-541.

6. Nieper HA (1973) The clinical applications of lithium orotate. A two years study. Agressologie 14: 407-411.

7. Ebenezer IS (2015) Affective disorders 2: Bipolar disorder. In: Neuropsychopharmacology and Therapeutics. John Wiley & Sons Ltd, Chichester, UK.

8. Burr HS (1957) Bibliography of Harold Saxton Burr. Yale J Biol Med 30: 163-167.

9. Hammerschlag R, Jain S, Baldwin AL, Gronowicz G, Lutgendorf SK, et al. (2012) Biofield research: a roundtable discussion of scientific and methodological issues. J Altern Complement Med 18: 1081-1086.

10. Movaffaghi Z, Farsi M (2009) Biofield therapies: biophysical basis and biological regulations? Complement Ther Clin Pract 15: 35-37.

11. Rivera-Ruiz M, Cajavilca C, Varon J (2008) Einthoven's string galvanometer: the first electrocardiograph. Tex Heart Inst J 35: 174-178.

12. Trivedi MK, Patil S, Tallapragada RM (2012) Thought intervention through bio field changing metal powder characteristics experiments on powder characteristics at a PM plant. Future Control and Automation LNEE 173: 247-252.

13. Trivedi MK, Patil S, Tallapragada RM (2015) Effect of biofield treatment on the physical and thermal characteristics of aluminium powders. Ind Eng Manage 4: 151.

14. Trivedi MK, Patil S, Tallapragada RM (2013) Effect of biofield treatment on the physical and thermal characteristics of vanadium pentoxide powder. J Material Sci Eng S11: 001.

15. Shinde V, Sances F, Patil S, Spence A (2012) Impact of biofield treatment on growth and yield of lettuce and tomato. Aust J Basic & Appl Sci 6: 100-105.

16. Lenssen AW (2013) Biofield and fungicide seed treatment influences on soybean productivity, seed quality and weed community. Agricultural Journal 8: 138-143.

17. Sances F, Flora E, Patil S, Spence A, Shinde V (2013) Impact of biofield treatment on ginseng and organic blueberry yield. Agrivita J Agric Sci 35.

18. Trivedi MK, Patil S, Shettigar H, Gangwar M, Jana S (2015) Antimicrobial sensitivity pattern of Pseudomonas fluorescens after biofield treatment. J Infect Dis Ther 3: 222.

19. Trivedi MK, Patil S, Shettigar H, Bairwa K, Jana S (2015) Phenotypic and biotypic characterization of Klebsiella oxytoca: An impact of biofield treatment. J Microb Biochem Technol 7: 202-205.

20. Nayak G, Altekar N (2015) Effect of biofield treatment on plant growth and adaptation. J Environ Health Sci 1: 1-9.

21. Dercz G, Prusik K, Prusik L (2008) X-ray and SEM studies on zirconia powders. JMME 31: 408-414.

22. Chauhan A, Chauhan P (2014) Powder XRD technique and its applications in science and technology. J Anal Bioanal Tech 5: 212.

23. Speakman SA: Basics of X-Ray Powder Diffraction.

24. Olinger B, Shaner JW (1983) Lithium, compression and high-pressure structure. Science 219: 1071-1072.

25. Hanfland M, Syassen K, Christensen NE, Novikov DL (2000) New high-pressure phases of lithium. Nature 408: 174-178.

26. Trivedi MK, Tallapragada RM (2009) Effect of super consciousness external energy on atomic, crystalline and powder characteristics of carbon allotrope powders. Mater Res Innov 13: 473-480.

27. Yeste M, Alvira D, Verdaguer E, Tajes M, Folch J, et al. (2007) Evaluation of acute antiapoptotic effects of Li+ in neuronal cell cultures. J Neural Transm 114: 405-416.

28. Grafe P, Reddy MM, Emmert H, ten Bruggencate G (1983) Effects of lithium on electrical activity and potassium ion distribution in the vertebrate central nervous system. Brain Res 279: 65-76.

29. Rashad MM, El-Shaarawy MG, Shash NM, Maklad MH, Afifi AF (2015) Controlling the composition, microstructure, electrical and magnetic properties of LiFe5O8 powders synthesized by sol gel auto-combustion method using urea as a fuel. J Magn Magn Mat 374: 495-501.

30. Gaber A, Abdel-Rahim MA, Abdel-Latief AY, Abdel-Salam MN (2014) Influence of calcination temperature on the structure and porosity of nanocrystalline SnO2 synthesized by a conventional precipitation method. Int J Electrochem Sci 9: 81-95.

31. Trivedi MK, Tallapragada RM (2008) A transcendental to changing metal powder characteristics. Met Powder Rep 63: 22-28, 31.

32. Moore JW (2010) Chemistry: The molecular science. 4th edn, Brooks Cole.

33. Trivedi MK, Patil S, Tallapragada RM (2013) Effect of biofield treatment on the physical and thermal characteristics of silicon, tin and lead powders. J Material Sci Eng 2: 125.

34. Ghasemi M, Sadeghipour H, Mosleh A, Sadeghipour HR, Mani AR, et al. (2008) Nitric oxide involvement in the antidepressant-like effects of acute lithium administration in the mouse forced swimming test. Eur Neuropsychopharmacol 18: 323-332.

35. Ghasemi M, Sadeghipour H, Poorheidari G, Dehpour AR (2009) A role for nitrergic system in the antidepressant-like effects of chronic lithium treatment in the mouse forced swimming test. Behav Brain Res 200: 76-82.

36. Grunwald E, Chang KC, Leffler JE (1976) Effects of molecular mobility on reaction rates in liquid solutions. Annu Rev Phys Chem 27: 369-385.

37. Dhabade VV, Tallapragada RM, Trivedi MK (2009) Effect of external energy on atomic, crystalline and powder characteristics of antimony and bismuth powders. Bull Mater Sci 32: 471-479.

38. Brooker MH, Bates JB (1971) Raman and infrared spectral studies of anhydrous Li2CO3 and Na2CO3. J Chem Phys 54: 4788-4796.

39. Simonov AP, Shigorin DN, Tsareva GV, Talalaeva TV, Kocheshkov KA (1965) The infrared absorption spectra and structure of some simpler alkoxldes of lithium, sodium and potassium. J Appl Spectros 3: 398-403.

40. Trivedi MK, Nayak G, Patil S, Tallapragada RM, Latiyal O (2015) Studies of the atomic and crystalline characteristics of ceramic oxide nano powders after bio field treatment. Ind Eng Manage 4: 161.

16

Conceptual Design of a Separation Process for Higher Alcohols Made by Catalytic Condensation of Ethanol

Venkat K. Rajendran, Andreas Menne and Axel Kraft*

Fraunhofer Institute for Environmental, Safety and Energy Technology UMSICHT, Germany

Abstract

A downstream process for the separation of n-butanol from a product mixture containing unreacted ethanol, higher alcohols, aldehydes, water and traces of other chemical species was studied and therewith a conceptual design for the separation train has been devised. A novel approach and a newly developed catalyst were introduced to produce n-butanol (or iso-butanol) from ethanol as a raw material through an alternative path. The product stream from the reactor outlet consists of various chemical species ranging from saturated alcohol mixture, to aldehydes, to traces of aromatics and high boilers, and is ought to be separated into individual components based on their commercial/industrial applicability. Nine azeotropes of which one being ternary and the remaining eight binary azeotropes were identified between the various product components. Due to the chemical complexity, a multi-column downstream separation unit is needed therefore the schema containing several distillation units is likely to be energy intensive. The goal of this work was primarily to assess the technical and commercial feasibility of such separation technology; further process intensification however, is a subject for later studies.

Keywords: Ethanol; Butanol; Distillation; Higher alcohols; Biofuels; Chemicals

Introduction

World demand for fuels, chemicals, energy, materials and feedstock are increasing exponentially day-by-day. Existing technologies and production methods are failing to keep abreast with the current consumption pattern in avoiding any turbulence to the demand-supply chain and simultaneously dealing with environmental issues. Moreover, the global impact on climate change has led to stringent environmental regulations, which makes the operating standards quiet arduous. High oil prices, market concerns regarding the subject of its availability and an increasing demand from fast growing economies like Brazil, Russia, India and China (BRICs) have propelled the global community to a new juncture to address the security of energy and material supplies and eventually decentralize our dependency on traditional resources. Therefore, there is a collective sense of urgency to deal with this challenge engulfing the 21st century in order to secure our energy and materials interests through sustainable means.

Today, biofuels account to 3.5% of the world's transportation and the investments into biofuels exceeded production capacity of 4.5 billion litres worldwide in 2012, and is steeply rising [1]. According to IEA, biofuels can provide up to 27% of world transportation fuel by 2050 [2]. Biofuels can also contribute to the improved development of the agro-industry and generate many new jobs [3]. In addition, many commodity materials like plastics, solvents and plasticizers are becoming greener at growth rather exceeding 10% per year. The trade association for European bioplastics predicted global bioplastics production capacity will reach 1.7 million tonnes by 2015, more than doubling 2010 capacity [4]. This trend has created a whole new market, the so called "green economy". Nevertheless, unfortunately, even after the huge subsidies, incentives and promotional schemes raised by the governments worldwide, bio-based materials and fuels still have some significant setbacks. The main issues are the cost-competitiveness to fossil-based products, the difficulties in policy making with regard to energy, transport, agriculture and environment, the *food vs. fuel* debate

and finally the social perception on the genetically modified crops for biomass production. Although the bio based fuels and materials have their own disadvantages and cannot totally replace fossil products, its contribution to the energy demand and their transitional role in cushioning the foreseeable oil driven supply and demand crisis is highly significant.

In order to fully realize the potential of bio-based products, a *sustainable business model* is required apart from low-carbon technologies. Such a model must comprise a mixture of value added products from ideally more than one independent value chain and supply chain respectively. Embarking on such a strategy helps rendering the economic and ecological balance sheet towards positive numbers, liberated from fluctuations of supply and demand in the crude oil market.

Ethanol as Raw Material

In recent years, ethanol has become a commodity chemical, which will be being produced at a scale of about 90 million tonnes worldwide [5]. According to NOVOZYMES large scale production of second-generation "2G" bioethanol is only a few years away, clearly indicating that ethanol prices will stay moderate and its availability will strongly increase in the future. Application of heterogeneous catalysis for the exploitation of ethanol as raw material for producing value added products is of very high interest. Currently BRASKEM is in the planning phase of building large-scale ethylene units. The same holds

*****Corresponding author:** Axel Kraft, Fraunhofer Institute for Environmental, Safety and Energy Technology UMSICHT, Osterfelder Str. 3, D-46047 Oberhausen, Germany, E-mail: axel.kraft@umsicht.fraunhofer.de

true for the bulk chemical ethylene oxide according to constructing company Scientific Design [6].

It is also well established that ethanol and methanol can be converted to higher alcohols in a gas phase reaction. This pathway will be further elucidated in this contribution. Converting ethanol to higher alcohols, which sell for about double the price of ethanol is clearly attractive from a commercial point of view. In such a system, a multiple mixture of various alcohols, aldehydes and hydrocarbons are considered. In spite of the potential for catalytic condensation of ethanol to higher alcohols only very little is understood about the technology for achieving the product separation of such complex mixtures. The goal of this study consequently was to assess the technical and commercial feasibility of the separation technology.

Higher Alcohols as 2G-Biofuels and Bio-based Chemicals

Second-generation biofuel production processes use biomass from a variety of non-food crops. These include waste biomass, hays stacks, wood chips, straw, miscanthus etc. Furthermore, the production of cellulosic ethanol from lignocelluloses as a feedstock has gained more attention in recent times. Historically the transformation of these raw materials is discussed under Biomass-to-Liquid (BTL) processes. Unlike first generation feedstock 2G feedstock does not compete with food production, are more environment friendly and preferably require less arable land. In the renewable fuel context, higher alcohols are seen as biofuels where one refers to butanol, ethers of n-hexanol and n-octanol, but potentially also other higher alcohols might be considered to be 2G-Biofuels and bio-based chemicals.

Butanol as a potential automotive fuel

Biobutanol seems to emerge as a potential motor fuel of the future and could partially replace ethanol. Bio-butanol refers to *n-* or iso-butanol. Both can be produced by microbial fermentation [7], alike the production of ethanol from very similar raw materials. Bio-butanol however displays several advantages over ethanol [8] as listed below:

(i) Lower vapour pressure compared to that of ethanol; (ii) Energy density of butanol is close to gasoline; (iii) Butanol has octane value closer to gasoline; (iv) Solubility of water to butanol is lower than that of ethanol; (v) Butanol can easily be transported through existing pipelines and (vi) Can be used in existing IC engines in higher concentrations than ethanol

The worldwide fuel market for butanol is about 1.3 billion litres per year. n-butanol (or iso-butanol) is widely produced via the so-called "Oxo-process" from propylene and syngas [9]. About 80% of the production costs are linked to the raw material price of propylene, which is usually manufactured in parallel to ethylene by steam cracking of naphtha. Hence, the butanol pricing is closely tied up with the oil market, and one can observe frequent cost fluctuations.

Ethers of n-hexanol and n-octanol as cetane number enhancer

Addition of ethers of hexanol and octanol to diesel fuel results in significantly increased cetane numbers. The addition of such ethers, at concentrations of 2 and 5 percent by weight, results in cetane number well above 100 [10]. Ethers could replace toxic, unstable and explosive alcohol nitrates like ethyl-hexyl-nitrate and in particular nitrates of n-hexanol and n-octanol.

Higher alcohols as diesel fuel

Likewise neat alcohols can also serve as diesel fuel. Practically suitable cetane numbers in the 40s are displayed by n-octanol, whereas numbers around 80 are reported for n-tetradecanol [11]. However, in general relatively moderate cetane numbers and the high price make alcohols economically unattractive as fuels for now.

Applications of higher alcohols as chemicals

Table 1 shows existing applications of higher alcohols for commodity and specialty products. Own estimations show that about 20 million tons of derivatives of higher alcohols are produced annually. The market and applications for n- and iso-butanol are already very diverse [12]. Without claiming completeness for all existing and potential applications, it becomes obvious that a large market can potentially be exploited by replacing the fossil based oxo-process with a green alternative, once costs are already competitive or at least could become competitive.

Applications of higher aldehydes as chemicals

In case aldehydes as by products cannot be avoided, it is then important to consider potential commercial outlets for these compounds. Table 2 lists existing applications of higher aldehydes for commodity and specialty chemicals.

Catalytic Technologies for Synthesis of Higher Alcohols

Several materials have already been investigated and tested as catalyst for butanol-synthesis. Yang et al. [13] reported that zeolites, alumino-silicates modified by ion exchange with alkaline earth metals such as Lithium or Potassium are suitable catalysts. In addition, aluminium oxides which have been doped with the transition metals nickel, iron and cobalt, were used by Yang and Meng [14] as catalysts for butanol-synthesis. Based on the experimental results they suggested that the synthesis of butanol occurs by a direct condensation of two molecules of ethanol. According to this, the carbon-hydrogen-bond of an activated basic molecule reacts with another molecule by elimination of water. Ndou analysed the catalytic production of butanol from ethanol and used pure alkali or transition metals modified magnesium or aluminium oxide [15].

The oxidation from ethanol to acetaldehyde, followed by aldol condensation to beta methyl acrolein with subsequent hydrogenation is described as a further mechanism. However, Ndou verified that the predominant part of the butanol is synthesised by the mechanism of direct condensation as postulated by Yang and Meng. That is, the main reaction path goes over direct condensation as illustrated in Figure 1.

Tsuchida et al. [16-18] investigated the butanol synthesis by the use of hydroxylapatites, a mineral of the category of anhydrous phosphates, with the stochometric formula $Ca_{10}(PO_4)_6(OH)_2$. This mineral can also be synthesised non-stoichiometrically out of calcium nitrate and ammonium phosphate, whereas the relation of calcium to phosphor defines the allocated acid and basic active sites. Following butanol, also higher linear and alpha-branched alcohols were produced in significant amounts.

Also Cosimo [19], who analysed the impact and the explanations for branching during the condensation reaction to higher alcohols, affirms that the butanol-synthesis could not only proceed by direct condensation but also by alcohol condensation and in the presence of

Chemical Composition	Alcohol	Branching	Applications
Phosphoric acid ester	C4, C6	Optional	Hydraulic fluid, solvent, plasticiser, lubricant
Citric acid ester	>C4	Optional	Environmentally benign plasticiser
Phthalic acid ester	>C4	Optional	Plasticizer (mostly banned, e.g. DEHP)
Trimellitate ester	>C6	Yes	TOTM/TEHTEM – alternative DEHP
Maleic acid ester	C4	Optional	Plasticiser
Adipic acid	>C8	Optional	Plasticiser
Acetic acid ester	≥C4	No	Solvent for coatings
Butyric acid ester	≥C4	No	Solvent
Fatty acid ester	≥C4	Optional	Lubricant, wax, herbicide additive
Ethoxylates	>C4	Optional	Cosmetics, drilling fluid additive
Neat alcohols	≥C4	Optional	Solvent, coal floatation frothing agent
Nitric acid ester	≥C6	Yes	Cetane enhancer
Sugar ethers	>C6	Optional	Low and high foaming surfactant
Dialkylether	≥C4	Optional	Cosmetics, solvent, heat storage material
Guerbet alcohols	>C6	Optional	Cosmetics, fatty alcohols, surfactant
Potential new applications in the pipeline			
Lactic acid ester	>C4	Optional	Solvent
Dialkylether	>C4	Optional	Cetane enhancer
Neat alcohols	>C4	Optional	Intermediate for alkenes

Table 1: Commodity and specialty products from higher alcohols.

Chemical Name	Applications
Acetal	Intermediate, acetic acid, 2-ethyl-1-butanol (solvent)
Butanal	Intermediate for n-butanol, 2-ethyl-hexanol, 2-ethyl-1-butanol
Hexanal	Intermediate for hexanol
Octanal	Intermediate for n-octanol, fatty alcohols, surfactants

Table 2: Commodity products from by-product aldehydes.

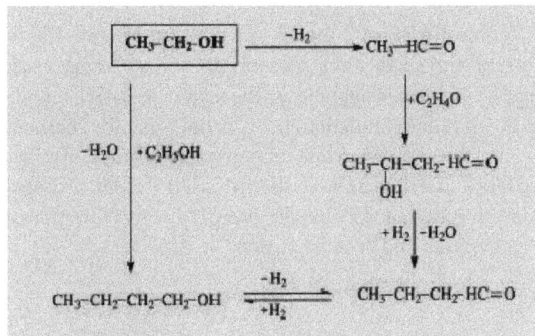

Figure 1: Reaction path of butanol synthesis.

acid and basic sites. Synthetic hydrotalcite, which are layered minerals out of magnesium and aluminium oxides that belong to the group of "Layered Double Hydroxides" and hold both acid and basic active sites, were employed as catalysts. Several patents have been filed by DuPont, using different types of hydrotalcite crystals [20].

The synthesis of higher alcohols out of a methanol-ethanol-mixture was analysed by Olson [21]. It turns out that activated carbon which is impregnated with magnesium oxide was active too. However, the manufactured impregnated carbons lost their catalytic activity within a very short time on the onset of the reaction.

As input for assessing the potential of catalytic condensation of higher alcohols and possibly also for the related aldehyde by-products with respect to downstream processing, the results from a continuously operated mini-plant have been used as an input for the calculation.

Details about the applied catalyst were published separately [22].

Downstream Processing Strategy for Higher Alcohols

Separations are the core to any physical, chemical or biochemical processes that requires purification, recovery or the elimination of certain components from the product mixture. Separation technologies play a vital role in terms of engineering as well as process economics. Separation systems constitute roughly 40-70% of both fixed and working capital. Distillation is the most widely used industrial separation process despite its low thermodynamic efficiency. It is estimated that around 40,000 columns operate in the U.S., handling 90-95% of all separations for product recovery and purification. The capital investment in distillation alone is at least $8 billion [23]. Despite this, distillation has a notable efficiency for large throughputs. Therefore, distillation has been chosen as the preferred method for product separation.

Structure and process model of the downstream units

The conceptual design work [24] was carried out in a hierarchical manner, starting from collection of thermodynamic data, understanding the VLE, LLE and VLLE behaviour and then setting up of a column sequence for the separation train using Residue Curve Maps. These data were used to establish an initial base case design and the subsequent simulation studies were carried out using Aspen Plus®. The best possible design after careful evaluations and sensitivity studies has been proposed as an outcome of this investigation. However, this publication will mainly focus on conceptual modelling and not on technical details.

An overview of the process model for the separation process for the purification of Biobutanol is illustrated in Figure 2. The chosen separation strategy for this work was to separate any low boiling aldehydes, i.e. acetaldehyde and butyraldehyde, from butanol upfront. Furthermore, it is necessary to separate ethanol as well, as unreacted ethanol will be recycled back to the reactor. All other higher alcohols are collected together as one fraction (Figure 3) for input/output flow structure. The chosen plant capacity for this study is 260,000 t/a which

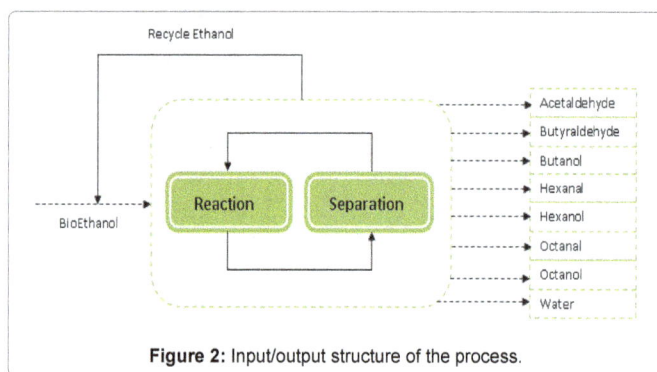

Figure 2: Input/output structure of the process.

breaks down to 129,000 t/a of bio-butanol and 131,000 t/a of other chemicals, mainly higher alcohols. Since the target products are future transportation fuels as well as chemicals, the specifications and purity of these fuels are not regulated so far. Therefore, the product purity for butanol was fixed to 95 mole percent in the present analysis. Aldehydes are considered to be technically pure.

The production unit is modelled to be operated for 8000 h per year. The reactor product stream which has been used as calculation basis for the following separation unit is shown in Table 3. A complex mixture of nine azeotropes, of which one being ternary and the remaining eight are binary azeotropes, are to be dealt with. Furthermore, the high water content of the product stream complicates the separation additionally.

The whole concept of distillation lingers around the boiling point of the chemical species. So, one need to be prudent while working with the vapor-liquid envelope which gives vital information regarding the feasibilities where a distillation column could probably operate. Complexities like azeotropes or regions of immiscibility act as a major hindrance for desired separation. i.e., an azeotrope creates a condition where equilibrium is reached in terms of separation with respect to boiling point. A detailed insight of the thermodynamics was necessary during the conceptual phase. One such graphical tool used was the *residue curve map*. These generated maps not only provide the separation possibilities and constraints of a ternary azeotropic system, but also help to validate the models used to predict equilibrium data.

Selection of a suitable thermodynamic model

In this work the thermodynamic data were primarily collected for physical properties and phase equilibria in terms of VLE, LLE and VLLE data, azeotropic points, distillation boundaries and finally residue curve maps (RCMs) for ternary systems. The gathered information has paved a generalised means to understand the underlying thermodynamics and also to visualise both the constraints as well as the feasibility for the separation train in a qualitative manner. Accurate phase equilibrium data of the component system to be separated is most important for the design, simulation and optimization of the separation process. Accuracy of a process simulation strongly depends on the chosen thermodynamic model. Therefore, a heuristic approach was employed to choose the appropriate property method [25] in order to establish a good base case scenario. Several factors like the nature of the components, the composition of the mixture, temperature and pressure range, availability of the binary interaction parameters etc., were carefully examined before setting up the simulation.

In this simulation UNIQUAC *[UNIversal QUAsi Chemical]* activity

coefficient model with binary parameters built-in Aspen Plus databank was used for the simulation after careful consideration of the available data and the literature. This choice was made due to the polar and non-electrolytic nature of the reaction products.

Column sequencing and composition profile mapping

A residue curve [26] represents the liquid composition against time as the result of a single stage batch distillation. The results, when plotted on the triangular graph are known as residue curves because the plot follows the liquid residue composition in the still. As shown in Figure 3 an example of an RCM, the ternary systems involving azeotropes contain different split regions and these regions are separated by distillation boundaries. These boundaries cannot be crossed by conventional distillation hence making certain splits extremely difficult if not impossible. However, various techniques have been devised for overcoming these distillation boundaries using pressure shifting, exploiting boundary curvature or volatility or kinetics using liquid-liquid decantation, membrane separation etc. respectively. Here, RCMs have been used extensively for composition profile mapping which in turn contributed a vital role in column sequencing. As an example the RCM for butanal, ethanol and water is depicted in Figure 3.

Connecting downstream units

Figure 4 presents the process flow sheet of the separation unit. The downstream processing unit consists of the distillation columns D101, D102 and D103 for the recovery of acetaldehyde and gaseous mixture, recovery of ethanol and the recovery of butanol respectively. A stripper S101, fed to a decanter, acts as an extraction unit for the recovery of butyraldehyde. All the columns were operated at atmospheric pressure.

The routine RADFRAC model was used to simulate the process conditions of the distillation columns it is a rigorous model for multistage vapour-liquid operations. The stage convention is numbered from top down in the simulation, reflux drum being the stage one. The operating variables such as reflux ratio, number of trays, distillate rate, feed tray location etc. were user specified. The design and operating variables were defined and eventually varied to observe its influence on energy requirement and product purity.

Design methodology for simulation

For a multicomponent mixture with single feed and two product outlets, there are $n+6$ degrees of freedom, n being the number of components for which the following variables are specified: (a) Feed flow rate, F; (b) Feed composition, Z_i; (c) Distillation composition, X_D; (d) Bottom composition, X_B; (e) Nature of the feed, q; (f) Temperature/Enthalpy, T_i/h_i; Reflux ratio, L/D; Boil-up ratio, V/B

Consequently, the following equations are solved for:

Overall material balance

$F=B+D$

Component balance

$Z_i F = X_{i,dist} D + X_{i,bot} B$

Energy Balance

$Q_R + Q_C + h_F F = h_D D + h_B B$

Qr/Qc: Heat duty of the reboiler and of the condensor

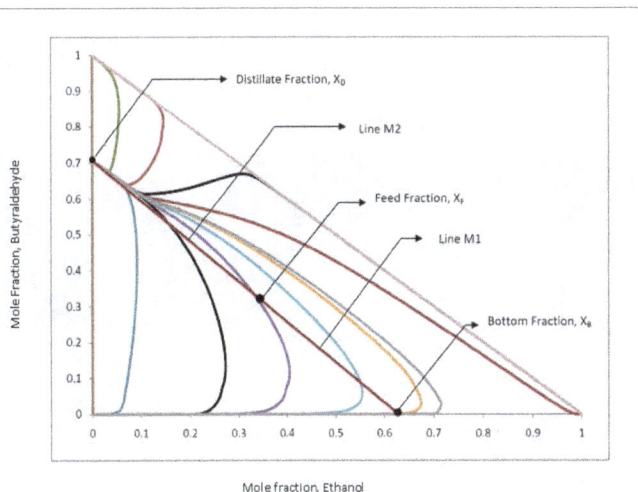

Figure 3: Residue curve map of butyraldehyde/ethanol/water.

With the increase in number of components the number of independent equations also increases, and so does the number of unknown variables. To solve this kind of problem a trial and error technique is employed, starting with an initial guess of the number of moles or mole fraction of one unknown component in the distillate or bottom fraction. The equations are then solved to convergence, and finally the solution is checked for conservation of mass and energy.

For a base case of 10 trays, the effect of feed tray location on the fractional recovery of the desired product was studied. The minimum reflux ratio, R_{min} for the separation and the minimum reboiler duty were thereby calculated. Now having known the minimum reflux ratio and reboiler duty, the effect of stripping stages, effect of rectifying stages and the effect of increase in reboiler duty were further calculated and optimized.

The design methodology can be explained by the following step-by-step algorithm:

a) Definition of a base case and fixing the product purity or distillate rate

b) Determination of the minimum reflux ratio

c) Studying the effect of feed tray location

d) Studying the effect of increase in rectifying stages

e) Studying the effect of increase in stripping stages

f) Studying the effect of increase in reboiler duty

Process simulation of the downstream separation unit

Distillation unit D 1001: The first step in separation process is the separation of the gaseous components and acetaldehydes from the reactor exit stream (Figure 5). The feed component mixture, stream *S01*, enters the distillation column *D-101* at a feed inlet temperature of 40°C. The reason for entering the column at very low temperature is merely to condense any gases present and ease the separation; moreover, since the boiling point of acetaldehyde is 21.06°C, the column is operated at these conditions. Since there are no azeotropes in the distillate, all the acetaldehyde and the trace gases are completely recovered as the top product, stream *S02*. Therefore, the distillate rate of this column is 39

kmol/hr (Table 4), for further specifications.

Distillation unit D102: The liquid stream from the bottom product of column 1 enters into the stripping column *D-102*, with a flow rate of 961 kmol/hr. The main function of this column is to separate butyraldehyde from the resulting mixture (Figure 6). Butyraldehyde forms binary low boiling heterogeneous azeotrope with water at 69°C and with ethanol at 73.27°C. At the same time, Ethanol forms an azeotrope with water at 78.17°C. This complex mixture makes the separation extremely difficult. The stream S03 is therefore fed into the column at a temperature of 82.5°C, which is the exit temperature of stream S03 from *D-101* andthe distillate rate of the column, is set quiet high to recover maximum water along with all the butyraldehyde as the top product. This explains the reason to limit the flow of water into other columns, which not only dilutes the stream but also complicates the separation process by the formation of azeotropes. The overhead product of the stream S05 should now contain no or only marginal traces of ethanol. All the remaining components are driven down the column to be collected as a bottom product through stream S09 (Table 5) for feed flows and product composition.

As seen from the RCM analysis, the butyraldehyde-water heterogeneous azeotrope consists on a molar basis of 0.71 butyraldehyde and 0.29 water. The mixture's heterogeneity it can be taken to one's advantage by using a simple decanter, *DECANT*. The distillate rate is initially set for 28.16 kmol/hr which is the azeotropic composition required to recover all the butyraldehyde. This top product is cooled down to a temperature of 30°C and sent into the decantation unit to be separated into organic and the aqueous phases. The organic phase which is completely rich in butanal is further sent into a feed splitter where a part of butanal is fed back as recycle into the column along with a make-up stream containing additional butyraldehyde to recover additional water.

Distillation unit D-103: The bottom product from the column *D-102* enters as a feed to column *D-103*. The feed stream S09 is at a temperature of 86.4°C. The task of this column is to completely recover all the ethanol along with traces of butanal (if any) from the remaining stream (Figure 7). The idea here was to ideally recover only ethanol and water, which could directly be fed back to the reaction unit as the unreacted stream. However, since ethanol also forms a ternary azeotrope with water and hexanal at 82.16°C. If a mixture of ethanol-water alone is desired to be separated without any hexanal as distillate, unavoidably high water content has to be tolerated in the bottom product. This complicates the process and increases the operation costs. The reason to this is, that hexanal also forms a binary azeotrope with water at 79.13°C which necessitates the requirement of two more columns for separation, one where hexanal-ethanol azeotrope is obtained as the distillate and another for separating butanol from the remaining stream. In order to avoid more complex separations at further downstream, high operating costs have to be tolerated in this unit. Refer to Table 6 for feed flows and product composition.

Distillation unit D-104: The bottom product from the column *D-103* enters as a feed to final column *D-104*. The feed stream S11 is at a temperature of 112.4°C. The column is used for the recovery of the desired commercial product butanol (Figure 8). Butanol forms a heterogeneous azeotrope with water at 92.5°C at a composition of 0.25 moles. Since the distillate composition in this column is azeotropic, in order to recover pure butanol one may invest in techniques such

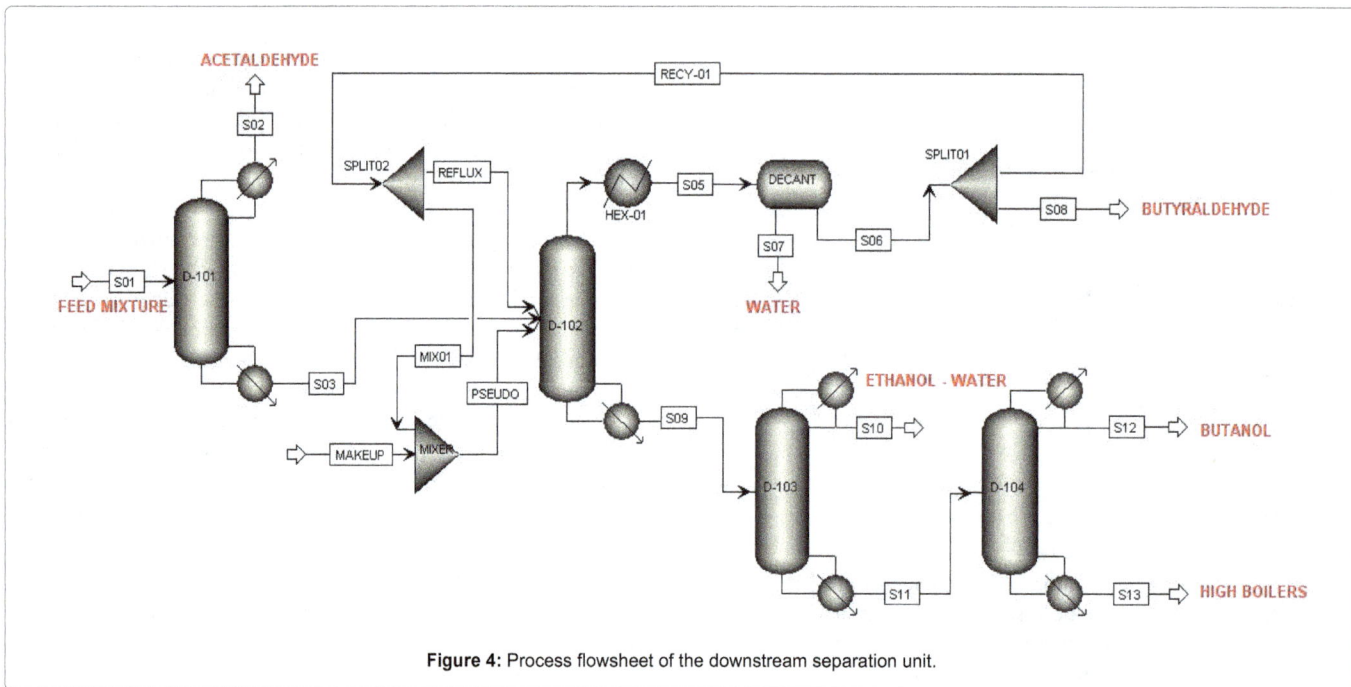

Figure 4: Process flowsheet of the downstream separation unit.

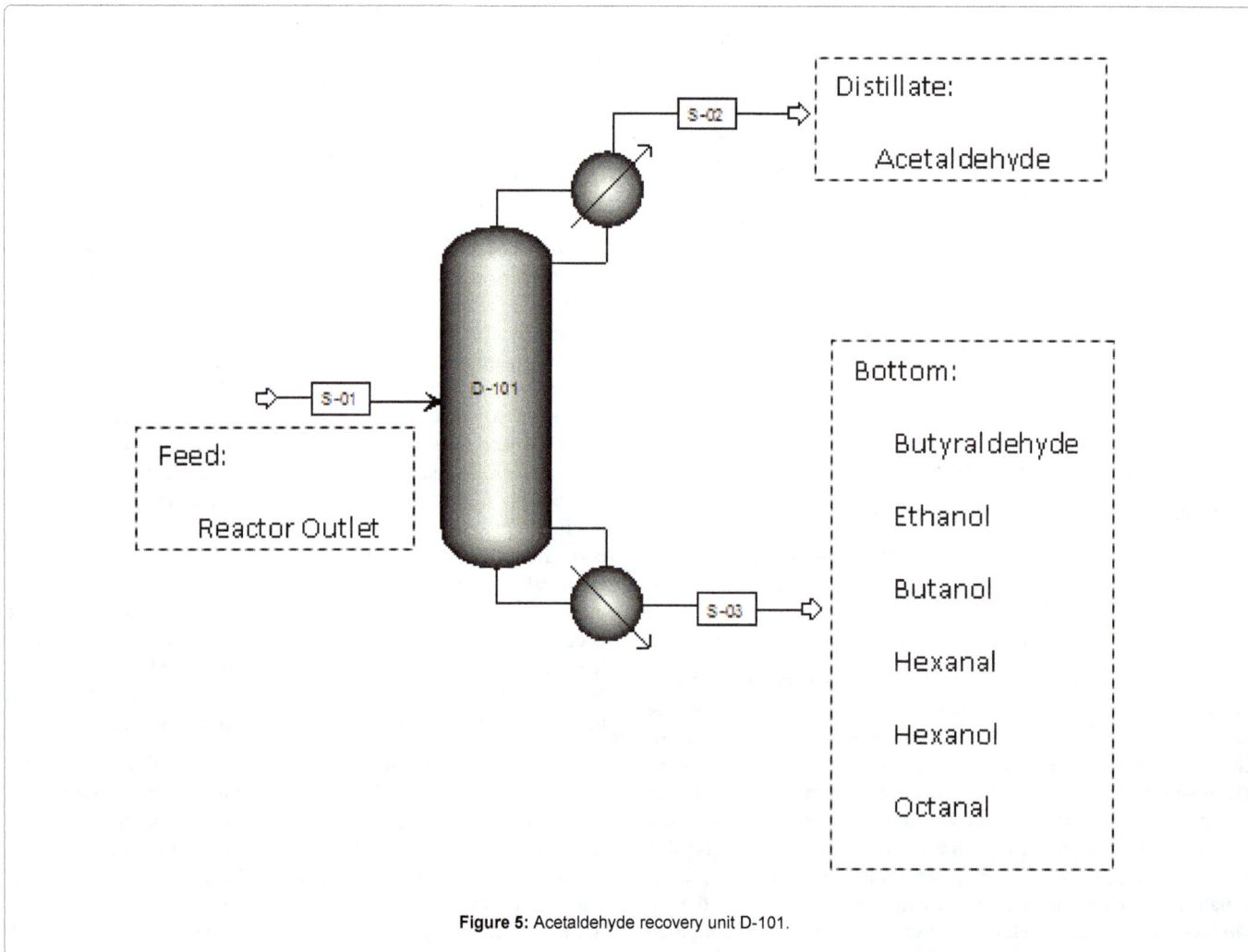

Figure 5: Acetaldehyde recovery unit D-101.

Component	Mole flow, kmol/hr
Acetaldehyde	39
Ethanol	421
Butyraldehyde	20
Butanol	114
Hexanal	5
Hexanal	19
Octanal	1
Octanol	4
Water	377

Table 3: Product stream from reactor outlet.

Unit: D-101	
Number of trays	40
Feed tray location	19
Feed temperature, °C	40
Distillate rate, kmol/hr	39
Product recovery, kmol/hr	39
Heat duty, MW	5
Reboiler type	Kettle
Condenser type	Partial condenser

Table 4: Specifications: unit D-101.

Unit: D-102	
Number of trays	60
Feed tray location	30
Feed temperature, °C	82.1
Distillate rate, kmol/hr	290
Product recovery, kmol/hr	184.5
Heat duty, MW	20
Reboiler type	Kettle
Condenser type	None

Table 5: Specifications: unit D-102.

Unit: D-103	
Number of trays	55
Feed tray location	27
Feed temperature, °C	82.7
Distillate rate, kmol/hr	620
Product recovery, kmol/hr	420.9
Heat duty, MW	13
Reboiler type	Kettle
Condenser type	Total condenser

Table 6: Specifications: unit D-103.

Unit: D-104	
Number of trays	40
Feed tray location	20
Feed temperature, °C	98
Distillate rate, kmol/hr	184
Product recovery, kmol/hr	113.97
Heat duty, MW	13
Reboiler type	Kettle
Condenser type	Total condenser

Table 7: Specifications: unit D-104.

as pervaporation. Refer to Table 7 for feed flow rates and product composition.

Conclusion

This study describes a preliminary feasibility study of a conceptual downstream processing plant manufacturing alcohols and aldehydes. Based on the experimental data conducted by Fraunhofer UMSICHT, the results with high compositions of alcohols, water and aldehydes were taken as a basis for this study. A generic design to handle complex multicomponent separation was developed. Therefore, the proposed design makes it highly flexible to produce both fuels and high value chemicals like.

It was found that the separation scheme is energy intensive involving four distillation columns. The main complexity in designing the separation unit was with the amount of water present in the feed entering in to the downstream section. The reason to this is that the units are all operated at atmospheric pressure and the top product compositions in the columns D102 and D103 equals their azeotropic composition, thereby carrying a large amount of water as a bottom product into the next successive column. Moreover, this necessitates the increase in the number of columns. In order to deal with this, a stripper followed by a condenser and a decanter has been designed to extract the maximum amount of water by recycling a split fraction of recovered butyraldehyde back into the column. The decanter was used to phase separate butyraldehyde-water azeotrope, which is heterogeneous in nature. In the Column D103, the design to recover ethanol-water, at its azeotropic compositions was found to be the best option, as it eliminates the requirement of an additional column. This distillate obtained will to be recycled back into the reactor as unreacted ethanol.

In order to make higher alcohols according to the proposed processing route feasible, further R&D is required. The following list outlines some important topics to be addressed in the future:

•   Alternative technologies for separation of water in terms of energy consumption, in particular evaporation should be investigated. This is already the state-of-the art for separation of ethanol and water.

•   Alternative technologies for water separation could include esterification or etherification of the crude products mix with acids, since up-front water separation would not be required or strongly reduced. Significant energy savings can be expected, and the number of overall unit operations would be reduced.

•   The commercial value of aldehydes versus alcohols should be evaluated. In particular the economics of hydrogenating aldehydes into alcohols versus separation of aldehydes, either stand-alone or coupled to an existing oxo-process, should be scrutinized.

•   If acetaldehyde cannot be avoided, its conversion to acetic acid via known technology, ideally coupled with esterification with the higher alcohols, should be considered.

•   The optimal product mix of chemicals versus fuel, if any, should be carefully evaluated to maximize economic returns and to minimize economical risks.

•   Life-cycle analysis tools would provide helpful to sort out the greenest processing window.

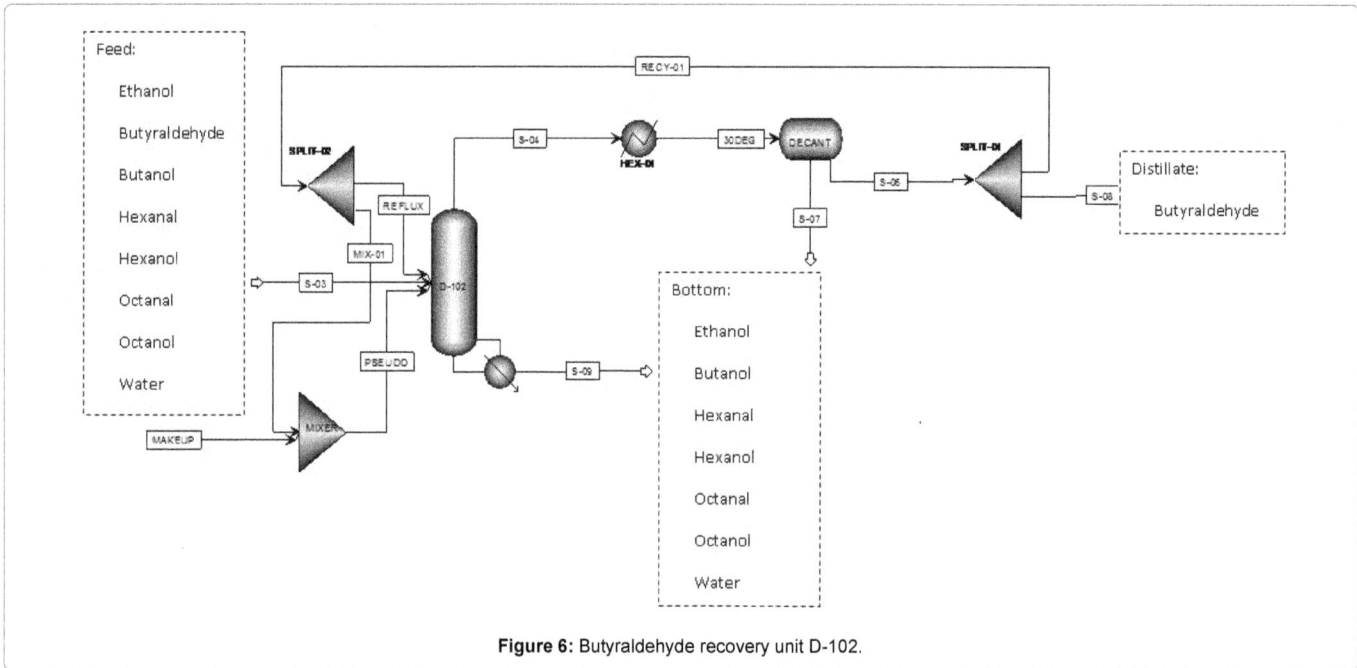

Figure 6: Butyraldehyde recovery unit D-102.

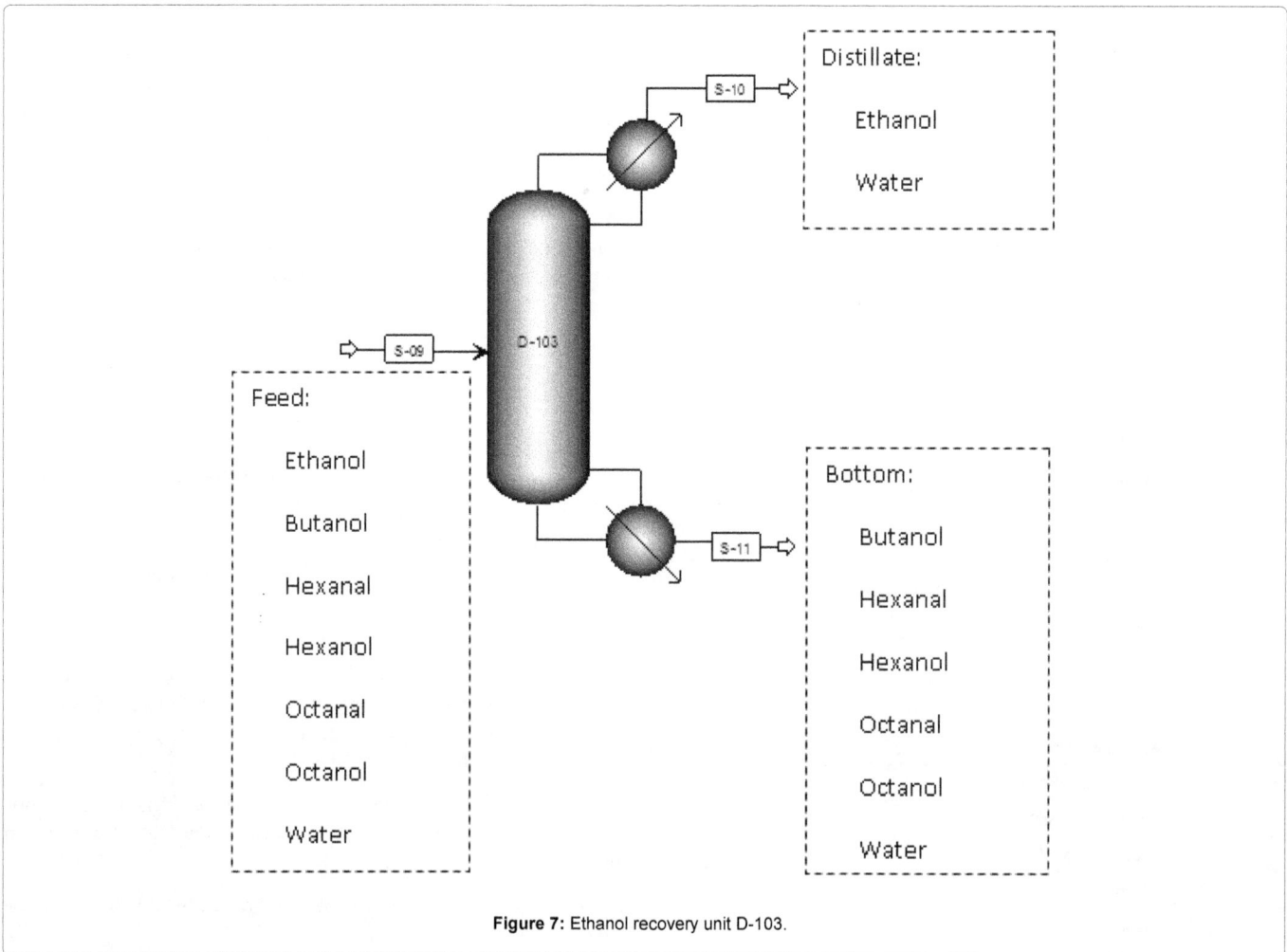

Figure 7: Ethanol recovery unit D-103.

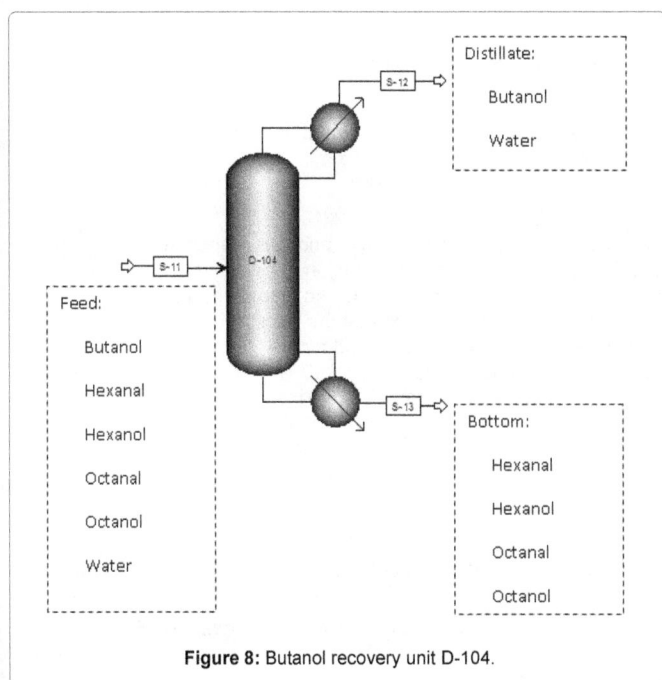

Figure 8: Butanol recovery unit D-104.

- Condensation of methanol and ethanol mixtures to branched alcohols is possible with the same technology. Opportunities arising from such a process should be evaluated.

Considering the existing market of higher alcohols and the broad mixture of existing applications of higher alcohols and derivatives thereof, it becomes conceivable that such products can become an important part of existing refineries or even chemical production sites without refineries, to introduce bio-based products into the portfolio in an affordable way. Ethanol as raw material could drastically lower the complexity of chemical production sites allowing further affordable and distributed back-ward integration, since capital intense unit operations of a refinery are not required anymore.

References

1. British Petroleum (2012) BP Statistical review of world energy.

2. IEA (2011) International Energy Agency - Roadmap.

3. UNEP (2011) Biofuel Vital Graphics - Powering a Green Economy.

4. Compounding world (2012) Why bioplastics are here to stay.

5. Wang Y, Sun J (2014) Recent Advances in Catalytic Conversion of Ethanol to Chemicals. Appl Catal 4: 1078-1090.

6. Braskem (2015) Braskem finishes the ethanol drying unit.

7. Duerre P (2007) Biobutanol: An attractive biofuel. Biotechnol J 2: 1525-1534.

8. Du Pont (2015) Butanol factsheet.

9. Falbe J, Bahrmann H, Lipps W, Mayer D (2000) Alcohols, Aliphatics. Ullmann's Encyclopedia of Industrial Chemistry.

10. Olah GA (1996) Cleaner burning and Cetane enhancing diesel fuel supplements. Patent US 5520710 A.

11. Mascal M (2012) Chemicals from Biobutanol: Technologies and Markets. Biofuels, Bioproducts and Biorefining.

12. Freedman B, Bagby MO (1990) Predicting Cetane Numbers of n-Alcohols and Methylesters from their physical properties. Journal of American Oil and Chemical Society 67: 565-571.

13. Yang KW, Jiang XZ, Zhang WC (2004) One-Step Synthesis of n-Butanol from Ethanol Condensation over Alumina-supported Metal Catalysts. Chin Chem Lett 15: 1497-1500.

14. Yang C, Meng ZY (1993) Bimolecular Condensation of Ethanol to 1- Butanol catalysed by Alkali Cation Zeolites. J Catal 142: 37-44.

15. Ndou AS, Pint N, Coville NJ (2003) Dimerisation of ethanol to butanol over solid-base catalysts. Appl Catal A Gen 251: 337-345.

16. Tsuchida T, Sakuma S, Takeguchi T, Ueda W (2006) Direct Synthesis of n-Butanol over non-stoichiometric hydroxyl-apatite. Ind Eng Chem Res 45: 8634-8642.

17. Tsuchida T, Kuboa J, Yoshioka T, Sakumaa S, Takeguchib T (2008) Reaction of ethanol over hydroxyapatite affected by Ca/P ratio of catalyst. J Catal 259: 183-189.

18. Tsuchida T, Sakuma S (2007) Method of synthesizing higher- molecular alcohol. EU Patent application: EP 1 829 851 A1.

19. Di Cosimo, Apesteguıaa CR, Ginésb MJL, Iglesiab E (2000) Structural requirements and reaction pathways in cond-ensation reactions of alcohols on MgyAlOx catalysts. J Catal 190: 261- 275.

20. Kourtakis K, Ozer R, and D'Amore MB (2010) Process for producing Guerbert alcohols using water tolerant basic catalysts. US Patent 20100160693 A1.

21. Breitkreuz K, Menne A, Kraft A (2014) New Process for sustainable fuels and chemicals from bio-based alcohols and acetone. Biofuels, Bioproducts & Biorefining 8: 504-515.

22. Olson E, Sharma RK, Aulich TR (2004) Higher Alcohols Biorefinery - Improvement of Catalysts for Ethanol Conversion. Appl Biochem Biotechnol 113-116: 913-932.

23. Humphrey JL (1995) Separation Processes: Playing a critical role. Chem Eng Prog 91.

24. Rajendran VK (2010) Thesis - Conceptual design for the downstream process synthesis of higher alcohols and aldehydes from ethanol. Fraunhofer Institute UMSICHT and Max-Planck-Institute Magdeburg.

25. Choosing Thermodynamic property models in Aspen Plus - User Guide.

26. Villiers WE, French RN, Koplos GJ (2002) Navigating phase equilibria via residue curve maps. Chem Eng Prog, pp. 66-71.

17

Single Gas Permeation on γ-Alumina Ceramic Support

Mohammed Nasir Kajama*, Habiba Shehu, Edidiong Okon and Ify Orakwe

Centre for Process Integration and Membrane Technology (CPIMT), School of Engineering, The Robert Gordon University, Aberdeen, AB10 7GJ, United Kingdom

Abstract

This study examines the characterization (SEM-EDXA observation, BET measurement) and gas transport through a commercial tubular alumina mesoporous (20 and 500 Å) support. Single gas permeation of helium (He), hydrogen (H_2), nitrogen (N_2) and carbon dioxide (CO_2) was measured at a temperature of 450°C and feed pressures between 0.85 up to 1.0 bar. Observation of the permeance of the alumina support revealed that the transport of the gases under these conditions is governed by Knudsen diffusion. Selectivity of 2.7 was obtained for He/N_2 at 1 bar. The selectivity obtained is comparable to the theoretical Knudsen value (2.65) for He/N_2.

Keywords: Alumina support; Gas permeation; Gas selectivity; Knudsen diffusion

Introduction

Applications of inorganic membranes continue to receive tremendous growth in the last three decades due to their improved permeances and thermal stabilities. In many cases, membrane technology has been efficiently applied in the industrial sectors in order to replace the conventional energy-demanding as well as environmentally polluting separation systems [1]. Inorganic membranes have been successfully applied for carbon dioxide separation [2], hydrogen separation [3], high-purity water and recovery of toxic or valuable components from industrial effluents [4]. Membranes can be classified into organic and inorganic systems. The organic ones are further divided into biological and polymeric constituents, while the inorganic membranes can be divided into metallic and ceramic (porous and non-porous) membranes [3]. In recent time, membranes were fabricated from polymeric materials but these are exposed to chemical attack and cannot withstand high temperature. It is for these reasons that inorganic membrane technology is receiving increasing attention. Inorganic membranes are commonly made from ceramic, metal oxide or sintered metal, palladium metal, zeolite among others [3].

Gas transport through porous ceramic membranes depends on pore diameter [5]. According to the International Union of Pure and Applied Chemistry (IUPAC) definition; macro pores with pore diameter >500 Å, where basically viscous flow and Knudsen diffusion occur; mesopores with pore diameter between 20 and 500 Å, where basically Knudsen diffusion is the dominant; and micro pores with pore diameter <20 Å, where molecular sieving is expected [5]. The so-called Knudsen number is used to differentiate between viscous and Knudsen flow which is written as [5,6]:

$$K_n = \frac{\lambda}{d_p} \qquad (1)$$

Where, λ is the mean free path of gas molecules, and d_p is the pore diameter.

Basically, the mean free path is the average distance travelled by the molecule between collisions. Therefore, mean free path is expressed as [6]:

$$\lambda = \frac{RT}{\sqrt{2}\pi d^2 N_A P} \qquad (2)$$

Where, R is the gas constant (8.314 J.K⁻¹.mol⁻¹), T is the temperature (K), d is the diameter (m), N_A is the Avogadro's number (mol), and P is the pressure (Pa).

Viscous flow is determined if the mean free path is smaller than the pore diameter, the flow characteristics are determined primarily by collisions among the molecules and can be written as [3,5,6]:

$$P_v = (\frac{\varepsilon r^2 P_{av}\Delta P}{8\pi\mu RTL}) \qquad (3)$$

Where, P_v is the viscous permeance (mol m⁻² s⁻¹ Pa⁻¹), ε is the porosity of the membrane, r_p is the mean pore radius (m), $P_{av}=P_1+P_2/2$ is the average pressure (Pa), ΔP is the pressure drop across the membrane (Pa), μ is the viscosity (Pas), and L is the thickness of the membrane (m).

Knudsen diffusion occurs if the mean free path is effectively larger than the pore diameter. The separation is based on molecular weight [3,5,6]. Thus, Knudsen permeance states that the permeation flux is proportional to the inverse square root of the molecular weight of the gas and temperature which can subsequently be written as [6]:

$$P_{kn} = \frac{\varepsilon 8 r_p \Delta P}{3\tau L(2\pi RTM)^{0.5}} \qquad (4)$$

Where, P_{kn} is the Knudsen permeance (mol m⁻² s⁻¹ Pa⁻¹), τ is the tortuosity and M is the molecular weight of the diffusing gas (g/mol).

However, if the mean free path of the gas molecule is equal to the pore diameter, then; the flow mechanism is governed by the combination of both mechanisms (i.e., Equations 3 and 4) which is written as:

$$P_t = \frac{\varepsilon}{\tau L}[\frac{r_p^2(p_{av})\Delta P}{8\mu RT} + \frac{8r_p\Delta P}{3(2\pi RTM)^0}] \qquad (5)$$

Where P_t is the total permeance (mol m⁻² s⁻¹ Pa⁻¹).

In this study, a commercially γ-alumina support was characterized and single gas permeation and selectivity was carried out at feed pressures between 0.85 up to 1.0 bar and a temperature of 450°C in order to elucidate their respective gas transport mechanism.

Corresponding author: Mohammed Nasir Kajama, Centre for Process Integration and Membrane Technology (CPIMT), School of Engineering, The Robert Gordon University, Aberdeen, AB10 7GJ, United Kingdom
E-mail: m.n.kajama@rgu.ac.uk

Experimental

A commercial tubular gamma alumina support supplied by Ceramiques Techniques et Industrielles (CTI SA) France was employed in this study. The gamma alumina support was mesoporous (20 and 500 Å) consisting of 7 and 10 mm internal and outer diameter respectively. The symmetric alumina support consisted of a permeable length of 348 mm (Figure 1) and a porosity of 45%. The gases used were helium (He), hydrogen (H_2), nitrogen (N_2) and carbon dioxide (CO_2). Table 1 shows the detailed characteristics of the gases.

The experimental set-up consisted of a membrane reactor, gas delivery system for pure gases, a permeate and retentate exit, a flow meter and K-type thermocouples fixed on the reactor (Figure 2). However, prior to permeation experiments the reactor and all connections were tested for leaks by means of a soap solution. The permeation tests involved passing the gas into the shell-side and directed to permeate across the alumina support at different pressures and a temperature of 450°C. The shell is made from stainless steel material and has 28 mm I.D., 36 mm O.D., 395 mm long, 5 mm thick that can withstand high temperatures. The stainless steel shell was covered with heating tapes in order to maintain the heating of the reactor system. The two ends were removable for membrane replacement purpose. Gas tightness between the shells was maintained by graphite O-rings. Two graphite rings (one at each end) were used as sealing for the alumina tube ends which withstand high temperature as well as allowing for thermal expansion of the alumina membrane. The inlet pressure of the reactor were measured with highly accurate and versatile digital pressure measuring gauges (Keller Druckmesstechnik, Winterthur, Switzerland) with an accuracy of 0.1% factory setting at room temperature. The permeate was connected to a digital flow meter to measure the flow rates. The permeance was calculated as:

$$F = q/A\Delta P \tag{6}$$

Where, F is the Permeance (mol/m^{-2} s^{-1} Pa^{-1}), q is the molar flow (mol/sec), A is the surface area of the support (0.0062 m^2), ΔP is the pressure difference across the support (Pa).

Scanning electron microscopy (SEM) and energy diffraction X-ray analysis (EDXA) were obtained from Zeiss EVO LS10 electron microscope. Figure 3 depicts the cross-section SEM micrographs of the alumina support. Approximately 10 μm thick alumina layer of highly intergrown alumina crystals was observed. Gas permeation

result reveals that the alumina support is defect free. Figure 4 depicts the EDXA result of the alumina support which clearly shows the Al_2O_3 peaks.

Nitrogen adsorption-desorption isotherms were measured using an automated gas sorption analyzer (Quantachrome instrument version 3.0) at liquid nitrogen temperature (77 K). The specific surface area was evaluated using the Brunauer-Emmett-Teller (BET) method (Quantachrome instrument version 3.0). The pore diameter was also obtained using the Barret-Joyner-Halenda (BJH) method. He, H_2, N_2 and CO_2 with at least 99.999 (%v/v) purity was used for permeation characterization. Permeation tests were carried out at feed pressures between 0.85 up to 1.0 bar and a temperature of 450°C.

Results and Discussion

Membrane characterization

The N_2 adsorption-desorption isotherm of the alumina support was obtained. The BET surface area and the BJH pore size distribution methods of the alumina support are depicted in Figures 5 and 6. It can be seen in Figure 5 that the isotherm exhibits a drop in the desorption branch at P/Po=0.5 (dotted line). However, the meniscus curve is not closed, this could be as a result of contaminants in the material. The BET surface area of the alumina support is 0.364 m^2/g. Figure 6 depicts the measurement of the pore-size distribution of the alumina support.

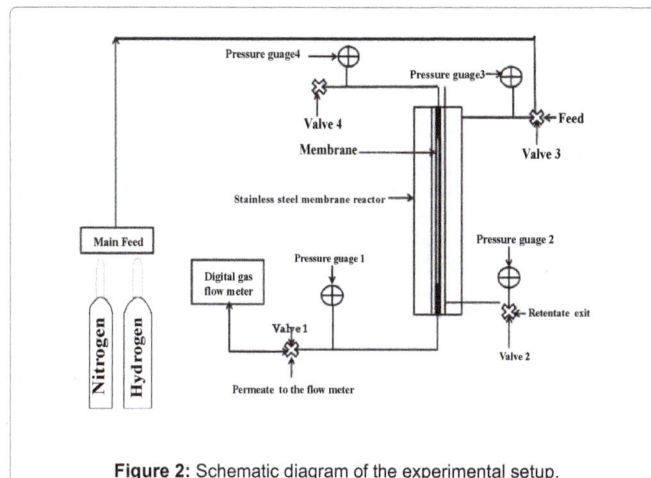

Figure 2: Schematic diagram of the experimental setup.

Gas	Kinetic Diameter (Å)	Molecular Weight (g/mol)
He	2.6	4
H_2	2.89	2
N_2	3.64	28
CO_2	3.3	44

Table 1: Gas kinetic diameter and molecular weight.

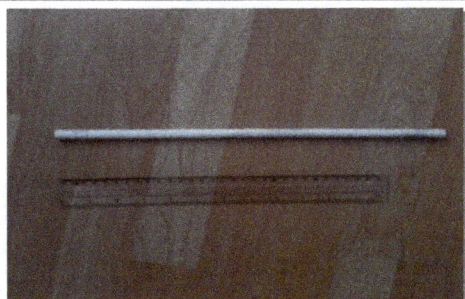

Figure 1: Pictorial view of tubular gamma alumina support.

Figure 3: Cross-section SEM image of the alumina support.

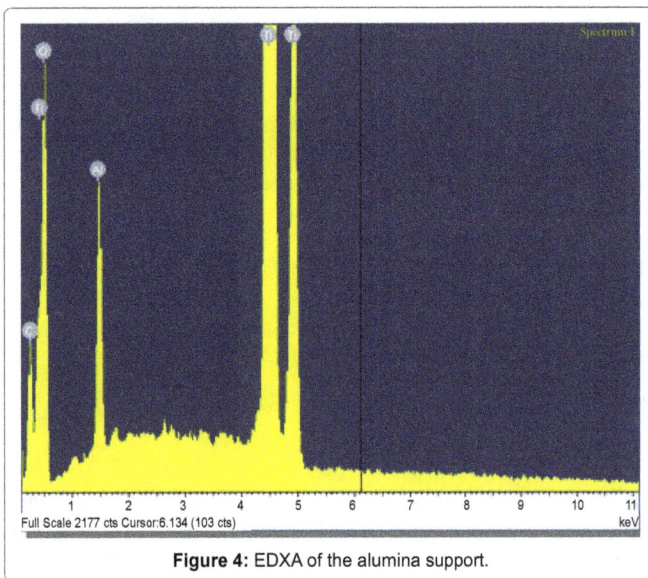

Figure 4: EDXA of the alumina support.

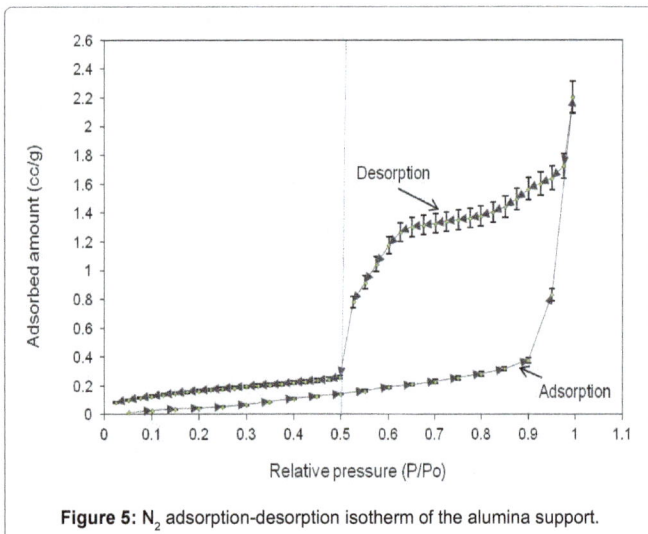

Figure 5: N_2 adsorption-desorption isotherm of the alumina support.

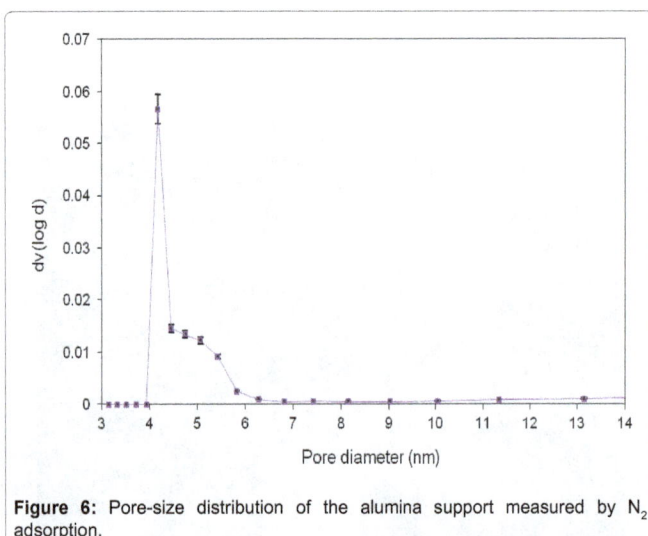

Figure 6: Pore-size distribution of the alumina support measured by N_2 adsorption.

An average pore diameter of 4.171 nm is calculated from the pore-size distribution graph, and the pore volume is 0.005 cm^3/g.

The adsorption-desorption isotherm exhibits a characteristics of mesopores solids (especially ceramics) resulting in Type IV physisorption isotherm according to the IUPAC recommendations which revealed the presence of mesoporous (20<pore size<500 Å) in the membrane undergoing capillary condensation and hysteresis during desorption [7-9].

Gas permeation

The variation of He, H_2, N_2 and CO_2 single gas permeances against permeation pressure across the alumina support was examined at 450°C. Figure 7 depicts gas permeances as a function of feed pressure across the alumina support. The permeances did not change with pressure increase which reveals that viscous flow contribution is not significant. From Equation (5); the first term can therefore be neglected

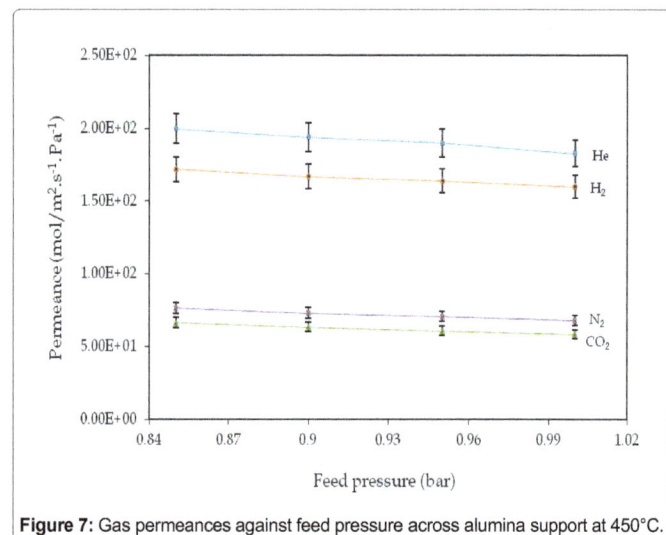

Figure 7: Gas permeances against feed pressure across alumina support at 450°C.

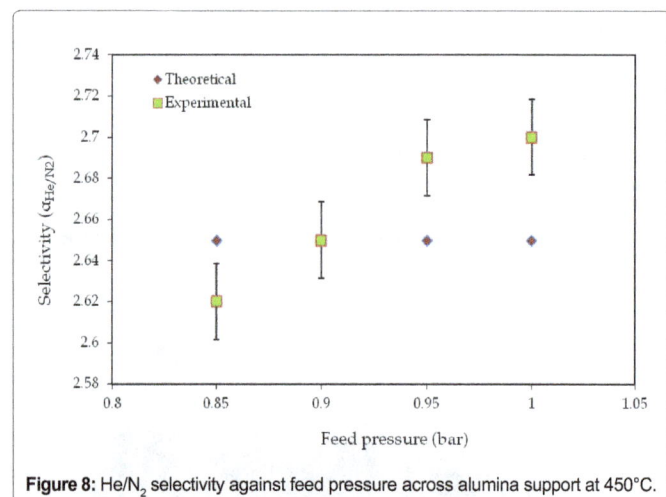

Figure 8: He/N_2 selectivity against feed pressure across alumina support at 450°C.

Pressure (bar)	Permeances (mol/m².s⁻¹. Pa⁻¹)		Selectivity ($\alpha_{He/N2}$)	
	He	N_2	Theoretical	Experimental
0.85	2.00×10^2	7.64×10^1	2.65	2.62
0.90	1.94×10^2	7.31×10^1	2.65	2.65
0.95	1.90×10^2	7.06×10^1	2.65	2.69
1.00	1.83×10^2	6.79×10^1	2.65	2.70

Table 2: He and N_2 permeances and $\alpha_{He/N2}$ at 450°C.

from the governing transport through the alumina support, owing to the fact that Knudsen diffusion dominates the flow regime. The transport of the gases with respect to their kinetic diameter and molecular weight behaves differently. It can be seen on Figure 7 that, He and H_2 permeation followed their respective kinetic diameter, whereas, N_2 and CO_2 permeation followed their respective molecular weight. Figure 8 depicts He/N_2 selectivity against feed pressure at 450°C. It can be seen that He/N_2 selectivity of 2.7 (Table 2) at 1 bar is obtained. The selectivity obtained is comparable to the theoretical Knudsen value (2.65).

Conclusions

The characterization (SEM-EDXA observation, BET measurement, permeation assessment) of a commercial tubular gamma alumina support was carried out. SEM result reveals that the alumina support is defect free. BET surface area, average pore diameter and the pore volume of the alumina support was also obtained (0.364 m^2/g, 4.171 nm and 0.005 cm^3/g). The adsorption-desorption isotherm exhibits a characteristics of mesopores solids. Single gas permeances were measured at feed pressures between 0.85 up to 1.0 bar and a temperature of 450°C. The permeances were influenced by Knudsen diffusion transport mechanism. He/N_2 selectivities obtained (2.7) were comparable to the theoretical Knudsen value (2.65).

Acknowledgements

The authors gratefully acknowledge Petroleum Technology Development Fund (PTDF) Nigeria for funding this research, and School of Pharmacy and Life Sciences RGU Aberdeen for the SEM and EDXA results.

References

1. Labropoulos AI, Athanasekou CP, Kakizis NK, Sapalidis AA, Pilatos GI, et al. (2014) Experimental investigation of the transport mechanism of several gases during the CVD post-treatment of nanoporous membranes. Chemical Engineering Journal 255: 377-393.

2. Kajama MN, Nwogu NC, Gobina E (2014) Experimental study of carbon dioxide separation with nanoporous ceramic membranes. WIT Transactions on Ecology and the Environment 186: 625-633.

3. Kajama MN, Nwogu NC, Gobina E (2016) Preparation and characterization of inorganic membranes for hydrogen separation. International Journal of Hydrogen Energy.

4. Castricum HL, Qureshi HF, Nijmeijer A, Winnubst L (2015) Hybrid silica membranes with enhanced hydrogen and CO2 separation properties. Journal of Membrane Science 488: 121-128.

5. Finol C, Coronos J (1999) Permeation of gases in asymmetric ceramic membranes. Chemical Engineering Education 33: 58-61.

6. Wall Y, Braun G, Brunner G (2010) Gas transport through ceramic membranes under super-critical conditions. Desalination 250: 1056-1059.

7. Smart S, Liu S, Serra JM, Diniz da Costa JC, Iulianelli A, et al. (2013) Porous ceramic membranes for membrane reactors. Handbook of membrane reactors: fundamental materials science, design and optimisation. Woodhead Publishing, Cambridge, United Kingdom 1: 298-336.

8. Weidenthaler C (2011) Pitfalls in the characterization of nanoporous and nanosized materials. Nano scale 3: 792-810.

9. Li J, Zhang Y, Hao Y, Zhao J, Sun X, et al. (2008) Synthesis of ordered mesoporous silica membrane on inorganic hollow fiber. Journal of colloid and interface science 326: 439-444.

18

Template-Assisted Synthesis of Metal Oxide Hollow Spheres Utilizing Glucose Derived-Carbonaceous Spheres As Sacrificial Templates

Haitham Mohammad Abdelaal [1]* **and Bernd Harbrecht** [2]

[1]*Ceramics Department, The National Research Centre, Al-Buhouth St. Dokki, Cairo, Egypt*

[2]*Department of Chemistry and Centre of Materials Science Philipps University, 35032 Marburg, Germany*

Abstract

A series of metal oxides hollow spheres (Cr_2O_3, α-Fe_2O_3, Co_3O_4, NiO and ZnO) have been fabricated using the glucose derived-carbonaceous spheres as sacrificial templates and the metal chlorides as precursors for the metal oxides in a sacrificial templating process. Heating of an aqueous solution of the metal chloride and glucose in an autoclave at 180 °C affords - as indicated by transmission electron microscopy (TEM) - a nanospherical composite consisting of a metal precursor shell sheathing a carbonaceous core. Consequently, hollow crystalline oxides spheres are obtained by removal of the carbonaceous cores through calcination in air. Correlations between the particle size and the various synthesis conditions such as glucose concentration, the molar concentration ratio between glucose and metal chloride, temperature, reaction time and the addition of acetic acid as a catalyst are uncovered. The obtained metal oxides hollow spheres were characterized by means of scanning electron microscopy (SEM), transmission electron microscopy (TEM), x-ray powder diffraction (XRD), infrared spectroscopy (IR), and nitrogen adsorption/desorption isotherms (BET).

Keywords: Oxides; Chemical techniques; Electron microscopy; Microstructure; Hollow materials

Introduction

Recently, in both academic and technological studies, inorganic oxides hollow spheres are attracting great attention due to their enhanced properties such as hollow cores, large specific surface area and low density a long with the distinct functions of oxides. Therefore, they present a class of distinct materials that provoke new options in the development of diverse potential applications such as protection of sensitive components (as enzymes and proteins), catalysis, coatings, water treatment, encapsulation of chemicals - for controlled-release applications - and adsorption. Moreover, other interesting applications may be proposed based on the chemical and physical characteristics of the inorganic hollow particles [1-12]. A variety of chemical and physicochemical methods such as sol-gel process [13], spray pyrolysis methods [14], surface polymerization processes [15], sonochemical route [16], colloidal templating methods and template free approaches have been used for the fabrication of hollow micro and nanomaterials [17,18]. Among the various synthesis methods, sacrificial templating approaches are considered as the most efficient and often used method for the fabrication of hollow structured micro and nanoparticles. In this way the fabrication of hollow particles usually involves the preparation of core-shell composite particles. These hybrid particles can be formed by precipitation of inorganic precursors of the target oxide hollow particles onto the surface of the core particles followed by etching of the cores by soaking the core in appropriate solvent in case of inorganic templates or thermal treatment in case of organic templates [12].

Various templating agents have been introduced for the fabrication of inorganic hollow materials including silica [19], gold [20], calcium carbonate [21], silver nanoparticles [22], hematite [23], polystyrene (PS) latex [17], polymethylmethacrylate (PMMA) [24], chitosan-polyacrylic acid (CS-PAA) [25], and n-propyle amine [26]. Latterly, carbohydrates such as glucose and sucrose have been used successfully as sacrificial templates for the synthesis of hollow inorganic particles [27-28]. Glucose is considered one of the most promising carbohydrates that can be used as sacrificial templates for the synthesis of inorganic hollow structures as it is one of the most inexpensive and widely available

carbohydrates. They have surface functionalities such as -C=O and -OH groups facilitating adsorption of the desired materials onto their reactive surfaces, as have been reported previously for different hollow materials [7,27-29].

In previous work, we have demonstrated the fabrication of silica hollow nanospheres via a facile one pot hydrothermal strategy by applying glucose-derived carbonaceous spheres as sacrificial templates. We further demonstrated that the shell size of hollow spheres can be varied by the variation of the starting materials [30]. Ta_2O_5 hollow nanospheres have been prepared by using glucose as sacrificial templates as well [31]. Further, fructose-derived carbonaceous spheres have been applied as sacrificial templates for the fabrication of some metal oxides hollow submicrospheres via a hydrothermal approach [32].

In this work we report the use of a facile route, hydrothermal hydrolysis, to fabricate porous crystalline metal oxides hollow spheres using glucose derived-carbonaceous spheres as sacrificial templates and metal chlorides as precursors for the metal oxides. In addition, the correlations between the particle size and the various synthesis conditions such as glucose concentration, the molar concentration ratio between the glucose and the metal chloride, temperature, reaction time and the addition of catalyst are investigated.

The interest in the use of monosaccharides, as sacrificial templates to fabricate the hollow metal oxides, arises from the reactive surface

---

*Corresponding author: Haitham Mohammad Abdelaal, Ceramics Department, The National Research Centre, Al-Buhouth St. Dokki, Cairo, P.O. Box 12622, Egypt, E-mail: hmaa_77@yahoo.com

of the glucose-derived carbonaceous materials - with its richness in O-functionalities - that facilitate the precipitation of the oxide precursor onto their surface layers without any further surface modifications. The former mentioned fact is probably the key to success in fabrication of the hollow spheres. In addition, the synthesis route is simple and can be readily analyzed and manipulated compared with complex multistep strategies including many procedures and a variety of chemical additives.

Experimental

Materials

D-(+)-Glucose monohydrate ($C_6H_{12}O_6 \cdot H_2O$), Chromium (III) chloride hexahydrate ($CrCl_3 \cdot 6H_2O$), cobalt (II) chloride hexahydrate ($CoCl_2 \cdot 6H_2O$), iron (III) chloride hexahydrate ($FeCl_3 \cdot 6H_2O$), zinc(II) chloride ($ZnCl_2$) and nickel(II) chloride ($NiCl_2$) were obtained from Merck (Darmstadt, Germany). All chemicals were analytical grade and employed without further purification. Distilled water (conductivity ~ 1.7 μS cm^{-1}) was used.

Fabrication of the metal oxide hollow spheres

The major process steps applied in the present work involved heating the metal chloride with glucose in closed system results in in-situ formation of hybrid particle due to the adsorption of the metal ions on the surface layers of the glucose-derived carbonaceous spheres. Finally, calcination of the hybrid spheres lead to the formation of hollow metal oxide.

For the formation of the metal oxide hollow spheres, glucose is applied as the sacrificial template, while metal chlorides [chromium (III) chloride hexahydrate ($CrCl_3 \cdot 6H_2O$), iron (III) chloride hexahydrate ($FeCl_3 \cdot 6H_2O$), cobalt(II) chloride hexahydrate ($CoCl_2 \cdot 6H_2O$), nickel(II) chloride ($NiCl_2$), and zinc(II) chloride ($ZnCl_2$)] are the precursors for the desired Cr_2O_3, α-Fe_2O_3, Co_3O_4, NiO and ZnO hollow spheres, respectively.

In a typical synthesis experiment, 1902 mg of glucose was dissolved in 100 mL distilled water. The water soluble metal chloride was added to satisfy the glucose: metal chloride molar ratio 10:1. The mixture was heated in a 100 mL Teflon-lined stainless steel autoclave at 180 °C for 24 h. The products were filtered off; washed three times, first with distilled water and then ethanol, and finally dried in a vacuum oven at 60 °C for 5 h. After synthesis, the core @ shell composites were calcined in air at 500 °C (heating rate 2 °C min^{-1}, 5 h) to remove the carbon core eventually leading to the target metal oxide hollow particles.

Characterization

The resulting hollow metal oxides and their corresponding composites were characterized by infrared (IR) spectra using IFS 88 from Bruker and XRD patterns obtained by using X-ray powder diffraction (X'Pert MPD, Pananalytical) operating in Bragg–Brentano geometry. The diffractometer was equipped with a graphite monochromator at the detector side. The sample holder was a single-crystal silicon plate. The XRD patterns, following Rietveld refinement procedures, were performed with the X'Pert software package supplied by the Pan Analytical Company.

The surface area was studied by nitrogen- sorption measurements performed with use of a Micromeritics ASAP 2020 gas sorptometer. The samples were degassed in vacuum at a pressure of 0.4 Pa for at least 3 h at 200 °C prior to measurements at 77 K over a range of relative pressures from 0.01 to 0.995. Specific surface areas were calculated

by assuming Brunauer-Emmet-Teller (BET) conditions. The particles morphology was visualized using a JEOL JSM-7500F field emission scanning electron microscope at an accelerating voltage of 5 kV. Therefore, the ground samples were mounted on an aluminium stub covered with a conductive carbon tape. To avoid surface charging the samples were coated with an ultra thin layer of platinum coating before SEM analyses. Transmission electron microscopy (TEM) was conducted with use of an electron microscope (JEM-3010, JEOL) operating at 300 kV. The samples were crushed to a powder and mounted by drop-drying of a chloroform suspension onto TEM cupper grids before TEM analyses.

Results and Discussion

The formation of the porous metal oxides hollow spheres

Figure 1 illustrates schematically the proposed mechanism of the formation of the porous metal oxides hollow spheres. The formation of hollow oxides spheres through the hydrothermal hydrolysis method involves adsorption of the metal ions dispersed in the solution mixture into the hydrophilic surface layers of the glucose-derived carbonaceous spheres which are rich in oxygen functionalities such as -OH and C=O [28-29]. This results in the in-situ formation of a core @ shell composite consisting of glucose-derived carbonaceous spheres with the metal ions bound to the oxygen functionalities in the outer shell; these are finally densified and cross-linked in a subsequent pyrolytic treatment eventually leading to free standing porous hollow metal oxide spheres after the removal of the carbonaceous core materials. The as-obtained hollow metal oxide spheres are replicas of the carbonaceous core spheres though about 60-80% smaller in size than the original corresponding composite granules.

The significant shrinkage in size in the course of the thermal treatment indicates that the composite with the metal cations bound to the carbonaceous spheres (CSs) transformed into a dense network of nanocrystalline metal oxide grains composing the shells of the hollow spheres. Frequently, the hollow spheres products exhibit a ball in ball (bnb) hollow structure which is obtained without any extra step.

The mechanism of the formation of the bnb structure is still vague and a challenge to the material scientists. However, it was postulated that some nano-islands of metal oxide nanoparticles in the shell may migrate to or be stuck on the surface of the shrinking CSs cores in the course of the calcination process at moderate temperature. Additional heating at elevated temperature, these nano-islands finally aggregate into a small ball in the interior when the CSs cores are wholly burnt off [33-34].

Figure 2 shows the XRD patterns of the metal oxide hollow spheres obtained through the typical experimental procedures after calcination at 500 °C for 5 h. It shows that, the metal oxide hollow spheres consist of the well crystalline single phase metal oxide as shown by Rietveld refinement of the XRD patterns [35]. The average crystallite size was calculated by applying the Scherrer equation using the full width at half maximum (FWHM) of the most intense peaks [36]. The mean size was determined to be 21, 23, 12, 11, 18 nm for Cr_2O_3, α-Fe_2O_3, Co_3O_4, NiO and ZnO, respectively. No crystalline peaks were observed before calcination (insets in Figure 2) disclosing that the metal ions are evenly adsorbed onto the hydrophilic shell of the carbonaceous cores or disseminated in the shell as amorphous cluster after hydrothermal treatment.

Figure 3 shows TEM micrographs of the products before

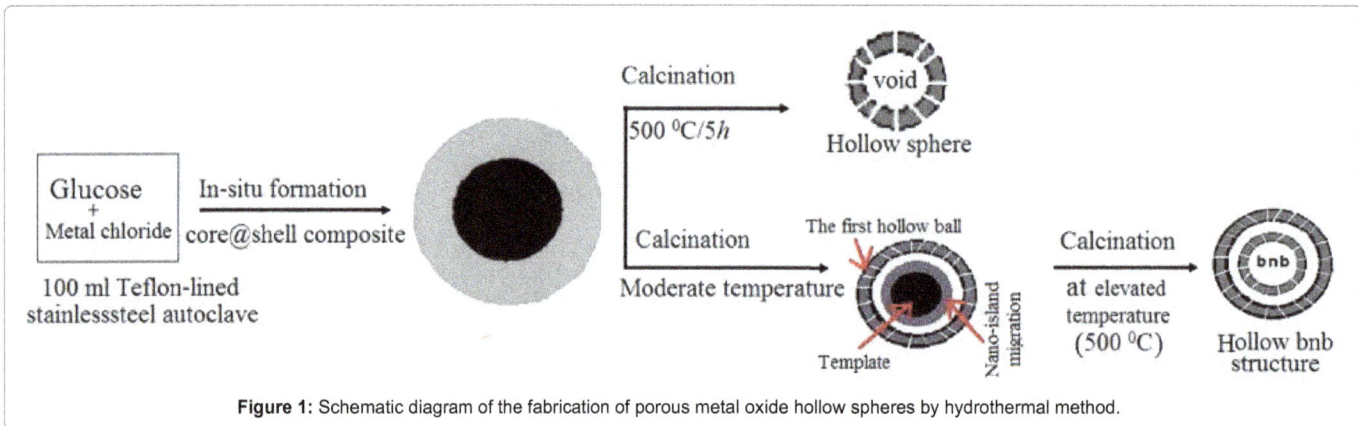

Figure 1: Schematic diagram of the fabrication of porous metal oxide hollow spheres by hydrothermal method.

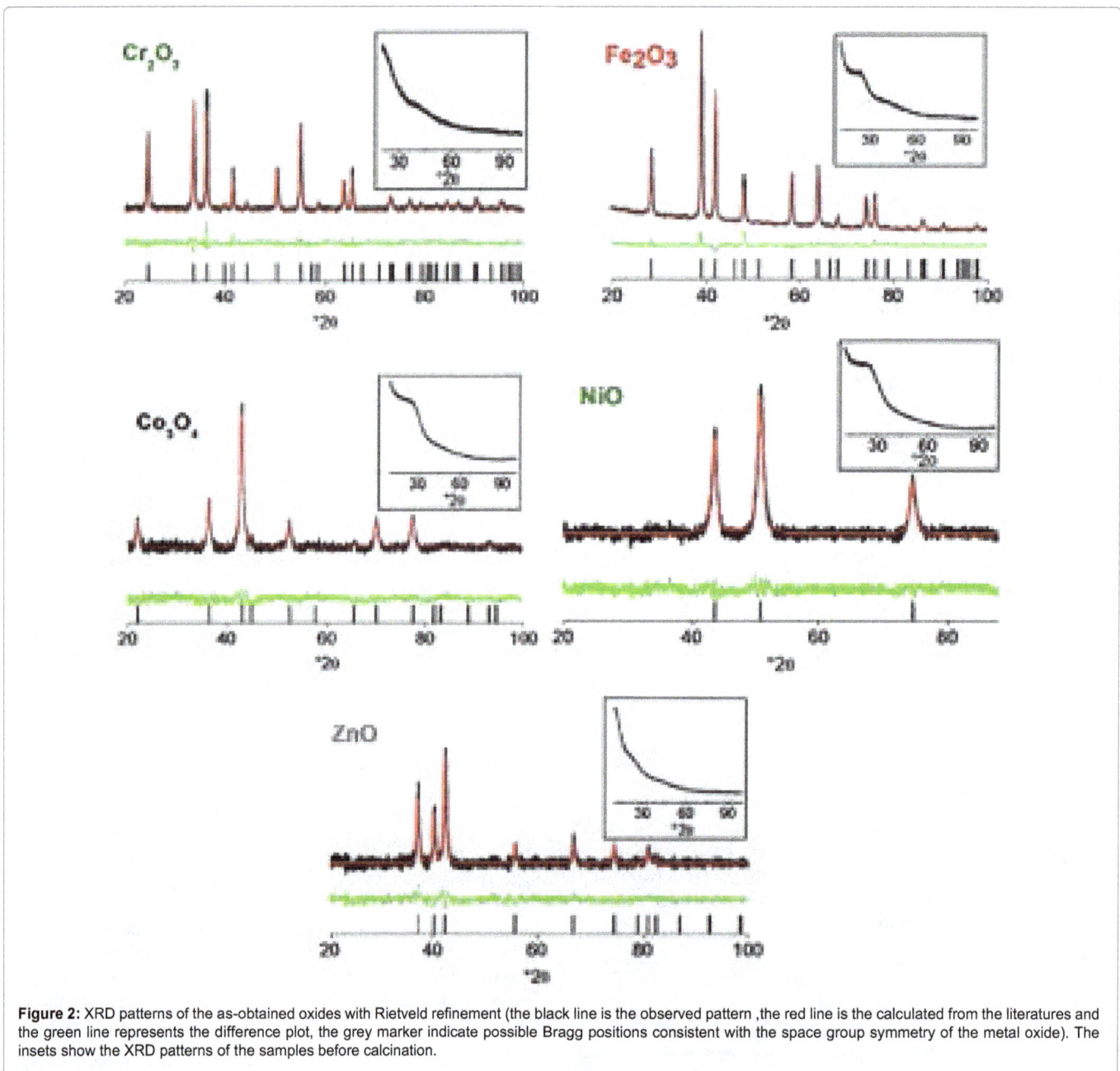

Figure 2: XRD patterns of the as-obtained oxides with Rietveld refinement (the black line is the observed pattern ,the red line is the calculated from the literatures and the green line represents the difference plot, the grey marker indicate possible Bragg positions consistent with the space group symmetry of the metal oxide). The insets show the XRD patterns of the samples before calcination.

calcination. They depict hybrid nature of the core @ shell composite of Cr_2O_3, α-Fe_2O_3 and ZnO samples. We can see that a contrast appears in the micrographs between the shell material and the core material. This provides support for the assumption of the spatial separation of the metal ions rich shell and the carbonaceous cores.

Comparison between IR spectra before and after calcination at 500 °C for 5 h, evidence the removal of the carbonaceous cores and the formation of the metal oxide hollow spheres as shown in Figure 4 for α-Fe_2O_3 hollow spheres. The IR spectrum before calcination displays a broad peak at 3400 cm^{-1}, which is attributed to be the stretching vibration of O-H groups. The peak at ~ 2900 cm^{-1} arises from the stretching vibrations of C-H bonds. The modes at 1701 cm^{-1} and 1630 cm^{-1} can be assigned to C=O and C=C, respectively. The C=C double bonds indicate that dehydration has taken place during the hydrothermal carbonization of glucose [37-38].

After calcination the carbonaceous templates and most peaks related to the functionalities, like carboxylic or aromatic groups disappeared and the observed peaks are typically related to M-O stretching vibrations as shown in Figure 4 which represents the IR spectrum of α-Fe_2O_3 hollow spheres. The observed bands at 570 and 480 cm^{-1} are typical for Fe-O modes of hematite α-Fe_2O_3 [39].

SEM micrographs in Figure 5 a, b and d, e display the metal oxide hollow particles before and after calcination of Cr_2O_3 and α-Fe_2O_3 respectively. SEM micrographs of the products Figure 5 b, e and Figure S2 (Supporting Information) evidence the formation of the spherical hollow metal oxides particles. The surface of the hollow spheres shows that the hollow metal oxides spheres walls are composed of many small nanoparticles of the metal oxide. From the broken shell, marked with a red arrow, we can notice the hollow porous nature of the metal oxides hollow spheres.

From Figure 5 a, b, d, e and also Figure S2 we can notice that after calcination the spheres preserve the three dimensional spherical shape of particles after removing of the carbonaceous core. In addition, about 60-80% shrinkage in size occurs after calcination as can be seen from the particle size distribution of the metal oxide hollow spheres and their corresponding composites (Figure S3). Apparently, the metal ions incorporated in the surface layer of the template densify and cross-link in the course of the pyrolysis to form the metal oxide hollow spheres replicas of the carbonaceous spheres template with significantly reduced size.

TEM micrographs (Figure 5 c, f, g, h, i) of Cr_2O_3, α-Fe_2O_3, Co_3O_4, NiO and ZnO, respectively, further confirm clearly the hollow interior as indicated by variation of the contrast between the dark shell and the pale core. The wall thickness of the porous hollow metal spheres can be estimated according to the cross sectional view obtained by TEM micrographs to be approximately 80 nm, 40 nm, 20 nm, 30 nm, and 18 nm for Cr_2O_3, α-Fe_2O_3, Co_3O_4, NiO and ZnO hollow spheres, respectively.

In general, SEM and TEM micrographs disclosed the formation of uniform hollow metal oxide spheres. Moreover, they reveal the formation of a ball in ball (bnb) hollow structure (Figures 5 and S4), as well. The high resolution TEM micrographs of Cr_2O_3, Co_3O_4, and ZnO porous hollow spheres (Figure S5) show that the size of the small nanoparticles composing the wall of the hollow spheres are in good agreement with the size calculated by the Scherrer equation for the as-obtained metal oxide hollow spheres.

The nitrogen adsorption/desorption isotherms were applied to study specific surface area of the hollow metal oxides spheres

(Figure S6). The observed hysteresis loops in the curves of all samples demonstrate the presence of mesoporous structures. They are typical isotherms characteristic of mesoporous materials according to the International Union of Pure and Applied Chemistry (IUPAC) [40]. The surface areas of the hollow oxides were 77, 54, 35, 33 and 65 m^2g^{-1} for Cr_2O_3, α-Fe_2O_3, Co_3O_4, NiO and ZnO hollow spheres, respectively. The specific surface area of the hollow metal oxides results from the sum of the areas of the outer and interior surface of the hollow spheres and the surface of the primary pores. The large surface areas and the spherical hollow shape that can be manipulated with respect to size, wall thickness and porosity makes this kind of materials interesting for various potential applications in several fields.

The impact of the synthesis conditions

Temperature (T), reaction time (t), concentration of glucose, the concentration ratio between glucose and metal chloride, and addition of acetic acid as catalyst are found to be five significant parameters that affect the outcome of the hydrothermal process. To study the impact of each synthesis parameter the experimental conditions were systematically varied whereby the parameter under investigation was varied while the other parameters remained unchanged according to the optimized synthesis procedures.

The results show that each parameter of the previously mentioned parameters has, to a large degree, a similar impact on the formation of the different types of oxides. The similar influence of each parameter on the formation of the hollow oxide spheres reported here might open the door for an improved understanding of the formation of the hollow metal oxide spheres with variable size.

The optimal temperature (T) for the formation of the hollow oxides using glucose as sacrificial templates is 180 °C. When decreasing T to 170 °C or raising it to 200 °C, no significant precipitates were seen at the former and no hollow spheres were formed at the latter temperature in all oxides under investigation. The optimal time (t) for the formation of metal oxide hollow spheres was 24 h. In case of increasing time to 36 h no hollow materials were observed except for Cr_2O_3 which formed fused hollow particles. While decreasing t to 12 h, the only observed hollow spheres were those for Cr_2O_3 with average size ~210 nm (Figures 6 a1, a2 and Figures S7 a).

Increasing glucose concentration from 96 $mmolL^{-1}$ - the typical glucose concentration- to 240 $mmolL^{-1}$ resulted in the formation of fused hollow particles for the oxides under investigation (Figures 6 b1 and b2; Figures S7 b, c and d). On the other hand, decreasing the glucose concentration to 64 $mmolL^{-1}$ resulted in small hollow spheres dispersed in nanoparticles of the metal oxide (Figure S7 e).

Figure 7 illustrates the impact of adding 0.5 mL of acetic acid to the reaction mixture solution. It is obvious that acetic acid catalyzes the reaction and increases the rate of the hydrothermal reaction and as a result the average size of the metal hollow oxides spheres increases by about 40-50%.

When the concentration ratio between glucose and metal chloride is increased from 10:1 to 20:1, we can anticipate that the amount of metal oxide forming the shell of hollow spheres will decrease. Figure 8 shows TEM micrographs for as-obtained hollow Cr_2O_3 samples through applying concentration ratio of 10:1 and 20:1. We notice that the wall thickness of the hollow spheres is inversely proportional to the molar ratio between the reactants. The wall thickness decreases from 80 nm to 33 nm when the concentration ratio increases. The particle size of the Cr_2O_3 particles forming the wall of the hollow spheres is found

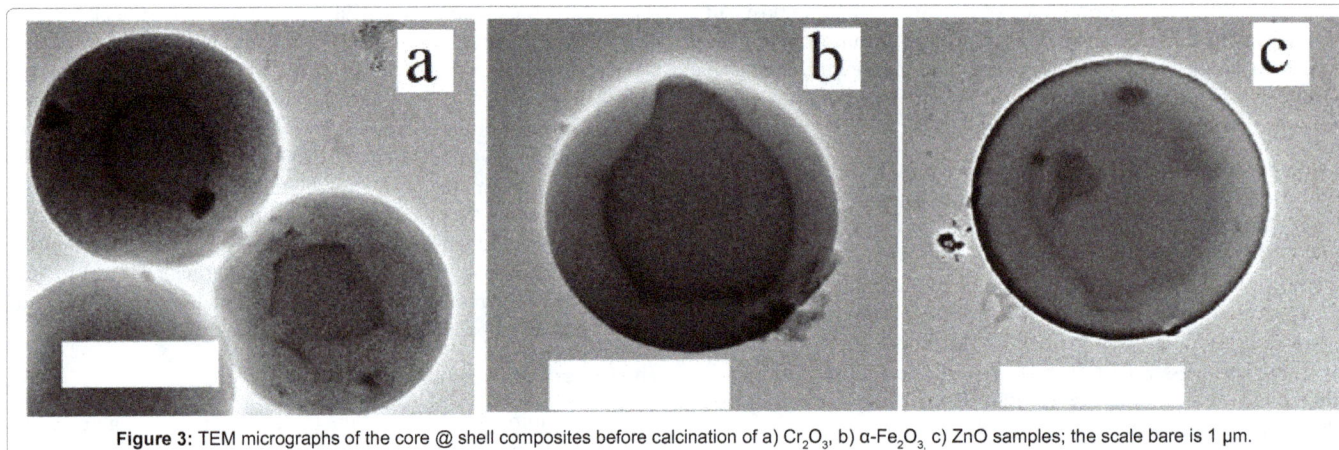

Figure 3: TEM micrographs of the core @ shell composites before calcination of a) Cr_2O_3, b) α-Fe_2O_3, c) ZnO samples; the scale bare is 1 μm.

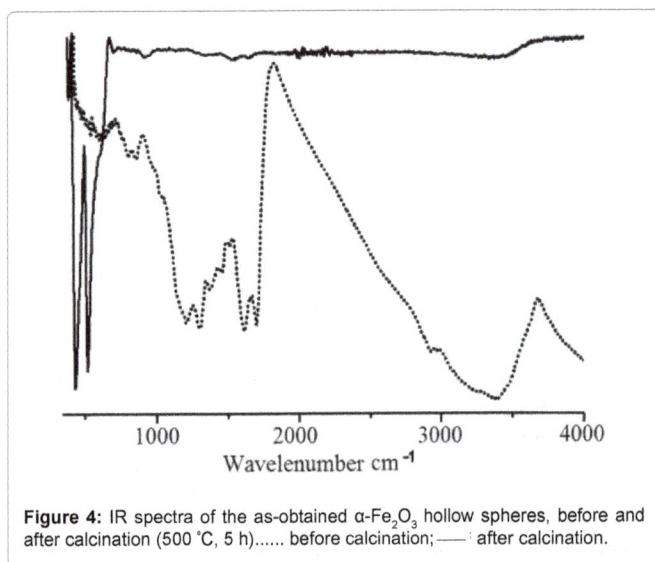

Figure 4: IR spectra of the as-obtained α-Fe_2O_3 hollow spheres, before and after calcination (500 ˚C, 5 h)...... before calcination; —— after calcination.

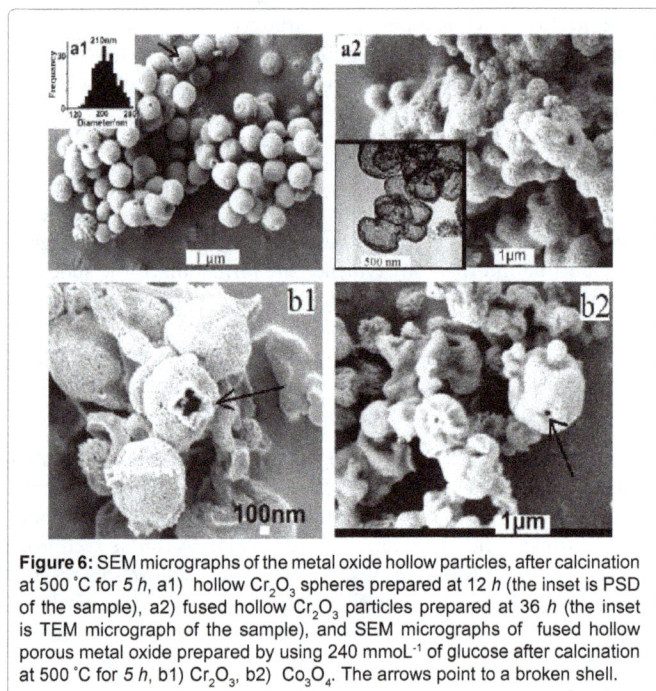

Figure 6: SEM micrographs of the metal oxide hollow particles, after calcination at 500 ˚C for *5 h*, a1) hollow Cr_2O_3 spheres prepared at 12 *h* (the inset is PSD of the sample), a2) fused hollow Cr_2O_3 particles prepared at 36 *h* (the inset is TEM micrograph of the sample), and SEM micrographs of fused hollow porous metal oxide prepared by using 240 $mmoL^{-1}$ of glucose after calcination at 500 ˚C for *5 h*, b1) Cr_2O_3, b2) Co_3O_4. The arrows point to a broken shell.

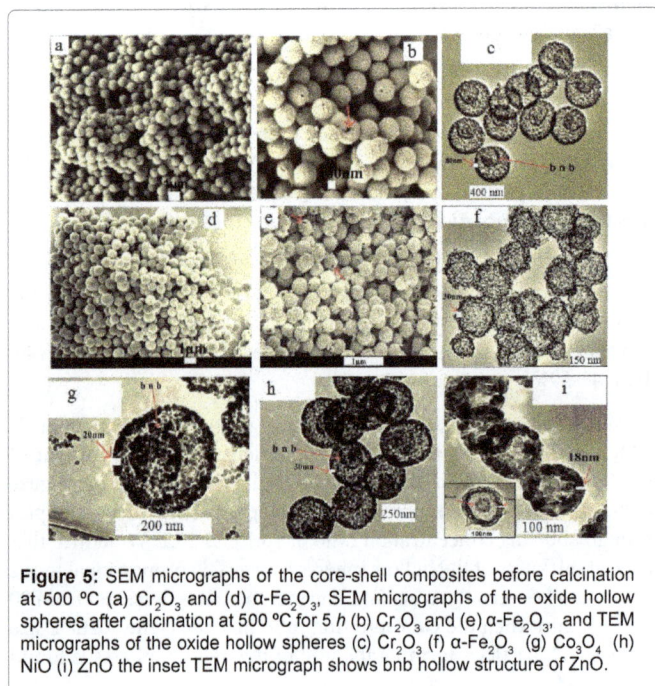

Figure 5: SEM micrographs of the core-shell composites before calcination at 500 ºC (a) Cr_2O_3 and (d) α-Fe_2O_3, SEM micrographs of the oxide hollow spheres after calcination at 500 ˚C for 5 h (b) Cr_2O_3 and (e) α-Fe_2O_3, and TEM micrographs of the oxide hollow spheres (c) Cr_2O_3 (f) α-Fe_2O_3 (g) Co_3O_4 (h) NiO (i) ZnO the inset TEM micrograph shows bnb hollow structure of ZnO.

to be approximately 21 nm. Hence, the 80 nm large walls of the hollow spheres are built up by about four layers of aggregated Cr_2O_3 grains for samples with higher Cr content (ratio 10:1). In contrast, lower Cr content (ratio 20:1) yields hollow spheres with walls consisting of nearly 2 layers of aggregated grains. Accordingly, the hollow spheres appear to be more robust in case of lower molar ratios (glucose/metal chloride). This is likely due to lower metal concentration leads to light packing of metal oxide nanoparticles and a thin wall, while an increase in the metal ions concentration yields a much denser packing and the formation of a robust thicker shell [29].

In general, the size of the hollow spheres is directly proportional to the reaction time and the addition of acetic acid (0.5 mL/100 mL solution) promoting the growth of the spherical core @ shell composite. While the wall thickness of the hollow spheres is inversely proportional with the increase of the molar ratio between glucose and metal chloride. The shape of the hollow materials is affected by the glucose concentration. The typical parameters given by the synthesis protocol procedures reported in experimental section are the optimized conditions for the fabrication of porous hollow metal oxides by using

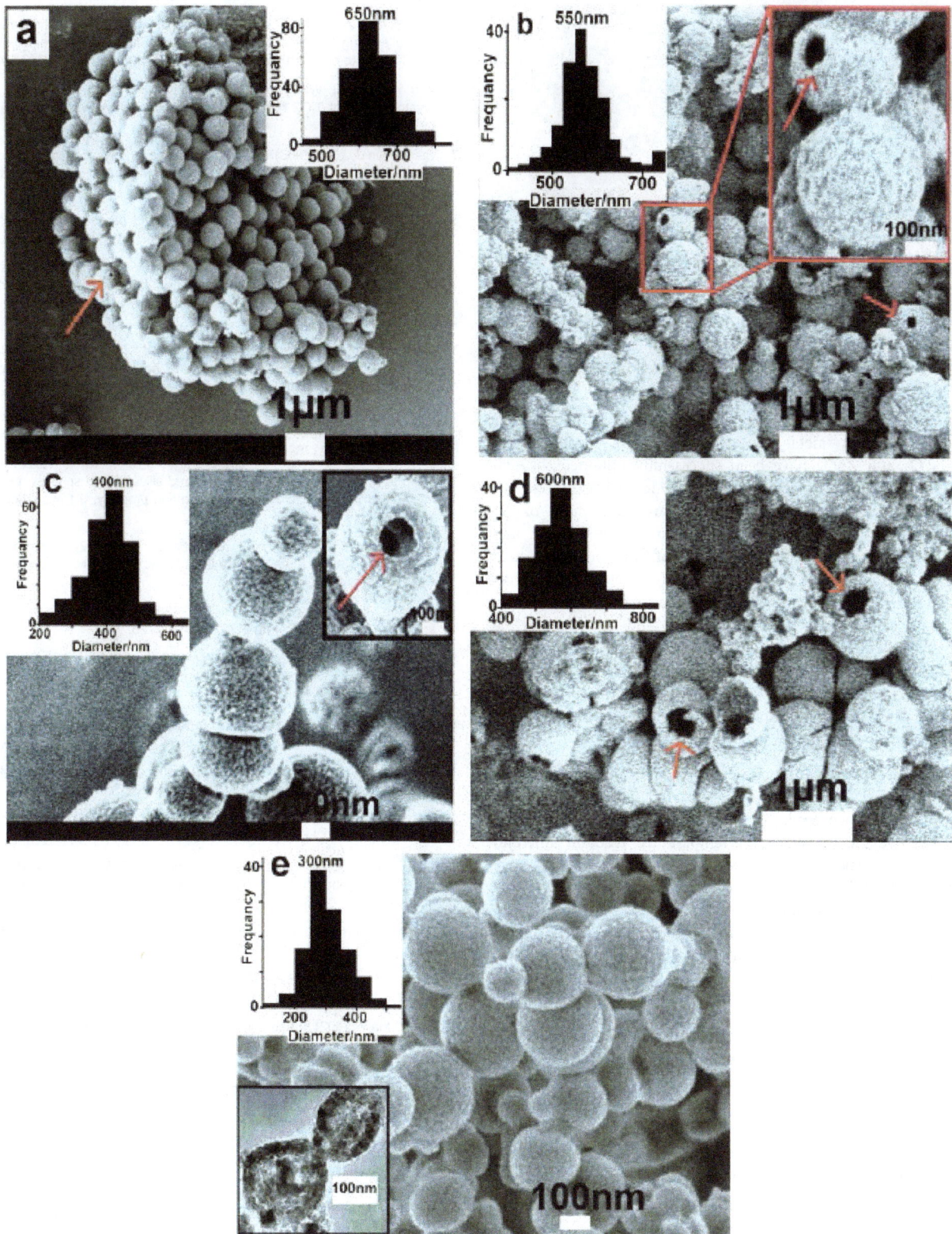

Figure 7: SEM micrographs of porous metal oxide hollow spheres prepared by applying 0.5 mL acetic acid as catalyst, after calcination at 500 °C for 5 h, a) Cr_2O_3, b) α-Fe_2O_3, the inset is a large view of the chosen area, c) Co_3O_4, the inset is SEM micrograph of a single hollow spheres with broken shell, d) NiO, and e) ZnO, the inset is TEM micrograph of hollow ZnO spheres; the particle size distribution of each sample is included in each micrograph. The red arrow refers to broken shells.

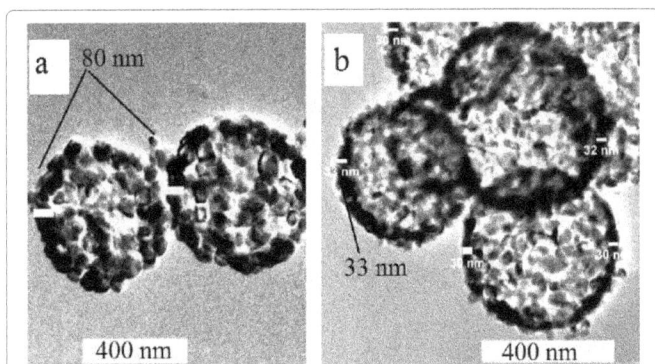

Figure 8: TEM micrographs of the as-obtained Cr_2O_3 hollow spheres by applying concentration ratio [c (glucose):c(metal chloride)] a) 10: 1 b) 20:1.

glucose as a sacrificial template. Table S2 in the Supporting Information summarizes some relationships between the synthesis parameters and the size and shape of the as-obtained hollow metal oxides spheres.

Conclusion

A series of porous metal oxides hollow spheres (Cr_2O_3, Co_3O_4, NiO, α-Fe_2O_3 and ZnO) have been successfully obtained through hydrothermal method by utilizing glucose-derived carbonaceous spheres as sacrificial templates and metal chlorides as metal oxides precursors. The key of success probably stems from the fact that the surface of the carbonaceous core is rich of functionalities which facilitate the adsorption of the metal ions into their surface layers without any surface modifications. The glucose-derived carbonaceous spheres used as templates have an integral and uniform surface functional layer which makes an additional modification of the surface of the shape-controlling carbonaceous template dispensable and provides the homogeneity of the shell. Though the pyrolytic treatment of the composites results in a drastic shrinkage in size, the spherical shape is preserved.

Correlations between the particle size and the concentration of glucose, as well as the ratio of metal precursor and the sugar concentrations are uncovered. Crucial factors, critical to fine-tune the final particle size and the shape, are temperature, reaction time and addition of acetic acid promoting particle growth. The results have shown that each of the varied parameters have similar impact on the various oxides. The similar impact of each parameter on the formation of hollow oxides reported here might open the door for improved understanding of the formation of the hollow metal oxides spheres with variable size.

Metal oxide particles of this type exhibit unique properties such as large specific surface areas and, in particular, hollow interior cores, and mesoporous walls of various size and thickness. In view of the rich and diverse property profiles of such nanoparticulate oxides the accruing properties arising from the specific shape and constitution of such hollow particles may offer improved and new useful applications in various fields. Catalysis, water treatment, photonic devices, chemical sensors and controlled release are just some of those.

References

1. Ludtke S, Adam T, Unger KK (1997) Application of 0.5μm porous silanized silica beads in electrochromatography. J Chromatogr A 786: 229-235.

2. Caruso F (2000) Hollow capsule processing through colloidal templating and self-assembly. Chemistry 6: 413-419.

3. Abdelaal HM (2014) Facile hydrothermal fabrication of nano-oxide hollow spheres using monosaccharides as sacrificial templates. ChemistryOpen.

4. Yuan J, Laubernds K, Zhang Q, Suib SL (2003) Self-assembly of microporous manganese oxide octahedral molecular sieve hexagonal flakes into mesoporous hollow nanospheres. J Am Chem Soc 125: 4966-4967.

5. Zhu Y, Shi J, Shen W, Dong X, Feng J, et al. (2005) Stimuli-responsive controlled drug release from a hollow mesoporous silica sphere/polyelectrolyte multilayer core-shell structure. Angew Chem Int Ed Engl 44: 5083-5087.

6. Zhu Y, Chen H, Wang Y, Li Z, Cao Y, et al. (2006) Mesoscopic photonic crystals made of TiO2 hollow spheres connected by cylindrical tubes. Chem Lett 35: 756-757.

7. Sun X, Liu J, Li Y (2006) Use of carbonaceous polysaccharide microspheres as templates for fabricating metal oxide hollow spheres. Chemistry 12: 2039-2047.

8. Yu J, Liu S, Yu H (2007) Microstructures and photoactivity of mesoporous anatase hollow microspheres fabricated by fluoride-mediated self-transformation. J Catal 249: 59-66.

9. Yu J, Yu H, Guo H, Li M, Mann S (2008) Spontaneous formation of a tungsten trioxide sphere-in-shell superstructure by chemically induced self-transformation. Small 4: 87-91.

10. Chen C, Abbas SF, Morey A, Sithambaram S, Xu L, et al. (2008) Controlled synthesis of self-assembled metal oxide hollow spheres via tuning redox potentials: versatile nanostructured cobalt oxides. Adv Mater 20: 1205-1209.

11. Yuan J, Zhang X, Qian H (2010) A novel approach to fabrication of superparamagnetite hollow silica/magnetic composite spheres. J Magn Magn Mater 322: 2172-2176.

12. Yuan J, Zhou T, Pu H (2010) Nano-sized silica hollow spheres: preparation, mechanism analysis and its water retention property. J Phys Chem Sol 71: 1013-1019.

13. Tissot I, Reymond J, Lefebvre F, Bourgeat-Lami E (2002) SiOH-functionalized polystyrene latexes. A step toward the synthesis of hollow silica nanoparticles. Chem Mater 14: 1325-1331.

14. Messing GL, Zhang SC, Jayanthi GV (1993) Ceramic powder synthesis by spray-pyrolysis. J Am Ceram Soc 76: 2707-2726.

15. Emmerich O, Hugenberg N, Schmidt M, Sheiko SS, Baumann F, et al. (1999) Molecular boxes based on hollow organosilicon micronetworks. Adv Mater 11: 1299-1303.

16. Dhas NA, Suslick KS (2005) Sonochemical preparation of hollow nanospheres and hollow nanocrystals. J Am Chem Soc 127: 2368-2369.

17. Shiho H, Kawahashi N (2000) Iron Compounds as Coatings on Polystyrene Latex and as Hollow Spheres. J Colloid Interface Sci 226: 91-97.

18. Mao LJ, Liu CY, Li J (2008) Template-free synthesis of VOx hierarchical hollow spheres. J Mater Chem 18: 1640-1643.

19. Salgueirino-Maceira V, Spasova M, Farle M (2005) Water-stable, magnetic silica–cobalt/cobalt oxide–silica multishell submicrometer spheres. Adv Funct Mater 15: 1036-1046.

20. Zhang R, Hummelgard M, Olin H (2010) Carbon nanocages grown by gold templating. Carbon 48: 424-430.

21. Zhang S, Li X (2004) Synthesis and characterization of CaCO 3@SiO2 core-shell nanoparticles. Powder Tech 141: 75-79.

22. Sun Y, Xia Y (2002) Shape-controlled synthesis of gold and silver nanoparticles. Science 298: 2176-2179.

23. Han YS, Jeong GY, Lee SY, kim HK (2007) Hematite template route to hollow-type silica spheres. J Sol State Chem 180: 2978-2985.

24. Kato T, Ushijima H, Katsumata M, Hyodo T, Shimizu Y, et al. (2002) Fabrication of hollow alumina microspheres via core/shell structure of polymethylmethacrylate/alumina prepared by mechanofusion. J Mat Sci 37: 2317-2321.

25. Tasi M, Li MJ (2006) A novel process to prepare a hollow silica sphere via chitosan-polyacrylic acid (CS-PAA) template. J Non Cryst Sol 352: 2829-2833.

26. Song L, Ge X, Wang M, Zhang Z (2006) Direct preparation of silica hollow spheres in a water in oil emulsion system: The effect of pH and viscosity. J Non Cryst Sol 352: 2230-2235.

27. Sun X, Li Y (2004) Colloidal carbon spheres and their core/shell structures with noble-metal nanoparticles. Angew Chem Int Ed Engl 43: 597-601.

Template-Assisted Synthesis of Metal Oxide Hollow Spheres Utilizing Glucose Derived-Carbonaceous...

117

28. Abdelaal HM (2014) Fabrication of hollow silica microspheres utilizing a hydrothermal approach. Chin Chem Lett. 25: 627-629.

29. Titirici MM, Antonietti M, Thomas A (2006) A Generalized synthesis of metal oxide hollow spheres using a hydrothermal approach. Chem Mater 18: 3808-3812.

30. Abdelaal HM, Zawrah MF, Harbrecht B (2014) Facile one-pot fabrication of hollow porous silica nanoparticles. Chemistry 20: 673-677.

31. Abdelaal HM, Pfeifer E, Grünberg C, Harbrecht B (2014) Synthesis of tantalum pentoxide hollow spheres utilizing a sacrificial templating approach. Matt Lett 136: 4-6.

32. Abdelaal HM, Harbrecht B (2014) Fabrication of metal oxide hollow spheres using fructose derived-carbonaceous spheres as sacrificial templates. C R Chim.

33. Suh WH, Jang A, Suh Y, Suslick KS (2006) Porous, hollow, and ball-in-ball metal oxide microspheres: preparation, endocytosis, and cytotoxicity. Adv Mater 18: 1832-1837.

34. Qian H, Lin G, Zhang Y, Gunawan P, Xu R (2007) A new approach to synthesize uniform metal oxide hollow nanospheres via controlled precipitation. Nanotech 18: 355602.

35. X'Pert Plus (1999) Program for Crystallography and Rietveld analysis, Philips Analytical, Almelo (The Netherlands).

36. Klug HP, Alexander LE (1974) X-ray Diffraction procedures. Wiley: New York, USA.

37. Ni D, Wang L, Sun Y, Guan Z, Yang S, et al. (2010) Amphiphilic hollow carbonaceous microspheres with permeable shells. Angew Chem Int Ed Engl 49: 4223-4227.

38. Sakaki T, Shibata M, Miki T, Hirosue H, Hayashi N (1996) Reaction model of cellulose decomposition in near-critical water and fermentation of products. Bioresour Technol 58: 197-202.

39. Pradhan GK, Parida KM (2011) Fabrication, growth mechanism, and characterization of $α$-Fe(2)O(3) nanorods. ACS Appl Mater Interfaces 3: 317-323.

40. Sing KSW, Everett DH, Haul RAW, Moscou L, Pierotti RA, et al. (1985) Reporting physisorption data for gas/solid systems with special reference to the determination of surface area and porosity (Recommendations 1984). Pure Appl. Chem. 57: 603-619.

19

Studies on Swelling and Absorption Properties of the γ – Irradiated Polyvinyl Alcohol (PVA)/Kappa-Carrageenan Blend Hydrogels

MD Tariqul Islam[1,2], NC Dafader[3], Pinku Poddar[1], Noor MD Shahriar Khan[1] and AM Sarwaruddin Chowdhury[1]*

[1]Department of Applied Chemistry and Chemical Engineering, Faculty of Engineering and Technology, University of Dhaka, Dhaka, Bangladesh
[2]Department of Chemistry, American International University of Bangladesh, Dhaka, Bangladesh
[3]Nuclear and Radiation Chemistry Division, Institute of Nuclear Science and Technology, Atomic Energy Research Establishment, Dhaka, Bangladesh

Abstract

A series of hydrogels were prepared from an aqueous mixture of Poly vinyl alcohol (PVA) and Kappa-carrageenan (KC) and irradiated the mixture at 25 kGy radiation dose with γ-radiation from 60Co γ source at room temperature (25°C). The effects of KC on the properties, such as gel fraction, swelling ratio (e.g., in distilled water, in NaCl solution with different concentration, buffer solution with different pH), water absorption, water desorption, moisture absorption and uptake of metal ion from aqueous solution of the prepared hydrogels were investigated. Incorporation of KC into the PVA obviously influences the properties of hydrogels. It is found that the gel fraction of the prepared hydrogel decreased but swelling ratio increased with increase in concentration of Kappa-carrageenan. Swelling properties in NaCl decreased with increased concentration of NaCl in aqueous solution. Swelling of the blend gel in buffer increased with the increase in pH. Water absorption properties showed that maximum absorption occurred within 24 hrs and then increasing trend of water absorption was insignificant. Water desorption is very fast upto 48 hrs and then attained a plateau value. The maximum moisture absorption occurred within 48 hrs and then the absorption was insignificant. Kappa-carragenan influences to uptake of metal (Cu+2) by PVA / KC blend hydrogel with time.

Keywords: Poly (vinyl alcohol); Hydrogels; Kappa-carrageenan; γ-Irradiation

Introduction

Hydrogels are two or multi component systems consisting of a three dimensional network of polymer chains and water that fills the space between macromolecules. The importance of hydrogels in biomedical applications was first realized in the late 1950s with the development of poly (2-hydroxyethyl methacrylate) (PHEMA) gels as a soft contact lens material. They are widely applicable in numerous biomedical [1] applications including ophthalmological devices, biosensors, biomembranes and carriers for controlled delivery of drugs or proteins. Although hydrogels have a number of non-biomedical applications (e.g., in agriculture), it seems that their use in the field of medicine and pharmacy is the most successful and promising [2]. Nowadays a new class of hydrogels, capable of reacting to various environmental stimuli as temperature, pH, ionic strength, solute concentration, electric field, light, sound etc., is tested for use in the so-called "intelligent biomaterials" [3].

The swollen state results from a balance between the dispersing forces acting on hydrated chains and cohesive forces that do not prevent the penetration of water into the network. Cohesive forces are most often due to covalent cross-linking [4]. Others are electrostatic, hydrophobic, or dipole-dipole in character [5]. The degree and nature of cross-linking and the crystallinity of the polymer are responsible for its characteristics in the swollen state. The ability to imbibe water and ions without the loss of shape and mechanical strength is valuable in many natural hydrogels, such as those found in muscle, tendons, cartilage, intestines, and blood.

Natural hydrogels are used in pulp and paper production, artificial silk, cellulosic membranes, and biomedical applications [6]. Synthetic hydrogels are used in prosthetic materials, soft lenses, and membranes for controlled drug release because of their compatibility with living tissue. Synthesis conditions such as temperature, monomer concentration and initiator level have significant effects on gel properties as well as impurities [7].

A simple cross-linking reaction is exemplified by polymer chains with several functional groups that are capable of reacting themselves to form chemical bonds. In principle, due to cross-linking, the polymer differs in many respects from linear and branched polymers. For example, they swell in a good solvent to form a gel but do not dissolve to form a solution. At elevated temperature, cross-linked polymers generally behave like soft but elastic solids rather than viscous liquids. Cross-linking can be carried out through the use of a cross-linking agent or by radiation energy. At some degree of conversion, the polymer chains began to form a cross-linked structure [8].

The gels are fixed to supports by coating, grafting, or chemical modification. Ion-exchange and separation membranes (qv) are hydrogels, although the term hydrogel is usually restricted to a synthetic, water-swelling polymer of soft, rubbery consistency. Its character is determined by the hydrophilic monomers and the density of the polymer network [9]. Ionogenic or charged gels [10] form a special group, with swelling and strength properties dependent on the pH of the environment.

There is a continuous search for materials that have the capability to improve the properties of the hydrogels. Kappa-carrageenan (KC) [11,12], agar [13] etc., are added to enhance the properties of poly (vinyl pyrrolidone) (PVP) hydrogel. By the addition of PVA [14] the

---

*Corresponding author: AM Sarwaruddin Chowdhury, Department of Applied Chemistry and Chemical Engineering, Faculty of Engineering and Technology, University of Dhaka, Dhaka-1000, Bangladesh
E-mail: profdrsarwar@gmail.com

gel strength of PVP hydrogel is improved. Poly (vinyl alcohol) (PVA) is a hydrophilic polymer with unique properties. It absorbs water, swells easily and it has extensively been used in wound dressing. In this research work it is found the effect of γ- irradiation on swelling and absorption properties of the PVA / KC blend hydrogels.

Experimental

Materials

Poly (vinyl alcohol) (PVA), medical grade obtained from BASF, Germany was used as received and molecular weight of PVA was 145000. Kappa-carrageenan (KC), commercial grade obtained from SIGMA, USA was used to improve properties of hydrogel. Distilled water was used to prepare the sample solution.

Preparation of PVA solution without and with KC

A nine percent aqueous solution of PVA was prepared by dissolving it in distilled water. KC of different concentrations (0-4%) was added to the 9% PVA solution and treated the solution in an autoclave at 120°C for 30 minutes. Then the solution is heated in a water bath at 70°C for 1 hour. The hot solution of PVA and KC were poured into test tubes of diameter 13 mm and cooled at room temperature (22-25°C).

Irradiation of sample

The samples that were poured into test tubes were irradiated at room temperature by Co-60 γ-sources with 25 kGy radiation dose at the dose rate of 0.79 kGy/hr (Figure 1).

Preparation of dry gel sample and measurement of gel fraction

The irradiated hydrogel samples from test tubes were cut into a number of equal sized test pieces. Pieces of gel samples were dried to constant weight in a vacuum oven at 50°C. The dried samples were immersed in distilled water at room temperature for 24 hrs to remove sol fraction (water soluble parts). Then the extracted samples were dried again to a constant weight at 50°C in a vacuum oven. In this way we get dry gel. The experiment was repeated for the hydrogel prepared from different concentration of KC.

The gel fraction was calculated as follows:

Gel Fraction(%)=$W_g/W_o \times 100$

Where W_g is the weight of dry gel after extraction in water and W_o is the initial weight of dry gel.

Determination of swelling ratio in distilled water

The gel samples dried to a constant weight were again immersed in distilled water at room temperature for 24 hours. Then the samples were taken out from distilled water, surface water was removed gently by soft tissue paper and weighed. The experiment was repeated for the

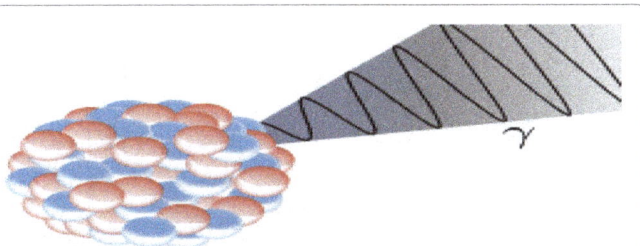

Figure 1: Emission of gamma ray (γ) from an atomic nucleus.

hydrogel prepared from different concentration of KC. The Swelling ratio was calculated as follows:

Swelling ratio=W_1/W_o

where W_1 is the weight of water absorbed by the gel sample and W_o is the weight of dry gel.

Determination of swelling ratio in NaCl solution with different concentration

The gel samples dried to a constant weight were immersed in NaCl solution having different concentration (0.1, 0.3, 0.5, 0.7 and 1.0%) for 24 hours at room temperature. Then the samples were taken out from solution, surface water was removed gently by soft tissue paper and weighed. The experiment was repeated for the hydrogel prepared from different concentration of KC. The Swelling ratio in NaCl solution was calculated as follows:

Swelling ratio in NaCl Solution=W_1/W_o

where W_1 is the weight of NaCl solution absorbed by the gel sample and W_o is the weight of dry gel.

Determination of swelling ratio in buffer solution with different pH

The gel samples dried to a constant weight were immersed in buffer solution of pH 4, 7 and 9 for 24 hours at room temperature. Then the samples were taken out from solution, surface water was removed gently by soft tissue paper and weighed. The experiment was repeated for the hydrogel prepared from different concentration of KC. The Swelling ratio in buffer solution was calculated as follows:

Swelling ratio in buffer=W_1/W_o

where W_1 is the weight of buffer solution absorbed by the gel sample and W_o is the weight of dry gel.

Determination of water absorption

The gel sample dried to a constant weight was immersed in distilled water at room temperature. This sample was taken out from distilled water at different time interval (1 to 48 hours) and weighed after soaking with a soft tissue paper to remove adhering water from the surface of gel sample. The experiment was repeated for the hydrogel prepared from different concentration of KC.

The percentage of water absorption was calculated as follows:

Water absorption (%)=$(W_a/W_o) \times 100$

where W_a is the weight of water absorbed by the gel sample at time interval and W_o is the weight of dried gel.

Determination of water desorption

The pre-weighed gel samples were kept at room temperature (23-25°C) and weighed at different time interval up to 76 hours. The experiment was repeated for the hydrogel prepared from different concentration of KC. Temperature and average humidity of the experimental environment were 23-25°C and 50-55% respectively. The percentage of water desorption was calculated as follows:

Water desorption (%)=$(W_d/W_1) \times 100$

where W_d is the weight lost of gel sample after a particular time and W_1 is the initial weight of hydrogel.

Determination of moisture absorption

The hydrogel samples were dried in oven at 50°C to a constant weight. The dried samples were then placed in an open environment for moisture absorption and weighed at different time interval up to 76 hours. The experiment was repeated for the hydrogel prepared from different concentration of KC. Temperature and average humidity of the experimental environment were 23-25°C and 50-55% respectively. The percentage of moisture absorption was calculated as follows:

Moisture absorption (%)=$(W_{ma}/W_{i0}) \times 100$

where W_{ma} is the weight of gel sample after a particular time and W_{i0} is the initial weight of dry gel sample.

Determination of uptake of metal ion (Cu^{+2})

Apparatus:

1. Spectrophotometer, Wavelength 435 mm, matched absorption cell 10 to 15 mm.

2. Separating funnel.

Reagents:

1. Zinc diethyldithiocarbamate: Dissolve 1 gm of sodium diethyldithiocarbamate in 100 ml water and 2 gm of $ZnSO_4$. Extract the resulting Zinc diethyldithiocarbamate by shaking with 100 ml $CHCl_3$ and separate the chloroform solution. Dilute to 1 litre. Store in amber colored bottle. This reagent is stable for at least six months.

2. Copper Standard Solution: 0.393 gm of copper sulphate pentahydrate ($CuSO_4 \cdot 5H_2O$) is taken into a small beaker and dissolve in water. Then 3 ml conc. H_2SO_4 is added and transfer in a 1000 ml volumetric flask and make up to the mark (100 ppm). From this solution 10 ml is taken in a 100 ml volumetric flask and added water up to the mark. This solution is 10 ppm (0.01 mg of Cu per milliliter or 10 microgram per gram).

Procedure: The gel samples dried to a constant weight were immersed in Copper solution at room temperature. These samples were taken out from Copper solutions at different time intervals (1 to 48 hours). The aqueous solution is then shaken with a solution in chloroform diethyldithiocarbamate to form and extract the yellow copper complex and optical density is measured by using spectrophotometer at wavelength 435 mm. in respect of a blank solution.

Preparation of calibration curve: 1, 3, 5, 7 and 10 ml solution is taken individually in five small beakers from the 10 ppm of standard solution of $CuSO_4 \cdot 5H_2O$ and made to 40 ml each by adding water. Take this solution in separating funnel and added 25 ml Zinc diethyldithiocarbamate in chloroform and shaked for 2 min. and separated chloroform layer and optical density is measured by using spectrophotometer at wavelength 435 mm. in respect of a blank solution (Figure 2).

Results and Discussion

Gel fraction of PVA/KC blend hydrogel

The gel is the insoluble part of a cross-linked polymer. It is measured after removing the sol (soluble portion) in a suitable solvent. The gel content (%) of Poly (vinyl alcohol) (PVA) / Kappa-carrageenan (KC) blend hydrogel prepared by γ-radiation at 25 kGy dose and different concentrations of KC are determined by gravimetric procedure.

From the Figure 3, it is found that the gel content of PVA / KC blend hydrogel is decreased with increase in concentration of KC. This

is occurred by the reason that KC is a radiation degradable polymer, when PVA with KC is subjected to radiation dose, KC degraded to low molecular weight compounds that causes the reduction of the gel fraction of PVA hydrogel with KC due to the degraded KC molecules may prevent to cross-link of PVA molecules [1]. The cross-link density in PVA/sago and PVP/sago blend hydrogels decreases due to degradation of sago by radiation [15]. Starch also reduces the gel content of PVA/starch blend hydrogel [16].

Swelling ratio of PVA/KC blend hydrogel in distilled water

Swelling ratio reflects the cross-linking of a polymer. The changes of swelling ratio of hydrogel without and with various concentration of Kappa-carrageenan are shown in the Figure 4.

Swelling is a result of the balance between two forces. One is osmotic force due to covalent or non covalent bond in the gel and another force is dispersing force. Osmotic force pushes water into polymer network. Whereas dispersing force exerted by polymer chains resists it. So with the increase of cross-link density a limited scope is available for free water enters into the vacant spaces of cross-link network.

From the Figure 4, it is shown that the degree of swelling increases with increase in concentration of the Kappa-carrageenan. The swelling ratio increases from 16.15 to 20.49 for the concentration of KC from 0.0% to 4.0% at the radiation dose of 25 kGy. Kappa-carrageenan is an anionic polymer. The water absorption capacity of kappa-carrageenan molecule is high due to the presence of strong hydrophilic group, $-OSO_3^-$ [17]. As a result, the swelling ratio of PVA/KC blend hydrogel is increased with increase in concentration of Kappa-carrageenan.

Swelling ratio of PVA/KC blend hydrogel in NaCl Solution

The degree of swelling for a hydrogel at equilibrium is characteristics of the polymer composition and cross-linking density, the temperature, P^H and ionic strength of the medium in which the gel is stored and the

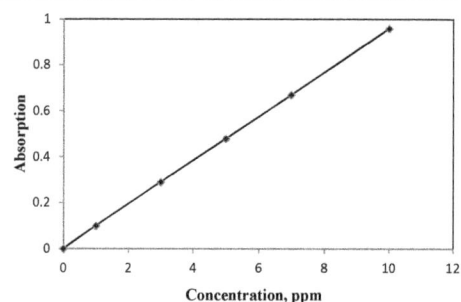

Figure 2: Calibration curve for uptake of metal concentration.

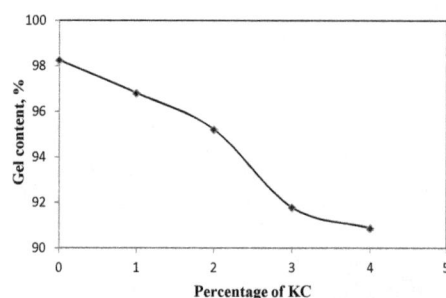

Figure 3: Effect of concentration of KC on the gel content for 9% PVA / KC blend hydrogel.

existing hydrostatic pressure. The degree of swelling can be altered by some environmental factors. Adding some an electrolyte can change the swelling ratio.

From the Figure 5 it is observed that swelling ratio in NaCl solution is lower than that of distilled water. Swelling ratio decreases with the increased the concentration of NaCl in aqueous solution. This is due to the fact that the concentration of counter ions increases, the effective ionization of the polymer is reduced and the electrostatic free energy of the system decreases. There is a reduction of volume i. e. shrinkage of the gel material. So the swelling ratio decreases with the increase concentration of the NaCl solution. It is also found from Figure 4 that the swelling ratio increases with increase in concentration of the kappa-carrageenan in PVA / KC blend hydrogel.

Swelling ratio of PVA/KC blend hydrogel in buffer solution

The first responsive gel was a pH sensitive gel made from poly (methacrylic acid) described by Katchalsky in 1949. This gel absorbed hundred times of its dry weight in water at high pH but shrink progressively as the pH was reduced. Figure 6 represents the effect of pH on the swelling ratio of polyvinyl alcohol/ kappacarragenan blend hydrogel.

From the Figure 6 it is found that the swelling ratio increases with the increase in pH. The lower swelling ratio in acidic region may be due to protonation and thereby resisting the association of water through hydrogen bonding. The increasing in the pH value of buffer, the deprotonation may be occurred and increases the swelling ratio.

From the Figure 6 it is found that the swelling ratio also increases with increasing of the concentration of Kappa-carrageenan. This may be happened due to reduction of cross-linking and increased the presence of –OSO$_3^-$ group with increase in concentration of KC in PVA / KC blend hydrogel.

Water absorption (%)

Water absorption is a measurement of swelling. The changes of water absorption (%) of PVA hydrogel without and with various concentrations of KC as well as different times are shown in the Figure 7.

It is found from the Figure 7 that the absorption of water of the hydrogel increases with time and increases with the increase of Kappa-carragenan ratio in the blend gel. It is also found that maximum absorption occurs within 24 hours of dipping in water. Then the absorption of water is insignificant.

Water desorption (%)

The changes of water desorption (%) of PVA hydrogel without and with various concentrations of KC as well as different times are shown in the Figure 8.

It is found from the Figure 8 that the desorption of water of the blend gel increases with time but decreases with the increase of Kappa-carragenan ratio in the blend gel. This decrease in the water desorption may be due to the formation of more cross linked density in the hydrogel as increase PVA concentration cause increase cross linking in the PVA network. It is also found that maximum desorption occurs within 48 hours in normal environmental condition. Then the desorption of water is insignificant.

Moisture absorption (%)

The changes of moisture absorption (%) of PVA hydrogel without and with various concentrations of KC as well as different times are shown in the Figure 9.

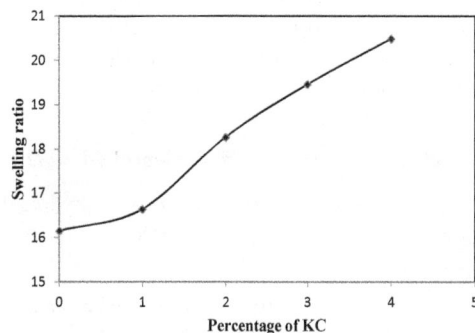

Figure 4: Degree of swelling in water with different concentration of KC in 9% PVA / KC blend hydrogel.

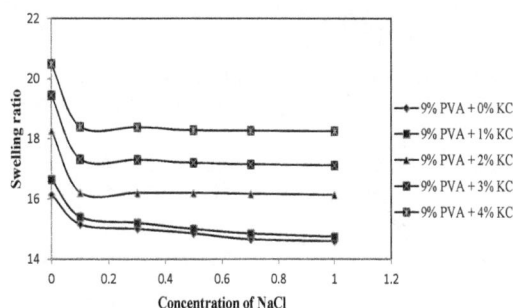

Figure 5: Effect of the concentration of KC on the ionic strength of NaCl solution on the swelling ratio of PVA / KC blend hydrogel.

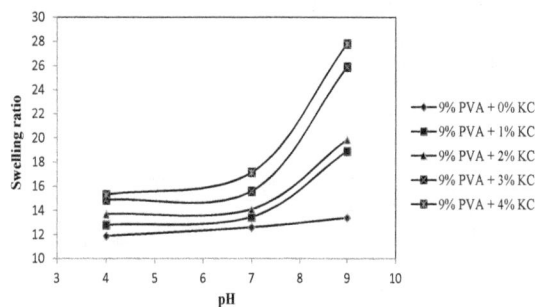

Figure 6: Effect of the concentration of KC on the swelling ratio of the PVA / KC blend hydrogel with different pH.

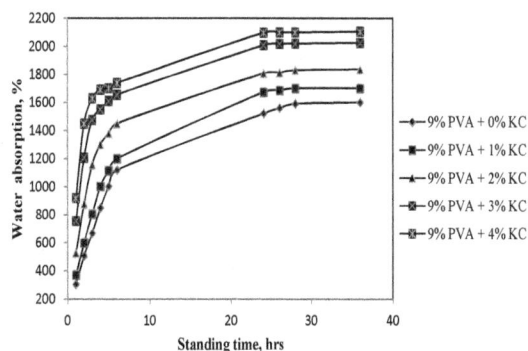

Figure 7: Effect of Kappa-carragenan ratio on the water absorption of PVA / KC blend hydrogel prepared at 25 kGy.

It is found from the Figure 9 that the absorption of moisture of the blend gel increases with time and increases with the decrease of Kappa-carragenan ratio in the blend gel. It is also found that maximum absorption occurs within 48 hours and then the absorption of moisture is insignificant.

Uptake of metal (Cu^{+2}) by PVA / KC blend hydrogel

The changes of uptake of metal (Cu^{+2}) of PVA hydrogel without and with various concentrations of KC as well as different times are shown in the Figure 10.

It is found from the Figure 10 that metal uptake capacity of PVA / KC blend hydrogel increases with time and maximum uptake occurs within 24 hours and then the uptake of metal (Cu^{+2}) is insignificant. It is also found that metal uptake capacity of PVA / KC blend hydrogel increases with increasing concentration of KC. This may be due to increase the content of strong hydrophilic group, $-OSO_3^-$ with increase KC in the blend hydrogel.

From the research work it is found that swelling and absorption properties of Poly vinyl alcohol (PVA) / Kappa-carragenan (KC) blend

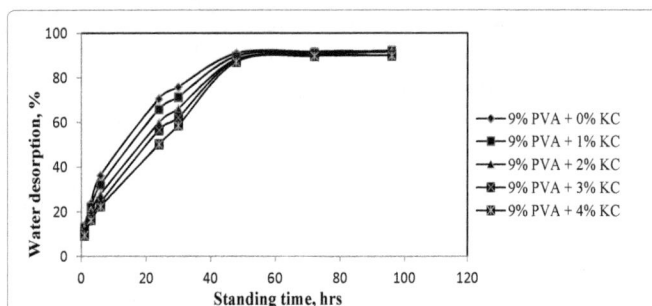

Figure 8: Effect of Kappa-carragenan ratio on the desorption in the PVA / KC blend hydrogel prepared at 25 kGy.

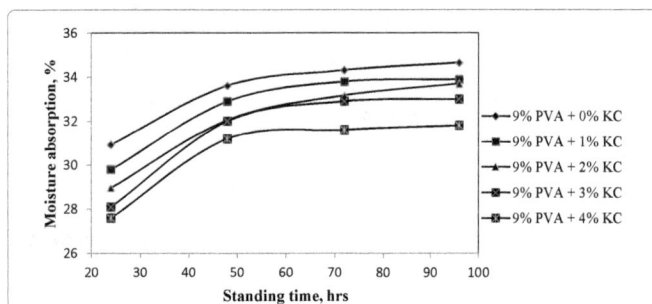

Figure 9: Effect of Kappa-carragenan ratio on the absorption in the PVA/KC blend hydrogel prepared at 25 kGy.

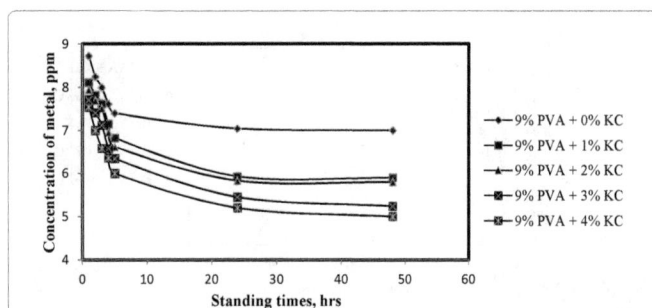

Figure 10: Effect of concentration of KC on the uptake of metal (Cu^{+2}), with PVA and KC blend hydrogel.

hydrogel prepared at 25 kGy increased with increase in concentration of KC. Since KC is a radiation degradable polymer, when PVA with KC is subjected to a radiation dose, KC degrades to a low molecular weight compound. The degraded KC molecules may prevent to cross-link of PVA molecules and may cause the reduction of the gel content of PVA hydrogel with increased concentration of KC. Again swelling and water absorption properties are increased with the decreased in gel content of hydrogel. Finally required properties are enhanced with the use of KC.

Conclusion

Poly vinyl alcohol (PVA) is a cross linked polymer as it forms gel when subjected to radiation. The gel fractions of the Poly vinyl alcohol (PVA) hydrogel and Poly vinyl alcohol (PVA)/Kappa-carragenan (KC) blend hydrogel increases with increased concentration of KC but it never become 100%. The gel fraction of PVA hydrogel is higher than that of PVA/KC blend hydrogel and with the increase of KC concentration in the blend hydrogel the gel fraction decreases. The swelling ratio in distilled water of PVA is higher than that of PVA/KC blend hydrogel and with the increase KC concentration the degree of swelling decreases. The metal ion (Cu^{+2}) uptake capacity increases with the increased concentration of KC in the PVA/KC blend hydrogel. The metal ion (Cu^{+2}) uptake properties showed that maximum uptake occurred within the first 24 hours and then the uptake is insignificant. PVA/KC blend hydrogel can be used for water purification because of its uptake capacity of metal ion from aqueous solution as well as for biomedical applications such as all types of wound dressing and covers can act also as slow-release drugs.

References

1. Dafader NC, Manir MS, Alam MF, Swapna SP, Akter T, et al. (2015) Effect of kappa-carrageenan on the properties of poly(vinyl alcohol) hydrogel prepared by application of gamma radiation. SOP Transactions on Applied Chemistry 2: 1-12.

2. Singh H, Vasudevan P, Ray AR (1980) Polymeric hydrogels: preparation and biomedical applications. J Scient Ind Res 39: 162.

3. Kaetsu I, Uchida K, Morita Y, Okubo M (1992) Synthesis of electro-responsive hydrogels by radiation polymerization of sodium acrylate. Radiat Phys Chem 40: 157-160.

4. Smidsrød O, Pernas BLAJ, Haug A (1967) The effect of alkali treatment on the chemical heterogeneity and physical properties of some Carrageenans. Acta Chem.Scand 21: 2585-2598.

5. McCandless EL, Gretz MR (1984) Biochemical and immunochemical analysis of carrageenans of the Gigartinaceae and Phyllophoraceae. Hydrobiologia 116: 175-178.

6. Lund S, Bjerre-Petersen E (1952) Industrial utilization of Danish seaweeds. Proc Int Seaweed Symp 1: 85-87.

7. Ukida J, Naito R, Kogyo KZ (1955) Chem Abstr 50: 8245h.

8. Kelly K (2005) Radiation may have positive effects on health: study -Low, chronic doses of gamma radiation had beneficial effects on meadow voles. University of Toronto, Canada.

9. Czechowicz-Janicka K, Romaniuk I, Piekarniak A, Wicha-Brazuchalska A, Galan S, et al. (1992) Polymer ocular implants for controlled release of drugs. I. Animal testing of the materials. Klin Oczna 94: 41.

10. Haehnel W, Herrmann WO (1924) Ger Pat 450 286.

11. Maolin Z, Hongfei H, Yoshii F, Makuuchi K (2000) Effect of kappa-carrageenan on the properties of poly (N-vinyl pyrrolidone)/kappa-carrageenan blend hydrogel synthesized by g-radiation technology. Radiation Physics and Chemistry 57: 459-464.

12. Dafader NC, Haque ME, Akhtar F (2005) Effect of kappa-carrageenan on the properties of poly (vinyl pyrrolidone) hydrogel prepared by the application of radiation. Polymer-Plastics Technology and Engineering 44: 1339-1346.

13. Dafader NC, Haque ME, Akhtar F (2005) Synthesis of hydrogel from aqueous solution of poly (vinyl pyrrolidone) with agar by gamma-rays irradiation. Polymer-Plastic Technology and Engineering 44: 243-251.

14. Hossen MK, Azim MA, Chowdhury SA, Dafader N, Haque M, et al. (2008) Characterization of poly (vinyl alcohol) and poly (vinyl pyrrolidone) co-polymer blend hydrogel prepared by application of gamma radiation. Polymer-Plastics Technology and Engineering 47: 662-665.

15. Hashim K, Mohid N, Bahari K, Dahlam KZ (2000) Radiation crosslinking of starch/water-soluble polymer blends for hydrogel. Jaeri-conf. 2000-03, JAERI, Takasaki, Japan, p: 23.

16. Zhai M, Yoshii F, Kume T, Hashim K (2002) JAERI-conf. 2002-03, JAERI, Takasaki, Japan, p: 54.

17. Makuuchi K, Yoshii F, Aranilla CT, Ahai M (2000) JAERE-conf. 2000-01. p: 192.

20

Subcritical Water Extraction of Xanthone from Mangosteen (*Garcinia Mangostana Linn*) Pericarp

Siti Machmudah[1*], Qifni Yasa' Ash Shiddiqi[1], Achmad Dwitama Kharisma[1], Widiyastuti[1], Wahyudiono[2], Hideki Kanda[2], Sugeng Winardi[1] and Motonobu Goto[2*]

[1]Department of Chemical Engineering, Sepuluh Nopember Institute of Technology, Kampus ITS Sukolilo, Surabaya 60111, Indonesia

[2]Department of Chemical Engineering, Nagoya University, Furo-cho, Chikusa-ku, Nagoya 464-8603, Japan

Abstract

Subcritical water extraction of phenolic compounds from mangosteen pericarps was examined at temperatures of 120-180°C and pressures of 1-5 MPa using batch and semi-batch extractor. This method is a simple and environmentally friendly extraction method requiring no chemicals other than water. Under these conditions, there is possibility for the formation of phenolic compounds from mangosteen pericarps from decomposition of bounds between lignin, cellulose, and hemicellulose via autohydrolysis. In both of systems, the total phenolic content inclusive xanthone increased with increasing extraction temperature. In batch-system, the maximum yield of xanthone was 34 mg/g sample at 180°C and 3 MPa with 150 min reaction time. The total phenolic content could approach to 61 mg/g sample at 180°C and 3 MPa with 150 min extraction time. The results revealed that subcritical water extraction is applicable method for the isolation of polyphenolic compounds from other types of biomass and may lead to an advanced plant biomass components extraction technology.

Keywords: Mangosteen; Phenolic compound; Xanthone; Subcritical water; Extraction

Introduction

Phenolic compounds are secondary plant metabolites, which are important determinants in the sensory and nutritional quality of fruits, vegetables and other plants. These compounds, one of the most widely occurring groups of phytochemicals, are of considerable physiological and morphological importance in plants. As a large group of bioactive chemicals, they have diverse biological functions. Phenolics may act as phytoalexins [1,2], antifeedants, attractants for pollinators, contributors to plant pigmentation, antioxidants and protective agents against UV light, amongst others [3]. These bioactive properties made these compounds play an important role in plant growth and reproduction, providing an efficient protection against pathogens and predators [2,4], besides contributing to the color and sensory characteristics of fruits and vegetables [2,5].

Recently, the consumption of fruits which contained high in antioxidant properties has become popular due to the increasing public awareness of the health. Pericarps of the fruit have been used in folk medicine for the treatment of many human illnesses [6]. Mangosteen (*Garcinia mangostana Linn*) is one of the fruits which used as an ingredient in commercial products including nutritional supplements, herbal cosmetics, and pharmaceutical products. This fruit belongs to the family of Guttiferae and is known the queen of the fruit. Mangosteen is a tropical tree and cultivated for centuries in South East Asia rainforests, and can be found in many countries worldwide. The major bioactive compounds found in mangosteens are phenolic acid, prenylated xanthone derivatives, anthocyanins, and procyanidins [6,7].

Xanthone is a kind of polyphenolic compounds that contain a distinctive chemical structure with a tricyclic aromatic ring. This compound had a variety of biological activity, for instance antioxidant, antibacterial, antiinflammatory, and anticancer effects [6,8]. Traditionally, xanthone is commonly obtained by extraction with organic solvents such as ethanol, acetone, hexane and methanol [6,9-12]. This extraction methods had several drawbacks; they are time consuming, laborious, have low selectivity and/or low extraction yields. Moreover, this technique employed large amounts of toxic solvents. In this work, water under subcritical conditions (100 to 200°C; 10 MPa) would be used to extract xanthone from mangosteen via autohydrolysis. This technique has received much attention in past several years, especially in food, pharmaceuticals and cosmetic industry, because it presents an alternative for conventional processes such as organic solvent extraction, steam distillation and the low temperature separation process prevents the degradation of chemical compounds [11]. Under subcritical conditions, water may extract polar organic compounds or decompose lignocellulosic materials to produce valuable compounds such as saccharides and aromatic organic acids. This technique has been applied to recover protein and amino acids [13], and phenolic compounds [14]. This treatment has also been demonstrated by several studies to effectively convert cellulosic [15] and lignocellulosic biomass [16] into useful products.

Experimental Section

Materials

The fruits of mangosteen were purchased from the market in Surabaya, Indonesia. They were cleaned and the pericarps of mangosteen were separated and cut into small pieces by using mechanical device. Then, the pericarps were dried in oven at 60°C for one or two days until it reached a constant weight. Next, the dried of pericarps was ground into fine homogeneous powder (around ± 0.65 mm) using millser.

*Corresponding authors: Siti Machmudah, Department of Chemical Engineering, Sepuluh Nopember Institute of Technology, Kampus ITS Sukolilo, Surabaya 60111, Indonesia, E-mail: machmudah@chem-eng.its.ac.id

Motonobu Goto, Department of Chemical Engineering, Nagoya University, Furo-cho, Chikusa-ku, Nagoya 464-8603, Japan E-mail: mgoto@nuce.nagoya-u.ac.jp

After sieving process, they were stored in refrigerator until next step experiments. Xanthone ($C_{13}H_8O_2$, 98.0%) and methanol (CH_2O, 99.7%) were obtained from Wako Pure Chemical Industries Inc. (Tokyo, Japan). They were used without further purification. The chemical structure of xanthone is shown in Figure 1.

Experimental setup and procedure

In this work, the subcritical water extraction was conducted in two processes: batch and semi-batch process. In batch process, the experiments were carried out with teflon lined autoclaves (Parr Instruments-model 4749) as a reactor. The maximum of temperature and pressure are 200°C and 10 MPa, respectively. 2.5 g of feed and pure water corresponding to 0.89-0.95 g cm^{-3} water density were loaded in the teflon lined. Then, it was placed in an autoclave cover (stainless steel) and closed firmly with screw cup. For operation safety, the stainless steel springs were placed at top and bottom of teflon lined autoclave. The reactor was placed into an electric furnace (Linn High Therm GmbH, model VMK 1600, Germany) and quickly heated to the desired temperature. The temperature in the reactor was measured by a thermocouple (K-type) inserted in the reactor cover. The time required to heat up the reactor from room temperature to the desired temperature was around 12-17 min and after that the reactor temperature was constant. The difference of the furnace temperature and the reactor temperature was around 8°C. Pressures were calculated from the water densities and steam tables. After 30-150 min (include the heating time about 12-17 min), the reactor was turned out from an electric furnace and quickly quenched in a water bath at room temperature. After cooling, the reactor was opened and then liquid and solid fractions were collected with washing inside the reactor by pure water so that the total volume of the product solution became 30 ml. Each experiment was conducted in duplicates/triplicates.

Figure 2 shows the schematic diagram of subcritical water extraction apparatus. The main apparatus consists of a high-pressure pump (200 LC Pump, Perkin Elmer, Germany), heater (Linn High Therm GmbH, model VMK 1600, Germany), reactor (10 ml in volume; Thar Design Inc., USA) and back-pressure regulators (BPR; AKICO, Japan). Both sides of the reactor were equipped with removable threaded covers included stainless-steel filters (0.1-1.0 μm). The pre-heater was fabricated from 1/8 inch stainless-steel tubing (SUS316) with a volume of 50 mL. The 1/16 inch stainless-steel tube was used to introduce hot water from the pre-heater to the reactor, which was located in the heater. After the reactor inclusive of 2.5 g of feed was installed to the system, distilled water at room temperature was pumped through the reactor inclusive pre-heater for a few minutes to purge air and completely wet the mangosteen pericarp; the system was then pressurized to the set pressure of 3 MPa through the back-pressure regulator, monitored by a pressure gauge (P, Migishita, Japan). These pressures are selected to keep the water in the liquid state at temperatures above its normal boiling point. In all experiments, feeds were placed between two layers of glass beads (the bottom and top) in the extraction container. The glass beads were used in order to distribute the solvent flow uniformly and reduce the dead space in the container. Therefore, the residence time was less than about 30 seconds. Glass beads (1.5-2.5 mm in diameter) were obtained from Oshinriko Co. Japan. When the system reached the desired pressure and a steady state was achieved, the electric heater was applied to heat the water. The reactor temperature was maintained at 120-180°C. The temperatures of the pre-heater, reactor and the electric heater were measured using K-type thermocouples and monitored using temperature controller (OMRON E5CJ, Japan). The time required to heat the reactor from room temperature to the desired temperature

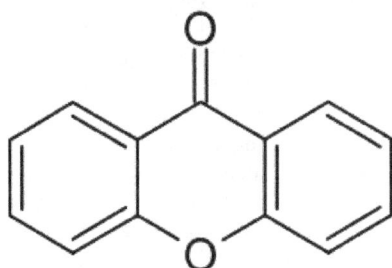

Figure 1: Chemical structure of xanthone.

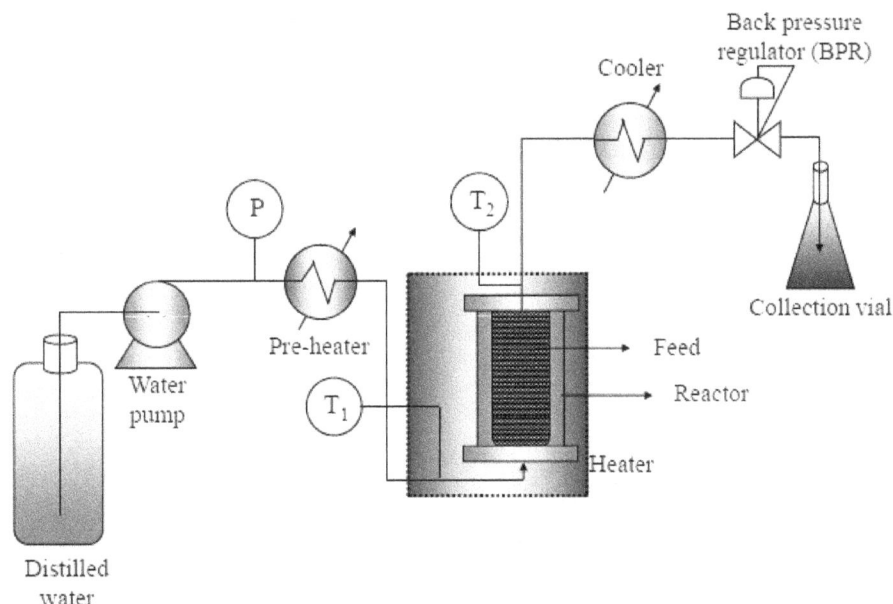

Figure 2: Schematic diagram of subcritical water extraction apparatus.

was 5-8 min, after which the reactor temperature equaled the electric heater temperature. After the temperature at the reactor has reached a preset temperature, the pump was used to feed water at 1.0 ml min⁻¹. Next, the outlet water was passed through the double-tube-type heat exchanger to quench the reaction. The time of experiment was 150 min, which, at 1.0 ml min⁻¹, produced a collected extract volume of 150 mL. Extracted solution was collected every 30 min.

Both of the processes (batch and semi-batch), the solid residues and liquid extracts were totally transferred to a petri dish and sealed bottles, respectively. During extraction process, the light exposure was avoided. The solid residues were dried in an oven at 60°C for 1 day and stored in a desiccator at room temperature. Extracted solution were directly stored in a refrigerator. These processes were maintained until analysis.

Analytical methods

Analysis of phenolic compounds and xanthone content in the extracts was conducted using UV-visible (UV-vis) spectrophotometry V-550 (Jasco Corporation, Japan), allowing spectra of between 190 and 800 nm with 10 nm min⁻¹ of bands in the fast scan mode. Liquid products were analyzed in a quartz cuvette with a 1 cm path length. UV-vis absorption is an effective tool for chemical characterization and may provide important information on the chemical structure of an analyte. The solid residues collected at each operating temperature were analyzed by a Spectrum One FT-IR spectrophotometer (Perkin-Elmer, Ltd., England) to determine the structure of the solid residues after the subcritical water treatment. The samples were placed directly in the diffuse reflectance attachment sample holder. Pre-flattening of the sample in a diamond cell was necessary prior to mounting. The spectra were measured in ATR (attenuated total reflectance) mode (golden single reflection ATR system, P/N 10500 series, Specac) at 4 cm⁻¹ resolution. The scanning wavenumber ranged from 4000 to 650 cm⁻¹.

Determination of total phenolic contents

The total phenolic content of the extracts was determined using the Folin-Ciocalteau's reagent. Initially, an aliquot of the extracts (0.1 mL) was diluted to a concentration (2 mL pure water) that was measurable using UV-Vis spectrophotometer prior to the addition of Folin-Ciocalteau's reagent and sodium carbonate. Then the Folin-Ciocalteau's reagent (0.5 mL) was added and mixed thoroughly. After shaking for 1 min, 2.0 ml of sodium carbonate (7.5% w/v) was added and mixed thoroughly. The mixtures were then allowed to stand for 2 h in the dark room before measuring its absorbance in a single beam UV-Vis spectrophotometer. A blank solution was required for initial calibration and it was prepared using methanol (pure solvent). The absorbance values of the extracts were referred to a standard calibration curve produced with five points of known gallic acid concentrations at 0 to 50 ppm to obtain its value in milligrams of gallic acid equivalents (GAE)/g of extract. Measurements were in triplicates.

Results and Discussion

In this work, mangosteen pericarps remained after subcritical water extraction was referred to as solid residue; this residue was characterized by infrared spectroscopy in the wavenumber region of 4000-650 cm⁻¹. Infrared spectroscopy is an analytical technique that allows identification of unknown substances and of the types of chemical bonds the compounds in those substances contain. Figure 3 shows the FT-IR spectra of mangosteen pericarps before and after treatment by subcritical water at temperature of 120°C and pressure of 3 MPa in batch process. Based on our previous work [17], the characterization of solid residues was only carried out for solid residue obtained at temperature of 120°C with reaction time 30 and 60 min. Generally, mangosteen pericarps consisting of cellulose, hemicellulose and lignin as three components of wood biomass is most likely composed of alkene, esters, aromatics, ketone and alcohol, with different oxygen-containing functional groups observed [17-19]. As a reference, the peak positions of all infrared bands and their functional groups are summarized in Table 1. Each molecule is composed of many different chemical bonds which are slightly elastic: they can stretch, bend, or vibrate. Therefore, some differences exist at each FT-

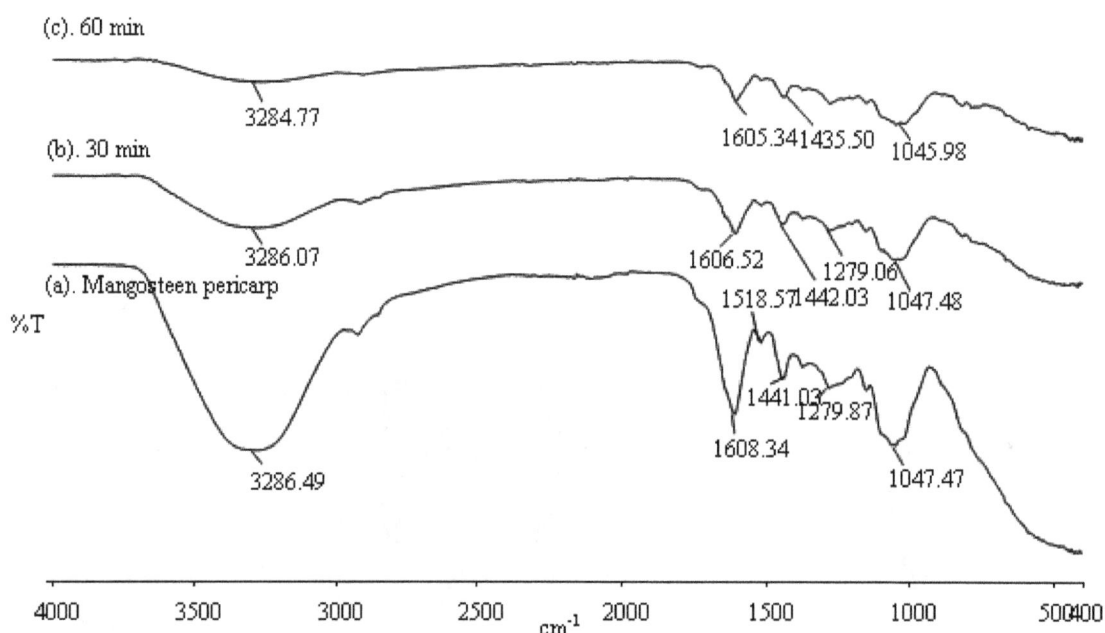

Figure 3: FT-IR spectrum of mangosteen pericarps before and after subcritical water treatment.

Wave number [cm⁻¹]	Functional groups	Compounds
3600-3000	O-H stretching	Acid, methanol
2860-2970	C-H$_n$ stretching	Alkyl, aliphatic, aromatic
1700-1730, 1510-1560	C=O stretching	Ketone and carbonyl
1632	C=C	Benzene stretching ring
1613, 1450	C=C stretching	Aromatic skeletal mode
1470-1430	O-CH$_3$	Methoxyl-O-CH$_3$
1440-1400	O-H bending	Acid
1402	C-H bending	
1232	C-O-C stretching	Aryl-alkyl ether linkage
1215	C-O stretching	Phenol
1170, 1082	C-O-C stretching vibration	Pyranose ring skeletal
1108	O-H association	C-OH
1060	C-O stretching and C-O deformation	C-OH (ethanol)
700-900	C-H	Aromatic hydrogen
700-650	C-C stretching	

Table 1: Main functional groups of the major constituents of plant biomass.

IR spectra due to their structural properties. Absorbance intensity due to hydrogen bonded O-H stretching (3600-3000 cm⁻¹) could be found in each spectrum. This intensity (3284.77-3286.49 cm⁻¹) decreased with increasing temperature, possibly due to the loss of alcoholic groups as further decomposition occurs at higher temperatures. The bands in the 1045.98-1047.48 cm⁻¹ and 1279.06-1279.87 cm⁻¹ regions are assigned to the stretching and deformation of aromatic C-O groups and the stretching of aryl-alkyl ether linkage C-O-C groups, respectively. In these regions, the peaks in (a and b) were sharper than in (c), showing that the C-O and C-O-C bonds in mangosteen pericarps were more reacted and consumed in (c). The same result also occurred at 1435.50-1442.03 cm⁻¹ and 1605.34-1608.34 cm⁻¹ due to the acid modes of O-H bending and the aromatic stretching modes of C=C, respectively. The intensity of the absorbance in these regions is mostly stable, indicating that methylene groups, syringyl units and ketone groups in mangosteen pericarps have difficulty cleaving under subcritical water conditions. Judging from these results, the extraction of xanthone from mangosteen pericarps by water at subcritical conditions was started at 120°C probably proceeded through hydrolysis reaction.

Although apparently a simple matter, xanthone determination is one of the least satisfactory of the analyses commonly performed on plants biomass. Generally, the methods used involve the solution and hydrolysis of all other plant constituents and the simple assumption that the extract after such treatment is target compound. Here, the determination of xanthone content in an extracts mangosteen pericarp was conducted by using UV-visible spectrophotometry at 243 nm [20,21]. The strong absorption at this wavelength in the UV is typical of a xanthone. Figure 4 shows the effect of extraction temperature on the yield of xanthone when the extraction was carried out at pressure of 3 MPa in batch system. As explained before, autohydrolysis could be applied for lignocellulosic materials lead to the solubilisation of phenolic compounds under subcritical water conditions, leaving a solid phase in both cellulose and hemicellulose. These interactions need to be broken to release of the antioxidant compounds. It was well known that the temperature of the extraction autoclave is a key variable of the extraction process under subcritical water conditions. As shown in Figure 4, the effect of temperature extraction on the yield of xanthone was significant. The yield of xanthone increased with the rise of extraction temperature from 120°C to 160 and 180°C at the same reaction time. The amount of xanthone extract at 160 and 180°C was almost 2-folds of that obtained at 120°C. It could be explained

that xanthone was more soluble in subcritical water at relative higher temperature. When temperature increased from 120 to 180°C, the dielectric constant of water significantly decreases from 50 to 38, which is closed to the dielectric constant of methanol (ε=33) or ethanol (ε=25). Therefore, the solubility of the compounds inclusive xanthone could be increased due to the decrease in water polarity. Kumar et al. [22] explained that a low dielectric constant allows liquid water to dissolve organic compounds, while a high ionization constant provides an acidic medium for the hydrolysis of biomass components via the cleavage of ether and ester bonds. Rangsriwong et al. [23] reported that the high amount of gallic acid and ellagic acid extracted with subcritical water at 150 and 180°C could possibly be due to the effect of hydrolysis reaction caused by the increase in the ionization constant of water at subcritical conditions. Nevertheless, they also explained that when the extraction temperature increased to 220°C, the amount of gallic acid and ellagic acid recovered was decreased, possibly due to further degradation at such high extraction temperature.

In subcritical water extraction, the water temperature as an extraction solvent is raised above the atmospheric boiling point, and pressure is applied to maintain the water in liquid state. The viscosity and surface tension of the water decrease, and the solubility and diffusion rate of the target compounds increase. The penetration of the water into the matrix and the transfer of the compounds out from the matrix are faster than in a similar extraction process performed at room temperature. Hence, compared to conventional extraction methods, the subcritical water extraction technique attains more rapid and efficient extraction. That is the main reasons why extraction at elevated temperatures and pressures give enhanced performance compared to extraction at lower temperature and atmospheric pressure [24]. Figure 5 illustrates the effect of extraction pressure on the yield of xanthone in flow type extraction process when the extraction was conducted at temperature of 160°C. As shown in this figure, the yield of xanthone increased with expanding extraction time. The yield of xanthone could approach to 20 mg/g sample at 3 MPa with 150 min extraction time. With the same extraction time, the yield of xanthone was around 18 mg/g sample at 1 and 5 MPa. As seen in this extraction process, the pressure did not result in significantly changes in the yields for the extraction of xanthone from mangosteen pericarps. It was also found that pressure does not have significant effect other than to keep the extraction solvent liquid at these range of temperatures [17,24-26]. Ko

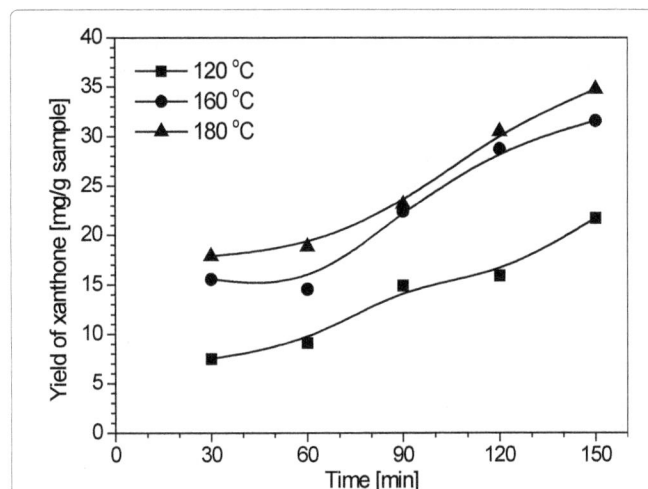

Figure 4: Effect of extraction temperature on the yield of xanthone in batch process.

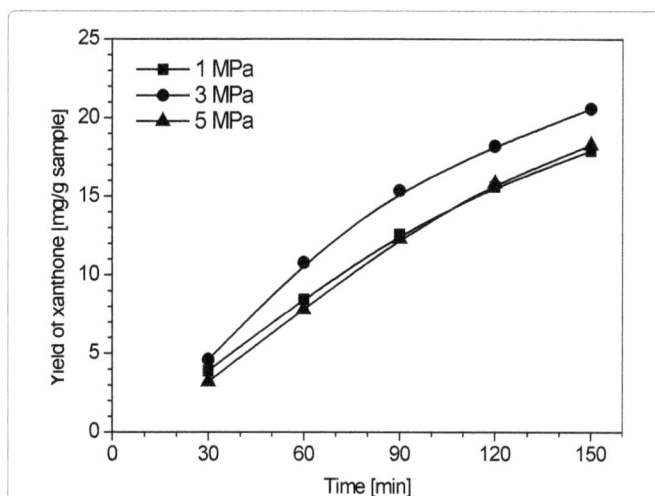

Figure 5: Effect of extraction pressure at 160°C on the yield of xanthone in semi-batch process.

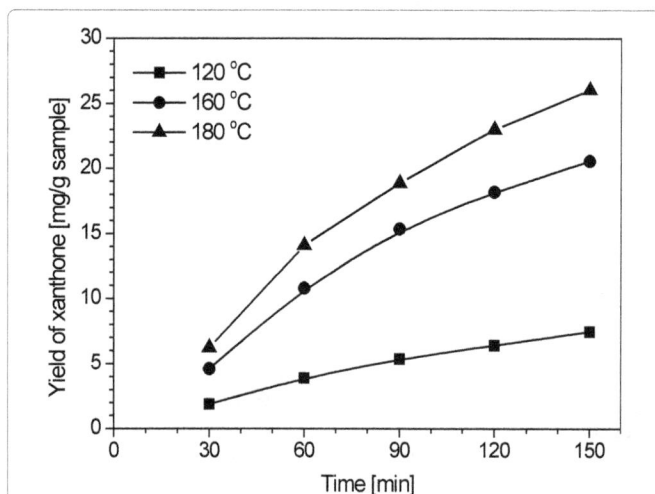

Figure 6: Effect of extraction temperature at 3 MPa on the yield of xanthone in semi-batch process.

et al. [26] performed subcritical water extraction of phenolic compound from onion skin at temperatures of 100-190°C and pressures of 90-131 bars with semi-batch processes. They reported that the changing the pressure has little effect on the extraction efficiency. Based on the result, it could be said that the pressure of extraction process is not a key variable of the extraction process under subcritical water conditions. However, a high pressure can help to enhance the subcritical water extraction efficiency by forcing the organic solvent into the matrix pores.

Figure 6 shows the effect of extraction temperature at a constant pressure on the yield of xanthone in semi-batch process. This figure shows the increase in the cumulative amount of xanthone with the increasing subcritical water extraction temperature from 120 to 180°C at the same extraction time. Similar to subcritical water extraction in batch process (Figure 4), here, subcritical water extraction in semi-batch process has also demonstrated its ability to extract xanthone from mangosteen pericarps depending on the temperature used. The amount of xanthone extract increased significantly when the extraction temperature was increased from 120°C to 160°C or 180°C. The yield of xanthone extract at 160 and 180°C could reach 3-folds of that obtained

at 120°C. Same to previous reasons, the increase in the xanthone content in the extract with the increasing extraction temperature could be explained by the fact that the increasing extraction temperature decreased the polarity or dielectric constant of water and that the decrease contributed to the increase in the solubility of most phenolic acids in the water at subcritical conditions. Even though it is known that increasing the temperature of subcritical water could dissolve several compounds similar to the organic solvents (methanol, ethanol), meanwhile the subcritical water particularly also promotes many reactions, such as hydrolysis and decomposition, which can degrade the compounds in the raw materials. As shown in Figure 6, at 180°C the yield of xanthone increased with longer extraction time and higher than that of 120 and 160°C. It indicated that xanthone was still stable at 180°C and there was no degradation reaction. Empikul et al. [27] carried out subcritical water extraction for defatted rice bran to extract phenolic compounds at temperatures of 120, 160, 200, and 250°C They reported that the subcritical water extraction at 200 and 250°C did not cause degradation of the phenolic substances but it increased the extraction efficiency by 2-3 times for the phenolic substances versus that at 120 and 160°C.

Phenolic compounds are known important due to their antioxidant activities. They possess aromatic structure along with hydroxyl substituents which enable them to protect human tissues from damages caused by oxygen or free radicals, and consequently reduce the risk of different diseases, and offer beneficial effects against cancers, cardiovascular disease, diabetes, and Alzheimer's disease [28]. Pourali et al. reported that there is possibility for the formation of total phenolic content from plant biomass at subcritical water conditions from decomposition of bounds between lignin, cellulose, and hemicellulose via autohydrolysis [28]. The result of phenolic compounds extraction from mangosteen pericarps by water at subcritical conditions would be presented. The Folin-Ciocalteu reagent is used to obtain a crude estimate of the amount of total phenolic content present in extract mangosteen pericarps. Blainski et al. suggested that after optimization of the conditions for the spectrophotometric determination of phenolics using the Folin-Ciocalteu reagent, all parameters analyzed showed adequate results [29]. Figure 7 describes the effect of subcritical water extraction temperature on total phenolic content when the extraction was performed at a constant pressure (3 MPa). At 120°C, the amount of total phenolic content increased with extending extraction time and

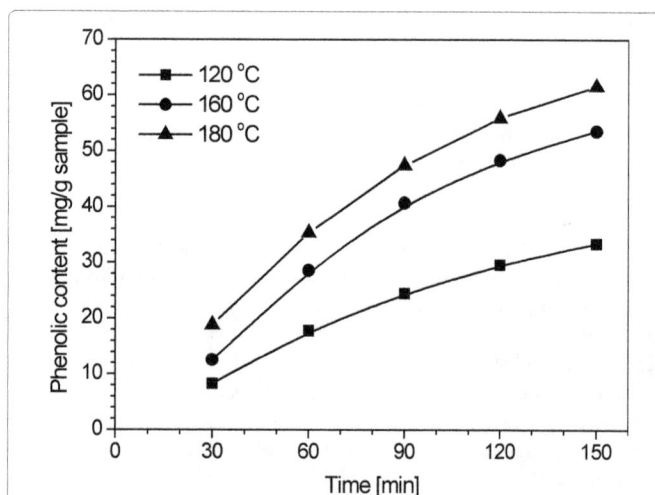

Figure 7: Effect of extraction temperature on total phenolic content of mangosteen extracts.

reached to 30 mg/g sample at 150 min extraction time. At 180°C, it increased significantly to 60 mg/g sample with the same extraction time. Again, this phenomenon could be explained that mangosteen pericarps phenolic compounds were more soluble in subcritical water at relative higher temperature. At such high temperature, phenolic compounds could dissolve in subcritical water as much as they dissolve in the organic solvents [27,14].

Conclusions

Subcritical water extraction of phenolic compounds from mangosteen pericarps was examined at temperatures of 120-180°C and pressures of 1-5 MPa using batch and semi-batch system. Under these conditions, there is possibility for the formation of phenolic compounds from mangosteen pericarps from decomposition of bounds between lignin, cellulose, and hemicellulose via autohydrolysis. In both of systems, the total phenolic content inclusive xanthone increased with increasing extraction temperature. In batch-system, the maximum yield of xanthone was 34 mg/g sample at 180°C and 3 MPa with 150 min reaction time. The total phenolic content could approach to 61 mg/g sample at 180°C and 3 MPa with 150 min extraction time. The FTIR spectrum of solid residues indicated that the liquefaction of mangosteen pericarps was started at 120°C and 3 MPa in batch process. This method could be an excellent alternative medium for extracting phenolic compounds from other types of biomass due to its temperature-dependent selectivity and environmental acceptability.

Acknowledgement

A part of this research was supported by a grant from Directorate General of Higher Education, Ministry of Education and Art of Indonesia through a research grant Desentralisasi - Penelitian Unggulan Perguruan Tinggi contract no. 016457.8/IT2.7/PN.01.00/2014.

References

1. Popa VI, Dumitru M, Volf I, Anghel N (2008) Lignin and polyphenols as allelochemicals. Ind Crops Products 27: 144-149.

2. Ignat I, Volf I, Popa VI (2011) A critical review of methods for characterisation of polyphenolic compounds in fruits and vegetables. Food Chem 126: 1821-1835.

3. Naczk M, Shahidi F (2006) Phenolics in cereals, fruits and vegetables: Occurrence, extraction and analysis. J Pharm Biomedic Anal 41: 1523-1542.

4. Bravo L (1998) Polyphenols: Chemistry, Dietary Sources, Metabolism, and Nutritional Significance. Nutr Rev 56: 317-333.

5. Alasalvar C, Grigor JM, Zhang D, Quantick PC, Shahidi F (2001) Comparison of volatiles, phenolics, sugars, antioxidant vitamins, and sensory quality of different colored carrot varieties. J Agric Food Chem 49: 1410-1416.

6. Aisha AFA, Abu-Salah KM, Ismail Z, Majid AMSA (2012) In vitro and in vivo anti-colon cancer effects of Garcinia mangostana xanthones extract. BMC Complement. Altern Med 12: 1-10.

7. Chaovanalikit A, Mingmuang A, Kitbunluewit T, Choldumrongkool N, Sondee J, et al. (2012) Anthocyanin and total phenolics content of mangosteen and effect of processing on the quality of mangosteen products. Int Food Res J 19: 1047-1053.

8. Ahmat N, Azmin NFN, Ghani NA, Aris SRS, Sidek NJ, et al. (2010) Bioactive Xanthones from the Pericarp of Garcinia mangostana. Middle-East J Sci Res 6: 123-127.

9. Asai F, Tosa H, Tanaka T, Iinuma M (1995) A xanthone from pericarps of garcinia mangostana. Phytochem 39: 943-944.

10. Garcia-Salas P, Morales-Soto A, Segura-Carretero A, Fernandez-Gutierrez A (2010) Phenolic-Compound-Extraction Systems for Fruit and Vegetable Samples. Molecules 15: 8813-8826.

11. Zarena AS, Sankar UK (2009) Supercritical carbon dioxide extraction of xanthones with antioxidant activity from Garcinia mangostana: Characterization by HPLC/LC-ESI-MS. J Supercrit Fluids 49: 330-337.

12. Negi JS, Bisht VK, Singh P, Rawat MSM, Joshi GP (2013) Naturally Occurring Xanthones: Chemistry and Biology. J Appl Chem ID 621459.

13. Zhu GY, Zhu X, Wan XL, Fan Q, Ma YH, et al. (2010) Hydrolysis technology and kinetics of poultry waste to produce amino acids in subcritical water. J Analyt Appl Pyrol 88: 87-91.

14. He L, Zhang X, Xu H, Xu C, Yuan F, et al. (2012) Subcritical water extraction of phenolic compounds from pomegranate (Punica granatum L.) seed residues and investigation into their antioxidant activities with HPLC-ABTS radical dot+ assay. Food Bioprod Process 90: 215-223.

15. Wang FF, Liu CL, Dong WS (2013) Highly efficient production of lactic acid from cellulose using lanthanide triflate catalysts. Green Chem 15: 2091-2095.

16. Zhou CH, Xia X, Lin CX, Tong DS, Beltramini J (2011) Catalytic conversion of lignocellulosic biomass to fine chemicals and fuels. Chem Soc Rev 40: 5588-5617.

17. Matsunaga Y, Wahyudiono, Machmudah S, Sasaki M, Goto M (2014) Hot compressed water extraction of polysaccharides from Ganoderma lucidum using a semibatch reactor. Asia-Pacific J Chem Eng 9: 125-133.

18. Xiao LP, Sun ZJ, Shi ZJ, Xu F, Sun RC (2011) Impact of hot compressed water pretreatment on the structural changes of woody biomass for bioethanol production. BioResources 6: 1576-1598.

19. Demirbas A (2000) Mechanisms of liquefaction and pyrolysis reactions of biomass. Energy Conv Manag 41: 633-646.

20. Ghazali SAISM, Lian GEC, Ghani KDA (2010) Chemical Constituent from Roots of Garcinia Mangostana (Linn.). Int J Chem 2: 134-142.

21. Aisha AFA, Abu-Salah KM, Ismail Z, Majid AMSA (2013) Determination of total xanthones in Garcinia mangostana fruit rind extracts by ultraviolet (UV) spectrophotometry. J Med Plant Res 7: 29-35.

22. Kumar MSY, Dutta R, Prasad D, Misra K (2011) Subcritical water extraction of antioxidant compounds from Seabuckthorn (Hippophae rhamnoides) leaves for the comparative evaluation of antioxidant activity. Food Chem 127: 1309-1316.

23. Rangsriwong P, Rangkadilok N, Shotipruk A (2008) Subcritical Water Extraction of Polyphenolic Compounds from Terminalia chebula Fruits. Chiang Mai J Sci 35: 103-108.

24. Peterson AA, Vogel F, Lachance RP, Froling M, Antal Jr MJ, et al. (2008) Thermochemical biofuel production in hydrothermal media: A review of sub- and supercritical water technologies. Energy Environ Sci 1: 32-65.

25. Kronholm J, Hartonen K, Riekkola ML (2007) Analytical extractions with water at elevated temperatures and pressures. Analyt Chem 26: 396-412.

26. Ko MJ, Cheigh CI, Cho SW, Chung MS (2011) Subcritical water extraction of flavonol quercetin from onion skin. J Food Eng 102: 327-333.

27. Viriya-Empikul N, Wiboonsirikul J, Kobayashi T, Adachi S (2012) Effects of Temperature and Flow Rate on Subcritical-water Extraction from Defatted Rice Bran. Food Sci Technol Res 18: 333-340.

28. Pourali O, Asghari F, Yoshida H (2010) Production of phenolic compounds from rice bran biomass under subcritical water conditions. Chem Eng J 160: 259-266.

29. Blainski A, Lopes GC, Palazzo de Mello JC (2013) Application and Analysis of the Folin Ciocalteu Method for the Determination of the Total Phenolic Content from Limonium Brasiliense L. Molecules 18: 6852-6865.

21

Technoeconomics and Sustainability of Renewable Methanol and Ammonia Productions Using Wind Power-based Hydrogen

Michael Matzen, Mahdi Alhajji and Yaşar Demirel*

Department of Chemical and Biomolecular Engineering, University of Nebraska Lincoln, Lincoln NE 68588, USA

Abstract

This study analyzes and compares the economics and sustainability aspects of two hydrogenation processes for producing renewable methanol and ammonia by using wind-power based electrolytic hydrogen. Carbon dioxide from an ethanol plant is used for producing methanol, while the nitrogen is supplied by an Air Separation Unit (ASU) for producing ammonia. The capacities are 99.96 mt/day methanol and 1202.55 mt/day anhydrous ammonia. The methanol plant requires 138.37 mt CO_2/day and 19.08 mt H_2/day. The ammonia is synthesized by using 217.72 mt H_2/day and 1009.15 mt N_2/day. The production costs and the carbon equivalent emissions (CO_2e) associated with the methanol and ammonia processes, electrolytic hydrogen production, carbon capture and compression, and ASU are estimated. The integral facilities of both the methanol and ammonia productions are evaluated by introducing a multi-criteria decision matrix containing economics and sustainability metrics. Discounted cash flow diagrams are established to estimate the economic constraints, unit product costs, and unit costs of hydrogen. The hydrogen cost is the largest contributor to the economics of the plants. For the methanol, the values of emissions are -0.85 kg CO_2e/kg methanol as a chemical feedstock and +0.53 kg CO_2e/kg methanol as a fuel with complete combustion. For the ammonia, the value of emission is around 0.97 kg CO_2e/kg ammonia. The electrolytic hydrogen from wind power helps reduce the emissions; however, the cost of hydrogen at the current level adversely affects the feasibility of the plants. A multi-criteria decision matrix shows that renewable methanol and ammonia with wind power-based hydrogen may be feasible compared with the nonrenewable ones and the renewable methanol may be more favorable than the ammonia.

Keywords: Electrolytic hydrogen production; Methanol production; Ammonia production; Technoeconomic analysis; Sustainability metrics; Multi-criteria decision matrix

Introduction

Electrolytic hydrogen using wind power may serve as a feedstock for hydrogenation processes and hence chemical storage for renewable electricity [1-5]. Hydrogen is a clean fuel; its burning causes no harmful emissions; however the cost to produce, store, compress, and transport of the hydrogen is still high [6-12]. Methanol may be used as a fuel and a valuable feedstock for producing methyl t-butyl ether, dimethyl ether, dimethyl carbonate, formaldehyde, acetic acid and other chemical secondary intermediates which are used in producing plywood, particleboard, foams, resins and plastics [13-21].

Methanol production using fossil fuels, mainly from natural gas and coal, is a mature technology [13]. Renewable hydrogen-based methanol as an alternative fuel is widely investigated by researchers worldwide [1,2,13,22-24]. CO_2 may come from flue gas, gasification of biomass, or ethanol plants [1,13,25]. Energy analysis of recycling CO_2 and reaction mechanisms of hydrogenation of CO_2 are some of the efforts toward non-fossil fuel-based methanol as a renewable energy storage and carrier [26-32]. Rihko-Struckmann et al. [33] carried out an energetic evaluation in order to assess the overall efficiency of methanol and hydrogen-based storage systems for renewable electric energy; the efficiency of the system using hydrogen is higher compared with that of using methanol as the storage medium; however, storage and handling of methanol as chemical storage is favorable when compared with H_2 [18-20,33-36].

The utilization of CO_2 as carbon source for chemical synthesis could have a positive but only marginal impact on the global carbon balance [1,14]. Because, we add 3500 million mt CO_2/year worldwide, while we use only 110 million mt CO_2/year to produce other chemicals (mainly urea); this is only around 3% usage of the CO_2 as feedstock [1,2,14,20,30]. On the other hand, the utilization of CO_2 in the fuel production or as a chemical storage of energy, such as methanol, could make a significantly larger impact, as only 16.8% of the world oil consumption was used in 2007 for non-energy purposes [14,17,36].

Like methanol, ammonia is a feedstock for manufacturing fertilizers such as urea, and may be considered as a chemical storage medium of renewable electricity [37-42]. Pure nitrogen for ammonia synthesis is produced using an air separation unit. In the U. S., about 98% of ammonia is produced by catalytic steam reforming of natural gas, while about 77% of world ammonia capacity is based on natural gas. The total energy consumption for the production of ammonia in a modern steam reforming plant is 40-50% above the thermodynamic minimum [40-42].

This study evaluates and compares the economics and sustainability aspects of the hydrogenation processes for renewable methanol and ammonia productions. A multi-criteria decision matrix is introduced in the feasibility evaluations of these productions. The cost and emissions for hydrogen, nitrogen, and CO_2 feeds used in these productions are estimated and the renewable hydrogen, methanol and ammonia economics are reassessed.

*Corresponding author: Yaşar Demirel, Department of Chemical and Biomolecular Engineering, University of Nebraska, Lincoln 68588, NE, USA
E-mail: ydemirel2@unl.edu

Sustainability in Chemical Processes and Energy Technology

The following sustainability metrics are applicable to specific chemical processes and energy systems [43-46]:

- Material intensity (nonrenewable resources of raw materials, solvents/unit mass of products)

- Energy intensity (nonrenewable energy/unit mass of products)

- Potential environmental impact (pollutants and emissions/unit mass of products)

- Potential chemical risk (toxic emissions/unit mass of products)

This study uses a comparative assessment of the renewable methanol and ammonia plants with the sustainability metrics of material intensity, 'energy intensity' and 'potential environmental impact' as emissions of CO_2e by using the 'Carbon Tracking' and the 'Global Warming Potential' options of Aspen Plus [47]. The costs/unit mass of products are also considered in these metrics.

Table 1 shows the U.S. average Levelized Costs of Electricity (LCE) for generating technologies entering service in 2019 projected in the Annual Energy Outlook 2014 [2]. LCEs are estimated in 2012 \$/MWh and measures of the overall competitiveness of different generating technologies over an assumed financial life and duty cycle. The cost of Carbon Capture and Storage (CCS) accounts for 21% of the total LCE in Integrated Gasification of Combined Cycle (IGCC). For natural gas-based advanced combined cycle, the cost of CCS accounts for 29% of the total LCE. Wind-based electricity is becoming comparable with the hydropower, IGCC with CCS, and natural gas-based advanced combined cycle with CCS, although the cost of offshore-based wind power still remains high [2,12,48-50].

This study employs the CO_2e emission factor data source of US-EPA-Rule-E9-5711 and the fuel source of natural gas [47,51,52]. Carbon equivalent emission, CO_2e, indicates the global warming potential of GHGs; this study uses US-EPA with a predetermined cost for CO_2 fee/tax of \$/mt CO_2e.

Hydrogen Production

Currently, 96% of H_2 is produced directly from fossil fuels and about 4% is produced indirectly by using electricity generated through fossil fuels [53]. The conventional technologies are steam reforming of natural gas, coal gasification, and partial oxidation of hydrocarbons such as biomass. Renewable hydrogen comes from the electrolysis of water using hydropower, wind power, and solar photovoltaic power [54-56].

Hydrogen Production from Syngas

Commercial processes for H_2 production are based on syngas feedstock produced from natural gas steam reforming (Figure 1) and coal (or biomass) gasification (Figure 2) with carbon capture and storage. These processes are complex, sensitive to the feedstock quality, and require large investments for larger units. The generated CO can also be used in the water-gas shift reaction to yield more hydrogen. In these processes, however, at least 20% of the energy of the fossil fuel is lost as waste heat.

Energy efficiency for biomass-based H_2 production is around 60% and likely become competitive in the future [55]. A representative

gasification reaction of biomass is $C_aH_b + O_2 \rightarrow H_2 + CO + CO_2 + H_2O$. Here the biomass reacts with oxygen supplied by an air separation unit (ASU) at 1150°C-1400°C and 400-1200 psig. Most modern plants purify the crude H_2 to 99.99-wt% by removing methane, CO_2, N_2, and CO using multi-bed pressure swing adsorption [53-59].

Current production of H_2 from natural gas and coal accounts for 48% and 18% of the total production, respectively. The emission of CO_2 varies between 7.33 kg CO_2/kg H_2 and 29.33 kg CO_2/kg H_2 using conventional fuels at about 75% energy efficiency. CO_2 emission (beside SO_x and NO_x) associated with producing H_2 from coal is about two-three times higher than that of the H_2 produced from natural gas [2,5-8,11,12].

Hydrogen Production from Water Electrolysis

Renewable option is electro-chemical conversion by water electrolysis using electricity from renewable sources or nuclear power [49,50,53-58]. Figure 3 shows the schematic of wind power-based hydrogen production. Alkaline electrolysis technologies are the most mature commercial systems. The electrolyzer units use process water for electrolysis, and cooling water for cooling. KOH is needed for the electrolyte in the system. The system includes the following equipment: transformer, thyristor, electrolyzer unit, feed water demineralizer, hydrogen scrubber, gas holder, two compressor units to 30 bar, deoxidizer, twin tower dryer (Figure 3) [5,49]. These electrolyzers have the energy efficiencies (57%-75%) based on higher heating value- HHV and 50-60% based on the lower heating value-LHV. The typical current density is 100-300 mA/cm² [12,49].

The amount of total water used is 26.7 kg/kg H_2; electrolysis uses approximately 45%, while manufacturing the wind turbines and the hydrogen storage consume around 38% and 17% of the total water used, respectively. The total greenhouse gas emission is 0.97 kg CO_2e/kg H_2, which is distributed as 0.757 kg CO_2e/kg H_2 (78%) for the wind turbine production and operation (because of steel and concrete used in its construction), 0.043 kg CO_2e/kg H_2 (4.4%) for the electrolyzer construction and operation, and 0.17 kg CO_2e/kg H_2 (17.6%) for the

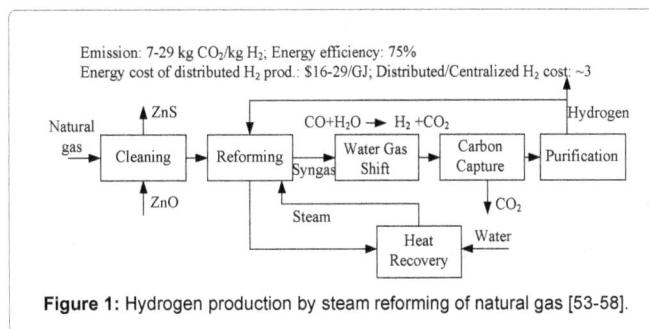

Figure 1: Hydrogen production by steam reforming of natural gas [53-58].

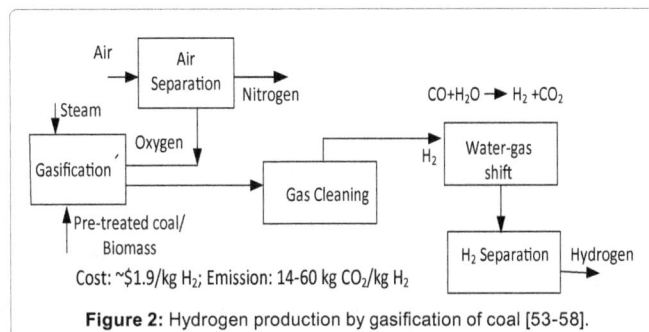

Figure 2: Hydrogen production by gasification of coal [53-58].

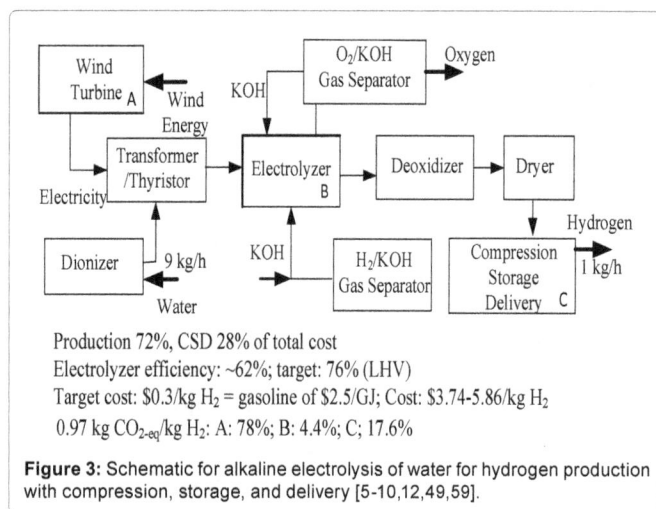

Production 72%, CSD 28% of total cost
Electrolyzer efficiency: ~62%; target: 76% (LHV)
Target cost: $0.3/kg H_2 = gasoline of $2.5/GJ; Cost: $3.74-5.86/kg H_2
0.97 kg CO_{2-eq}/kg H_2: A: 78%; B: 4.4%; C; 17.6%

Figure 3: Schematic for alkaline electrolysis of water for hydrogen production with compression, storage, and delivery [5-10,12,49,59].

hydrogen compression and storage (mainly due to the production of steel used in the storage tanks) [59].

M-Langer et al. [54] evaluated hydrogen production processes based on natural gas steam reforming, coal and biomass gasification, and water electrolysis. H_2 production cost is around $65/GJ using wind electricity, $30/GJ using nuclear power, and $600/GJ using photovoltaic electricity based on 2007 $. Large-scale processes, using natural gas and coal, are the most economical processes while biomass gasification still needs technological improvements. The operating cost of an electrolyzer is driven by the energy efficiency and the cost of electricity. Energy efficiency needs to be increased to 76% from the current average of about 62%. The capital costs of wind-based H_2 are $2086/kW (2011) and $2067/kW (2012) for 50000 kg H_2/day for a centralized production plant. New classes of materials could be designed at the nanoscale to produce catalysts that are more selective, less prone to poisoning, and able to operate at lower temperatures [5-9]. High-temperature solid oxide electrolysis can use lower cost energy (in the form of steam) for water-splitting to decrease electricity consumption [12,49].

Economics of Wind Power-Based Hydrogen

Wind power-based electrolysis production cost estimates are limited geographically and the base cost of H_2 ranges from $3.74/kg H_2 to $5.86/kg H_2. Capacities of H_2 productions range from 1,000 to 50,000 kg H_2/day [2-5]. Other factors such as large-scale storage, compression, pipeline transport, and dispensing economics need separate analyses [49]. Currently, the production of H_2 by electrolysis using renewable electricity is not competitive with chemical production methods based on fossil fuels. However, using the off-peak power could increase plant load factor and improve the economics [49,53]. Electrolytic H_2 may be more attractive for regions without access to natural gas or if H_2 is used as an energy storage medium [33,49].

The current capital equipment cost for advanced electrolysis is between $600/kW and $700/kW. This cost needs to be reduced to $200/kW to achieve $2.75/GGE (untaxed gasoline gallon equivalent) by 2015 [49,50]. This shows around 60% of the improvement needed. Table 2 shows some electrolyzer types with their efficiencies. Higher efficiencies are possible with Polymer Electrolyte Membrane (PEM) and Solid Oxide Electrolytic Cell (SOEC) electrolyzers, which are still under development. Table 3 shows a typical sensitivity analysis to determine how the availability of wind farms and the capacity of electrolyzer affect the electricity needed for the production of H_2 [53-56]. Capital cost of

electrolyzer increases considerably as the wind farms' availability and electrolyzer capacity decrease.

Integration with low-cost renewables and the flexibility to produce H_2 from the grid electricity during off-peak periods may help lower the production cost of H_2. A large alkaline (bipolar design) electrolyzer unit is the Norsk Hydro Atmospheric Type No. 5040, which can produce 1046 kg H_2/day (381,790 kg H_2/year) by using approximately 2.3 MW of electricity. Small systems however, are often built around Polymer Electrode Membrane (PEM) electrolyzer cell technology. Table 4 shows the streams of the Norsk hydro atmospheric type electrolyzer unit. The levelized cost is $6.63/kg H_2 (2007$) and the purchased electrolyzer system cost: $489/kW (2014$) [5-10]. Economic analysis shows that final production cost is around $4.97/kg H_2, which is much higher compared with the cost of $1.91/kg H_2 from coal gasification [9,10].

The gas output streams from the electrolyzer are assumed to be 100% pure (typical real outputs are 99.9 to 99.9998% for H_2 and 99.2 to 99.9993% for O_2). Electricity cost is typically 70 to 80% of the total cost of H_2 production. Table 5 shows the typical energy usage by the Norsk electrolyzer. The system energy requirement includes compression to bring the gas output to 33 bar (480 psi) [7,8]. The minimum power conversion system would require rectification of the variable ac output from the wind turbines to dc output for the electrolyzer cells. Future energy requirements are targeted at 50 kWh/kg H_2 [9,10,57-59].

Hydrogen production costs change approximately from $1.75/kg H_2 to $4.6/kg H_2 as the electricity prices change from $0.02/kWh to $0.08/kWh, for an advanced electrolyzer technology at 76% efficiency, and capital cost of $250/kW (current state of technology is 56%-75% efficiency and $700/kW) [12,49]. These costs represent distributed hydrogen production and include compression, storage, and delivery. The electrolyzer has a capacity factor of 70% to adjust for seasonal and weekend/weekday fluctuations in demand and a 97% availability of the equipment.

Production of H_2 is an energy-consuming process, and may not be environmentally friendly [18,56]. In addition, the low density and extremely low boiling point of H_2 increase the energy costs of compression or liquefaction and the investment costs of storage and delivery. Distributed electrolysis case may play a role in the transition to the hydrogen economy when there is little delivery infrastructure for hydrogen [12]. Underground gas storage of hydrogen and oxygen in connection with the electrolysis may enable the electrolyzer to accommodate the variations in the power produced by renewable resources. The output-input efficiency cannot be much above 30%, while the advanced batteries have a cycle efficiency of above 80%. Even the most efficient fuel cells may not recover these losses [56,58-61].

Methanol Production

Methanol synthesis needs carbon-rich feedstock (natural gas, coal or biomass), hydrogen, and a catalyst, mainly $Cu/ZnO/Al_2O_3$ [26-34]. Methanol is produced almost exclusively by the ICI, the Lurgi, and the Mitsubishi processes. These processes differ mainly in their reactor designs and the way in which the produced heat is removed from the reactor. To improve their catalytic performance, the CuO/ZnO catalysts have been modified with various metals, such as chromium, zirconium, vanadium, cerium, titanium, and palladium [30-33,62]. The long-term stability of the catalysts may be improved by adding a small amount of silica to the catalysts at reaction conditions of 5 MPa, 523 K [63]. A high catalyst activity is related to a high copper surface area or small crystallite size combined with intimate contact with the zinc promoter. Table 6 shows some of the experimental reactor operating

Plant type	Capacity factor (%)	LCE	O&M with fuel	Transmission investment	Total LCE	Emission* mt CO_2e/MWh
IGCC*	85	76.1	31.7	1.2	115.9	0.94-0.98
IGCC with CCS	85	97.8	38.6	1.2	147.4	0.94-0.98
NG-CC	87	15.7	45.5	1.2	64.4	0.55
NG-CC with CCS	87	30.3	55.6	1.2	91.3	0.55
Biomass	83	47.4	39.5	1.2	102.6	
Wind	35	64.1		3.2	80.3	
Wind-Offshore	37	175.4		5.8	204.1	
Solar PV	25	114.5		4.1	130.0	
Solar thermal	20	195.0		6.0	243.1	
Hydro	53	72.0	6.0	2.0	84.5	

*Steam-electric generators in 2012 for calculating the amount of CO_2 produced per kWhr[2];
IGCC: Integrated gasification combined cycle; O&M: Operations and Maintenance cost;
CCS: Carbon capture and storage; NG: Natural gas; PV: Photovoltaic

Table 1: Estimated U.S. average levelized cost of electricity (LCE) 2012 $/MWh for advanced generation resources entering service in 2019 [2].

Electrolyzer	Capacity (kW)	Efficiency % (HHV)	Efficiency % (LHV)
Alkaline	1-2,300	72	61
PEM	1-130	60	51
Solid Oxide	Pilot scale only	82	69

*Norsk Hydro's 30,000 Nm³/hr (~ 150 MW) connected to a hydroelectric power plant, generating about 70,000 kg H_2/day.
The higher heating values for hydrogen: HHV= 39.42 kWhr/kg and the lower heating value LHV= 33.31 kWhr/kg.
100% HHV efficiency translates into 84.5% efficiency based on LHV.

Table 2: Electrolyzer types* [9-11].

Wind turbine capital cost ($/kW)	1654	2067	2481
Electrolyzer energy use (kWh/kg H_2)	47.5	50	60
Electrolyzer capital cost ($/kW)	326	408	489
Wind farm availability (%)	90	88	86
Electrolyzer capacity factor (%)	99.5	98	96

Table 3: Sensitivity analysis changing the unit cost of H_2 with the production efficiency and electricity cost [9,10,55-57].

Water		Hydrogen		Oxygen		Water	
kg/hr	kmole/hr	kg/hr	kmole/hr	kg/hr	kmole/hr	kg/hr	kmole/hr
485	26.9	43.59	21.6	346.51	10.8	94.82	5.3

Table 4: Stream table of the norsk hydro atmospheric type electrolyzer unit [9,10].

System energy required (includes compression)		Hydrogen production		Electrolyzer energy required	System power required
kWh/(Nm³)	kWh/kg H_2	kg/h	kmole/hr	kWh/(Nm³)	kW
4.8	53.5	43.59	21.6	4.3	2330

Table 5: Energy usage for the Norsk electrolyzer [9,10].

Reactions	T, °C	P, bar
Based on all three reactions (1-3) [63]	250	50
Based on all three reactions (1-3) [65]	200-244	15-50
Based on reaction (1) and (2) [66]	215-270	50
Based on reaction (1) and (3) [67]	187-277	30-90
Based on reaction (1) and (3) [68]	180-280	51
Based on reaction (1) and (3) [69]	220-300	50-100

Table 6: Experimental conditions of methanol synthesis with the catalyst Cu/ZnO/Al_2O_3

temperatures and pressures with the catalyst Cu/ZnO/Al_2O_3. During the synthesis these following reactions occur [63-69]

$$CO_2 + 3H_2 = CH_3OH + H_2O \quad \Delta H°(298\ K) = -49.4\ kJ/mol \quad (1)$$

$$CO + 2H_2 = CH_3OH \quad \Delta H°(298\ K) = -90.55\ kJ/mole \quad (2)$$

$$CO_2 + H_2 = H_2O + CO \quad \Delta H°(298\ K) = +41.12\ kJ/mole \quad (3)$$

Only two of these reactions are linearly independent and two reaction rate equations can describe the kinetics of the all reactions.

Methanol from Natural Gas

Figure 4 shows the main blocks of natural gas-based methanol production. Three fundamental steps are: (i) natural gas reforming to produce syngas with an optimal ratio of $[(H_2\ CO_2)/(CO + CO_2)] = 2$, (ii) conversion of syngas into crude methanol, and (iii) distillation of crude methanol. Methanol synthesis from natural gas has a typical energy efficiency of 75% and emits around 1.6 kg CO_2/kg methanol [13]. Specific energy consumption for natural gas-based methanol is around 8.0 GJ/mt methanol [22]. Captured CO_2 is commonly reused internally in ammonia and some methanol plants.

Table 7 compares the cost of methanol production and emissions from fossil fuel resources. Coal-based syngas process has the highest emission of GHGs, which is around 2.8-3.8 kg CO_2/kg methanol. The typical energy efficiency for the coal-based methanol is in the range of 48% to 61% [13,22]. Technical and economic analyses of methanol production from biomass-based syngas show that overall energy efficiency is around 55% based on HHV. The level of emission is around 0.2 kg CO_2/kg methanol, which is mainly from biomass growing, harvesting, and transportation. Methanol from biomass or flue gas CO_2 is at least 2-3 times more expensive than the fossil-fuel based methanol [13,64-70].

Methanol from CO_2 and H_2

Converting CO_2 into chemicals is thermodynamically challenging, and inherently carries costs for the energy and hydrogen supply [22]. The conversions of reactions (1) to (3) with catalyst of Cu/ZnO/Al_2O_3 are limited by the chemical equilibrium of the system. The temperature rise must be minimized in order to operate at good equilibrium values. However selectivity for methanol is high with a value of 99.7% at 5 MPa and 523 K with a H_2/CO_2 ratio of 2.82 [63]. The energy efficiency for the concentrated CO_2 and hydrogen based methanol is around 46%. Figure 5 shows a schematic of renewable hydrogen production.

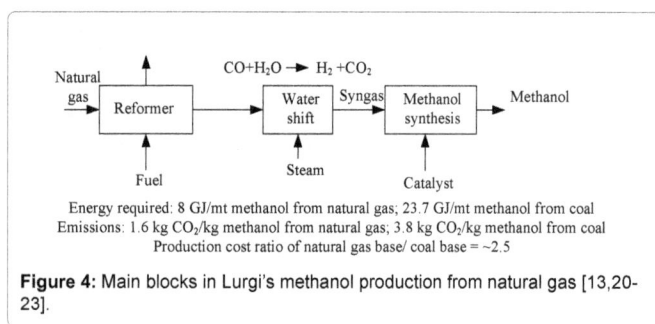

Energy required: 8 GJ/mt methanol from natural gas; 23.7 GJ/mt methanol from coal
Emissions: 1.6 kg CO_2/kg methanol from natural gas; 3.8 kg CO_2/kg methanol from coal
Production cost ratio of natural gas base/ coal base = ~2.5

Figure 4: Main blocks in Lurgi's methanol production from natural gas [13,20-23].

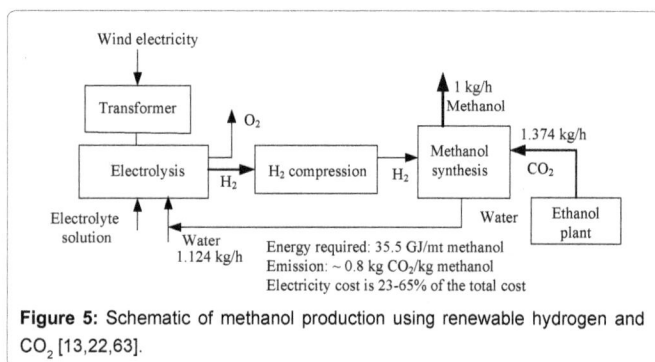

Energy required: 35.5 GJ/mt methanol
Emission: ~ 0.8 kg CO_2/kg methanol
Electricity cost is 23-65% of the total cost

Figure 5: Schematic of methanol production using renewable hydrogen and CO_2 [13,22,63].

Process	Production cost $/mt methanol'	Emissions kg CO_2/kg methanol	Energy efficiency %
Natural gas based syngas	170	0.5-1.6	75
Coal based syngas	432	2.8-3.8	48-61
Biomass based syngas	723	0.2	51
CO_2 from flue gas	973	0.8	46

'The cost data[13] for 2005 has been updated using: $Cost_{new} = Cost_{old}$ [CEPCI(2014)/CEPCI(2005)]
CEPCI (2014) = 576.1 and CEPCI (2005) = 468 [70].

"This emissions account for methanol production process and the emissions occurring with the utilization of methanol.

Table 7: Methanol costs and emissions" [13,22,71].

Methanol synthesis from water, renewable electricity, and carbon may lead to renewable energy storage, carbon recycle, fixation of carbon in chemical feedstock, as well as extended market potential for electrolysis. For methanol production with coal as carbon source, 23.7 GJ/mt methanol and with CO_2 as carbon source 35.5 GJ/mt methanol are required.

Currently the cost for hydrogen from electrolysis is roughly twice of that from natural gas steam reforming. Therefore, methanol production from renewable hydrogen would increase the energy consumption; however, a significant GHG reduction may be possible [22]. Clausen et al. [70] used electrolytic H_2 in methanol production using the post combustion captured CO_2. The alkaline electrolyzer is operated at 90°C and atmospheric pressure with an electricity consumption of 4.3 kWh/Nm^3 H_2 corresponding to an efficiency of 70% (LHV). With underground storage for hydrogen and oxygen and the electricity price during the off-pick hours of operation, the costs are estimated as $15.0/GJ, $20.0/mt CO_2, and $217/mt methanol (2010 $), respectively. The electricity cost is around 23%-65% of the methanol production cost because of high stoichiometric hydrogen demand in the synthesis [66-68].

CO_2 Capture and Compression

Some of the available sources for CO_2 are fermentation processes such as ethanol production plants, fossil fuel-based power stations, ammonia, and cement plants. Table 8 shows the equipment and operating costs to capture and liquefy 68 mt CO_2/day and 272 mt CO_2/day (the maximum capture rate for a typical 40 million gal/year ethanol plant). The estimated costs are for food grade CO_2 (99.98% minimum and <0.4 ppmv of sulfur) and also for less purified CO_2 suitable for enhanced oil recovery or sequestration [25].

Methanol Production Plant

We have designed and simulated a methanol plant using renewable electrolytic H_2 and CO_2 supplied from an ethanol plant. The RK-SOAVE equation of state is used. The plant uses 19.1 mt H_2/day and 138.4 mt CO_2/day, and produces 99.9 mt methanol/day at 99.7-wt% together with 57.3 mt/day 98.3-wt% of waste water. Table S1 in the 'Supporting Information' presents the stream table representing the energy and material balances of the plant.

Figure 6 presents the process flow diagram for the methanol plant using CO_2 and H_2. The feedstock is at the conditions associated with typical storage, with H_2 at 25°C and 33 bar and CO_2 at -25.6°C and 16.422 bar (liquid phase) [7,8,25]. The ratio of H_2 to CO_2 is held at of 3:1 to promote methanol synthesis. In the feed preparation block, the renewable H_2 and CO_2 are compressed to 50 bar in a multi-stage compressor and pump, respectively, and mixed with the recycle stream S9 in mixer M101. Stream S4 is the feed of the plug-flow reactor R101 where the methanol synthesis takes place. This multi-tube reactor has 15 tubes with a diameter 0.127 m and a length of 5 m, loaded with a total of 250 kg of catalyst. The reactor operates at 50 bar with a constant temperature of 235°C representing the Lurgi's low pressure isothermal system [66].

Langmuir-Hinshelwood Hougen-Watson (LHHW) kinetics formulations, with fugacities, are used for reactions (1) and (2). LLHW kinetics considers the adsorption of the reactants to the catalytic surface, the surface reactions to synthesize the methanol, and the desorption of the products from the catalytic surface [47,66]. The reactor output stream S5 is expanded in a turbine in order to cool down the outlet and produce power. This turbine produces 0.69 MW of electrical energy which can be fed back into the process or sold for revenue. In flash drum F101, stream S6 is separated into liquid (S6) and gas streams (S7). Stream S7 is crude methanol, which is separated from the water in the distillation tower T101. The product methanol is the distillate, while the wastewater is the bottoms flow of T101. The streams of methanol and water are cooled by the heat exchangers of E101 and E102, respectively, and are stored. Gas stream S8 is sent to a flow splitter SF101, in which 90% of S8 is recycled to the reactor after it is compressed in the multi stage compressor REC-COMP. Stream S9 is chosen as a tear stream. The mole fraction of methanol in the distillate is controlled by varying the reflux ratio and the ratio of bottoms flow to feed flow rate by using two design specifications in the Radfrac column T101. The column has 20 stages with a feed stage 17 and partial condenser. Methanol production has the potential for the best possible technology deployment ranging from 16% to 35% [65]. Therefore the design reflects that potential in a simple design delivering almost pure methanol and waste water containing less than 1% methanol.

Ammonia Production

Ammonia is synthesized by the catalytic reaction of H_2 and nitrogen gas at around 400-600°C and 200-400 atmospheres (Haber

Cost	68 mt CO_2/day beverage grade	272 mt CO_2/day beverage grade	272 mt CO_2/day Non-beverage grade
Capital cost, $	2,530,000	5,770,000	4,700,000
Capital cost, $/ mt CO_2	37205	21213	17279
Electricity*, $/ mt CO_2	19.46	18.8	18.9

*Electricity cost: $0.10/kWh

Table 8: Estimated cost of CO_2 recovery options from ethanol plant ($ 2006) [25]

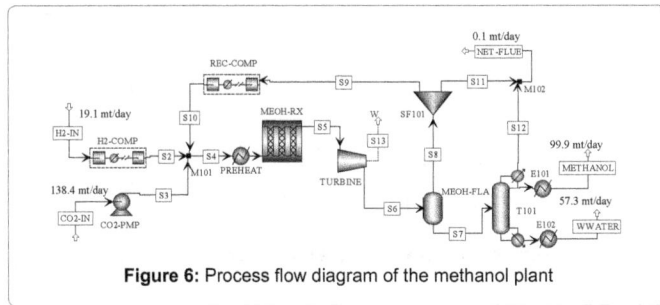

Figure 6: Process flow diagram of the methanol plant

and Bosch process).

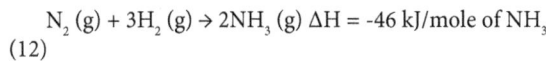

$$N_2 (g) + 3H_2 (g) \rightarrow 2NH_3 (g) \quad \Delta H = -46 \text{ kJ/mole of } NH_3$$
(12)

The sources of H_2 are steam reforming and/or water-gas shift from natural gas or gasification of coal, while an Air Separation Unit (ASU) supplies the nitrogen [71-73]. Figure 7 shows the both processes of renewable H_2 based and syngas-based NH_3 production.

Air Separation Unit

ASU can produce nitrogen (99.999% purity) and oxygen (98% purity) for synthesis of ammonia using the air [72,73]. Ambient air is compressed in multiple stages (accounting for 86% of the total energy consumption) with inter-stage cooling to 6.45 bar and sent into the molecular sieve to remove residual water vapor, carbon dioxide, and atmospheric contaminants. Table 9 shows typical power consumptions. A larger plant with efficiency improvements (energy consumption of less than 10%) and process optimization would deliver air liquefaction at around 0.4 MWh/mt liquid nitrogen. Operation and Maintenance (O&M) costs typically amount to between 1.5% and 3% of the plant purchase price per annum. Production cost is around $54/mt nitrogen for a 300 mt/day and $49/mt nitrogen for a 600 mt/day capacity [71-73].

Ammonia Production Plant

Figure 8 shows the process flow diagram for the ammonia plant. Production of ammonia is based on the Haber-Bosch synthesis process with a high pressure reactor in the presence of porous iron oxide. Typically for ammonia synthesis these conditions are about 150 atmospheres and 370-500°C. Under equilibrium conditions the proportion of reactants and the product of a chemical reaction are balanced and determined by the existing physical conditions such as pressure, temperature and concentrations. Since the reaction is exothermic, lowering the temperature in the reactor will increase the yield of ammonia. However, this also slows down the reaction therefore, for higher efficiency; the temperature is kept as high as possible. Increasing the pressure will increase the yield of ammonia but there is a limit in pressure for safety reasons [41,42].

The nitrogen is supplied by an air separation unit SEP 101, to produce 1202.66 mt/day anhydrous ammonia. The ammonia process is designed and simulated by using the RK-SOAVE equation of state property method. The ammonia plant uses 217.71 mt/day H_2 and 1009.15 mt/day nitrogen, and produces 1202.66 mt/day 99.9 wt % ammonia. The flow rate of ammonia is maximized to be 2943 kmol/hr and its composition to be 0.99wt% NH_3, using the constrained optimization option. There is a slight loss of ammonia in the stream BLEED. Air is separated in SEP 101, and the feeds of nitrogen and hydrogen at 20.27 bar are mixed in M101. This mixture is compressed to about 212 bar in compressors C101 and C102. Temperature of this mixture is adjusted in heat exchanger E201. In reactor R201 the ammonia synthesis takes place at around 556°C and 212 bar with a platinum group metal such as ruthenium [40-42]. The reactor R201 is a RGIBBS reactor and estimates the equilibrium composition of the reactor by Gibbs free energy minimization. The output of the reactor is conditioned in heat exchangers E202 and E203 and sent to adiabatic flash drums FL301 and FL302, which operate at 203 and 12 bar, respectively. The bottom flow of FL302 is the product ammonia at -26°C and 12.4 bar. Stream table and overall mass and energy balances for the ammonia plant are presented in Table S3 and S4 within the "Supporting Information." There is a large energy difference between the input and output, and must be compensated by utilities from outside in the form of cooling water, steam, electricity, and refrigeration.

Sustainability and Economic Analyses

Sustainability analysis

The integral methanol production facility consists of three units: an electrolytic hydrogen production, CO_2 capture and storage, and the methanol production. Similarly, the integral ammonia production facility consists of three units: an electrolytic hydrogen production, ASU, and the ammonia production. Figures 9 and 10 show these integral facilities subject to sustainability and economic analyses. Table 10 shows the main results of the material and energy usages, as well as the CO_2 emissions for the integral facilities. The energy costs are estimated by the unit cost of utilities listed in Table 11.

The integral methanol facility requires 19.08 mt H_2/day and 138.38 mt CO_2/day in total. The total emissions of CO_2 from each unit are -111.54 mt CO_2/day, 18.51 mt CO_2/day, and 8.77 mt CO_2/day for the methanol production, H_2 production, and CO_2 capture and storage, respectively. The net carbon fee is -$9.3/h for the methanol facility and $69.89/h for the ammonia facility based on a set value of $2/mt CO_2e. As Table 10 shows, the values of net duty and cost are the highest for the hydrogen production units used in methanol and ammonia productions.

The integral ammonia facility requires 217.72 mt H_2/day and 1009.15 mt N_2/day in total. The total emissions of CO_2 from each unit are 838.78 mt CO_2/day, 211.18 mt CO_2/day, and 111.47 mt CO_2/day for the ammonia production, H_2 production, and ASU, respectively.

Figure 11 presents an approximate energy balance with the energy required for the electrolyzer, carbon capture and storage, and total duty required in methanol production versus energy content in methanol as fuel combusted fully. The energy efficiency for the integral facility is around 58%.

Figure 12 shows an approximate energy balance with the energy required for the electrolyzer, nitrogen production through the ASU, and the total duty required for the ammonia production versus the energy content in ammonia as a fuel combusted fully. The total energy efficiency for the integral facility is around 35%.

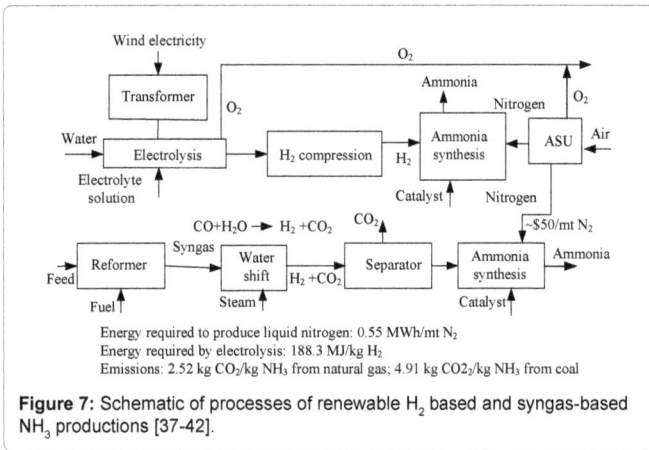

Figure 7: Schematic of processes of renewable H_2 based and syngas-based NH_3 productions [37-42].

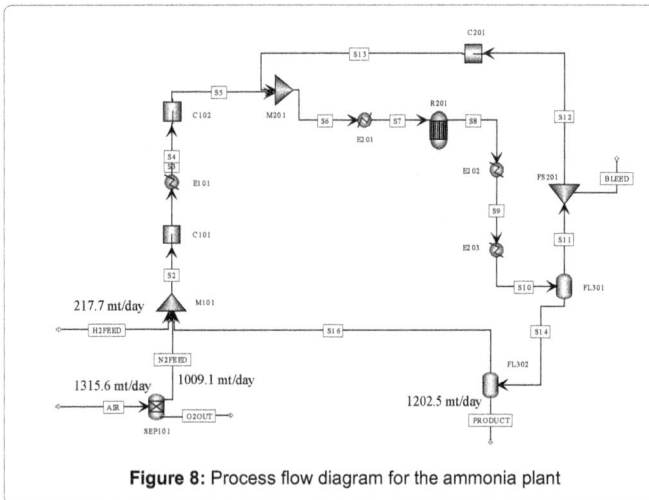

Figure 8: Process flow diagram for the ammonia plant

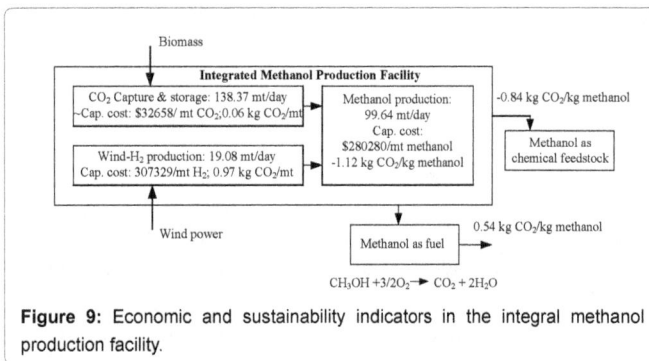

Figure 9: Economic and sustainability indicators in the integral methanol production facility.

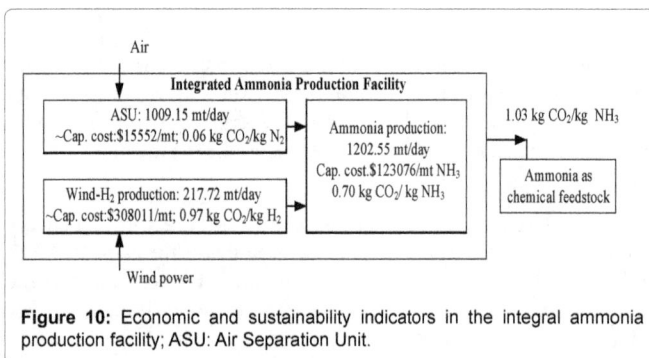

Figure 10: Economic and sustainability indicators in the integral ammonia production facility; ASU: Air Separation Unit.

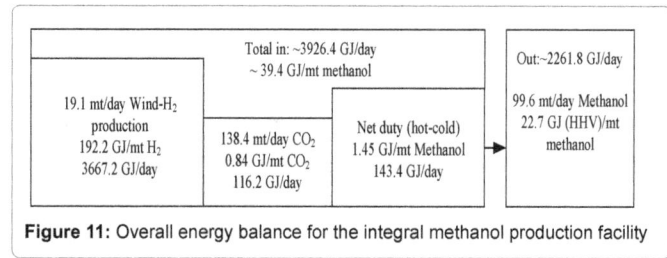

Figure 11: Overall energy balance for the integral methanol production facility

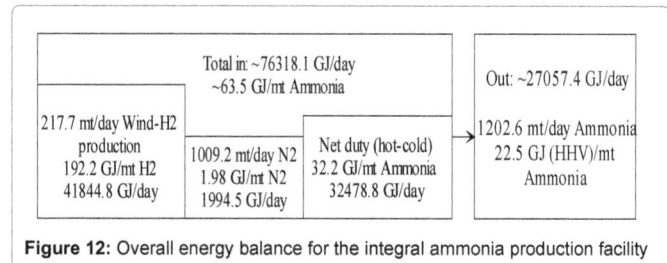

Figure 12: Overall energy balance for the integral ammonia production facility

Table 12 presents the following sustainability metrics that are estimated for the integral methanol and ammonia facilities:

• Material intensity (nonrenewable energy/unit mass of product)

• Energy intensity (nonrenewable energy/unit mass of product)

• Potential environmental impact (pollutants and emissions/unit mass of product)

The overall facility emissions of CO_2 are normalized with respect to methanol and ammonia capacities. The material intensity metrics show that the methanol facility requires 1.39 mt CO_2/mt methanol. The environmental impact metrics shows that the integral methanol facility reduces -0.84 kg CO_2/kg methanol when utilizing it as a chemical feedstock, and recycles 0.53 kg CO_2/kg methanol after its complete combustion, as seen in Figure 9. On the other hand, the environmental impact metrics for the integral ammonia facility is 1.03 kg CO_2/kg ammonia, as seen in Figure 10. The duty (heating-cooling) becomes negative due to excessive cooling required in the ammonia facility.

Economic analysis

The economics analyses of the integral methanol and ammonia plants are based on the Discounted Cash Flow Diagrams (DCFD) prepared for a ten-year of operation using the current economic data. Based on the equipment list from the process flow diagrams (Figures 6 and 8), bare module costs are estimated and used as Fixed Capital Investments (FCI). Chemical Engineering Plant Cost Index [48] (CEPCI-2014) (=576.1) is used to estimate and update the costs and capacity to the present date by

$$\text{Cost}_{\text{New}} = \text{Cost}_{\text{New}} \frac{CEPCI_{\text{New}}}{CEPCI_{\text{Old}}} \left(\frac{Capacity_{\text{New}}}{Capacity_{\text{Old}}} \right)^x \quad (5)$$

Where x is the factor, which is usually assumed to be 0.6. Working capital is 20% of the FCI. Depreciation method is the Maximum Accelerated Cost Recovery System (MACRS) with a 7-year recovery period [74]. After estimating the revenue and the cost of production, DCFDs are prepared. The details can be found within the 'Supporting Information.' DCFDs generate the three economic feasibility criteria that are Net Present Value (NPV), Payback Period (PBP), and Rate of Return (ROR). At least two out of three criteria should be favorable for the operation to be feasible. These criteria are favorable if NPV 0, PBP ≤ useful operational years; and ROR ≥ i, where i is the internal interest rate. In addition, the economic constraint (EC) and the unit product cost (PC) are also estimated

$$EC = \frac{\text{Average Discounted Annual Cost of Production}}{\text{Average Discounted Annual Revenue}} \quad (6)$$

$$PC = \frac{\text{Average Discounted Annual Cost of Production}}{\text{Capacity of the plant}} \quad (7)$$

The PC takes into account the Operating and Maintenance (O&M) costs. An operation with EC < 1 shows the opportunity to accommodate other costs and improve the cash flows of the operation toward a positive NPV.

The estimated approximate values of the FCIs are $5.87 million for the wind-based electrolytic H_2 production unit, $4.52 million for the CO_2 production unit, and $28 million for the methanol production unit. The H_2 production includes the compression, storage, and dispensing from a centralized production facility with an average electricity cost of 0.045/kWh. Therefore, the total value of the FCI is around $38.39 million.

The distribution of unit capital costs for the integral methanol production facility shows that the contribution from wind-based H_2 is the highest (Figure 9). The cost of H_2, which makes the NPV = 0, is $0.88/kg H_2 when the selling price of methanol is $600/mt with the corresponding values of EC = 0.85 (< 1) and PC = $518/mt methanol (< $600/mt). Global prices of methanol vary widely; the prices in 2014 are $435/mt in Europe, $482/mt in North America, $410/mt and in Asia Pacific [75]. Compared with natural gas-based methanol, the cost of renewable methanol production is almost five times higher. Only the biomass production cost is comparable, as seen in Table 13. The cost of renewable hydrogen and the selling price of methanol affect the economics of the renewable methanol.

The approximate value of FCI for the ammonia process is around $148.5 million, while the values of FCIs for the ASU and wind-based electrolytic H_2 production unit are around $15.6 million and $66.9 million, respectively. The capital cost of the integrated production, including the ammonia process, the ASU, and the H_2 production unit, becomes $231.0 million. An average selling price of ammonia is around $700.0/mt (2014 $) [76]. The cost of H_2, which makes the NPV = 0, is $2.33/kg H_2 when the selling price of ammonia is $700/mt with the corresponding values of EC = 0.95 (< 1) and PC = $662.9/mt methanol (< $700/mt). The details of the economic analysis of the ammonia plant are given in the 'Supporting Information.'

Assessment of Renewable Methanol and Ammonia Productions

Minimum and maximum current world-wide productions of methanol are around 55 to 5000 mt/day. Methanol has half of the volumetric energy density relative to gasoline or diesel; however, it can be used in the direct methanol fuel cell [13-15,20,21,29,34,36]. Renewable hydrogen-based methanol would recycle carbon dioxide as a possible alternative fuel to diminishing oil and gas resources [77-79]. It is also used as a chemical feedstock to ultimately fix the carbon. This would lead to a "methanol economy" [18,19]. There are already vehicles which can run with M85, a fuel mixture of 85% methanol and 15% gasoline [1,18-22]. Methanol can be used with the existing distribution infrastructure of conventional liquid transportation fuels. In addition, fuel cell-powered vehicles are also in a fast developing stage, although

they are not yet available commercially [1,2,19].

Table 13 shows the specific energy consumptions and emissions in producing methanol and ammonia by various feed stocks [22]. The coal-based process has the emissions of 3.8 kg CO_2/kg methanol, while natural gas-based process leads to 1.6 kg CO_2/kg methanol. Lifecycle CO_2 emission is around 0.8 kg CO_2/kg methanol for the flue gas based methanol. Around 50% of these emissions are due to the CO_2 capture processes [22,80].

Current capacities for ammonia vary from 1,000 to 2,000 mt/day or 360,000 to 720,000 mt/year. NH_3 can be used as fertilizers, industrial chemicals, and fuel. Ammonia cracking is endothermic and depends on the catalyst [37,38]. Ammonia has a capacity of 17.6 wt% for H_2 storage; however, considerable energy is required to release H_2 from ammonia. Ammonia synthesis coupled with hydrogen production may increase efficiency. Ammonia can burn directly in an internal combustion engine and can be converted to electricity directly in an alkaline fuel cell, or converted to H_2 for non-alkaline fuel cell. However, Polymer Electrolyte Membrane (PEM) fuel cell technology is incompatible in the presence of ammonia (>0.1 ppm) [37]. For sites in a remote island, ammonia fuel may become competitive around $10/gallon of diesel fuel [39-41].

When it is produced from natural gas, ammonia production cost depends on the price of natural gas; for example, for $4.5/MMBtu natural gas, NH_3 production cost is around $180/mt, while for $7.0/ MMBtu natural gas, NH_3 production cost becomes $260/mt at 2006 $. Only 60-65% of the energy input of natural gas to the process is contained in the product ammonia. Replacing natural gas with coal as the feedstock increases energy consumption and production costs 1.7 times and the investment cost 2.4 times [37-39,73]. The cost of ammonia from renewable hydrogen ranges between $660/mt and 1,320 $/mt, which is higher than both coal and natural gas based-ammonia production costs [22].

Emission for a natural gas-based ammonia is around 2.52 mt CO_2/ mt NH_3, while coal-based ammonia produces nearly 4.91 mt CO_2/ mt NH_3. The emission of CO_2 based on natural gas represents a lower limit for the GHG emissions from ammonia production. Some of the CO_2 emitted is captured and subsequently used for the production of urea [22,37,38,77]. Energy consumption, as well as the capital cost, in ammonia production is higher than of that for methanol production [22]. The best possible technique for NH_3 production uses H_2 from renewable energy sources. Hydrogen production is one of the largest energy-consuming steps in the production of ammonia and methanol. Capital cost for a centralized 20000 mt H_2/year plant is around $60 million (2011$) with operational cost estimated at $3.3 million/year. The investment costs of a centralized water electrolysis plant would be roughly one third of the investment costs of a conventional natural gas based plant of equivalent production capacity [22]. As Table 13 shows, this is by far the highest energy consuming process step in the overall scheme and dominates all subsequent steps, such as hydrogen compression and, in the case of ammonia production, the air separation unit for production of nitrogen from air [22].

Process Steps	kWh/Nm³	MJ/Nm³	MJ/kg	kg H_2/ Nm³	kg N_2/ Nm³	$/mt N_2
Electrolysis	4.7	17.0	188.3	0.09		
ASU	1.0	4.0	3.1		1.17	49(600 mt N_2/ day)

Table 9: Specific energy consumptions for hydrogen and nitrogen [22,71-73]

Material metrics	Integral methanol production			Integral ammonia production		
	Methanol production	H₂ prod.	CO₂ C&S	NH₃ prod.	H₂ prod.	ASU
CO_2 Input, mt/day	138.37					
H_2 Input, mt/day	19.08			217.72		
N_2 Input, mt/day				1009.2		
Methanol production, mt/day	99.66					
Ammonia production, mt/day				1202.6		
Energy intensity metrics						
Total heating duty, MW	4.60	42.49	1.05	103.54	484.89	23.08
Total cooling duty, MW	2.93	0.12	0.03	162.32	1.40	0
Net duty (heating - cooling), MW	1.67	42.37	1.02	-58.78	483.49	23.08
Total heating cost flow, $/h	59.18	3292.83	81.31	2648.9	37579	1789
Total cooling cost flow, $/h	2.24	0.09	0.02	1236	4.85	
Net cost (heating + cooling), $/h	61.42	3292.92	81.33	3885	37584.08	1789.04
Environmental impact metrics						
Net stream CO_2e, mt/day	-138.37	0	0	0	0	0
Utility CO_2e, mt/day	26.83	18.51	8.77	838.78	211.18	111.47
Total CO_2e, mt/day	-111.53	18.51	8.77	838.78	211.18	111.47
Net carbon fee, $/h	-9.29	1.54	0.73	69.89	17.60	9.29

˙US-EPA-Rule-E9-5711; natural gas; carbon fee: $2/mt.

Table 10: Sustainability indicators for the methanol and ammonia plants˙

Utilities	Energy price, $/MJ	T_{in} °C	T_{out} °C	Factor˙	U˙˙ kW/m² K
Electricity	$0.0775/kW h			0.58	
Cooling Water	$0.09/mt	20	25	1	3.75
Medium Pressure Steam	2.2×10^{-3}	175	174	0.85	6.00
High Pressure Steam	2.5×10^{-3}	250	249	0.85	6.00
Refrigeration	3.3×10^{-3}	-39	-40	-1	1.30

˙CO_2 energy source efficiency factor; ˙˙ Utility side film coefficient for energy analysis.

Table 11: Unit energy cost for various utilities with energy source of natural gas for 2014 [47].

Metrics	Integral methanol plant	Integral ammonia plant
Material metrics		
CO_2 used/Unit product	1.39	
N_2 used/Unit product		0.84
H_2 used/Unit product	0.19	0.18
Energy intensity metrics		
Net duty/unit product, MWh/mt	9.55	-1.17
Net cost/Unit product, $/mt	828.67	863.33
Environmental impact metrics		
Total CO_2e/Unit product	-0.85	1.03
Net carbon fee/Unit product, $/mt	-1.70	2.07

Table 12: Sustainability metrics for the integral methanol and ammonia plants

Process	kg H₂/ kg prod.	H₂ prod. /comp.	Average prod.	BPT	Theor min.	Average kg CO₂/kg prod.
Methanol from CO_2	0.189	37.06				
Syngas-coal methanol	0.126	24.20	24.0	20.1	5.1	2.83
Syngas-NG methanol			13.9	9.0-10	5.1	0.52
Ammonia	0.178	35.57				
syngas-NG Ammonia			15.4	7.2-9.0	5.8	2.52
Syngas-coal Ammonia			27.9	22.0	8.1	4.91

SEC: Specific energy consumption that includes fuel, steam and electricity for the process.
BPT: Best possible technology; GHG: greenhouse gas emissions as CO_2 equivalent per ton of product;
CO_2e includes CO_2, CH_4, and NO_x.

Table 13: Specific energy consumptions and emissions for ammonia and methanol productions [22]

Economics and sustainability indicators	Weighting factor:0-1	Fossil-methanol	Non-fossil-methanol	Fossil-ammonia	Non-fossil-ammonia
Economic indicators					
Net present value NPV	1	+	−	+	−
Payback period PBP	0.8	+	−	+	−
Rate of return ROR	0.8	+	−	+	−
Economic constraint EC	0.9	+	−	+	−
Impact on employment	1	+	+	+	+
Impact on customers	1	+	+	+	+
Impact on economy	1	+	+	+	+
Impact on utility	0.7	−	+	−	+
Sustainability indicators					
Material intensity	0.7	−	+	−	+
Energy intensity	0.8	+	−	+	−
Environmental impact GHG in production	0.8	−	+	−	+
Environmental impact GHG in utilization	0.8	−	−	+	+
Toxic/waste material emissions Process safety and Public safety	1	−	+	−	−
Potential for technological improvements and cost reduction	0.8	−	+	−	+
Security/reliability	0.9	−	+	−	+
Political stability and legitimacy	0.8	−	+	−	+
Quality of life	0.8	−	+	−	+
Total positive score		8	11	9	11
Total minus score		9	−6	−8	−6
Net score (positive-minus)		−1	+5	+1	+5
Weighted total score		**+0.2**	**+5.4**	**+2**	**+4**

Table 14: Multi-criteria decision matrix for feasibility evaluation of chemical processes and energy systems

Tallaksen and Reese [38] compared the renewable and with fossil-based ammonia productions in terms of energy use and carbon emissions using the Life Cycle Assessment (LCA) methods. Renewable ammonia production requires around 60 GJ of electricity/mt ammonia. This is considerably more total energy than conventional fossil fuel based produced ammonia, however it requires less fossil energy and results in less GHG emissions. The boundary of LCA for the wind to ammonia contains wind power, water electrolysis, hydrogen compression, nitrogen separation and compression, ammonia production and ammonia storage. LCA is more focused on environmental issues rather than raw material depletion [38].

Main chemical storage of electricity involves the production of hydrogen, synthetic natural gas, and chemicals, which are mainly methanol and ammonia. Combination of several storage applications together may help electricity storage to be more feasible. The initial investment requires a cost per unit of power ($/kW) and a cost per unit of energy capacity ($/kWh), which are technology dependent [77]. The economics of electricity storage are influenced by the type of storage technology, electricity price, the frequency of charging and discharging cycles, and the system in which the storage facility is located. Besides, one needs to consider direct and localized impacts of the technology and the generation source used [77].

Assessment of Chemical Processes by a Multi-Criteria Decision Matrix

Beside the economics analysis, sustainability metrics should also be used to evaluate the feasibility of chemical processes [81-84]. Table 14 shows a Pugh decision matrix [85] developed using '+' and '- 'for the ratings to assess the methanol and ammonia production plants. Four scores generated show the number of plus scores, minus scores, the overall total, and the weighted total. The weighted total adds up the

scores times their respective weighting factors. The totals are guidance only for decision making. If the two top scores are very close or very similar, then they should be examined more closely to make a more informed decision. Renewable energy-based systems may require the combined use of scenario building and participatory multi-criteria analysis for sustainability assessment [84].

Table 14 indicates the weighted decision matrix to compare the plants producing methanol and ammonia from fossil and non-fossil resources. The weight factor can be adjusted with respect the location, energy policies, and energy costs and security. With the weight factors and the combined economic and sustainability indicators, the decision matrix has estimated the highest weighted scores for the renewable methanol and ammonia production facilities. The positive weighted score for the renewable methanol (+5.4) is slightly better than the renewable ammonia production (+ 4). These scores indicate the overall impact of sustainability indicators beside the economics.

Conclusion

Renewable hydrogen, methanol, and ammonia productions may lead to renewable electricity storage and reduce the carbon emissions either by recycling and/or fixation of the carbon. The cost of hydrogen production plays an important role within the economics of the renewable methanol and ammonia productions and determines the scope of improvements necessary for feasible operations. The economic analysis shows that the cost of electrolytic hydrogen is critical in the economics of renewable methanol and ammonia plants at the capacities assumed in this study and using the currently available technologies. Despite its poor overall efficiency and high up-front capital costs, chemical storage may provide the large-scale and long-term storage requirements of a mixed renewable power generation. Multi-criteria decision matrix, containing the sustainability indicators, show that chemical processes that use non-fossil fuels may achieve better overall weighted scores. This helps accounting the cost of environmental damage from using fossil fuels in the overall assessment of feasibility for chemical process and energy systems. This is in line with the need for the development of low-carbon chemical processes and energy technologies in order to address the global challenges of energy security, climate change, and economic growth.

Acknowledgement

The authors acknowledge the partial financial support (4200001187) by the Nebraska Public Power District (NPPD) Columbus, Nebraska 68602-1740.

References

1. Demirel Y (2012) Energy: Production, Conversion, Storage, Conservation, and Coupling. Springer-Verlag London, London.

2. U.S. Energy Information Administration (2014): Annual Energy Outlook.

3. Parsons B, Milligan M, Zavadil B, Brooks D, Kirby B, et al. (2004) Grid Impacts of Wind Power. A Summary of Recent Studies in the United States. Wind Energy 7: 87-108.

4. Wiser R, Ryan H, Bolinger M (2011) Wind Technologies: Market Report. Golden, CO: NREL.

5. Solar and Wind Technologies for Hydrogen Production (2005) ESECS EE-3060.

6. Esmaili P, Dincer I, Naterer GF (2012) Energy and exergy analyses of electrolytic hydrogen production with molybdenum-oxo catalysts. Int J Hydrogen Energy 37: 7365-7372.

7. Turner J, Sverdrup G, Mann MK, Maness PC, Kroposki B, et al. (2008) Renewable hydrogen production. Int J Energy Research 32: 379-407.

8. Dincer I, Ratlamwala TAH (2013) Development of novel renewable energy based hydrogen production systems: a comparative study. Int J Hydrogen Energy 72: 77-87.

9. Dingizian A, Hansson J, Persson T, Ekberg HS, Tuna PA (2007) Feasibility Study on Integrated Hydrogen Production Presented to Norsk Hydro ASA Norway.

10. Norsk Electrolyzer.

11. U.S. Department of Energy (2007) Hydrogen, Fuel Cells & Infrastructure Technologies Program, Safety Planning Guidance for Hydrogen Projects.

12. James BD, Moton JM, Whitney G, Colella WG (2013) Guidance for Filling out a Detailed H2A Production Case Study. EERE, US.

13. Galindo CP, Badr O (2007) Renewable hydrogen utilization for the production of methanol. Energy Convers Manag 48: 519-527.

14. Olah GA, Goeppert A, Prakash GKS (2009) Chemical recycling of carbon dioxide to methanol and dimethyl ether from greenhouse gas to renewable environmentally carbon neutral fuels and synthetic hydrocarbons. J Org Chem 74: 487-498.

15. Demirel Y, Matzen M, Winters C, Gao X (2015) Capturing and using CO_2 as feedstock with chemical-looping and hydrothermal technologies and sustainability metrics. Int J Energy Research 39: 1011-1047.

16. Nguyen N, Demirel Y (2013) Biodiesel-glycerol carbonate production plant by glycerolysis. J Sustainable Bioenergy Systems 3: 209-216.

17. Jiang Z, Xiao T, Kuznetsov VL, Edwards PP (2010) Turning carbon dioxide into fuel. Phil Trans R Soc A 368: 3343-3364.

18. Olah GA, Goeppert A, Prakash GKS (2011) Beyond Oil and Gas: The Methanol Economy. 2nd edition, Wiley, New York, USA.

19. European Parliamentary Research Service (2014) Methanol: a future transport fuel based on hydrogen and carbon dioxide. Economic viability and policy options, Science and Technology Options Assessment.

20. Specht M, Staiss F, Bandi A, Weimer T (1997) Comparison of the renewable transportation fuels, liquid hydrogen and methanol, with gasoline: energetic and economic aspects. Int J Hydrogen Energy 23: 387-396.

21. EPA (2002) Clean alternative fuels: Methanol. Technical Report-420-F-00-040. Washington DC, USA.

22. IEA (2013)Energy and GHG reductions in the chemical industry via catalytic processes: Annexes, Dechema/, ICCA.

23. Hugill JA, Overbeek JP, Spoelstra SA (2001) Comparison of the eco-efficiency of two production routes for methanol: Report ECN-I-01-003. Energy Research Centre of the Netherlands, Netherlands.

24. US Department of Energy (2003) Commercial-scale demonstration of the liquid phase methanol (LPMEOH) process: Report No-DOE/NETL-2004/1199. Washington DC, USA.

25. Finley R (2006) Illinois State Geological Survey: Evaluation of CO2 Capture Options from Ethanol Plants.

26. Martin O, Perez-Ramírez J (2013) New and revisited insights into the promotion of methanol synthesis catalysts by CO_2. Catal Sci Technol 3: 3343-3352.

27. Kansha Y, Ishizuka M, Tsutsumi A (2013) Development of innovative methanol synthesis process based on self-heat recuperation. Chem Eng Transac 35: 37-42.

28. Bandose A, Urukawa A (2014) Towards full one-pass conversion of carbon dioxide to methanol and methanol-derived products. J Catalysis 309: 66-70.

29. Hoekman SK, Broch A, Robbins C, Purcell R (2010) CO_2 recycling by reaction with renewably-generated hydrogen. Int J Greenhouse Gas Control 4: 44-50.

30. Wang W, Wang S, Ma X, Gong J (2011) Recent advances in catalytic hydrogenation of carbon dioxide. Chem Soc Rev 40: 3703-3727.

31. Lim HW, Park MJ, Kang SH, Chae HJ, Bae JW, et al. (2009) Modeling of the kinetics for methanol synthesis using Cu/ZnO/Al$_2$O$_3$/ZrO$_2$ catalyst: Influence of carbon dioxide during hydrogenation. Ind Eng Chem Res 48: 10448-10455.

32. Studt F, Sharafutdinov I, Abild-Pedersen F, Elkjaer CF, Hummelshoj JS, et al. (2014) Discovery of a Ni-Ga catalyst for carbon dioxide reduction to methanol. Nature Chem 6: 320-324.

33. Rihko-Struckmann LK, Peschel A, Hanke-Rauschenbach R, Sundmacher K (2010) Assessment of methanol synthesis utilizing exhaust CO_2 for chemical storage of electrical energy. Ind Eng Chem Res 49: 11073-11078.

34. Kauw M (2012) Recycling of CO_2, the perfect biofuel. Master report, University of Groningen, Netherlands.

35. National Energy Technology Laboratory (2013) Carbon dioxide transport and storage costs in NETL studies.

36. Yang CJ, Jackson RB (2012) China's growing methanol economy and its implications for energy and the environment. Energy Policy 41: 878-884.

37. Thomas G, Parks G (2006) Potential roles of ammonia in a hydrogen economy. US Department of Energy.

38. Tallaksen J, Reese M (2013) Ammonia production using wind energy: An early calculation of life cycle carbon emissions and fossil energy consumption. Tenth Annual NH3 Fuel Conference, University of Minnesota, West Central Research and Outreach Center.

39. Morgan E, Manwell J, McGowan J (2014) Wind-powered ammonia fuel production for remote islands: A case study. Renew Energy 72: 51-61.

40. European Fertilizer manufacturer Association (2000) Production of Ammonia: Belgium.

41. LeBlanc JR, Knez SA (1998) Ammonia production with enriched air reforming and nitrogen injection into synthesis loop. US5736116 A.

42. Whitlock DR (1999) Method for ammonia production. US5968232 A.

43. Martins AA, Mata TM, Costa CAV, Sikdar SK (2007) Framework for sustainability metrics. Ind Eng Chem Res 46: 2962-2973.

44. Sikdar SK (2003) Sustainable development and sustainability metrics. AIChE J 49: 1928-1932.

45. Center for Waste Reduction Technologies (2004) Focus Area: Sustainability Metrics.

46. IChemE (2004) Sustainable development progress metrics recommended for use in the process industries.

47. Aspen Technology Inc. Burlington, MA, USA.

48. Chemical Engineering (2015) June: 80.

49. Wind-To-Hydrogen Project (2008) Electrolyzer Capital Cost Study. NREL, Technical Report NREL/TP-550-44103.

50. Saur G, Ainscough C, Harrison K, Ramsden T (2013) Hour-by-Hour Cost Modeling of Optimized Central Wind-Based Water Electrolysis Production. National Renewable Energy Laboratory.

51. European Commission Decision (2007) 2007/589/EC: Official Journal of the European Commission, L229 1-4.

52. EPA Rule E9-5711 (2009) Federal Register, Proposed Rules 74: 16639-16641.

53. Kothari R, Buddhi D, Sawhney RIL (2008) Comparison of environmental and economic aspects of various hydrogen production methods. Renewable Sustainable Energy Reviews 12: 553-563.

54. Mueller-Langera F, Tzimas E, Kaltschmitt M, Peteves S (2007) Techno-economic assessment of hydrogen production processes for the hydrogen economy for the short and medium term. Int J Hydrogen Energy 32: 3797-3810.

55. Barranon DCC (2006) Methanol and hydrogen production: energy and cost analysis. Lulea University of Technology, Lulea, Sweden.

56. Committee on Alternatives and Strategies for Future Hydrogen Production and Use (2004) The Hydrogen Economy: Opportunities, Costs, Barriers, and R&D Needs 99.

57. Dodds PE, McDowall W (2012) A review of hydrogen production technologies for energy system models, UCL Energy Institute, University College London.

58. Holladay JD, Hu J, King DL, Wang Y (2009) An over view of hydrogen production technologies. Catal Today 139: 244-260.

59. Spath PL, Mann MK (2004) Life Cycle Assessment of Renewable Hydrogen Production via Wind/Electrolysis. Milestone Completion Report-NREL/MP-560-35404.

60. International Energy Agency (2006) Hydrogen Production and Storage. R&D Priorities and Gaps.

61. Penev M (2013) Hybrid hydrogen energy storage. NREL, All-Energy, Aberdeen, UK.

62. National Academy of Science (2004) The Hydrogen Economy: Opportunities, Costs, Barriers, and R&D Needs. National Academies Press, Washington DC, USA.

63. Toyir J, Miloua R, Elkadri NE, Nawdali M, Toufik H, et al. (2009) Sustainable process for the production of methanol from CO_2 and H_2 using Cu/ZnO-based multicomponent catalyst. Physics Procedia 2: 1075-1079.

64. Machado CFR, de Medeiros JL, Araujo OFQ, Alves RMB (2014) A comparative analysis of methanol production routes: synthesis gas versus CO2 hydrogenation. Proceedings of the 2014 International Conference on Industrial Engineering and Operations Management Bali, Indonesia 7-9.

65. Graaf GH, Stamhuis EJ, Beenackers AACM (1988) Kinetics of low pressure methanol synthesis. Chem Eng Sci 43: 3185-3195.

66. Weiduan S, Junli Z, Bingchen Z, Honfshi W, Dingye F, et al. (1998) Kinetics of methanol sysnthesis in the presence of C301 Cu-based catalyst (I) intrinsic and global kinetics. J Chem Ind Eng 39: 401-409.

67. Skrzypek J, Lachowska M, Moroz H (1991) Kinetics of methanol synthesis over commercial copper/zinc oxide/alumina catalysts. Chem Eng Sci 46: 2809-2813.

68. Bussche KMV, Froment GF (1996) A steady-state kinetic model for methanol synthesis and the water gas shift reaction on a commercial Cu/ZnO/Al2O3 catalyst. J Catalysis 161: 1-10.

69. Aksgaard TS, Norskov JK, Ovesen CV, Stoltze P (1995) A kinetic model of methanol synthesis. J Catalysis 156: 229-242.

70. Clausen LR, Houbak N, Elmegaard B (2010) Technoeconomic analysis of a methanol plant based on gasification of biomass and electrolysis of water. Energy 35: 2338-2346.

71. Nielsen SE (2007) Latest developments in ammonia production technology. FAI International Conference in Fertiliser Technology, New Delhi, India 12-13.

72. Yan L, Yu Y, Li Y, Zhang Z (2010) Energy Saving Opportunities in an Air Separation Process; International Refrigeration and Air Conditioning Conference, Lafayette, Indiana, USA.

73. Liquid Air Energy Network (2013)Liquid air production and cost.

74. Turton R, Bailie RC, Whiting WB, Shaeiwitz JA, Bhattacharya D (2012) Analysis, Synthesis, and Design of Chemical Processes. 4th edition, Upper Saddle River, Prentice Hall.

75. Methanex (2014) Methanol Price.

76. Knorr B (2014) Weekly Fertilizer Review.

77. Electricity Storage Association (2009) Power Quality, Power Supply.

78. Ting LH, Man LH, Yee NW, Yihan J, Fung LK (2012) Techno-economic analysis of distributed hydrogen production from natural gas. Chinese J Chem Eng 20: 489-496.

79. Zoulias EI, Lymberopoulos N (2007) Techno-economic analysis of the integration of hydrogen energy technologies in renewable energy-based stand-alone power systems. Renewable Energy 32: 680-696.

80. Arons JDS, Kooi HVD, Sankaranarayanan K (2004) Efficiency and Sustainability in the Energy and Chemical Industries. CRC Press, New York, USA.

81. Dincer I, Rosen MA (2007) Exergy: Energy, Environment and Sustainable Development. Burlington, Elsevier.

82. Patel AD, Meesters K, den Uil H, de Jong E, Blok K, et al. (2012) Sustainability assessment of novel chemical processes at early stage: application to biobased processes. Energy Environ Sci 5: 8430-8444.

83. Demirel Y (2013) Sustainable Operations for Distillation Columns. Chem Eng Process Techniq 1005: 1-15.

84. Kowalski K, Stagl S, Madlener R, Oman I (2009) Sustainable energy futures: Methodological challenges in combining scenarios and participatory multi-criteria analysis. Europ J Operat Res197: 1063-1074.

85. Pugh S (1981) Concept selection, a method that works. In: Hubka V (ed) Review of design methodology. Proceedings international conference on engineering design, Rome, Zürich: Heurista 497-506.

22

Statistical Optimization, Kinetic and Isotherm Studies on Selective Adsorption of Silver and Gold Cyanocomplexes Using Aminoguanidyl-Chitosan Imprinted Polymers

Ahamed MEH[1], Marjanovic L[3] and Mbianda XY[1,2]*

[1]*Department of Applied Chemistry (Doornfontein Campus), University of Johannesburg, South Africa*
[2]*Centre for Nanomaterial Science Research, University of Johannesburg, South Africa*
[3]*Department of Chemistry, Faculty of Science, University of Johannesburg, South Africa*

Abstract

Aminoguanidyl-chitosan imprinted polymers (AGCIPs) were synthesized and applied to the selective extraction of silver and gold cyanocomplexes from aqueous solutions. Batch adsorption parameters for the recovery of silver and gold cyanocomplexes from aqueous solutions by the AGCIPs viz., contact time, solution pH, initial metal concentrations and temperature, were optimized by a two-level fractional factorial design and the Box-Behnken matrix. The equilibrium data correlated well with Langmuir isotherm model; and the maximum adsorption capacities for silver cyanide calculated from the Langmuir equation were 429.2 mg Ag g^{-1} and 319.5 mg Ag g^{-1} at pH 6.9 and 10, respectively; whereas they were 319.5 mg Au g^{-1} and 312.5 mg Au g^{-1} for gold cyanide in the same order. Adsorption kinetics suggested that these materials predominantly display a pseudo-second-order kinetic mechanism, while thermodynamic parameters revealed that the adsorption process was spontaneous and of exothermic nature. Investigation on the adsorption selectivity showed that the selectivity coefficients of AGCSIP (gold cyanide) with respect to $Ag(CN)_2^-$, $Fe(CN)_6^-$, and $Hg(CN)_2^-$ were 8.675, 26.005 and 5694.667 respectively whereas for AGCIP (Silver cyanide) they were 3.017, 75.478 and ∞ for $Au(CN)_2^-$, $Fe(CN)_6^-$, and $Hg(CN)_2^-$ respectively. This indicates that AGCSIPs have excellent selectivity for silver and gold cyanide complexes. Regeneration and reusability studies also revealed that 2M solution of KNO_3 at pH 10.5 could be used to regenerate the AGCIPs; and these materials could be recycled up to five times without significantly diminishing their adsorption capacity.

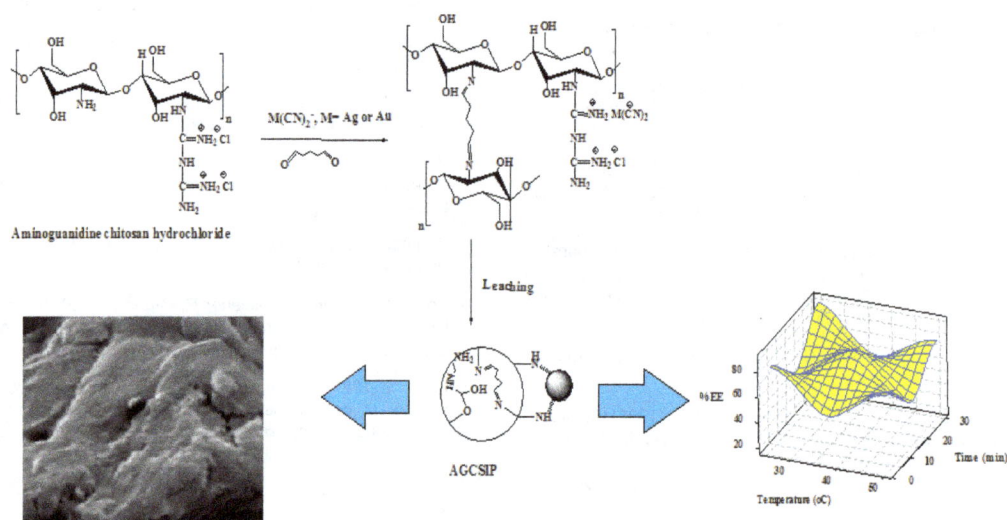

Aminoguanidine chitosan hydrochloride

$M(CN)_2^-$, M = Ag or Au

Leaching

AGCSIP

Keywords: Aminoguanidyl-chitosan imprinted polymers; Silver and gold cyanocomplexes; Selective adsorption; Experimental design; Equilibrium isotherm; Kinetics

Introduction

The recovery of precious metals, particularly gold and silver, from their primary and secondary sources has always attracts a great deal of attention due to their scarcity. Since both sources contain various coexisting metals, selectivity towards target species plays a crucial role in their extraction.

Nowadays hydrometallurgical processes are extensively used to recover precious metals. In these processes, the cyanidation method

*Corresponding author: XY Mbianda, Department of Applied Chemistry (Doornfontein Campus)/Department of Chemistry, Faculty of Science, University of Johannesburg, South Africa, E-mail: mbianday@uj.ac.za

is still the dominant technique for gold and silver ores as well as metal containing waste treatments [1]. The recovery of gold and silver from the alkaline cyanide leachate is usually accomplished through precipitation, solvent extraction, adsorption on activated carbon and ion-exchange [2]. Among these, adsorption on activated carbon and ion-exchange has been extensively employed and proved to be more effective compared to other separation methods [3,4]. However, these adsorbents are costly, not selective and require extensive labor and time [5]. This prompted the conception of this study which aims to develop efficient, cheaper and environmentally friendly sorbents for the recovery of silver and gold in the cyanidation process.

In the last decade chitosan-based materials have received increased attention as a versatile class of adsorbent that can be used in the hydrometallurgical processes [6]. The main advantage in using chitosan-based materials is their availability at low cost and their versatility. But native chitosan could not be used as anion exchanger for precious metals in the cyanidation process, since its cationic behavior is limited to acidic condition (pK_a near 6.2) [7]. However it can be modified, in a number of ways, with various ligating groups to produce chitosan-based materials which display higher specific affinity towards the target species [2,8-10].

Nowadays ion imprinted polymers (IIPs), have received much attention as sorbents for solid phase extraction due to several potential reasons viz. high affinity and selectivity for the target ion, high adsorption reproducibility without loss of recognition memory, low cost and stability [11]. IIPs are prepared by cross-linking a polymer derivative containing a metal ion template with a bifunctional reagent; then the metal ion is removed from the polymer matrix thus generating a specific bonding site that is complementary in size and shape to the target metal ion [12].

In this paper two aminoguanidyl modified chitosan ion imprinted polymers (AGCIPs), are proposed as low cost environmentally friendly biopolymers, for the selective recovery of gold and silver cyanocomplexes from aqueous solutions. The synthesis of these new materials was done using aminoguanidyl modified chitosan as a biopolymer, glutaraldehyde (GLA) as a crosslinker and silver and gold cyanide as ion templates. The characterization was done by analytical techniques such as FT-IR, XRD, SEM, BET and zeta potential. It is noteworthy mentioning that aminoguanidyl-chitosan imprinted polymers (AGCIPs) have never been used before, in the extraction of gold and silver cyanide. A number of studies involving guanidyl based organic synthetic polymers for the recovery of gold and silver cyanocomplexes ions from aqueous solutions, can be found in the literature [8,13,14]; But to the best of our knowledge, such studies involving the synthesis and uses of aminoguanidyl-chitosan imprinted biopolymers have never been reported. In order to optimize parameters (contact time, solution pH, initial metal concentrations and temperature) affecting the extraction efficiency of silver and gold cyanocomplexes anions from aqueous media, a response surface design combined with an advance multivariable optimization method was used in this study. The multivariable approach, when compare to the single variation method, has the advantage of reducing the processing costs by saving time and chemicals [15,16]. The adsorption isotherms, kinetics, thermodynamics as well as selectivity and reusability studies of the AGCSIPs materials were also investigated.

Experimental

Chemicals and solutions

Chitosan (medium M_n ~ 9000, N- deacetylation degree > 75%) and glutaraldehyde (50%) were obtained from Sigma -Aldrich Chemicals, Saint Louis, Mo, USA. Potassium dicyanoargentate and potassium aurocyanide were supplied by South Africa Precious Metal, Ltd, South Africa. 1-Cyanoguanidine and hydrochloric acid were purchased from Riedel-de Haën (Germany). Acetic acid (97%), ethanol, acetone, sodium hydroxide, thiourea, sodium nitrate, sulphuric acid, metal salts of $Hg(CN)_2$ and $K_3Fe(CN)_6$ were purchased from Merck South Africa. Working solutions, as per the experimental requirements, were freshly prepared from the stock solution for each experimental run. A Spectra scan silver and gold standard solutions (1000 mg L^{-1}) (Industrial Analytical Pty, Ltd, South Africa) were used to prepare working standard solutions at concentrations of 10-100 mg L^{-1} for Ag and Au. Reagents used were of analytical grade, and ultrapure water (18.3 $\mu\Omega$ cm^{-1} at 25°C) was obtained from an Elix/Milli- Q Element system (France).

Instrumentation

Inductively coupled plasma optical emission spectroscopy (ICP-OES, Spectro Arcos, model Arcos FH512- Germany) was used to determine the concentration of metal ions. Temperature controlled water bath-shaker (LABCON, shaking water bath-25 L, USA) was used to agitate the sample solutions. pH of the solutions was adjusted using 0.1M HCl or NaOH using a Hanna pH meter (Italy). FT-IR Attenuated Total Reflection (ATR) (Perkin Elmer, USA) was used to analyze the functional groups in the adsorbent. ^{13}C NMR spectrum was recorded at 75 MHz, on a Bruker AV- 400 spectrometer (USA). Zeta potentials of the polymers were measured by a Malvern Nanosizer (ZENN 3600, UK). The surface morphology of polymers was characterized by Scanning Electron Microscope (SEM Thermo Scientific, model 6658A-1NUS-SN, USA). X-ray diffraction (XRD) analysis was conducted using a *Philips PAnalytical Xpert PRO* powder diffractometer, employing Cu K_α radiation of wavelength 1.54. Nitrogen adsorption/desorption experiments were carried out using Micrometric ASAI 2020 Surface Area and Porosity Analyzer (USA).

Procedure

Synthesis of aminoguanidyl-chitosan hydrochloride: Aminoguanidyl chitosan hydrochloride was prepared by dissolving 2.0 g of chitosan (12.00 mmol NH_2) and 2.04 g of 1-cyanoguanidine (24.24 mmol) in 100 mL of (0.15M) HCl solution. The solution was heated for 2 h at 100°C. After cooling to room temperature, the aminoguanidyl chitosan hydrochloride was precipitated in acetone. Then, the wet solid was purified with ethanol in a Soxhlet extractor for 24 h. Finally, the product was dried under vacuum to constant mass. The preparation of aminoguanidyl chitosan hydrochloride is illustrated in Scheme 1 (step 1).

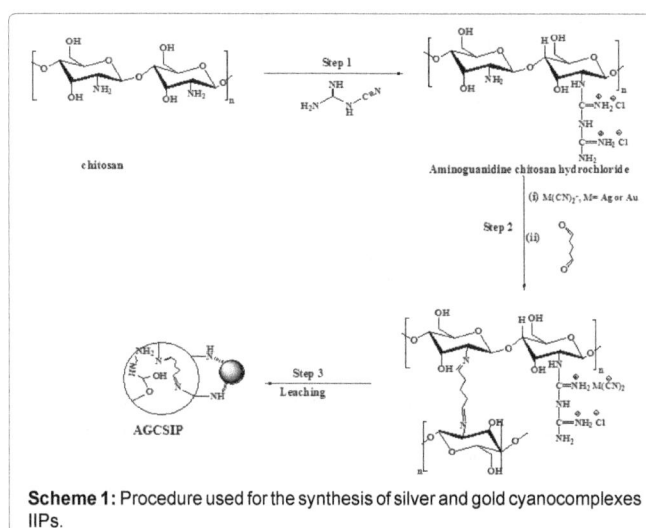

Scheme 1: Procedure used for the synthesis of silver and gold cyanocomplexes IIPs.

Synthesis of aminoguanidyl-chitosan imprinted polymer (AGCIP) and control polymers (CP): The aminoguanidyl chitosan hydrochloride (3.00 g) was dissolved with constant sonication in 50 mL acetic acid solution (2%V/V) and then transferred into a 250 mL round bottom flask containing 50 mL of potassium dicyanoargentate or potassium aurocyanide solutions, to give metal cyanide solution of 2000 mg L⁻¹. The mixture was stirred continuously at 60°C for 12 h. Then, 3.42 mL glutaraldehyde solution (1.81 g, 18.11 mmol) was added to the mixture to form a gel. The process was followed by filtering and intensive washing of the precipitate with acetone to remove any unreacted glutaraldehyde. The precipitate was again stirred, this time with a 2M solution of NaNO$_3$ at pH 10.5, to strip the template anion. This stage was monitored with an ICP-OES spectrophotometer. The residue was subsequently mixed with a 0.1M solution of hydrochloric acid for 5 h to remove non-crosslinking aminoguanidyl chitosan. The precipitate obtained was filtered out and washed with distilled water and acetone; then dried inside a vacuum oven at 60°C for 12 h. The resulting material was grounded and sieved to collect the particles which were later used for this study. The preparation process is shown in Scheme 1 (step 2 and 3). The aminoguanidyl-chitosan control polymer (CP) was similarly synthesized in the absence of templates

Design of experiments

Aiming to achieve the highest adsorption uptake for silver and gold cyanocomplexes using a batch system, several variables in the adsorption process should be optimized. However, in this study, the variables chosen for optimization were: kind of adsorbents (e.g., imprinted polymers versus control) (A), contact time (B), initial metal concentration (C), pH (D) and temperature (E). The optimization was conducted using multivariate method. Firstly, the half full 2^5 factorial design with two central points was carried out to screen the influential variables on the extraction efficiency. The extraction efficiency (% EE) of silver and gold cyanocomplexes was taken as the response of the design experiments. The factor levels were coded as -1 (low), 0 (central point), and +1 (high) and presented in Table S1 (Supplementary data). The experimental data were processed by using the MINITAB Statistical Software program release 16.1 (Trial version, USA).

After identifying the significant variables according to the half factorial design, the response surface analysis (RSM) was applied in order to optimize the silver and gold cyanocomplexes extraction. The experimental sets were evaluated using Box-Behnken design with five replicates at centre points (Table 1). The design of experiments was carried out using MINITAB Statistical Software program release 16.1 (Trial version, USA).

Adsorption experiments

Batch tests based on fractional factorial and Box-Behnken designs were conducted at random to study the effect of the pre-selected five operating variables on the silver and gold cyanocomplexes adsorption capacity of the AGCIPs. All equilibrium adsorption experiments were individually conducted for precious metal cyanides in a thermostatic water bath at agitation rate 200 rpm. The adsorption of silver or gold cyanocomplex was tested by shaking 10 mg of the adsorbent and 10 mL of silver or gold cyanocomplex solutions at various pHs, initial metal concentration, contact time, temperature (Table S1, Supplementary data). Aliquots for analysis were filtered, and the residual Ag/ Au concentration was measured by ICP-OES. The extraction efficiency (%EE) was determined as given in equation 1.

Isotherm data were carried out by adding 10 mg adsorbent to 10 mL of 50-1000 mg L⁻¹ silver or gold metal ions at pH 6.9 for 60 min.

Experiment	Time (min) B	pH D	X E or C	%EE Ag(CN)$_2^-$	Au(CN)$_2^-$
1	0	-1	-1	85.126	87.042
2	+1	-1	0	85.175	93.244
3	0	0	0	84.370	89.230
4	-1	0	+1	76.9616	54.988
5	0	+1	+1	20.228	49.702
6	0	+1	-1	23.575	64.787
7	+1	0	-1	89.820	90.126
8	-1	+1	0	19.128	44.056
9	0	0	0	81.103	89.541
10	0	-1	+1	78.425	82.899
11	-1	-1	0	81.358	69.012
12	+1	0	+1	77.002	75.030
13	0	0	0	80.472	89.224
14	0	0	0	85.565	89.641
15	0	0	0	76.839	87.194
16	+1	+1	0	25.428	63.000
17	-1	0	-1	82.099	59.987
Levels		-1	0	+1	
Time (min) B		10 (2)*	50 (16)*	90 (30)*	
pH D		6	8	10	
X (mg L⁻¹) C		105	157.5	210	
or X: (°C) E		25	37.5	50	

Note: The bracket star ( )* denote to the condition for adsorption silver cyanide by the polymer.

X stands for the factors temperature (°C) or initial concentration (mg L⁻¹) for adsorption silver and gold cyanocomplexes by the imprinted polymers, respectively.

Table 1: Optimization of silver and gold cyanocomplexes adsorption AGCSIPs using Box- Behnken design with five-central points.

Further, experiment were also performed at pH 10 (typical pH of mining operating of gold and precious metal cyanides leach liquor). Sorption kinetic experiments of silver or gold cyanocomplex were performed with 10 mg adsorbent and an initial concentration (100 mg L⁻¹) of metal ions. For the evaluation of the thermodynamic parameters, the experiment was carried out under similar mentioned condition at 25, 35 and 50°C for 30 min. After filtration, the residual silver or gold concentrations were analyzed by ICP-OES

Calculation of extraction efficiency and other constants

The extraction efficiency (%EE) was calculated by relating the obtained concentration (C_f) of the analyte to the original concentration (C_i) of the metal ion in the model solution (Equation 1)

$$\%EE = \frac{C_i - C_f}{C_i} \times 100 \qquad (1)$$

The amount adsorbed per unit mass of polymer (q, mg g⁻¹) was calculated by mass balance equation.

$$q = \frac{C_i - C_f}{m} \times V \qquad (2)$$

where m (g) and V (L) is the mass of the polymer and solution volume, respectively.

The competitive sorption experiments were conducted by preparing binary-mixture of Ag(CN)$_2^-$/Au(CN)$_2^-$, Ag(CN)$_2^-$ or Au(CN)$_2^-$/Fe(CN)$_6^{-3}$ and Ag(CN)$_2^-$ or Au(CN)$_2^-$/Hg(CN)$_2$. Ag(CN)$_2^-$, Au(CN)$_2^-$ and the two competitors had an initial metal concentration of 100 mg L⁻¹. The pHs of the solutions were then adjusted accordingly to pH 6.9. These were placed in sealed containers and mechanically stirred for 30 min. The experiments were performed in triplicates.

The effect of imprinting on selectivity was defined by:

$$k_d = \frac{C_i - C_f}{C_f} \times \frac{V}{m} \quad (3)$$

where K_d is the distribution coefficient and V, the volume of the solution used for the extraction and m, the mass of the polymer used for extraction. The selectivity coefficient (β), for the binding of a particular metal ion in the presence of a competing ion can be obtained by;

$$\beta_{Ag(CN)_2^- \text{ or } Au(CN)_2^-} = \frac{k_{Ag(CN)_2^- \text{ or } Au(CN)_2^-}}{k_{competing\ ion}} \quad (4)$$

The relative selectivity coefficient k`;

$$k` = \frac{k_{Ag(CN)_2^- \text{ or } Au(CN)_2^-}}{k_{non-imprinted}} \quad (5)$$

The results allow an estimation of the effect of imprinting on selectivity.

Effect of different desorbents on the desorption of silver and gold cyanocomplexes-imprinting ions

Batch sorption/desorption experiments were conducted using different desorption solutions of Thiourea (0.5M)/ H_2SO_4 (2M), NaOH (1.0M) and $NaNO_3$ (2M) at pH 10.5. The adsorbed silver and gold cyanocomplexes (100 mg L^{-1} metal ions) (20 mg dry adsorbents) were washed with deionized water several times, dried and transferred into stoppered reagent bottles. To this, the desorption agent (10 mL) was added, and the bottles were shaken in a shaker (200 rpm) at room temperature (25-27°C) for 2.0 hrs. The concentration of Ag or Au ions desorbed from the imprinted polymers into aqueous phase was quantified by ICP-OES. The desorption ratio (%) could be calculated from the following equation:

$$\text{Desorption ratio} = \frac{\text{amount of ions desorbed to the elution medium}}{\text{amount of ions adsorbed onto the sorbent}} \times 100 \quad (6)$$

Results and Discussion

Synthesis and characterization of the polymers

Synthesis and characterization of the aminoguanidyl-chitosan hydrochloride: In this work, aminoguanidyl-chitosan hydrochloride was successfully synthesized by condensation of chitosan with 1-cyano guanidine in acidic media (Scheme 1 step 1). The reaction is believed to follow the addition mechanism, where the primary amino groups on chitosan backbone can undertake direct addition to C≡N bond. The reaction requires acid catalysis in order to proceed at a practical rate [17]. Evidence of the successful grafting of the aminoguanidyl groups on the chitosan molecules was provided by ^{13}C NMR spectrum, where the appearance of a signal at 165 ppm is assigned to the carbon attached to active hydrogen atom of guanidium salt groups (Figure S1, Supplementary data). This signal does not exist in the spectral data of native chitosan [18]. Additional confirmation of the chemical modification of chitosan was provided by the Fourier transform infrared (FT-IR) spectra, where new peaks characteristics of guanidyl groups are observed. These new features include: a strong peak at 1653 cm^{-1} associated to the stretching vibration of C=N of guanidyl group; and two peaks at 1358 cm^{-1} and 1590 cm^{-1} which are assigned to the stretching vibration of C–N and the scissoring bending of N-H, respectively [19].

Synthesis and characterization of the AGCIPs: For the synthesis of AGCSIPs, we adopted a three step strategy, which involved the formation of electrostatic interaction between imprinted cyano-anions and guanidyl groups of chitosan followed by the freezing of the complex configuration by a cross-linking process with glutaraldehyde, and finally the removal of the templated anions to give the aminoguanidyl chitosan -imprinted polymers (AGCIPs) (step 2 and 3 in Scheme 1). The FT-IR was also used as an important tool to ascertain the binding of metal cyanide ions to AGCIP as well as their complete removal from the imprinted polymers. The spectra of aminoguanidyl chitosan control polymers (CPs), AGCIPs before and after the ion template removal (unleached IIP and leached IIP respectively) are presented in Figures 1c-2f. When comparing the spectra of these three materials, it can be observed that CP and leached AGCIP have similar spectrum, while noticeable differences can be observed before and after the leaching process on the IIP spectra. For instance the N-H stretching vibrations peaks at 3242 cm^{-1} and 3290 cm^{-1} observed in the spectra of unleached silver and gold cyanocomplexes imprinted polymers, respectively moved to higher wave number 3334 cm^{-1} after the leaching process.

Another difference is found in the 1540 to 1655 cm^{-1} region, where the aminoguanidyl characteristic peaks observed for the silver and gold cyanocomplexes unleached polymers at 1637 cm^{-1} and ~1633 cm^{-1} respectively appeared at 1643 cm^{-1} (st. vib for C=N) in the leached polymers. Furthermore the bending vibration peak of N-H that appeared at 1540 cm^{-1} and ~1543 cm^{-1} for silver and gold cyanocomplexes unleached polymers respectively is found at 1557 cm^{-1} in the leached IIPs. These results clearly confirm that the aminguanidyl groups are strongly involved in the binding of metal cyanide complex ions. Also noteworthy is the stretching vibration of C≡N groups that are observed near 2160 cm^{-1} and 2154 cm^{-1} respectively (Figures 1d-2e) in the unleached polymers, but disappeared after the leaching process, thus indicating the complete removal of the metal cyanide ion templates from the polymer.

Another confirmation of the binding of metal cyanide complex ions is given by XRD patterns of control, unleached and leached polymer particles (Figure 2a-e). In the XRD pattern of the unleached

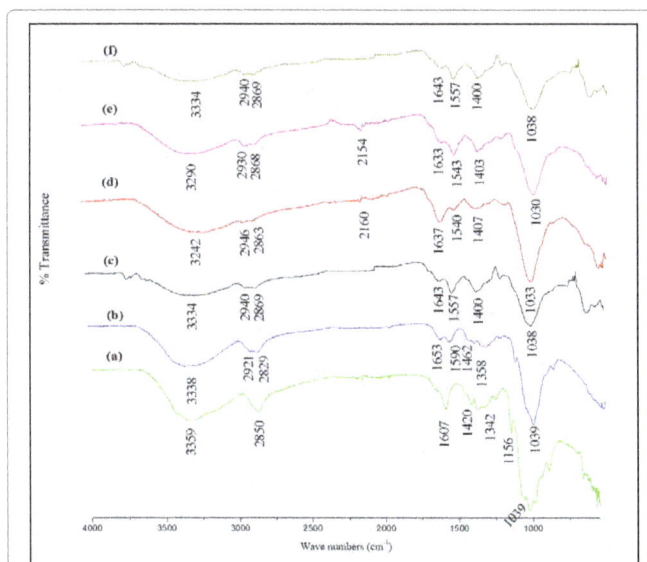

Figure 1: FTIR spectra of native chitosan (a), aminoguanidine chitosan hydrochloride (b), CP (c), silver cyanide AGCIP before leaching (d) gold cyanide AGCIP before leaching (e) and silver cyanide/gold cyanide AGCIPs (f) after leaching.

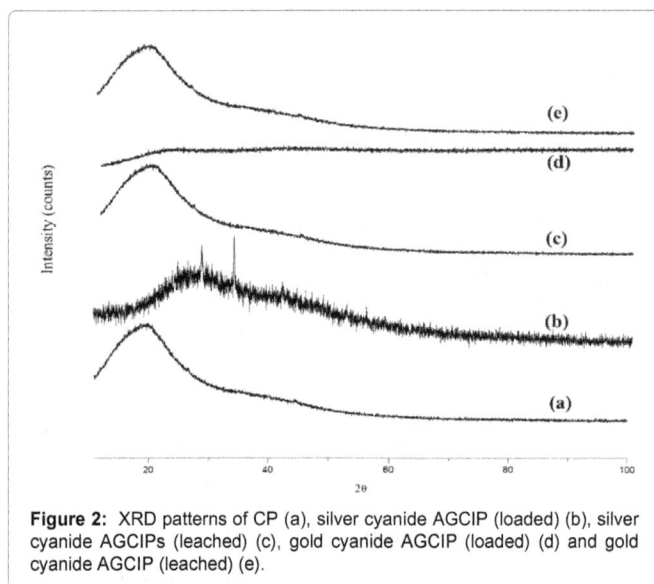

Figure 2: XRD patterns of CP (a), silver cyanide AGCIP (loaded) (b), silver cyanide AGCIPs (leached) (c), gold cyanide AGCIP (loaded) (d) and gold cyanide AGCIP (leached) (e).

silver cyanide AGCIP particles (pattern 3b), the peaks at $2\theta = 38.5°$ (1 1 1) and 59.2° (2 2 0) is assigned to silver metal which may be formed due to the reduction of silver cyanide to crystalline metallic silver [20]. In contrast, the diffractogram of unleached gold cyanide AGCIP (pattern d) show a significant decrease in the crystallinity at $2\theta = 20.42$ without indication of the reduction to Au metal, thus suggesting that gold cyanide disturb the inter-polymer bond and form stable complex with functional groups on the polymer. After leaching, the peaks related to silver cyanide (in pattern 3b) were absent in pattern (3c), while for leached gold cyanide AGCIP particles, the polymer retains its crystallinity (in pattern 3e). This indicates that the template anions were completely removed during leaching. The XRD pattern of leached AGCIP was found to be similar to that of the CP particles. Thus confirming the results obtained in the FTIR analysis.

Surface morphology, surface area and zeta potential measurements of the AGCSIPs and CPs

The surface morphology of the AGCIPs and CPs were studied using scanning electron microscopy (Figure 3a-c). The micrographs reveal that the leached imprinted polymers exhibit a rougher and porous surface which may be due to the imprinting of template anions on the polymer matrices.

The specific surface area of the particles were determined using BET method and were found to be 0.320 m² g⁻¹, 0.372 m² g⁻¹and 0.001 m² g⁻¹ for the leached silver cyanide AGCIP, the leached gold cyanide AGCIPs, and CP, respectively. Furthermore, the pore sizes for leached silver and gold cyanocomplexes AGCIPs were also found to be greater than that of CPs and were 38.908 Å, 44.603 Å and 14.166 Å, respectively. The BET result indicates an increase in surface area for template anions-AGCIPs with a porous patterned surface due to imprinting effect. This result is consistent with the conclusion obtained from SEM and again confirms that the porous texture observed in SEM images of IIPs is due to the template anions imprinting.

Zero point charge (pH_{zpc}) for the polymers was determined to get a better understanding of the adsorption mechanism for adsorbing silver and gold cyanocomplexes on the sorbent. The pH_{zpc} for all AGCIPs and their CPs were found to be at 8.8. From the electrostatic interaction point of view, we expected that guanidyl groups in the polymer will be protonated at $pH<pH_{zpc}$ (10.5) to afford positively charged sorbent,

which was beneficial for anion adsorption. Whereas, at $pH > pH_{zpc}$ solution, a relatively lower number of positively charged sites on the adsorbent surface do not favour the adsorption of anionic metal cyanides due to the electrostatic repulsion.

Statistical design of experiments

Screening of factors for silver and gold cyanocomplexes uptake by AGCIPs and control CPs polymers (Half factorial design): Table S1 (Supplementary data) depicts the results of RSM optimization where 5-factors fractional factorial was used. The regression analysis was performed to fit the response (%EE). A good fit model should have a correlation coefficient value above 0.8 [21]. The correlation coefficients obtained in our study were 0.998 and 0.979 for the extraction of silver and gold cyanocomplexes, respectively. These values were higher than 0.8. Thus indicating a good agreement between the predicted and observed results [22].

Analysis of Variance (ANOVA) was used to investigate the significance of the variables and their interaction in the extraction of the silver and gold cyanocomplexes. The information acquired from the ANOVA is presented in a Pareto chart (Figure 4A and 4B). Bar lengths are proportional to the absolute value of estimated effect, which helps in comparing the significance of effects. If the bar exceed the vertical reference line (P 0.05), the effect of the variable or interaction is significant [23]. On the other hand, when the value of the factor is positive it implies that increasing the factors from minimum to maximum maximizes the extraction. In contrast; the negative values mean that the factors must be kept at low levels to enhance the extraction [24].

Figure 4A and 4B showed that the type of polymer (A) and the pH of the solution (D) used for the extraction of silver cyanide (+7.30 and -7.02, respectively) and gold cyanide (+13.96 and -5.27, respectively) were statically significant at 95% confidence level. This result demonstrated that the IIPs (high level) were better than CPs (low level) for both silver and gold cyanocomplexes. Therefore, it was expected that experiment with IIPs optimized system will present higher adsorption capacities for silver and gold cyanocomplexes than CPs. On the other hand, the negative value of the pH coefficient meant that the silver and gold cyanocomplexes uptake by IIPs and CPs were favoured at low pH values (pH 6.0). The increase in the pH led to a notable decrease of both cyanide anions uptake by the two kinds of adsorbents (IIPs and CPs). The other statically significant factors for the extraction of silver cyanide (Figure 4A) were solution temperature (-3.77) and interaction between BC (-4.93), CD (-3.11) and AD (-3.04). Contact time (+0.48) and initial concentration (+0.34) were not significant at 95% confidence level. In the case of gold cyanide extraction (Figure 4B), contact time (+4.44) and interaction between AB and AD were statically significant at 95% confidence level. However, the main effect contact time (-1.56) and temperature (+0.68) were statically significant at 95% confidence level.

For further optimization experiments on the adsorption of silver and gold cyanocomplexes, only imprinted polymers (IIPs) will be considered, and only the most three significant factors will be undertaken. These include: pH, temperature and contact time for the extraction of silver cyanide; and pH, contact time and initial metal concentration for the extraction of gold cyanide.

Surface analysis: After identifying the most significant variables using a half 2^5 factorial designs, a Box- Behnken response surface design (RSM) was used to find the optimum condition for the highest silver and gold cyanocomplexes extraction. The list of experiments

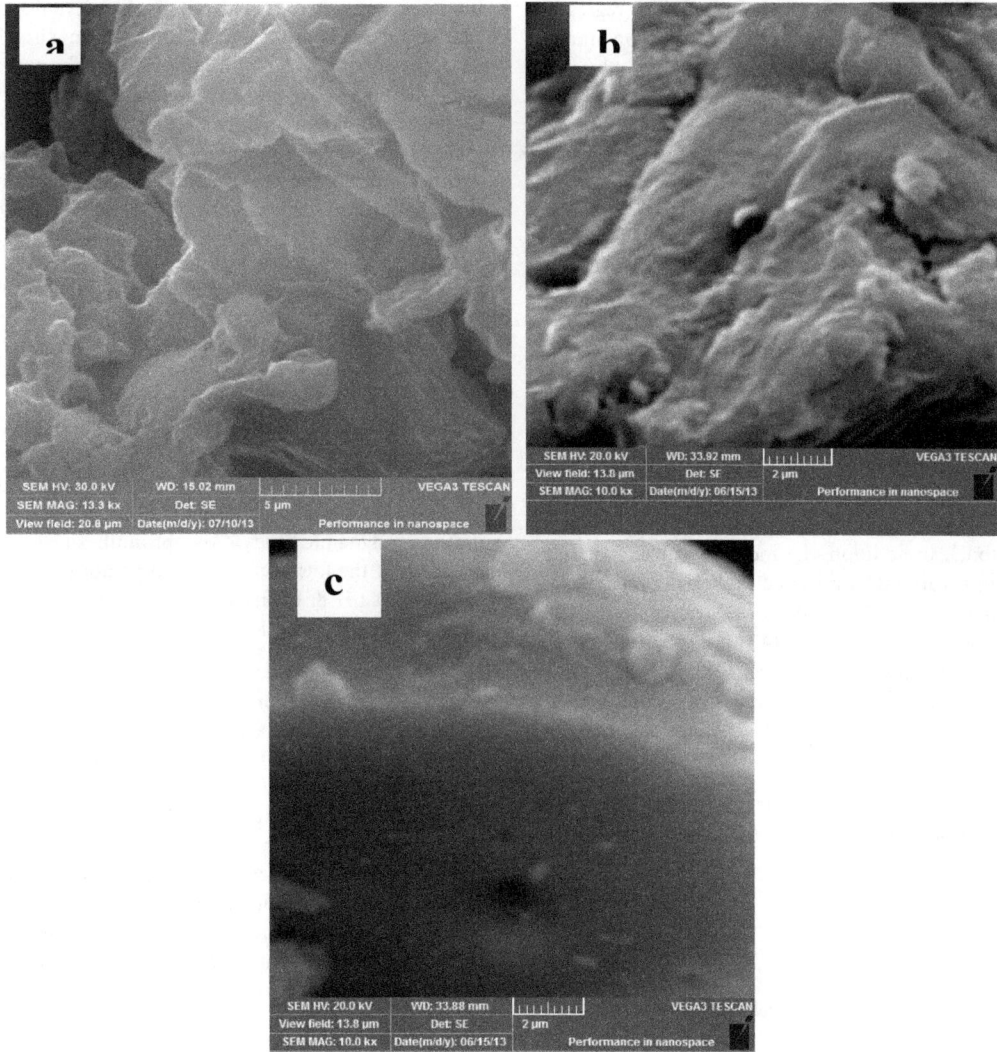

Figure 3: Scanning electron micrographs of silver cyanide AGCIP (a), gold cyanide AGCIP (b) and CP (c), magnification × 10000.

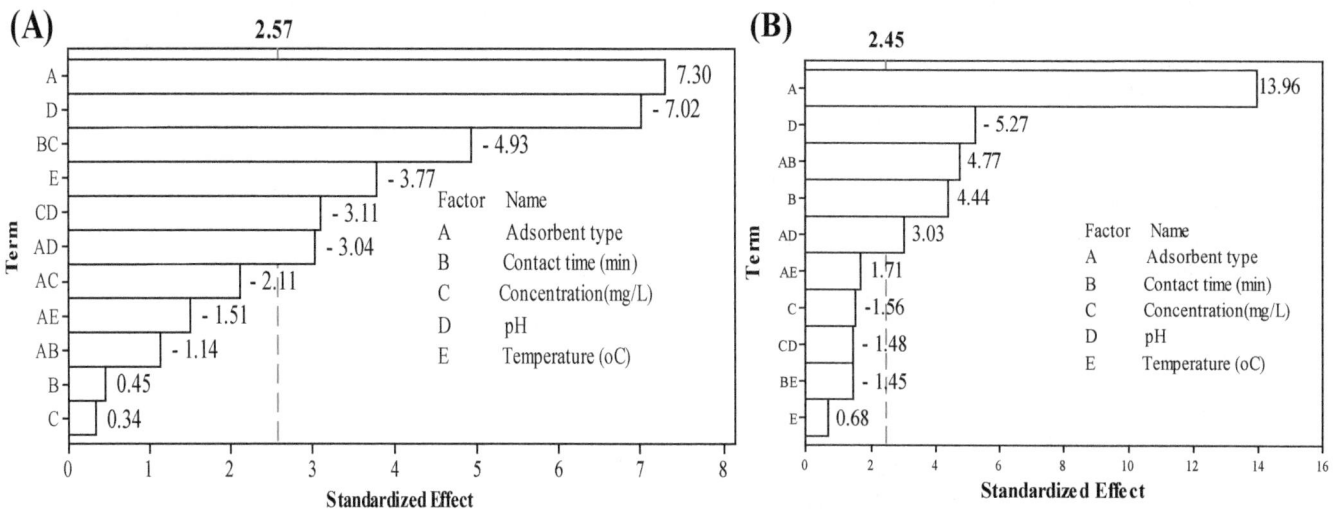

Figure 4: Pareto chart of standardized effects for variables related to the extraction of silver cyanide (A) and gold cyanide (B).

as designed by RSM and the values of the response (% EE) for each metallic cyanide anion obtained are given in Table 1. The regression analysis was carried out to fit the response. The final regression function of the extraction response (extraction efficiency, %EE) of silver and gold cyanocomplexes on imprinted polymers were expressed in terms of actual factors.

$$\text{\%EE (silver cyanide)} = -251.71 + 0.39B + 100.38D - 0.37E - 7.32D^2 + 0.02BD - 0.01BE + 0.03DE \tag{7}$$

$$\text{\%EE (gold cyanide)} = -106.15 + 1.19B + 1.01C + 25.74D - 0.01B^2 - 2.75 \times 10^{-3}C^2 - 1.65D^2 - 1.20 \times 10^{-3}BC - 0.02CD \tag{8}$$

The significance of each coefficient present in Equations (7 and 8), were determined by the student's t-test and p-values. The results of the quadratic model for % extraction efficiency in the form of analysis of variance (ANOVA) are given in Table 2. The values of R^2, $R^2_{adjusted}$ and $R^2_{predicted}$, for the extraction of both metallic cyanide anions were close to 1, thus demonstrating a qualitative agreement between the experimental values and the predicted values obtained for the three variables under study [15]. It also means that most of the data variations were explained by the regression model. In addition, the model is considered to be statistically significant because the associated Probability (p value) for the model is lower than 0.05, at 95% confidence. Meanwhile, the lack-of-fit for p-values were more than 0.05 (0.142 and 0.954 for recovery silver and gold cyanocomplexes, respectively), and thus statically insignificant at 95% confidence level. These results demonstrate that the quadratic model is statically significant for the response and therefore it will be used for further analysis [15].

Three-dimensional (3D) response surface plots: The three-dimensional (3D) response surface plots of the dependent variable as a function of two independent variables, maintaining all other variables at fixed levels can provide information on their relationships and can be helpful in understanding both the main and the interaction effects of these two independent variables. Therefore, to investigate the interactive effect of two factors on the extraction efficiency of silver and gold cyanocomplexes, the three dimensional response surface plots were constructed based on the quadratic model. The constructed plots under the combined effect of process parameters shown in Figure 5A-5F together with the inferences so obtained are discussed below.

The effect of pH: Figure 5A, 5B, 5D and 5F show the interactive effect of pH with time of contact (Figure 5A and 5D), pH with temperature (Figure 5B) or pH with initial concentration (Figure 5F) for the extraction of silver and gold cyanocomplexes from the aqueous phase. As can be seen, the influence of pH was found to be the highest amongst the studied variables. A major decrease was observed in the extraction efficiency as pH was increased from 6.0 to 10.0 in both figures. Accordingly, when pH was held below 9.0 (preferably at pH ~8.0), 80% and 90% of silver and gold cyanocomplexes were recovered respectively, regardless of contact time, temperature or initial concentration. The results implied that the adsorption capability of the adsorbent was nearly independent of pH in a range from 6.0 to 9.0, which was consistent with the pHzpc discussed above. While the observed reduction in adsorption ability of the adsorbent when the pH further increase (pH > 9) can be explained by the competition between the excessive OH⁻ ions (introduced by base adding) and template metal cyanide for the binding sites. There are also electrostatic repulsion between the negatively charged surface of adsorbent and the anionic metal cyanide. So the optimal pH range of the IIP CUCS-GLA polymer was 6.0–9.0, which was much wider than that of the protonated cross-linked chitosan (4.0-6.0) [7].

The effect of contact time: Figure 5A and 5B reveal that among all the three parameters considered for silver cyanide adsorption, contact time has the minimum impact since two maxima has been found; so from the 3D graphs the optimum value of the contact time was found to be 30 min. On the contrary, the effect of contact time was significant for gold cyanide extraction, and the extraction efficiency increases with increasing contact time between the metal cyanide and adsorbent, within its respective experimental range. The optimum condition of time, pH and initial concentration in the adsorption of gold cyanide were estimated to be 72 min, 8.5 and 170 mg.L⁻¹, respectively.

The effect of temperature and initial concentration: The conjugated effects of temperature with contact time and temperature with pH on the extraction efficiency of silver cyanide under predefined condition are visualized in Figures 5C and 6B, respectively. The solution temperature has no significant effect on the silver cyanide extraction. The extent of extraction was suppressed only by 8% when solution temperature was increased from 25 to 50°C for constant pH and contact time levels. Thus, it may be evident that the adsorption of silver cyanide is an exothermic process which will be verified later in thermodynamic study. An optimum silver cyanide extraction is observed at the lowest solution temperature (25°C) and contact time (30 min) at pH 8. Under these conditions, 80% extraction efficiency recovery for silver cyanide was obtained.

The combined effects of initial concentration and contact time, and initial concentration and pH on the extraction of gold cyanide by the imprinted polymer are visible in Figure 5E and 5F, respectively. As can be seen, the extraction efficiency increases linearly with increasing concentration from 100 to 150 mg L⁻¹; and the trend is reversed at higher concentration (above 150 mg L⁻¹), where the gold cyanide sorption decreases with increasing concentration. This behaviour can be understood as the increase in metal cyanide ions concentration with fixed adsorbent amount results in saturation of the binding sites on the surface of the polymer. The maximum adsorption of gold cyanide by the imprinted polymer was 90% at 150 mg L⁻¹ initial concentration at constant pH (~6.5) and time (72 min), which is accordance with the model.

Confirmation experiments: To check the validity and suitability of the data obtained from the software numerical modelling under optimized condition, confirmatory experiments were conducted with parameters suggested by the model. The optimum values of the independent process variables to achieve the maximum extraction efficiency (94.4%) of the silver cyanide from an aqueous solution of 100 mg L⁻¹ were; contact time 30.0 min; solution pH 6.9 and temperature 25°C. In the case of the gold cyanide, the theoretical maximum extraction efficiency (97.3%) was obtained at 25°C when the independent process variables were set as: contact time 72.2 min; solution pH 6.9 and the initial metal concentration 142.1 mg L⁻¹. The corresponding experimental value of the silver and gold cyanocomplexes adsorption under the optimum condition of the variables were determined respectively as 92.9% ± 1.2 and 95.4% ± 1.3 (mean ± SD of three replicates), which is very close to the theoretical value (error margin of less than 1.5). The good correlation between these two set of results, indicates that the Box-Behnken design coupled with desirability functions could be effectively used to optimize the adsorption parameters for the extraction of silver and gold cyanocomplexes.

Adsorption isotherms

In this study, the adsorption isotherms at two different pHs, 6.9 and 10 were investigated using two equilibrium models, which are Langmuir and Freundlich isotherm models with the quality of the

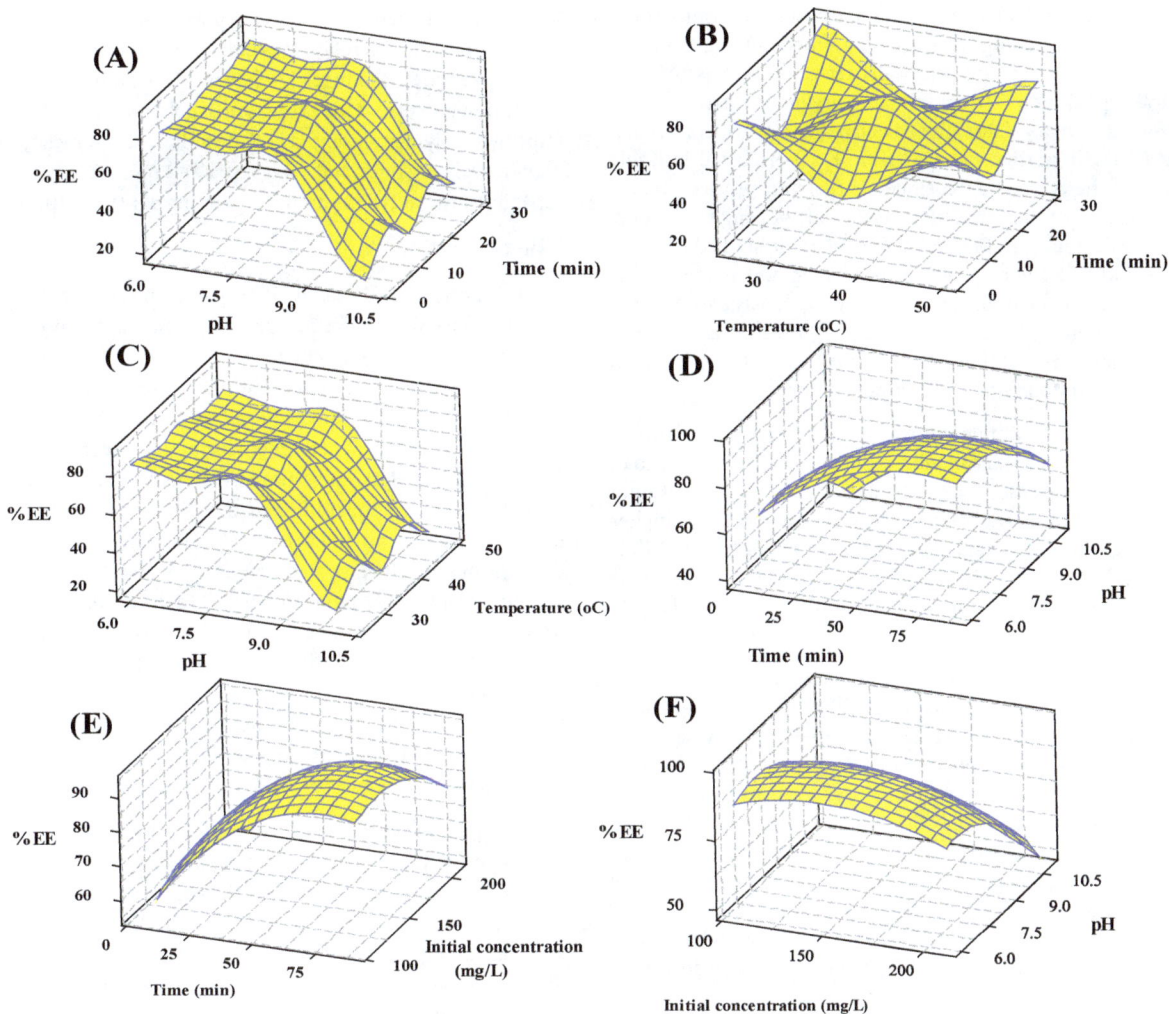

Figure 5: Three-dimensional response surface for extraction efficiency of silver cyanide (A-C) and gold cyanide (D-F) using AGCIPs.

Analysis of variance (ANOVA) for quadratic model for $Ag(CN)_2^-$ adsorption						
Source	DF	Seq SS	Adj SS	Adj MS	F	P
Regression	7	11091.000	11091.000	1584.420	243.710	0.000
Linear	3	7441.900	2638.000	879.340	135.260	0.000
Square	1	3630.000	3630.000	3629.950	558.340	0.000
Residual Error	9	58.500	58.500	6.500		
Lack-of-Fit	5	11.000	11.000	2.190	0.180	0.954
Pure Error	4	47.500	47.500	11.890		
Analysis of variance (ANOVA) for quadratic model for $Au(CN)_2^-$ adsorption						
Source	DF	Seq SS	Adj SS	Adj MS	F	P
Regression	8	4239.760	4239.760	529.970	244.660	0.000
Linear	3	2813.220	2813.22	937.740	432.900	0.000
Square	3	1371.130	1371.130	457.040	210.990	0.000
Residual Error	8	17.330	17.330	2.170		
Lack-of-Fit	4	13.230	13.230	3.310	3.220	0.142
Pure Error	4	4.100	4.100	1.030		
Models summary Statistic						
Source	$Ag(CN_2^-$			$Au(CN)_2^-$		
R^2	0.993			0.996		
R^2 (adjusted)	0.990			0.991		
R^2 (predicted)	0.988			0.977		

Table 2: Analysis of variance (ANOVA) and statistical summary for quadratic models for silver and gold cyanocomplexes adsorption.

fitting evaluated using the correlation coefficient and functional error function. It was reasonable to study the adsorption isotherm at pH 10 since a typical mining leaching solution containing gold and silver as cyanide complexes are practically recovered at this pH. The mathematical equations of Langmuir and Freundlich are given in Table S2 (Supplementary data). The calculated constants, correlation coefficients (R^2) and relative average error (F_{error}) are listed in Table 3. Inspection of the correlation coefficients (R^2) and relative average errors (F_{error}) show that the correlation coefficients of Langmuir model was stronger with respect to Freundlich model at the studied pHs, for both AGCIP and CP. The q_{max} calculated by Langmuir model (Table 3) at the studied pH for both IIP and CP were consistent with the experimental saturated adsorption of resins (q_{exp}), which also proved that the monolayer adsorption was dominant and the process is preferably chemisorption rather than physisorption.

Furthermore, the maximum sorbent capacities for silver cyanide AGCIP and CP decreased substantially with increasing pH (more than 25% at pH 10). But, the adsorption capacity for $Au(CN)_2^-$ AGCIPs did not change much with increasing pH and was found to be, at higher pH, comparable with that obtained at the optimum pH (pH 6.9) (less than 2.2% at pH 10), thus indicating the strong interaction between template gold anions and active sites of the adsorbent even at relatively high pH.

The essential degree of suitability of resin towards metal ions can be expressed in terms of separation factor R_L, which describes the type of isotherm and it is calculated by the equation $R_L = 1 + (1/K_L C_i)$ where, C_i is the initial concentration of template anions. The values of R_L calculated for both IIP and CP at the two studied pH were in the range between 0.004 and 0.7, thus indicating highly favourable adsorption of template anions onto the studied resins. Another support for the favourability of the adsorption process is given by the values of adsorption intensity (n) which were determined from Freundlich isotherm. In this study the calculated n values were greater than unity indicating favorable adsorption conditions and reflecting a high affinity between adsorbate and adsorbent which is indicative of chemisorption.

Comparison of the maximum adsorption capacities of AGCSIPs with different adsorbents reported in literature

For comparison study, the values of some adsorbent capacities towards silver and gold cyanocomplexes available in the literature are shown in Tables 4 and 5. The q_{max} values for adsorption silver cyanide (429.2 and 321.5 mg g^{-1}) and gold cyanide (319.5 and 312.5 mg g^{-1}) at pH 6.9 and 10, respectively are higher than those in most previous studies, except for the adsorption of gold cyanide by the Amberjet™

4400 and the biosorbent reported by In Seob et al. [25]. However, the adsorption pH used in the reported work was not practically feasible since the extraction was conducted in acidic medium and the adsorption time was longer (24 hours). This finding suggests that the adsorption process for the extraction of gold cyanide by the synthesized imprinted polymer presented in this study is economically more viable than the existing ones, since it is faster and presents a good adsorption capability even at typical mining operating pH (pH =10).

Kinetic studies

Kinetic analysis is required to get an insight of the mechanism that describes the adsorption process, which are mainly used in the modeling, and designing of continuous process. Four linearized form of kinetic models viz., pseudo-first order, pseudo-second order, intra-particle diffusion and liquid film diffusion have been used to analyze the collected experimental data during (Table S2, Supplementary data). The parameters, correlation coefficients (R^2) and relative average error (F_{error}) of the four different models were all listed in Table S3 (Supplementary data) and presented in Figure S2 a-d (Supplementary data). From the investigated models discussed above, the pseudo-second-order kinetics model fitted well to the adsorption data inferred from the high R^2 value of 0.9775 to 0.9999, as well as better predicted the value for q_e than those of first-order model given in Table S3 (Supplementary data). Thus it can be assumed that chemisorption is the rate-controlling step involving valence forces through sharing or exchange of electrons between the adsorbent surface and adsorbate ions with no involvement of a mass transfer in solution.

The plot of intra-particle diffusion which can be obtained by plotting q_t vs. $t^{0.5}$ (Figure S2c, Supplementary data) is linear over the entire time range but did not pass through the origin point indicating that the intra-particle diffusion may not be the rate controlling factor in determining the kinetics of the process. The role of liquid film diffusion in governing the adsorption process was also verified using the kinetic equation. The plot of −ln(1−F) versus time (presented in Figure S2d, Supplementary data) and the fitting parameter (Table S3, Supplementary data) gave linear line with the higher correlation coefficients for IIPs ranged from 0.9781 to 0.9878. However, the corresponding calculated correlation coefficients for CPs were low (0.9257 and 0.1301). This indicates that the liquid film diffusion model can be applied to predict the adsorption kinetic for the adsorption of template anions by AGCIPs and demonstrates that the surface areas of AGCIPs are higher compared to the CPs resin.

Isotherm	$Ag(CN)_2^-$ AGC				$Au(CN)_2^-$ AGCAGC			
	pH 6.9		pH 10		pH 6.9		pH 10	
	IIP	CP	IIP	CP	IIP	CP	IIP	CP
Langmuir q_{exp}	407.54	163.99	293..69	101.88	315..17	52.74	303.79	31.84
q_m (mg g^{-1})	429.185	150.830	321.543	104.822	319.489	52.826	312.500	35.199
K_L (L mg^{-1})	0.041	0.008	0.062	0.028	0.0914	0.079	0.084	0.024
R^2	0.999	0.887	0.998	0.990	0.999	0.998	0.999	0.909
F_{error}	0.041	0.158	0.042	0.058	0.026	0.084	0.041	0.163
Freundlich								
K_F ((mg.g^{-1}.(mg L^{-1})$^{1/n}$)	109.648	46.666	31.696	15.849	109.648	44.463	87.096	9.683
1/n	0.221	0.147	0.475	0.316	0.185	0.036	0.232	0.250
R^2	0.977	0.584	0.962	0.986	0.985	0.436	0.860	0.437
F_{error}	0.061	0.134	0.043	0.066	0.060	0.083	0.047	0.053

Table 3: Fitting parameters for the Langmuir and Freundlich to experimental adsorption isotherm of AGCSIPs and CPs.

Adsorbents	pH	q_e (mg Ag/g^{-1})	Reference
Conventional anion exchangers with modified amino alkyl group and S or S,N-modified polystyrene	9.0	50.0 -117.0	[30]
Weak base resin Wofatit AD-4	10.4	36.7	[31]
Weak base resin (Wofatit AD-42)	10.4	28.0	[31]
Weak base resin (Lewatit MP-64)	10.4	2.3	[31]
Weak base resin (Amberlite IRA-93)	10.4	25.9	[31]
Strong base resin (Wofatit RO)	10.4	83.1	[31]
Strong base resin (Wofatit SBW)	10.4	107.9	[31]
Strong base resin (Lewatit M 500)	10.4	129.4	[31]
AGCIP	6.9	429.2	This work
AGCSIP	10.0	321.5	This work

Table 4: Maximum adsorption capacities for the adsorption of silver cyanide ions onto various adsorbents.

Adsorbents	pH	q_e (mg Au/g^{-1})	Reference
Minix resin	9.0	26.0	[32]
Aminoguanidine of acrylonitrile/vinyl acetate/divinylbenzene	9.0	49.3	[9]
Aminoguanidine of vinyl benzyl chloride/ methyl methacrylat/ divinylbenzene resin	9.5	30.0	[8]
Aminoguanidine of poly(acrylonitril-co vinyl acetat-co-divinyl benzene) polymer	9.6	2.3	[14]
aminoguanidine of poly(vinylbenzyl chloride-co-divinylbenzene) polymer	9.6	23	[14]
Hypersol Macronet resin MN300	8.5	>40	[1]
Amberjet™ 4400	3.0	427.8	[25]
activated carbon	3.0	170.6	[25]
polyethylenimine modified biomass	3.0	361.8	[25]
AGCIPs	6.9	319.5	This work
AGCSIPs	10.0	312.5	This work

Table 5: Maximum adsorption capacities for the adsorption of gold cyanide ions onto various adsorbents.

Adsorption thermodynamic

Thermodynamic parameters were evaluated to confirm the nature of the adsorption process. The thermodynamic parameters, Gibbs free energy change (ΔG^o), enthalpy change (ΔH^o) and entropy change (ΔS^o), were calculated according to the equations given in Table S2 (Supplementary data) and the results are reported in Table S4 (Supplementary data). The negative value of ΔH^o indicates the exothermic nature of the adsorption reaction, while the negative values of entropy change (ΔS^o) reflect a decreased in randomness at the solid–solution interface due to the electrostatic attraction between the template anions and the active sites of the resins to form stable structure [26]. The negative values of ΔG^o (except for the adsorption of gold cyanide anions by CP) confirm the favourability of the process and the spontaneous nature of adsorption with a high preference for silver and gold cyanocomplexes onto the resins. This means that the energy input from outside is not necessary. The data given in Table S4 (Supplementary data) also show a slight change in the value of $T\Delta S^o$ at all temperatures and all $|\Delta H^o| > |T\Delta S^o|$. This indicates that the adsorption process is dominated by enthalpic rather than entropic changes. However, the adsorption of gold cyanide onto CP is an exceptional case since $|T\Delta S^o| > |\Delta H^o|$, meaning that the former process is dominated by entropic change [27].

Selectivity studies

AGCIPs and CP particles were tested for the separation of silver or gold cyanocomplexes from mixtures containing hexacyanoferrate(III) and mercuric cyanide. The distribution coefficient (K_d), the selectivity coefficient (β) and the relative selectivity coefficient (k′) values of the competing ions with respect to the target ions, silver and gold cyanocomplexes are summarized in Table 6. The data obtained clearly indicated that the distribution coefficient increased for target anion, silver and gold cyanocomplexes and decreased for other coexisting ions. AGCIPs showed maximum selectivity for the template anions over other anions. The relative selectivity order of AGCSIPs silver and gold cyanocomplexes increased from $Ag(CN)_2^-> Au(CN)_2^->>$ $Fe(CN)_6^{3-} >>>>$ $Hg(CN)_2$ and $Au(CN)_2^-> Ag(CN)_2^->> Fe(CN)_6^{3-} >>>>$ $Hg(CN)_2$, respectively, indicating quantitative separation of target template anions when it is present together with other anions. Though the silver and gold cyanocomplexes resemble each other in terms of (i) the weak hydration state compared with other cyanide-complexes anions; (ii) the large ionic size (Ag 12,6 nm; Au 13.7 nm) and (iii) their single anions charged [28], the IIP synthesized in this study has a higher selectivity for the specific targeted precious metal anion since, it perfectly fitted the fabricated recognition sites on the polymers.

Desorption and reusability studies

In order to reduce the cost of the extraction process, the reusability of polymer is a crucial factor for the adsorbent. Table S5 (Supplementary data) demonstrates the results of three different eluents used for the extraction of silver and gold cyanocomplexes from the metal anion adsorbed resin. It was observed that silver and gold cyanocomplexes could be quantitatively desorbed with 10 ml of 2M KNO_3 at pH 10.5 with extraction above 95% in the first elution cycle. The probable mechanism of the regeneration might be that the aminoguanidyl functional groups are deprotonated causing release of the precious metal cyanide anions [1,29]. The reason for using relatively high concentrations of nitrate counter ions is to replace the loaded silver and gold cyanocomplexes in the loaded resin. This synergic effect is required to establish a driving force for the elution of the cyanocomplexes in aqueous media. To assess the regeneration of the adsorbent, five consecutive adsorption-desorptions cycles were conducted using 2M KNO_3 at pH 10.5 as desorbing agent. It is shown (Figure S3, Supplementary data) that the uptake capacity of $Ag(CN)_2^-$ and $Au(CN)_2^-$ on the adsorbents decreased slowly with increasing number of cycles. At the fifth regeneration cycle, the adsorption remained above 85%. These results show that the adsorbents could be effectively recycled and reused for silver and gold cyanocomplexes adsorption with 2M KNO_3 at pH 10.5, and the adsorbents. And it can be concluded that silver and gold cyanocomplexes imprinted polymers could be used several times without significantly diminishing their adsorption capacity.

Conclusion

In summary, two aminoguanidyl-chitosan ion imprinted polymers have been successfully prepared and tested for the extraction of gold and silver cyanide ions in aqueous solution. High adsorption rates, high adsorption capacity and high selectivity have been observed for the two polymers. The optimum adsorption conditions were established at 72 mn, 6.9 and 142.1 mg L^{-1} for time, pH and initial concentration respectively. AGCSIPs also exhibit excellent stability and reusability (up to five cycles). When compare to the existing gold and silver cyanides adsorbents, these polymers were faster and presented good adsorption capability even at typical mining operation pH (pH=10).

Ag(CN)$_2^-$ AGCIP						
Couple mixture	K$_D$ Ag(CN)$_2^-$ (ml/g)		K$_D$ M(CN)$_x^{-n}$ (ml/g)		B Ag(CN)$_2^-$/M(CN)$_x^{-n}$		k`
	IIP	CP	IIP	CP	IIP	CP	
Ag(CN)$_2^-$/ Au(CN)$_2^-$	5.856	0.361	1.941	0.592	3.017	0.610	4.946
Ag(CN)$_2^-$/Fe(CN)$_6^{3-}$	21.813	0.039	0.289	0.0294	75.478	1.327	56.878
Ag(CN)$_2^-$/ Hg(CN)$_2$	∞	0.059	0	0	∞	∞	∞
Au(CN)$_2^-$ AGCIPs						
Couple mixture	K$_D$ Au(CN)$_2^-$ (ml/g)		K$_D$ M(CN)$_x^{-n}$ (ml/g)		β Au(CN)$_2^-$/M(CN)$_x^{-n}$		k`
	IIP	CP	IIP	CP	IIP	CP	
Au(CN)$_2^-$/ Ag(CN)$_2^-$	4.693	0.999	0.541	1.561	8.675	0.128	67.773
Au(CN)$_2^-$/Fe(CN)$_6^{3-}$	10.376	0.050	0.399	0.281	26.005	0.177	146.921
Au(CN)$_2^-$/ Hg(CN)$_2$	205.008	0.910	0.036	0.163	5694.667	5.583	1020.001

Table 6: Competitive adsorption properties of AGCSIPs (mixture: Ag(CN)$_2^-$/Au(CN)$_2^-$/Fe(CN)$_6^{3-}$/Hg(CN)$_2$), T 25 ± 1°C, time 30 min and pH 6.9 ± 0.1)

Thus suggesting that these new aminoguanidyl-chitosan imprinted polymers could be used effectively in practical applications for the selective recovery of silver and gold cyanide ions in aqueous solutions.

Acknowledgements

The authors acknowledge the Centre for Nanomaterial Science Research (CNSR), the Department of Applied Chemistry; and the University of Johannesburg (UJ) for their financial support.

References

1. Cortina JL, Kautzmann RM, Gliese R, Sampaio CH (2004) Extraction studies of aurocyanide using macronet adsorbents: Physico-chemical characterization. Reactive & Functional Polymers 60: 97-107.

2. Xie F, Lu D, Yang H, Dreisinger D (2014) Solvent extraction of silver and gold from alkaline cyanide solution with lix 7950. Mineral Processing & Extractive Metall Rev 35: 229-238.

3. Syed S (2012) Recovery of gold from secondary sources-a review. Hydrometallurgy 115: 30-51.

4. Soleimani M, Kaghazchi T (2008) The investigation of the potential of activated hard shell of apricot stones as gold adsorbents. Journal of Industrial and Engineering Chemistry 14: 28-37.

5. Fleming C, Cromberge G (1984) The extraction of gold from cyanide solutions by strong- and weak-base anion-exchange resins. J S Afr Inst Min Metall 84: 125-137.

6. Ahamed MEH, Mbianda XY, Mulaba-Bafubiandi AF, Marjanovic L (2013) Selective extraction of gold(iii) from metal chloride mixtures using ethylenediamine n-(2-(1-imidazolyl)ethyl) chitosan ion-imprinted polymer. Hydrometallurgy 140: 1-13.

7. Xie Y, Li S, Liu G, Wang J, Wu K (2012) Equilibrium, kinetic and thermodynamic studies on perchlorate adsorption by cross-linked quaternary chitosan. Chemical Engineering Journal 192: 269-275.

8. Bozena NK, Dorota JB, Andrzej WT, Wieslaw A (1999) Influence of the structure of chelating resins with guanidyl groups on gold sorption. Reactive and Functional Polymers 42: 213-222.

9. Boiena NK, Dorota B, Andrzej WT, Wieslaw A, Barbara P (1998) New selective resins with guanidyl groups. Reactive and Functional Polymers 36: 185-195.

10. Pan L, Wang F, Bao X (2013) Solvent extraction of gold(i) from alkaline cyanide solution with furfuryl thioalcohol. Separation Science and Technology 48: 2007-2012.

11. Yan H, Row KH (2006) Characteristic and synthetic approach of molecularly imprinted polymer. International journal of molecular Sciences 7: 155-178.

12. Nishad PA, Bhaskarapillai A, Velmurugan S, Narasimhan SV (2012) Cobalt (ii) imprinted chitosan for selective removal of cobalt during nuclear reactor decontamination. Carbohydrate Polymers 87: 2690-2696.

13. Cortina JL, Meinhardt E, Roijals, Marti V (1998) Modification and preparation of polymeric adsorbents for precious-metal extraction in hydrometallurgical processes. Reactive & Functional Polymers 36: 149-165.

14. Dorota JB, Bozena NK (2002) Gold sorption on weak base anion exchangers with aminoguanidyl groups. European Polymer Journal 38: 2239-2246.

15. Umesh KG, Kaur MP, Garg VK, Dhiraj S (2008) Removal of nickel(ii) from aqueous solution by adsorption on agricultural waste biomass using a response surface methodological approach. Bioresource Technology 99: 1325-1331.

16. Zare F, Ghaedi M, Daneshfar A, Agarwal S, Tyagi I, et al. (2015) Efficient removal of radioactive uranium from solvent phase using agoh–mwcnts nanoparticles: Kinetic and thermodynamic study. Chemical Engineering Journal 273: 296-306.

17. Yongbo S, Qiuxiao L, Yunling L (2012) Self-aggregation and antimicrobial activity of alkylguanidium salts. Colloids and Surfaces A: Physicochem Eng Aspects 393: 11-16.

18. Liu M, Zhou Y, Zhang Y, Yu C, Cao S (2013) Preparation and structural analysis of chitosan films with and without sorbitol. Food Hydrocolloids 33: 186-191.

19. Bozena NK, Dorota JB, Julia J, Wieslaw A (2001) Anion exchangers with alkyl substituted guanidyl groups gold sorption and cu(ii) coordination. Reactive & Functional Polymers 48: 169-179.

20. Liang Z, Shuwei Y, Tong H, Lvling Z, Cailian M, et al. (2012) Improvement of ag(i) adsorption onto chitosan/triethanolamine composite sorbent by an ion-imprinted technology. Applied Surface Science 263: 696-703.

21. Joglekar AM, May AT (1987) Product excellent through experimental design. General Food World 32: 857-868.

22. Camila GP, Fernanda SR, Nathália MS, Araci ASJ, Júlio CPV, et al. (2006) Use of statistical design of experiments to evaluate the sorption capacity of 7-amine-4-azaheptylsilica and 10-amine- 4-azadecylsilica for cu(ii), pb(ii), and fe(iii) adsorption. Journal of Colloid and Interface Science 302: 396-407.

23. Somera Bf, Corazza MZ, Yabe MMJS, Segatelli MG, Galunin E, et al. (2012) 3- mercaptopropyltrimethoxysilane- modified multi-walled carbon nanotubes as a new functional adsorbent for flow injection extraction of pb(ii) from water and sediment samples. J Water, Air and soil Pollution 223: 6069-6081.

24. Jorge LB, Ricardo RE, Caroline DM, Lucas CM, Flavio AP, et al. (2006) Statistical design of experiments as a tool for optimizing the batch conditions to cr(vi) biosorption on araucaria angustifolia wastes. Journal of Hazardous Materials B 133: 143-153.

25. In Seob K, Min AB, Sung WW, Juan Maob KS, Jiyeong P, et al. (2010) Sequential process of sorption and incineration for recovery of gold from cyanide solutions: Comparison of ion exchange resin, activated carbon and biosorbent. Chemical Engineering Journal 165: 440-446.

26. Jayakumar R, Rajasimman M, Karthikeyan C (2015) Optimization, equilibrium, kinetic, thermodynamic and desorption studies on the sorption of cu (ii) from an aqueous solution using marine green algae: Halimeda gracilis. Ecotoxicology and environmental safety 121: 199-210.

27. Zhou L, Shang C, Liu Z, Huang G, Adesina AA (2012) Selective adsorption of uranium (vi) from aqueous solutions using the ion-imprinted magnetic chitosan resins. Journal of Colloid and Interface Science 366: 165-172.

28. Xihui Y, Aleksandra O, Hao D, Jan DM (2011) Molecular dynamics simulations of metal–cyanide complexes: Fundamental considerations in gold hydrometallurgy. Hydrometallurgy 106: 64-70.

29. Adelia MO, Versiane AL, Carlos AS (2008) A proposed mechanism for nitrate and thiocyanate elution of strong-base ion exchange resins loaded with copper and gold cyanocomplexes. Reactive & Functional Polymers 68: 141-152.

30. Knothe M, Feistel L, Hauptmann R, Schwachula G, Schwachula Chemie H (1991) Studies on the adsoorption of silver from cyanide solutions by fuctional polymers. Solvent Extraction and Ion Exchange 9: 677-696.

31. Slavica P, Ivan P (1998) Electrolytic desorption of silver from ion-exchange resins. Wat Res 32: 2913-2920.

32. Conradie PJ, Johns MW, Fowles RJ (1995) Elution and electrowinning of gold from gold-selective strong-base resins. Hydrometallurgy 37: 349-366.

23

The Preparation of CNTs/PE Nanocomposites Particles with Coral Shape and Core-Shell Structure *In Situ* Produced via Nanotemplate Catalyst Based on MWCNTs

Jing Wang[1], Jiangping Guo[1], Yang Zhou[1], Qigu Huang[1*], Jianjun Yi[2], Hongming Li[2], Yunfang Liu[1], Kejing Gao[2] and Wantai Yang[1]

[1]*State Key Laboratory of Chemical Resource Engineering, Key Laboratory of Carbon Fiber and Functional Polymers, Ministry of Education, Beijing University of Chemical Technology, People's Republic of China*

[2]*Lab for Synthetic Resin Research Institution of Petrochemical Technology, China National Petroleum Corporation, People's Republic of China*

Abstract

A kind of nano template catalyst was prepared through loading the active compound (m-CH$_3$PhO)TiCl3 on carbon nanotubes (CNTs). This catalyst can catalyze (co)polymerization of ethylene to form CNTs/polyethylene (PE) nanocomposites particles. The nano template catalyst showed high catalytic activity up to 5.8 kg/(gTi.h.p) for the copolymerization of ethylene with 1-hexene. The results revealed that the nascent CNTs/PE nanocomposites particles looked like coral shape and featured with the core-shell structure which CNTs as the core and polyethylene as the shell.

Keywords: Carbon nanotubes; Nanocomposites; Microstructure

Introduction

Porous materials such as SiO$_2$, MgCl$_2$ and Al$_2$O$_3$ have been applied as supports for heterogeneous Ziegler-Natta catalysts in olefins polymerization. Many reports on the use of porous materials as supports for Ziegler-Natta catalysts, metallocene catalysts and nickel diimine catalysts in the olefin polymerization have appeared in the literature [1-8].

Carbon materials have been attracted much attention on promising fillers in various polymers because of their excellent mechanical, thermal and electrical properties. Carbon nanofibers [9] and graphene oxide [10] are treated with methylaluminoxane and then loaded the active species on the surface for ethylene/propylene polymerization. All most previous works report that the efficient active sites are directly anchored on the supports surface without chemical bond between the active site and the support. Zhu [11] utilized two methods for the impregnation of Ni-diimine complexes on silicate-based nanotubes, which were carried out for ethylene polymerization. Wanke [12] studied the morphology of polyethylene particle catalyzed by MgCl$_2$-supported Ziegler-Natta catalyst. He found that the polymer particle was a perfect replica of the catalyst particle. The obtained milli-scaled polyethylene particles were spherical. Young [13] used a thermally pretreated bimetallic MgCl$_2$/tetrahydrofuran (THF)/ TiCl$_4$ catalyst for producing polyethylene particles with regular and homogeneous globule. Mao [14,15] developed MgCl$_2$-supported Ziegler-Natta catalysts for ethylene and propylene polymerization to produce the polymer particles with good spherical morphology. Choi [7] synthesized MgCl$_2$-supported nickel diimine catalysts for ethylene polymerization with high catalytic activity, and the obtained polymer particles had good spherical morphology. Kanellopoulos [16] studied the single particle growth in heterogeneous olefin polymerization according to the random pore polymeric flow model. Our previous work [8] reported that MgCl$_2$-supported Ziegler-Natta catalysts with different structure ligands had efficient catalytic activity for the copolymerization of ethylene with 1-octene. The comonomer incorporation content of the copolymer was relative to the catalysts' ligand structure. Carbon nanotubes (CNTs) have been attracted much attention because of their excellent mechanical, thermal and electrical properties. Based on these distinctive advantages, they were considered

to be promising fillers in various polymers. Huang [10] prepared the polypropylene/graphene oxide nanocomposites by *in situ* Ziegler-Natta polymerization. The active site TiCl$_4$ was anchored to the support graphene oxide through MgCl$_2$. Milani [17] obtained polypropylene (PP)/graphene nanosheet (GNS) nanocomposites with good molecular weight, thermal properties, and tacticity by *in situ* polymerization using metallocene catalysts. Kaminsky [9] reported that carbon black (CB), carbon nanofibers (CNFs) and different types of CNTs were separated by ultra sound and then treated with methylaluminoxane (MAO). These catalytic active centers efficiently promoted ethylene/ propylene polymerization and polyolefin nanocomposites were obtained after mixing zirconocenes or other transition metal complexes with the fillers. Pinheiro [18] prepared linear low-density polyethylene nanocomposites containing different types of nanofiller (TiO$_2$, MWCNT, expanded graphite, and boehmite) by *in situ* polymerization using a Zirconium-Nickel tandem catalyst system. All most previous works reported that the efficient active sites were directly anchored on the supports surface without chemical bond between the active site and the support. The active sites included TiCl$_4$, metallocene catalysts, α- and β-diimine complexes, FI catalyst, Cr-based catalyst and et al. The supports included MgCl$_2$, SiO$_2$, carbon nanotubes; graphene oxide and et al. Recently, Wang [19] reveal that the active site TiCl$_4$ is directly bonded with the support oxidized nanosized carbon spheres through Ti-O bond. Coperet [20] investigated SiO$_2$-supported dinuclear CrIII sites to polymerize olefins forming polymers initiated by C-H bond activation, the active site is directly bonded with the support through Cr-O-Si bond.

---

*Corresponding author: Qigu Huang, State Key Laboratory of Chemical Resource Engineering, Key Laboratory of Carbon Fiber and Functional Polymers, Ministry of Education, Beijing University of Chemical Technology, Beijing 100029, People's Republic of China, E-mail: huangqg@mail.buct.edu.cn

In this work, we report that the CNTs/PE nanocomposites particles with the coral-shaped and core-shell structure *in situ* produced by the nano template catalyst (m-CH$_3$PhO)TiCl$_3$/CNTs which is directly bonded to the oxidized multiwalled carbon nanotubes (MWCNTs) through Ti-O bond, AlEt$_3$ used as a co-catalyst. The morphology and structure of the nano template catalyst and the obtained CNTs/PE nanocomposites particles, as well as the performance of the polymerization, are investigated.

Experimental Section

General remarks

All operations of air- and moisture-sensitive materials were performed using the rigorous repellency of oxygen and moisture in flamed Schlenk-type glassware on a dual manifold Schlenk line under a nitrogen atmosphere. CH$_3$MgCl with 22 wt% in tetrahydrofuran (THF) and triethylaluminum (AlEt$_3$) with 1.0 M in hexane were purchased from Acros Organics Agent in China. MgCl$_2$ (water included<0.01%), AlEt$_3$ (2.0 M in n-hexane) and m-cresol were purchased from Acros in China. 1-hexene were from Fluka in China. Other chemicals were from Beijing chemical agent company. 3-Chloro-1,2-epoxypropane and tributyl phosphate were treated with activated 5 Å molecular sieves under high-purity nitrogen for one week before use. Toluene, THF and n-hexane were further purified by refluxing over metal sodium under nitrogen for 48 h and distilled before use. Ethylene (polymerization grade) was from Sino-petrochemical Company, and used without further purification.

Synthesis of non-metallocene catalyst

To a stirred solution of TiCl$_4$ (11.0 mL, 0.10 mol) in toluene (30 mL), one equivalent of m-cresol (10.5 mL, 0.10 mol) was added with a syringe at -25°C. The mixture was gradually warmed to 30°C for 4 h. The mixture was filtrated and extracted by toluene. The solvent was removed, brown solid (m-CH$_3$PhO)TiCl$_3$ was obtained with yield 22.40 g (89.6%). (m-CH$_3$PhO)TiCl$_3$ (M$_w$=261.3) compound: ^{1}H NMR: δ 7.13 (tri, 1H), δ 6.75 (m, 1H), δ 6.65 (m, 2H), 2.30 (s, 3H); ELEM.ANAL, Calcd: C, 32.12; H, 2.68. Found: C, 32.20; H, 2.59. ^{13}C NMR: δ 169.54 (C1), 119.42 (C2), δ 139.57 (C3), δ 126.71 (C4), δ 128.92 (C5), δ 115.96 (C6), δ 21.18 (CH$_3$).

Preparation of oxidized MWCNTs

Two grams (2.0 g) MWCNTs and 100 mL of concentrated nitric acid were added into a 250 mL Schlenk flask. The mixture was subjected to ultrasonic washing for 5 min with an ultrasonic washer and was stirred for 5 h with a stirring bar at 80°C. The obtained residue after filtrating was washed twice with distilled water and anhydrous ethanol, respectively, and then it was further purified by vacuum for 1 h to obtain black bulk solid, oxidized MWCNTs with diameter of 50 nm and length of *ca* 20 μm. The yield was 1.70 g.

Preparation of MWCNTs supported catalyst

Oxidized MWCNTs (1.5 g) and n-hexane (100 mL) were added into a 250 mL schlenk flask. The mixture was stirred in high speed with a dripping of 21.0 mL of CH$_3$MgCl (0.06 mol). The reaction was carried out for 2 h at room temperature, the reactant was filtrated and the residue was washed twice with 50 mL of n-hexane to remove the remnant Grignard reagent. The obtained solid and 50 mL of n-hexane were added into a 250 mL schlenk flask, followed by adding 15.0 g of (m-CH$_3$PhO)TiCl$_3$ at 0°C. Then the reaction system was stirred for 1 h at the same temperature. Warming to 60°C slowly, the mixture was maintained for 2 h at 60°C. After that, the reaction system was filtrated.

The residue was washed in turn with toluene and n-hexane (50 mL each time) at 50°C until no titanium in the filtrate was determined by Inductively Coupled Plasma (ICP) and drying under vacuum for 2 h. Black powder product was obtained with yield of 1.83 g. The titanium content of the catalyst was 4.0 wt% confirmed by ICP, *i.e.* (m-CH$_3$PhO)TiCl$_3$ content was 21.8 wt%. The number of the active sites on CNTs presented with (m-CH$_3$PhO)TiCl$_3$ was 63%, estimated by the result of the poisoning experiment according to the literature [21].

Preparation of MgCl$_2$ supported catalyst

The supported catalyst (m-CH$_3$PhO)TiCl$_3$/MgCl$_2$ was prepared according to the literature [8]. To a mixing solution of MgCl$_2$ (solid, 1.0 g) dissolved in toluene (60 mL), 3-chloro-1,2-epoxypropane and tributyl phosphate, 11.0 g of (m-CH$_3$PhO)TiCl$_3$ in toluene (Mg:Ti=1:4 in mol) were added by a syringe at -10°C over a period of 1 h in a 300 mL Schlenk flask with a magnetic bar. Then the mixture was enhanced to 60°C and kept it for 3 h at the temperature. Adding n-hexane 20 mL, spherical precipitate was given. The mixture was filtrated and washed in turn with toluene and n-hexane for times (40 mL each time) at 50°C until no titanium in the filtrate was determined by ICP, then dried by vacuum. Spherical catalyst particles were obtained in yield of 1.31 g with brown color. The titanium content of the catalyst was 3.8 wt %, as determined by ICP analysis, *i.e.* (m-CH$_3$PhO)TiCl$_3$ content was 20.7 wt %. The number of the active sites on CNTs presented with (m-CH$_3$PhO)TiCl$_3$ was 51.6% which was also estimated according to the literature [21].

Polymerization procedure

All polymerizations were carried out in a 2000 mL stainless steel reactor equipped with a magnetic stirrer after purging all moisture and oxygen by a high-vacuum pump, the reactor was sealed under a nitrogen atmosphere. Freshly distilled hexane (1000 mL), along with the desired amounts of heterogeneous non-metallocene catalyst and AlEt$_3$, were added in the order. The mixture was stirred for 15 min for preactivation, and heating to 80°C, comonomer 1-hexene was charged into the reactor and the copolymerization was initiated by the introduction of ethylene and ethylene pressure was kept at 0.2 MPa for a desired time. The monomer pressure was kept constant during the polymerization by continuously charging with ethylene. The reaction was stirred for a desired time. Finally, the polymerization was terminated with 10 wt% HCl in alcohol. The obtained polymer was filtered, washing with alcohol and water, and then drying overnight in a vacuum oven at 50°C. Ethylene consumption was automatically recorded by a flowmeter. Ethylene polymerization followed the same procedure, but there was no comonomer added.

Characterization

^{13}C NMR spectra were recorded on an INOVA500 MHz instrument operating at 125 MHz. The condition used for quantitative ^{13}C NMR was of the copolymer content up to 15 wt% in solution, using *ortho*-dichlorobezene (d_4) as the solvent at 125°C. Tetramethylsilane was used as internal chemical shifts reference. The 1-hexene incorporation was estimated from ^{13}C NMR spectra according to the literature [22-24]. The average molecular weight and molecular weight distribution were measured by a PL-GPC220 instrument using standard polystyrene as a reference and 1,2,4-trichlorobenzene as a solvent at 150°C. DSC thermograms were recorded with a PA5000-DSC instrument at a rate of 10 K/min. Scanning electron microscopy (SEM), morphological observation for the catalyst particle and polymer particle was performed on SUPRA 55/55VP field emission scanning electron microscope. The samples (the catalyst particle and polymer particle) for SEM were obtained by immobiled on a film, and were sputter coated with gold to *c.a.* 15 nm thicknesses. Micrographs were taken at 20 kV HR-TEM. The titanium content in the catalyst was determined using a Shimadzu ICPS-5000 inductively coupled plasma emission spectrometer. ^{1}H NMR spectra was measured on an INOVA500MHz instrument. ^{1}H chemical shifts were reported in ppm relative to proton resonance in chloroform-d at δ 7.26 ppm. Elemental analyses were performed on a Perkin-Elmer 2400 microanalyzer at the College of Material Science and Technology, Beijing University of Chemical Technology.

Results and Discussion

Properties of the heterogeneous nano template catalyst

The kinetic curves of the catalyst (m-CH$_3$PhO)TiCl$_3$/CNTs, AlEt$_3$ used as a cocatalyst, for the copolymerization of ethylene with 1-hexene is shown in Figure 1. From Figure 1a, the catalyst (m-CH$_3$PhO)TiCl$_3$/CNTs exhibited much higher catalytic activity than that of catalyst (m-CH$_3$PhO)TiCl$_3$/MgCl$_2$. The catalytic activity of the catalyst system (m-CH$_3$PhO)TiCl$_3$/CNTs/AlEt$_3$ maintains nearly constant at 80°C over 2 h even at fairly low titanium concentration, such as 5.4×10^{-5} M in 1000 mL of solvent. The polymer mass increases linearly with increasing polymerization time within 2 h (Figure 1a'). The results indicate that the catalyst (m-CH$_3$PhO)TiCl$_3$/CNTs is stable even at 80°C for 2 h. Thus a low rate of deactivation of the catalyst (m-CH$_3$PhO)TiCl$_3$/CNTs might be attributable to the enhanced steric hindrance of the support CNTs bond with (m-CH$_3$PhO)TiCl$_3$. But the activity of catalyst (m-CH$_3$PhO)TiCl$_3$/MgCl$_2$ showed obviously decline for the copolymerization of ethylene with 1-hexene under the same conditions (Figure 1b). The polymer mass increases not linearity with increasing polymerization time within 2 h (Figure 1b). The discovery indicated that the support of CNTs is more efficient for non-metallocene catalyst for the copolymerization of ethylene with 1-hexene than MgCl$_2$. The data recording started at temperature 80°C (t=0 min), before the temperature, the product mass was 1.9 g for line a'.

The HR-TEM image of the catalyst (m-CH$_3$PhO)TiCl$_3$/CNTs and low magnification TEM image of MWCNTs are shown in Figure 2. As shown in Figure 2, the CNTs have an outer average diameter of 30-60 nm, multilayer arrays and many defects on the surface; exhibiting the nano template catalyst has a rough surface. It is believed that to some extent, the surface structure is damaged during the treatment with acid. The rough surface of the CNTs should be the result of the reaction between the functional group located on the surface and the transition metal compound (m-CH$_3$PhO)TiCl$_3$, which is in favor of the attachment for the active sites for ethylene polymerization.

Morphology of CNTs/PE nanocomposites particles

HR-TEM result reveals that the polymeric shell covers around the CNTs core of the CNTs/PE nanocomposites particles. As ethylene was charged continuously in feed, the polymerization didn't quit within the polymerization time, the particle size increased with the polymerization time. It is about 35 nm at 2 min of polymerization time (Figure 3a), 70-80 nm at 10 min (Figure 3b) and 500 nm at 20 min (Figure 3c). It is interesting, the SEM images confirm that the CNTs/PE composite particles produced by the nano template catalyst (m-CH$_3$PhO)TiCl$_3$/CNTs, look like coral shape with about 8-10 μm in diameter and about 30 μm in length when the polymerization time is 120 min (Figure 3d). The SEM image of the cryo-fracture surface of the CNTs/PE nanocomposites is shown in Figure 3d. The sample *in situ* produced by (m-CH$_3$PhO)TiCl$_3$/CNTs/AlEt$_3$ after 2 min of polymerization time exhibits a lot of nano fibers in PE matrix. The result reveals that CNTs of the CNTs/PE nanocomposites are well-distributed in PE matrix.

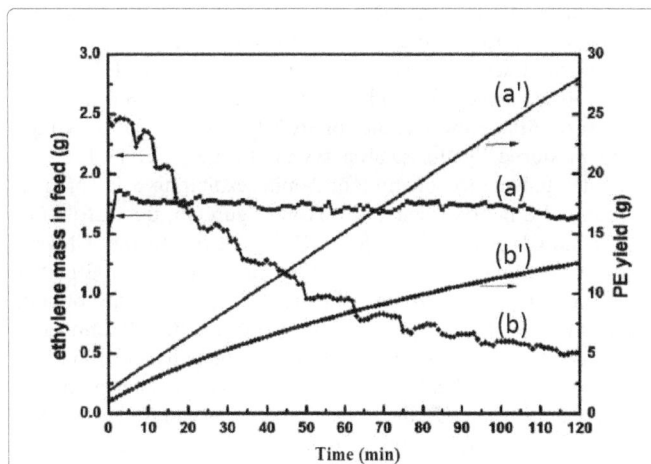

Figure 1: Ethylene mass in feed in polymerization catalyzed by catalyst systems of (a) (m-CH$_3$PhO)TiCl$_3$/CNTs and (b) (m-CH$_3$PhO)TiCl$_3$/MgCl$_2$, PE yield obtained by catalyst systems of (a') (m-CH$_3$PhO)TiCl$_3$/CNTs and (b') (m-CH$_3$PhO)TiCl$_3$/MgCl$_2$

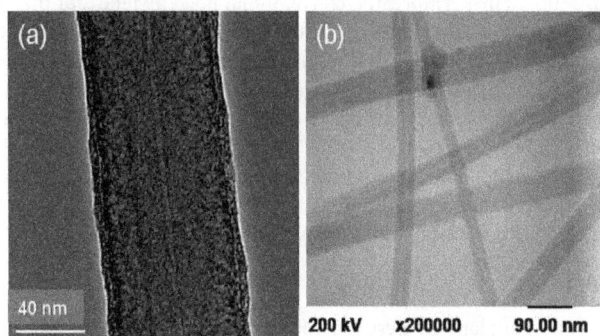

Figure 2: (a) HR-TEM image of the nano template catalyst of (m-CH$_3$PhO)TiCl$_3$/CNTs and (b) low magnification TEM image of MWCNTs

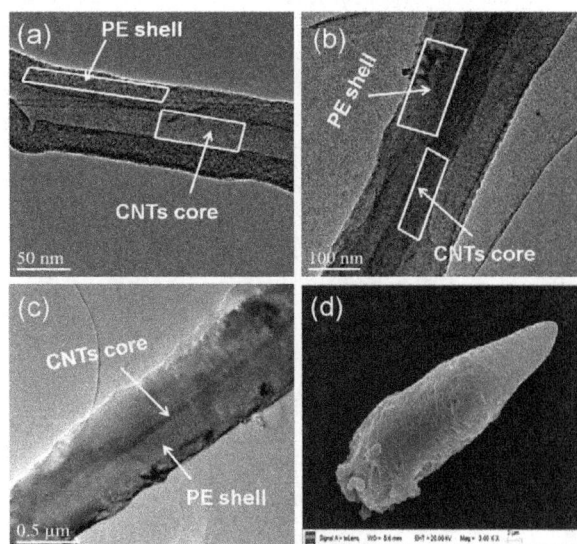

Figure 3: HR-TEM images of the CNTs/PE nanocomposites particles with polymerization time of (a) 2 min, (b) 10 min and (c) 20 min, and SEM image of the CNTs/PE nanocomposites particles with polymerization time of (d) 120 min

The SEM micrographs taken from the cryo-fracture surfaces of coPE/CNTs nanocomposites samples are shown in Figure 4. Figure 4a shows the fracture surface of pure polyethylene which has a flat crack propagation region in addition to some wrinkle. But no CNTs can be seen on it. However, the coPE/CNTs nanocomposites sample *in situ* produced by the catalyst system (m-CH$_3$PhO)TiCl$_3$/CNTs/ AlEt$_3$ after polymerization time of 2 min, exhibiting a lot of nano CNTs in coPE matrix (Figure 4b). From Figure 4b, the distribution of the nano CNTs is fairly uniform. The result reveals that CNTs of the coPE/CNTs nanocomposites are well distributed in coPE matrix. With increasing polymerization time up to 10 min, the size of coPE shell of the coPE/CNTs nanocomposites reaches about 75 nm, coPE component is aggregated (Figure 4c). When the polymerization time is 120 min, the size of coPE shell of the coPE/CNTs nanocomposites is about 5 μm (Figure 4d); the cryo-fracture surface of the coPE/CNTs nanocomposites sample is similar to that of pure PE.

TGA analyses were performed for both conventional polyethylene and PE/MWCNTs nanocomposites in Figure 5. From Figure 5, it shows that conventional polyethylene and PE/MWCNTs nanocomposites have the same starting decomposition temperature of ca. 300°C, the decomposition temperature of loss 5 wt% of them is ca. 360°C. Furthermore, the conventional polyethylene and the PE/ MWCNTs nanocomposites have the similar sharp decomposition temperature at ca. 450°C. The results indicate that the PE fraction of the PE/MWCNTs nanocomposites is thermally stable. From Figure 5, we can also find that for the conventional polyethylene, the complete decomposition temperature is 500°C (Figure 5a), while the PE/ MWCNTs nanocomposites has approximately 75°C higher complete decomposition temperature at ca. 575°C (Figure 5b). MWCNTs can be decomposed at the temperature range from 500°C to 600°C.

Conclusion

Nano template catalyst (m-CH$_3$PhO)TiCl$_3$/CNTs was prepared through (m-CH$_3$PhO)TiCl$_3$ bonded directly to the oxidized MWCNTs. The kinetics of the copolymerization of ethylene with 1-hexene catalyzed by the nano template catalyst revealed that the composite mass increased linearly with increasing of the polymerization time. CNTs/PE nanocomposites particles with coral-shaped and core-shell structure were formed *in situ*. CNTs of the CNTs/PE nanocompositess were well-distributed in the matrix confirmed by SEM.

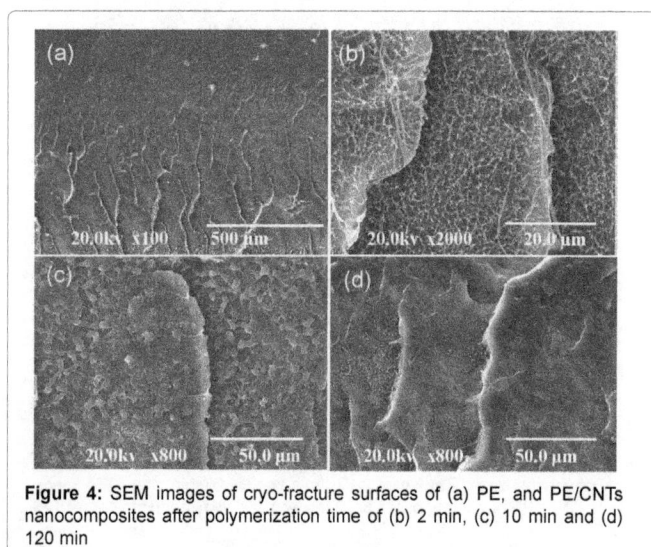

Figure 4: SEM images of cryo-fracture surfaces of (a) PE, and PE/CNTs nanocomposites after polymerization time of (b) 2 min, (c) 10 min and (d) 120 min

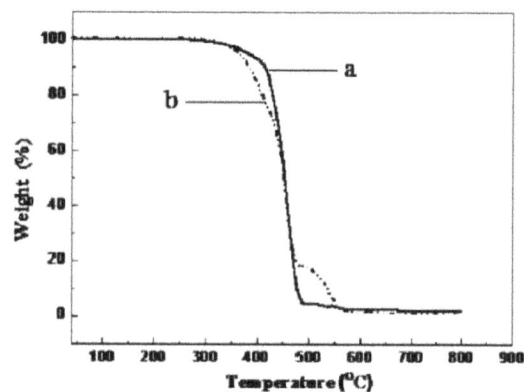

Figure 5: TGA weight loss curves of (a) conventional polyethylene and (b) polyethylene/MWCNTs nanocomposites.

Acknowledgment

We sincerely thank the National Natural Science Foundation of China (No. 21174011 and U1462102).

References

1. Fang Y, Xia W, He M, Liu BP, Hasebe K, et al. (2006) Novel SiO$_2$-supported chromium catalyst bearing new organo-siloxane ligand for ethylene polymerization. Mol Catal A: Chem 247: 240-247.

2. Fink G, Steinmetz B, Zechlin J, Przybyla C, Tesche B (2000) Propene Polymerization with Silica-Supported Metallocene/MAO Catalysts. Chem Rev 100: 1377-1390.

3. Liu WJ, Huang QG, Yi JJ, Yang WT, Ma LF (2010) The Progress of the Catalysts for Olefins Coordination Polymerization. Polymer Bulletin (Chinese) 6: 1-33.

4. Delferro M, Marks JT (2011) Multinuclear Olefin Polymerization Catalysts. Chem Rev 111: 2450-2485.

5. Dong XC, Wang L, Sun TX, Zhou JF, Yang Q (2006) Study on ethylene polymerization catalyzed by Cp2ZrCl2/carbon nanotube system. J Mol Catal A: Chem 255: 10-15.

6. Espinas J, Pelletier J, Szeto KC, Basset JM, Taoufika M (2014) Preparation and characterization of metallacalixarenes anchored to a mesoporous silica SBA-15 LP as potential catalysts. Microporous and Mesoporous Mater 188: 77-85.

7. Choi YY, Soares JBP (2010) Ethylene slurry polymerization using nickel diimine catalysts covalently-attached onto MgCl$_2$-based supports. Polymer 51: 2271-2276.

8. Kong Y, Yi JJ, Huang QG, Yang WT (2010) With different structure ligands heterogeneous Ziegler–Natta catalysts for the preparation of copolymer of ethylene and 1-octene with high comonomer incorporation. Polymer 51: 3859-3866.

9. Wiemann K, Kaminsky W, Gojny FH, Schulte K (2005) Synthesis and Properties of Syndiotactic Poly(propylene)/Carbon Nanofiber and Nanotube Composites Prepared by *in situ* Polymerization with Metallocene/MAO Catalysts. Macromol Chem Phys 206: 1472-1478.

10. Huang YJ, Qin YW, Zhou Y, Niu H, Yu ZZ, et al. (2010) Polypropylene/Graphene Oxide Nanocomposites Prepared by *In Situ* Ziegler−Natta Polymerization. Chem Mater 22: 4096-4102.

11. Ye ZB, Alsyouri H, Zhu SP, Lin YS (2003) Catalyst impregnation and ethylene polymerization with mesoporous particle supported nickel-diimine catalyst. Polymer 44: 969-980.

12. Wu L, Lynch DT, Wanke SE (1999) Kinetics of Gas-Phase Ethylene Polymerization with Morphology-Controlled MgCl2-Supported TiCl4 Catalyst. Macromolecules 32: 7990-7998.

13. Ko YS, Han TK, Park JW, Woo SI (1997) Copolymerization of ethylene–1-Hexene over a thermally pretreated MgCl2/THF/TiCl4 bimetallic catalyst. J Polym Sci Part A: Polym Chem 35: 2769-2776.

14. Huang R, Liu D, Wang S, Mao BQ (2005) Preparation of spherical MgCl2 supported bis(imino)pyridyl iron(II) precatalyst for ethylene polymerization. J Mol Catal A: Chem 233(1-2): 91-97.

15. Xu R, Liu D, Wang S, Wang N, Mao B (2007) Preparation of spherical MgCl2-supported bis(phenoxy-imine) zirconium complex for ethylene polymerization. J Mol Catal A: Chem 263: 86-92.

16. Kanellopoulos V, Dompazis G, Gustafsson B, Kiparissides C (2004) Comprehensive Analysis of Single-Particle Growth in Heterogeneous Olefin Polymerization: The Random-Pore Polymeric Flow Model. Ind Eng Chem Res 43: 5166-5180.

17. Milani AM, Quijada R, Basso NRS, Graebin AP, Galland GB (2012) Influence of the graphite type on the synthesis of polypropylene/graphene nanocomposites. J Polym Sci Part A: Polym Chem 50: 3598-3605.

18. Pinheiro AC, Casagrande ACA, Casagrande OL (2014) Linear low-density polyethylene nanocomposites by in situ polymerization using a zirconium-nickel tandem catalyst system. J Polym Sci Part A: Polym Chem 52: 3506-3512.

19. Wang J, Yu MS, Jiang WH, Zhou Y, Li FJ, et al. (2013) The Preparation of Nanosized Polyethylene Particles via Carbon Sphere Nanotemplates. J Ind Eng Chem Res 52: 17691-17694.

20. Conley MP, Delley MF, Siddiqi G, Lapadula G, Copéret C (2014) Polymerization of Ethylene by Silica-Supported Dinuclear CrIII Sites through an Initiation Step Involving C[BOND]H Bond Activation. Angew Chem Int Ed 53: 1872-1876.

21. Shen XR, Hu J, Fu ZS, Lou JQ, Fan ZQ (2013) Counting the number of active centers in MgCl2-supported Ziegler–Natta catalysts by quenching with 2-thiophenecarbonyl chloride and study on the initial kinetics of propylene polymerization. Catal Commun 30: 66-69.

22. Carman CJ, Harrington RA, Wilkes CE (1977) Monomer Sequence Distribution in Ethylene-Propylene Rubber Measured by ^{13}C NMR. 3. Use of Reaction Probability Model. Macromolecules 10: 536-544.

23. Kimura K, Yuasa S, Maru Y (1984) Carbon-13 nuclear magnetic resonance study of ethylene-1-octene and ethylene-4-methyl-1-pentene copolymers. Polymer 25: 441-446.

24. He T, Jia S, Jiang P, Chen Y (2006) Solid state 13c-nmr study of γ-irradiated poe. Acta Polymerica Sinica 4: 624-626.

24

Technical and Economic Evaluation of Phorbol Esters Extraction from *Jatropha curcas* Seed Cake using Supercritical Carbon Dioxide

Cristiane de Souza Siqueira Pereira[1*]**, Fernando Luiz Pelegrini Pessoa**[1]**, Simone Mendonca**[2]**, Jose Antonio de Aquino Ribeiro**[2] **and Marisa Fernandes Mendes**[3]

[1]Technology of Chemical and Biochemical Processes, Chemistry School, Federal University of Rio de Janeiro, Brazil
[2]Embrapa Agroenergy, Brasilia, Brazil
[3]Chemical Engineering Department, Federal Rural University of Rio de Janeiro, Brazil

Abstract

Jatropha curcas plant shrub of *Euphorbiaceae* family is a plant whose seeds are rich in oil that can be used for biofuel production. However the seeds contain many toxic compounds which the most important ones are known as phorbol esters (PEs). This study has as aim the study of the technical and economic feasibility of the supercritical fluid for the PEs extraction present in the *Jatropha* seed cake. The effect of temperature (40-100°C) and pressure (100-500 bar) on the phorbol yield was investigated using a central composite design methodology to determine the significance and interactions of these parameters. PEs in the extracted samples were analyzed and quantified by HPLC. The supercritical fluid extraction was effective in the recovery of PE extracted from *Jatropha curcas* cake varying from 23.0%, at 70°C and 500 bar to 2.6% at 90°C and 160 bar. The results showed that pressure had the most significant enhancing effect on the phorbol ester yield. Simulations of phorbol ester extraction from *Jatropha curcas* cake were carried out using SuperPro Designer 9.0 (Intelligen, Inc) to evaluate production costs of an industrial process to treat the necessary quantity of cake. It was possible to conclude that the supercritical extraction is viable to be applied.

Keywords: Detoxification; Screw press; Design experiment; Scale-up

Introduction

Jatropha curcas L. seed, known as "pinhão-manso" in Brazil, is an important oleaginous which has received great attention in recent years due to its utilization in biodiesel production [1]. *Jatropha curcas* oil is usually extracted by screw presses and for every thousand of liters of *Jatropha* oil, around 2 tonnes of press cake are produced [2].

The cake remaining after oil extraction is rich in proteins but it also contains toxic compounds [3]. The toxicity is due to the presence of high levels of toxic and anti-nutritional components as trypsin inhibitors, lectins, saponins, phytate and PEs. Although others compounds are present, PE is the main toxic component of *Jatropha curcas* and the concentration of them is the parameter that limits the utilization of the protein rich pressed cake for animal nutrition [4].

The phorbol esters molecules are tetracyclic diterpenoids with a tigliane skeletal structure [5]. These authors isolated six different types of PE of *Jatropha curcas* and all of these compounds possess the same diterpene core, namely, 12-deoxy-16-hydroxyphorbol. These compounds present in the seeds were designated as jatropha factors C1, C2, C3, epimers C4, C5 and C6, with the molecular formula $C_{44}H_{54}O_8$.

Different chemical, physical and biologic methods have been employed for the removal or inactivation of PEs from oil, kernel or pressed cakes in order to promote their use as a protein source for animal feed [6-10]. PEs are double-edged swords, having a lot of negative effects on human and livestock. They also possess some beneficial effects, because not all of the PEs are toxic and their activity and potency vary from one type of PE to another [11]. Many works have been studied in order to extract the PEs to add value to this by product [12-15]. The purified PEs could also be converted or transformed chemically into nontoxic compounds with beneficial activities such as the hydrolysis of 12-deoxy-16-hydroxyphorbol that results in the

synthesis of 12-deoxyphorbol-13-phenyl acetate, a compound that is considered as a promising adjuvant for antiviral therapy because of its anti-HIV properties [14].

There are several studies on the detoxification of *Jatropha curcas* cake using organic solvents and chemical treatments, for example, hexane, methanol, petroleum ether and potassium hydroxide [16,17]. However, authors reveal that these techniques are not economically feasible and the multiple steps involved in the processing of the biomass up to the detoxification stage are also costly and not environment friendly. Based on this fact, the supercritical extraction process has advantages compared to conventional methods with organic solvent. This technology implies in the use of the principles of green chemistry and engineering, from process inception in the research environment to process application on a commercial scale [18]. The carbon dioxide in its supercritical state is a promising solvent due to its characteristics like inertness, non-toxicity, no flammability, non-explosiveness, and availability with high purity at low cost [19].

There are no reports in the literature concerning the extraction of PEs from *Jatropha curcas* cake with supercritical carbon dioxide.

Corresponding author: Cristiane de Souza Siqueira Pereira, Technology of Chemical and Biochemical Processes, Chemistry School, Federal University of Rio de Janeiro, Brazil, E-mail: crispereirauss@gmail.com

Marisa Fernandes Mendes, Chemical Engineering Department, Federal Rural University of Rio de Janeiro, Brazil, E-mail: marisamf@ufrrj.br

Because of the scarce of data concerning PEs extraction with supercritical CO_2, this study has as aim the technical and economical evaluation of this type of extraction of PEs present in the *Jatropha curcas* seed cake.

Materials and Methods

Materials

Jatropha curcas seeds were kindly provided by Empresa de Pesquisa Agropecuária de Minas Gerais (EPAMIG) grown in the region of Janaúba city, located in the north of Minas Gerais state (Brazil). The *Jatropha curcas* press cake was obtained using a tubular radial screw press with capacity of 50 kg/h (Brand: SCOTTECH), which provided approximately 19.6% of oil. The press cake composition was detailed in Ref. [10]. After pressing, the material was stored in a plastic bag in the refrigerator for later experiments. Liquid CO_2 (99.9% pure) was from White Martins (Rio de Janeiro, Brazil).

Experimental design

Jatropha pressed cake was submitted to the extraction process following a central composite rotational design (CCRD) with two independent variables (pressure and temperature), commonly studied in a supercritical extraction process. Table 1 shows the coded and actual levels of variables and describes the 11 experiments that were carried out. The results were statistically analyzed using a statistical program.

Supercritical carbon dioxide extraction

The supercritical fluid extraction (SFE) experiments were performed in an apparatus, built in the Applied Thermodynamics and Biofuel Laboratory at Chemical Engineering Department/ UFRRJ, consisting of a stainless steel 316S extractor with 42 mL of capacity. The extractor contains two canvas of 260 mesh to prevent the entrainment of material. A high-pressure pump (Palm model G100), specific for pumping CO_2 was responsible for the solvent feeding into the extractor. A thermostatic bath (Fisatom model) was coupled in the extractor to control the temperature and a manometer was on line installed for pressure measurement. The flowsheet of the experimental apparatus is shown in Figure 1. The same apparatus has been used in numerous studies done by the research group [20] and the experimental procedure was done in a semi-batch way. Initially, the extractor was filled with the solid material, approximately, 10 g of *Jatropha* pressed

cake. The sampling was done using a micrometric valve, reducing the pressure, and the oily extract was recovered in a previously weighed polypropylene tube. Sampling occurred at each 10 minutes with the depressurization of the system.

The extraction yield was estimated according to equation 1:

$$Yield(\%) = \frac{mass\ of\ extract(g)}{mass\ of\ cake(g)} x100 \qquad (1)$$

with mass of extract as the oily fraction extracted at each 10 min and the mass of cake was 10 g, approximately, for all the experiments.

Phorbol ester analysis

The PEs from extracts of *Jatropha* cake with supercritical CO_2 and from cake pressed was analyzed in Embrapa Agroenergia (Brasília, Brazil).

Phorbol esters analysis from pressed cake

The methodology used in this work was adapted from Makkar et al. [4]. It was transferred, approximately, 4 g of *Jatropha curcas* cake to cells accelerated solvent extractor (ASE 350). The samples contained in the cells were extracted with methanol using the following conditions: temperature: 60°C; heat time: 5 min; static time: 2 min; number of cycles: 5; rinse volume: 150% and a purge time of 60 s. The extracts from the ASE tubes were evaporated under vacuum in a water bath at 60°C (rotaevaporator). It was added 2.5 mL of HPLC grade methanol to the ASE tubes and it was mixed for 20 to 30 seconds. The methanolic extracts were transferred to a test tube of 10 mL and centrifuged at 4000 rpm for 3 minutes. The clear supernatant was transferred to a volumetric flask of 5 mL with the aid of micropipette. This procedure was repeated with a further portion of 2.5 mL of HPLC grade methanol, bringing the mixture methanol plus residue at the same test tube. The methanol solution was filtered to vial (VertiPure PTFE Syringe, 13 mm, 0.2 μm) and 25 μL were injected into the chromatographic system.

HPLC analysis of PEs extracts from supercritical CO_2 process

It was added 3 mL of HPLC grade methanol to the falcon tube containing the extract obtained by supercritical fluid extraction. The tube was mixed for 20 to 30 seconds and centrifuged at 9000 rpm for 5 minutes. The clear supernatant was transferred to a 10 mL volumetric flask with the aid of micropipette. This procedure was repeated with two further portions of 3 mL of methanol, adding the methanolic extracts in the same 10 mL volumetric flask and completing the volume with methanol. The methanolic solution was filtered to vial (VertiPure PTFE Syringe, 13 mm, 0.2 μm) and 25 μL were injected into the chromatographic system. A gradient of (A) phosphoric acid 0.1% (V/V) and (B) acetonitrile was used as following described: start with 60% of B, increase B to 100% in the next 25 min, and keep 100% B for the next 3 min. Then the column was washed with 2-propanol in the next 5 min and equilibrated with the starting conditions (60% of B) for 10 min. The chromatographic conditions were also adapted from Makkar et al. [21]. The PEs were analyzed and quantified by HPLC (Agilent) on a reverse phase C18 SB-C18 250 × 4.6 mm (5 μm), maintained at 40°C.

Phorbol-12-myristate 13-acetate (PMA) was used as an external standard, which has a retention time around 23.5 min. PEs peaks were integrated at 280 nm, and the concentration was expressed as equivalent to PMA. The PEs peaks appeared between 17.5 and 21.5 min. The percentage of PEs present in the extracts were calculated as a ratio between the areas of the PEs analysed in the supercritical extracts and the areas of PEs presented in the original cake.

Run	Coded level of variables		Actual level of variables	
	Temperature (x1)	Pressure (x2)	Temperature (°C)	Pressure (Bar)
Factorial points				
1	-1	-1	50	160
2	+1	-1	90	160
3	-1	+1	50	440
4	+1	+1	90	440
Axial points				
5	-α (-1.41)	0	40	300
6	+α (+1.41)	0	98	300
7	0	-α (-1.41)	70	100
8	0	+α (+1.41)	70	500
Center points				
9	0	0	70	300
10	0	0	70	300
11	0	0	70	300

Table 1: Central composite rotatable design matrix applied for the supercritical fluid extraction.

Economic evaluation

According Moraes, Zabot and Meireles [22], studies involving economic aspects are needed to transfer the knowledge acquired at laboratory/pilot scales to industrial scale. Many of them have simulated the manufacturing cost (MC) of extracts, mostly obtained by supercritical fluid extraction from vegetal raw materials and reported the financial viability of the process such as antioxidant extracts from *Myrciaria cauliflor* [23] and production of phenolic rich extracts and extraction of carotenoids from Brazilian plants [24,25].

Due to the fact that the technical feasibility of PEs extraction with supercritical carbon dioxide was assured and because of an absence of works using this type of raw material, an economic evaluation was done to predict the extraction behaviour of a process that will be conducted in a pilot scale of 42 L, using the better operational conditions of temperature (70°C) and pressure (500 bar) obtained in the experimental unit, until the cake was considered detoxified. The scale up criterion adopted consisted in maintaining solvent mass to feed mass ratio (S/F) constant. The overall extraction curves obtained from laboratory scale experiments were used as reference, so that S/F = 53.6.

The commercial simulator SuperPro Designer v9.0 (Intelligen, Inc) was used to simulate the extraction process and to estimate the production cost of phorbol esters extracted with supercritical fluid until the cake was destoxified. This simulator has been used by other authors to simulate supercritical fluid extraction processes using different raw materials [25,26]. The method utilized by the software to estimate the manufacturing cost [27,28] which uses the summation of the fixed cost of investment (FCI), the cost of utilities (CUT), the cost of labor (COL), the cost of raw material (CRM) and the cost of waste treatment (CWT) involved in the studied chemical process to compose the manufacturing cost.

The extraction process was simulated in a batch mode using a solid-liquid extractor present in the database of the simulator. The scale-up criterion adopted consisted of maintaining a constant solvent to feed ratio (S/F). The cost of solid-liquid extractor of 42 L (size 2) of capacity was expressed by equation 2, based on the value US$ 500.000 for extractor 50 L (size 1) provided by Prado [29], considering an annual depreciation rate of 10%.

$$Cost\,2 = Cost\,1\left(\frac{Size\,2}{Size\,1}\right)^{0,6} \qquad (2)$$

The *Jatropha* cake press cost (US$ 75/ton) was based on those reported by Sriram [30]. The cost of CO_2 was considered US$ 19.76/Kg (White Martins) and it was adopted that the solvent can be recycled without further treatment. The utility cost was based from a simulation of the energy balance in SuperPro Designer: steam at US$ 4.2/MT and the electricity at US$ 0.092/kWh. The revenue of the plant may consist of phorbol ester. The cost of phorbol ester was considered based on the reference price of phorbol 12-myristate 13-acetate at US$ 212.85/mg [31]. This consideration was necessary due to the absence of this price in the literature.

Results and Discussion

The total yields, presented in equation 1, for each operational condition were shown in Table 2.

According to the results, the best yields were obtained at high pressures and the best extraction conditions were obtained at 70°C and pressure of 500 bar with the yield of 9.33%. The accumulated yield in function of the extraction time of *Jatropha* cake can be seen in Figure 2. The extraction time was different from one condition to another according to the saturation of the raw material: 200 minutes for 98°C at 300 bar, 300 minutes for 70°C at 100 bar, 90°C at 160 bar, 40°C at 300 bar, 70°C at 500 bar and 400 minutes for 50°C and 160 bar, 70°C at 300 bar, 50°C at 440 bar and 90°C at 440 bar.

It is important to note that Table 2 and Figure 3 show the total yield of the extraction, which not represents the PE yield for each operational condition. These behaviours cannot be represented because only the last fraction with the accumulated mass was analysed to know the quantity of PEs. It was observed that the yield increases with increasing pressure and higher pressures (300, 440 and 500 bar) had better yields when compared to lower ones (100 and 160 bar).

According to Figure 2, the increase in temperature, at constant pressure (300 bar), leads to an increase in yield. This may be related to the competitive effect of the vapor pressure of the solute and solvent density, due to the fact that when the vapor pressure of the solutes increased, the extraction process is favoured. At 160 bar, it was observed a cross-over behaviour between the curves of 50°C and 90°C, showing the competitive effects of density and vapor pressure of the solute in the extraction efficiency.

The extracts obtained with supercritical extraction in different conditions were analysed to better quantify the content of PEs by HPLC and shown in Table 3.

The PE content in untreated seed cake was 1.97 mg/g. This measure is the average of the HPLC analysis done in triplicate. Table 3 shows that the extraction at 70°C and 500 bar and 50°C and 440 bar were able to remove 22.19% and 23.03% of the PE present in the seed cake, respectively. The profile of the best condition of extraction (70°C and 500 bar) is shown in Figure 3 and indicates four peaks between retention times of 17 and 21.5 minutes, that represents the PE derivatives.

A central composite rotatable design (CCRD) was used to evaluate the PE yields extracted. The influence of the two variables (temperature and pressure) was statistically investigated at 95% of confidence level ($p \leq 0.05$). The linear and quadratic coefficients of the variables studied and their interactions, the standard error, the significance of each coefficient determined by the p-value and the values of the analysis of variance (ANOVA) are listed in Table 4. The p-values of the regression coefficients suggest that only pressure, as a linear factor, was significant ($p \leq 0.05$). This result corroborated the experimental results presented in Figure 3, in which the better operational conditions occurred at higher pressures. Moreover, the statistical analysis of the experimental data showed that the pressure had a significant effect on PE removal. A first order model was established (Equation 3) based on ANOVA which described the yield of PEs extracted as a function of the significant

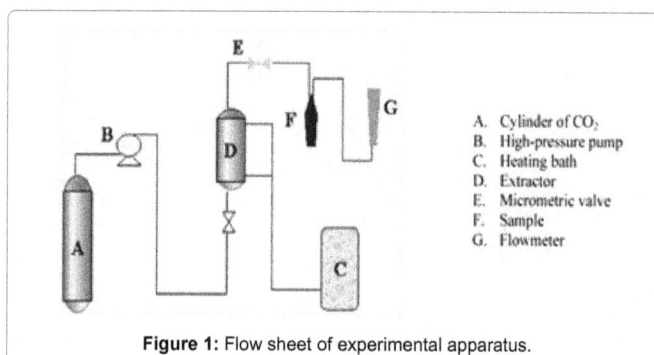

A. Cylinder of CO_2
B. High-pressure pump
C. Heating bath
D. Extractor
E. Micrometric valve
F. Sample
G. Flowmeter

Figure 1: Flow sheet of experimental apparatus.

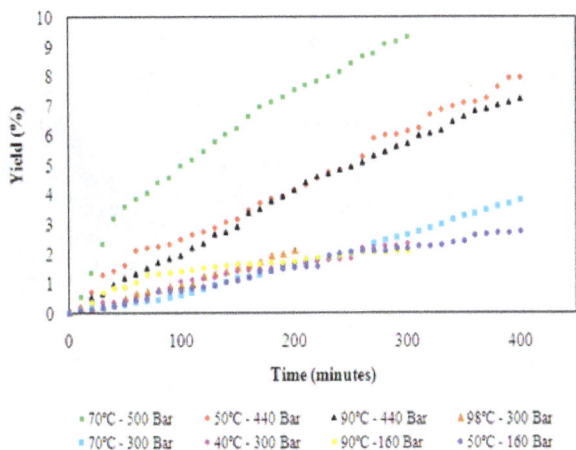

Figure 2: Extraction curves of *Jatropha* cake press with supercritical CO_2.

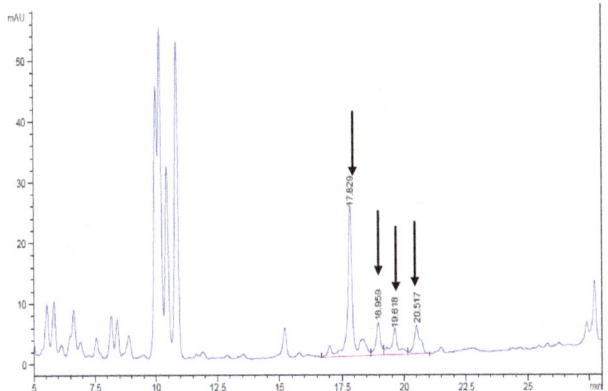

Figure 3: HPLC profile of the extract obtained at 70°C and 500 bar from *Jatropha* cake press.

independent variables (pressure).

$$Y = 9.19 + 7.29\,x_2 \qquad (3)$$

where Y is the yield of PEs and x_2 is the pressure.

It was observed that the calculated F factor for the regression was greater than the tabulated one, demonstrating that the model is representative and, because of that, a response surface can be obtained.

Figure 4 shows the response surface illustrating the extraction yield as a function of temperature and pressure.

Analysing the response surface generated by the first order model, it be can observed an optimized region at higher pressures. The data indicated that, at any temperature, the yield of the extraction and consequently PEs concentration will increase with pressure increase.

The results of efficiency and temperature and pressure influences on the PEs extraction cannot be compared with another work, because until now, it was not found in the literature works concerning the extraction of PEs from *Jatropha curcas* cake with supercritical carbon dioxide to compare data.

As the phorbol ester content in untreated seed cake is 1.97 mg/g and the percentage extracted was 23%, it can be concluded that the best operational condition (500 bar and 70°C) was not able to remove the

phorbol esters to a tolerable level of 0.11 mg/g according to Makkar and Becker [32]. Due to that, it was simulated a supercritical extraction process involving more than one batch to assure the total extraction of PEs, based on the best operational conditions obtained in the experimental step.

Economic evaluation

The better condition of temperature and pressure (500 bar and 70°C) was used to simulate the application of supercritical carbon dioxide in the PEs extraction until the cake was detoxified. A preliminary simulation using only one extractor was not sufficient and because of that, this study considered the industrial SFE containing 24 extractors (Figure 5), with a total of 7866 h of operation per year, cycle time of 60 hours which represents a total of 132 batch/year. The experimental data corroborated the experimental data.

The annual throughput of phorbol ester was 2.47 kg/yr and a concentration in PE of 0.11 mg/g could be reached with the simulation. The seed cake can be considered non-toxic with this level, confirming the efficiency of the methodology studied in this work.

Table 5 provides a summary of the overall material balances and material consumption of this process.

The direct fixed capital (DFC) refers to the fixed assets of an investment, such as plant and equipment. It is calculated at the process section level as the sum of direct, indirect and miscellaneous costs that are associated with a plant capital investment. The direct costs include cost elements that are directly related to an investment, such as the cost of equipment, process piping, instrumentation, buildings, facilities, etc. The indirect costs include costs that are indirectly related to an investment, such as the costs of engineering and construction. Additional costs such as the contractor's fee and contingencies are included in miscellaneous costs. In this study, the total DFC is around US$ 98 million approximately and the individual cost items that contributed to the DFC are shown in Table 6.

Figure 6 shows the breakdown of the operating cost. The facility dependent cost is the most important item, accounting for 90% of the overall operating cost. In cases of designs, where there is no prior experience on the use of equipments, this is typically calculated as the sum of the costs associated with equipment maintenance, depreciation of the fixed capital cost, costs such as insurance, local taxes and possibly other overhead-type of factory expenses. Raw materials account for around 5.7% of the overall cost. Carbon dioxide is the most expensive raw material, accounting for 99.99% of the raw materials cost, comparing to the *Jatropha* press cake. Labour costs were based on the sum of the labour requirements for each unit procedure multiplied by a fixed labour rate (which was based on a basic labour rate, plus adjustments for fringe benefits, administration, and others). Supercritical fluid extraction is an environmentally safe technology, thus the cost of waste treatment was considered null.

Table 7 shows the results of the economic evaluation. For a plant of this capacity (42 L), the total capital investment is US$ 108,417,000. Assuming a selling price of US$ 212.85/mg, the project yields an after-tax internal rate of return (IRR) of 132.58% and a net present value (NPV) around US$ 2 billion (assuming a discount interest of 7%). The NPV is an indicator of how much value an investment adds to the industry. In this case, NPV is positive, what means that the investment would add value to the industry and therefore the project is economically viable and may be accepted and implemented. Based on these results, this project represents an attractive investment. It is

Operational Conditions	70°C 100 bar	50°C 160 bar	90°C 160 bar	40°C 300 bar	70°C 300 bar	98°C 300 bar	50°C 440 bar	90°C 440 bar	70°C 500 bar
Yield (%)	0.65	2.76	2.01	2.35	3.83	2.12	7.97	7.25	9.33

Table 2: Yields of the extracts of *Jatropha* cake with supercritical CO_2.

Run	Temperature (°C)	Pressure (bar)	PE (mg/g cake)	Yields PE (%)
1	50	160	0.1639	8.32
2	90	160	0.0508	2.58
3	50	440	0.4536	23.03
4	90	440	0.3812	19.35
5	40	300	0.1761	8.94
6	98	300	0.1057	5.36
7	70	100	0.0640	3.25
8	70	500	0.4370	22.19
9	70	300	0.2230	11.32
10	70	300	0.1355	6.88
11	70	300	0.1860	9.44

Table 3: PEs yields from *Jatropha* cake extracted with supercritical CO_2.

Factors	Regression coefficient	Standard deviation	p - value
Constant	9.19	1.29	0.0190
(x_1) Temperature (L)	-1.81	0.79	0.1483
Temperature (Q)	-0.18	0.94	0.8654
(x_2) Pressure (L)	7.29	0.79	0.0115
Pressure (Q)	2.62	0.94	0.1085
Temperature by Pressure	0.51	1.11	0.6894

ANOVA				
	Sum of squares	Degrees of freedom	Mean sum of squares	F-value
Regression	429.73	1	429.73	37.32
Residue	103.62	9	11.51	
Total	533.35	10		

Table 4: Coefficients of regression of the CCRD and analysis of variance.

Bulk Material	Annual Amount (kg)	kg/batch	Unit Cost (US$)	Annual Cost (US$)
Carbon Dioxide	70,831.00	536.60	19.76	1,399,625.00
Jatropha Cake	1,320.00	10.00	0.075	99.00

Table 5: Overall Balance and Material Consumption.

A. Total Plant Direct Cost (TPDC)	
1. Equipment Purchase Cost	13,009,000.00
2. Installation	6,361,000.00
3. Process Piping	4,033,000.00
4. Instrumentation	1,691,000.00
5. Insulation	390,000.00
6. Electricals	3,773,000.00
7. Buildings	781,000.00
8. Yard Improvement	1,301,000.00
9. Auxiliary Facilities	7,155,000.00
TPDC	38,494,000.00
B. Total Plant Indirect Cost (TPIC)	
10. Engineering	12,318,000.00
11. Construction	13,088,000.00
TPIC	25,406,000.00
C. Total Plant Cost (TPC = TPDC + TPIC)	
TPC	63,899,000.00
D. Contractor's Fee and Contingency	
12. Contractor's Fee	11,502,000.00
13. Contingency	23,004,000.00
CFC	34,506,000.00
E. Direct Fixed Capital Cost (DFC = TPC + CFC)	
DFC	98,405,000.00

Table 6: Fixed capital estimate summary (2014 prices US$).

Direct Fixed Cost	98,405,000.00
Total Investment (US$)	108,417,000.00
Operating Cost (US$) /yr	24,572,000.00
Total Revenues (US$) /yr	525,425,000.00
Cost Basis Annual Rate mg ester/yr	37,197.00
Unit Production Cost US$/mg ester	660.60
Gross Margin (%)	95.32
Return On Investment (%)	285.35
Payback Time (years)	0.35
IRR After Taxes%	132.58
NPV at (7.00%) US$	2,094,841,000.00

Table 7: Key economic evaluation results.

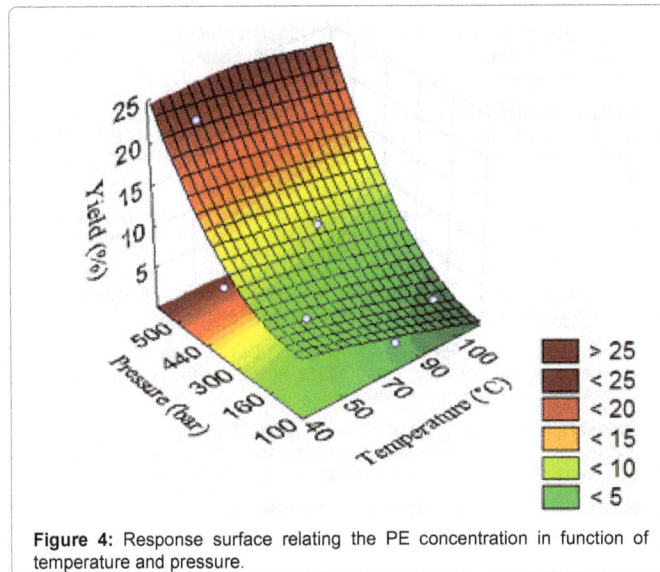

Figure 4: Response surface relating the PE concentration in function of temperature and pressure.

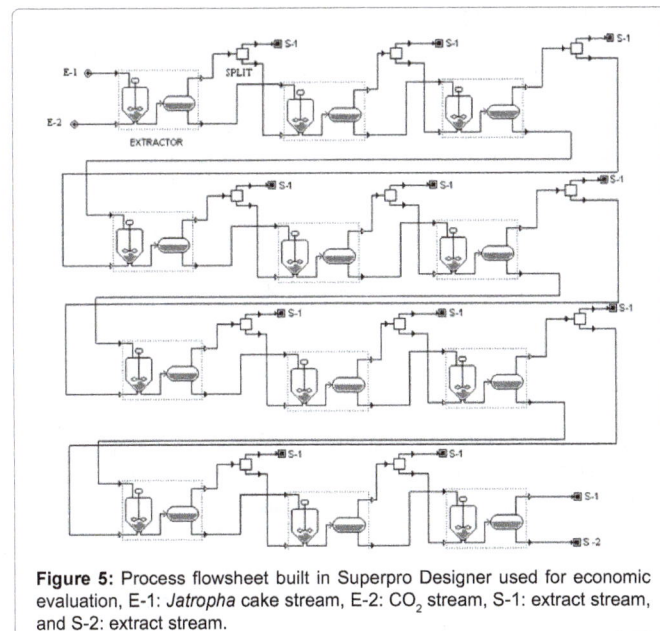

Figure 5: Process flowsheet built in Superpro Designer used for economic evaluation, E-1: *Jatropha* cake stream, E-2: CO_2 stream, S-1: extract stream, and S-2: extract stream.

important to note that this industrial plant can be used with other raw materials to extract different high aggregated components.

The analysis was based on the assumption that a new facility will be constructed for this process with a construction time of 30 months; start-up of 4 months and a 15 years project lifetime. The internal rate of return (IRR) is compared to the minimum attractive rate of return (MARR) or the cost of capital of the company. The decision criteria for accepting the project is to have an IRR greater than or equal to the minimum acceptable [20]. The criterion for evaluating the IRR in this study was that this is equal to or greater than 25% per year. A sensitivity study varying the selling price of phorbol ester in relation to total costs was conducted. The influences of the price in internal rate of return on total revenue and payback time are presented in Figure 7.

Figure 8 show the costs and profits in function of the rate of the production for phorbol ester. The break-even point was calculated in function of the direct fixed cost, variable costs and revenue. The break-even point was situated in the figure at, approximately, 20% of the rate of production. It means that if the plant is operated above this rate of production, the profits are higher than the total costs of production.

The Superpro Designer was also used to simulate the operating cost of the cake using chemical treatment described by Guedes et al. [10], to compare the process costs. The detoxification of seed cake was obtained with soxhlet apparatus, reducing the PE content in 97.30% (0.10 mg/g) using methanol, extraction time of 8 hr and solute/solvent ratio of 1:10 (w/v). Soxhlet simulation considered an industrial setup with three extractors of 42 L. The results showed that the Soxhlet extraction presented lowest operating cost US$ 16,195,000.00/yr

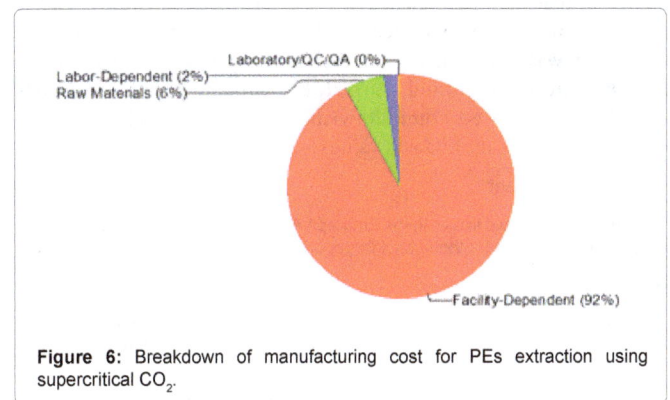

Figure 6: Breakdown of manufacturing cost for PEs extraction using supercritical CO_2.

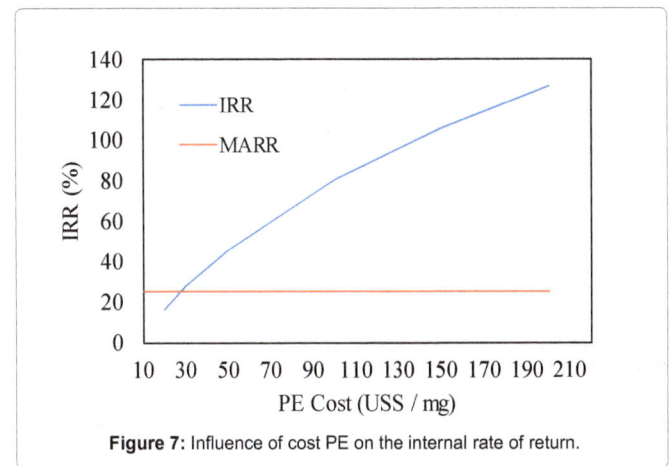

Figure 7: Influence of cost PE on the internal rate of return.

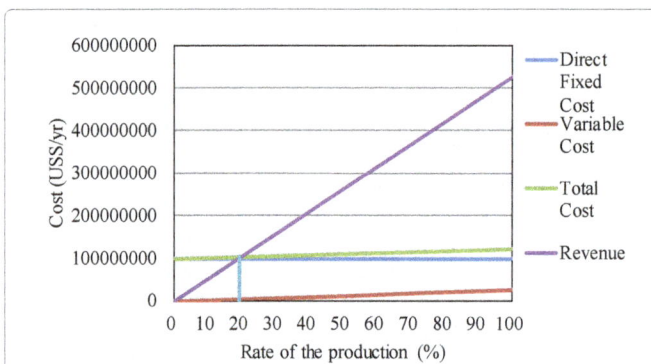

Figure 8: Variation of the costs and profit in function of the rate of the production of the plant.

compared with the operating cost of the supercritical fluid extraction (US\$ 24,572,000.00/yr). Probably, this behaviour is associated with the number of extractors used in the processes. On the other hand, the carbon dioxide in supercritical state is a promising solvent for green chemical processes due to its characteristics like non-toxicity, no flammability and non-explosiveness. Moreover, perhaps, it will be necessary more separation steps, using organic solvent, to eliminate the presence of solvent in the cake, able to be used in animal feed.

Conclusion

It was studied the technical evaluation of supercritical fluid to extract PEs from *Jatropha curcas* cake. The experiments were done applying a statistical experimental planning, showing that the best results occurred at 70°C and 500 bar with the yield of 9.33%. It was observed that the yield increases with increasing pressure and a statistical analysis corroborated this behavior. The supercritical fluid extraction was very effective in the recuperation of the PE from *Jatropha curcas* cake around 23.0%. The economic evaluation indicates that the process is economically viable and can be implemented to obtain detoxified cake and add value to phorbol ester.

Acknowledgement

The authors would like to thank Embrapa Agroenergy (Brasília, Brazil), for the analysis and FAPERJ for the financial support.

References

1. Ceasar SA, Ignacimuthu S (2011) Applications of biotechnology and biochemical engineering for the development of Jatropha and biodiesel: a review. Renew Sustain Energy Rev 15: 5176-5185.

2. Kootstra AMJ, Beeftink HH, Scott ES, Sanders JPM (2011) Valorization of Jatropha curcas: solubilization of proteins and sugars from the NaOH extracted de-oiled press cake. Industrial Crops and Products 34: 972-978.

3. Herrera JM, Martinez CJ, Ayala AM, Siciliano LG, Escobedo RM, et al. (2012) Evaluation of the Nutritional Quality of Nontoxic Kernel Flour From Jatropha Curcas in rats. J Food Qual 35: 152-158.

4. Makkar HPS, Becker K, Sporer F, Wink M (1997) Studies of nutritive potential and toxic constituents of different provenances of Jatropha curcas. J Agric Food Chem 45: 3152-3157.

5. Haas W, Sterk H, Mittelbach M (2002) Novel 12-deoxy-16-hydroxyphorbol diesters isolated from the seed oil of Jatropha curcas. J Nat Prod 65: 1434-1440.

6. Devappa RK, Makkar HPS, Becker K (2010) Optimization of conditions for the extraction of phorbol esters from Jatropha oil. Biomass Bioenergy 34: 1125-1133.

7. Joshi C, Mathur P, Khare SK (2011) Degradation of phorbol esters by Pseudomonas aeruginosa PseA during solid-state fermentation of deoiled Jatropha curcas seed cake. Bioresour Technol 102: 4815-4819.

8. Xiao J, Zhang H, Niu L, Wang X, Lu X (2011) Evaluation of detoxification methods on toxic and antinutritional composition and nutritional quality of proteins in Jatropha curcas meal. J Agric Food Chem 59: 4040-4044.

9. Phasukarratchai N, Tontayakom V, Tongcumpou C (2012) Reduction of phorbol esters in Jatropha curcas L. pressed meal by surfactant solutions extraction. Biomass and Bioenergy 45: 48-56.

10. Guedes RE, Cruz FDA, Lima MCD, Sant'Ana LDO, Castro RN, et al. (2014) Detoxification of Jatropha curcas seed cake using chemical treatment: Analysis with a central composite rotatable design. Industrial Crops and Products 52: 537-543.

11. Joshi C, Khare SK (2011) Utilization of deoiled Jatropha curcas seed cake for production of xylanase from thermophilic Scytalidium thermophilum. Bioresour Technol 102: 1722-1726.

12. Devappa RK, Malakar CC, Makkar HP, Becker K (2013) Pharmaceutical potential of phorbol esters from Jatropha curcas oil. Nat Prod Res 27: 1459-1462.

13. Roach JS, Devappa RK, Makkar HP, Becker K (2012) Isolation, stability and bioactivity of Jatropha curcas phorbol esters. Fitoterapia 83: 586-592.

14. Wender PA, Kee JM, Warrington JM (2008) Practical synthesis of prostratin, DPP, and their analogs, adjuvant leads against latent HIV. Science 320: 649-652.

15. Devappa RK, Angulo-Escalante MA, Makkar HPS, Becker K (2012) Potential of using phorbol esters as an insecticide against Spodoptera frugiperda. Industrial Crops and Products 38: 50-53.

16. Aregheore EM, Becker K, Makkar HPS (2003) Detoxification of a toxic variety of jatropha curcas using heat and chemical treatments, and preliminary nutritional evaluation with rats. South Pac J Nat Sci 21: 51-56.

17. Martinez-Herrera J, Siddhuraju P, Francis G, Davila-Ortiz G, Becker K (2006) Chemical composition, toxic/antimetabolite constituents, and effects of different treatments on their levels, in four provenances of Jatropha curcas L from Mexico. Food Chem 96: 80-89.

18. Machida H, Takesue M, Smith Jr RL (2011) Green chemical processes with supercritical fluids: Properties, materials, separations and energy. The Journal of Supercritical Fluids 60: 2-15.

19. Brunner G (1994) Gas extraction: an introduction to fundamentals of supercritical fluids and the application to separation process.

20. Mendes MF, Pessoa FLP, Uller AMC (2005) Optimization of the process of concentration of vitamin E from DDSO using supercritical CO_2. Brazilian Journal of Chemical Engineering 22: 83.

21. Makkar H, Maes J, De Greyt W, Becker K (2009) Removal and degradation of phorbol esters during pre-treatment and transesterification of Jatropha curcas oil. Journal of the American Oil Chemists' Society 86: 173-181.

22. Moraes MN, Zabot GL, Meireles MAA (2014) Applications of Supercritical Fluids in Latin America: Past, Present and Future Trends. Food and Public Health 4: 162-179.

23. Cavalcanti VMS, Aznar M, Melo SAV, Cruz FG (2013) Economic evaluation based on an experimental study of extraction of alkylamides from Spilanthes genera using supercritical CO_2. In: 10th Conference on Supercritical Fluids and Their Applications 1: 476.

24. Prado JM, Assis AR, Maróstica-Júnior MR, Meireles MAA (2010) Manufacturing cost of supercritical-extracted oils and carotenoids from amazonian plants. J Food Process Eng 33: 348-369.

25. Prado JM, Dalmolin I, Carareto NDD, Basso RC, Meirelles AJA, et al. (2012) Supercritical fluid extraction of grape seed: process scale-up, extract chemical composition and economic evaluation. J Food Engineering 109: 249-257.

26. Delgado B, Pessoa FLP (2014) Simulation of butanol production by an integrated fermentation process using supercritical fluid extraction with carbon dioxide. New Biotechnology.

27. Peters MS, Timmerhaus KD (1991) Plant design and economics for chemical engineers. McGraw-Hill, USA.

28. Turton R, Bailie RC, Whiting WB, Shaeiwitz JA (2003) Analysis, Synthesis, and Design of Chemical Processes. Prentice Hall, USA.

29. Prado IM, Albuquerque CLC, Cavalcanti RN, Meireles MAA (2009) Use of commercial process simulator to estimate the cost of manufacturing (COM) of carotenoids obtained via supercritical technology from palm and buriti trees. 9th International Symposium on Supercritical Fluids, Arcachon, France.

30. Sriram S (2012) Potential of New Jatropha Varieties.

31. Sigma Aldrich.

32. Makkar HP, Becker K (1999) Nutritional studies on rats and fish (carp Cyprinus carpio) fed diets containing unheated and heated Jatropha curcas meal of a non-toxic provenance. Plant Foods Hum Nutr 53: 183-192.

25

Thermal Properties of Polyurethane-Polyisocyanurate (PUR-PIR) Foams Modified with Tris(5-Hydroxypenthyl) Citrate

Joanna Liszkowska*, Bogusław Czupryński and Joanna Paciorek-Sadowska

Kazimierz Wielki University, Bydgoszcz, Poland

Abstract

New polyol (citrate) for the production of PUR-PIR foams was obtained. The hydroxyl number of the obtained compound is 496 mg KOH/g. The amount of water in the new compound was 0.98%. Due to this amount, the foam recipe does not need to be modified. The obtained foam series contain from 0.1 R to 0.5 R of the new compound. The $T_{5\%}$ values were higher by about 50% for the foams examined under nitrogen. The foams showed similar $T_{10\%}$ values under both atmospheres. The $T_{20\%}$ were higher by around 15°C for foams heated in nitrogen. The difference in the $T_{50\%}$ was circa 150°C and foams heated in oxygen registered better value. The foams showed slight decrease in the $T_{5\%}$, $T_{10\%}$, $T_{20\%}$ and $T_{50\%}$ values in both atmospheres, as well as a decrease in the softening temperatures, along with the increasing amount of the tris(5-hydroxypentyl) citrate compound in them. New compound is new product, doesn't described in literature. It has good thermal properties as industrial polyols but it is cheaper. Foams with this product have good thermal properties and they may be used in the building industry for thermal insulation.

Keywords: Polyurethane foam; Citric acid; Pentane glycol; Thermal properties; TG

Introduction

Constant technological development in the polymer field and widening the range of their use impose great requirements regarding their properties. One of the demands for macromolecular compounds is thermal stability during processing and their application. In the recent years it has become more important to obtain polymers with increased thermal stability.

The term "thermal stability" is often used to describe thermal durability or resistance, as well as heat resistance. Polymers with higher thermal stability are characterized by: higher temperatures of melting, softening and thermal decomposition, smaller mass loss during heating in higher temperature, higher heat deflection temperature under load, and less significant changes of their properties (physical, mechanical and chemical) during short-term and long-term exposure to higher temperature. Those parameters are usually considered as criteria for the assessment of specific polymer's thermal stability. The improvement of thermal stability can be achieved by various measures: by choosing the right structure, by increasing the crystallinity level, by incorporating strongly polar or stereoregular groups, by limiting the rotation range in the chain using high-volume groups, by cross-linking, by using cyclic monomers [1,2]. In the temperature above 400°C, almost all polymers decompose in the presence of oxygen. The literature sources also describe that foam's thermal stability decreases along with the lowering of foam's density. In lower temperatures, the speed of foam decomposition is comparable in vacuum, oxygen and nitrogen environments, however in higher temperatures the foam decomposes the quickest in vacuum, and the slowest in oxygen environment [3].

The method of synthesizing polyurethane foams with increased resistance was described by Ożóg and Lubczak [4-6]. The authors proved that by adding polyetherol produced with cyiamelure ring to foams, they increased their thermal resistance.

Polymers' thermal stability is connected to their chemical structure, the bond energy between individual atoms comprising the material (energy of dissociation into radicals). Multiple bonds (e.g., C=N, C=C) or bonds containing e.g., boron, nitrogen, silicon: C-F, B-O, B-N, Si-O, have high energy value. The thermostability of polymers, in which those bonds are present, is higher than in polymers with only C-C

polymers. The type of environment plays a significant role here as well, e.g., the substituents' character, the type of neighboring atoms, and so-called "macroscopic" factors, such as the structural unit and the structure of the entire macromolecule, especially the cyclic structure and the bridges between aromatic and heterocyclic links.

To determine the comparison scale of different polymers, the term "polymer's half-live temperature" was introduced. The sample heated for 45 minutes loses 50% of its initial mass (Th thermostability) [1,2].

When it comes to foams, thermal and heat resistance are significant parameters, especially when the foams are used in construction (e.g., as insulation) or automotive industries [7]. Thermal resistance is related to foam's physical changes taking place under increased temperature and applied force. They are often characterized by providing softening temperature and the method of determining it [8]. In polyurethane foams, the softening temperature is lower by couple dozens of degrees from the foams' decomposition temperature. Thermal resistance is related to the worsening of properties connected to mass loss. Those properties change due to changes in temperature during foam utilization in specific conditions. The material decomposes and molecules with smaller molecule weight are produced [9].

Thermal resistance of polyurethane foams is mainly related to the thermal dissociation temperature of the weakest bonds in them. Depending on the raw materials used and the equivalence ratio of –NCO to –OH groups, the urethane bonds (with dissociation temp. of 200°C) can be present along with ester bonds (with dissociation temp. of 260°C), ether bonds (with dissociation temperature of 350°C), allofanian bonds (with dissociation temp. of 106°C) and other bonds [9-14].

---

*Corresponding author: Joanna Liszkowska, Kazimierz Wielki University, Bydgoszcz, Poland, E-mail: liszk@ukw.edu.pl

As a result of a reaction with polyisocyanurate, when using water to produce CO_2 (porophor), urea and biuret bonds are created with the dissociation temperature of 250°C and from 130°C to 145°C respectively [15]. When there is too much polyisocyanurate in relation to polyol, isocyanurate bond with the dissociation temperature of circa 300°C and carbodiimide bond with the dissociation temperature of circa 240°C are introduced into the chain of the macromolecular compound.

Thermal properties of the polyurethanes are dependent on their mechanical properties [16]. Foams with closed cells show higher thermal stability than those partially comprising of open cells. The mass loss of polyurethane foams depends on the porophor used for their production [17].

Besides the abovementioned, the types of polyol used for the synthesis, as well as the conditions (nitrogen, oxygen atmosphere) have an influence on foams' thermal properties [12]. The study conducted under nitrogen showed that foams are less stable in higher temperatures (250-400°C) than the same foams examined under oxygen. The amount of polyol in rigid polyurethane foams also has an influence on their thermal properties [18]. The increase of, for example, bio-polyol, produced based on sugar cane, causes a shift of foam's first weight loss temperature towards higher temperatures.

Often the temperature at which 5% ($T_{5\%}$) or 10% ($T_{10\%}$) mass loss is reached (in relation to the initial mass) is taken as a measure of thermal stability. Thermal stability examinations are conducted under neutral atmosphere (nitrogen, oxygen), oxidizing atmosphere (oxygen, air) or under other gases, e.g., carbon dioxide [19].

Moreover, the term "thermal analysis", in accordance with ICTAC (*International Confederation for Thermal Analysis and Calorimetry*) recommendations, describes the analysis of the dependencies between substance's specific property and its temperature, towards which the sample is heated or cooled in controlled environment [20]. Thermogravimetric analysis (TGA), differential thermal analysis (DTA) and differential scanning calorimetry (DSC) are the most popular techniques of thermal analysis used in polymer analysis. The examination of polymer structures using thermal analysis techniques is rather widely applied [21]. However, some special attention needs to be directed towards such phenomena as: determining the crystallinity level of a polymer, the glass state phenomenon, determining the range of melting temperatures, and measuring the enthalpy of phase changes. Especially interesting seems to be the use of DTA and DSC techniques in the examination of polymers' thermal and oxidation stability. The TGA technique determines the relation of the mass of the sample to the steadily increasing temperature, and to the time of the sample heating in constant temperature [19].

Particularly interesting is the combination of thermogravimetry with thermal analysis because during single analytical procedure, simultaneously two parameters are being measured, i.e., changes in the mass and thermal effects. In the case of distilling or subtilizing products (e.g., enriching supplements), the decomposition temperature has the value of the temperature, at which the first changes in thermogravimetric curve (TG or DTG) can be observed [22]. Differential scanning calorimetry (DSC) enables the measurement of the difference in heat streams flowing among the model foam, examined sample and the sensor. The measurement is performed in isothermic conditions or with programed temperature increase. The DSC curve (as well as DTA) registers thermal effects using characteristic points, e.g.,

❖ Temperature of thermal effect start T_0/ end T_k

❖ Temperature of extreme point T_{max}

❖ Effect amplitude at the extreme point ΔT_{max}

❖ Square surface of thermal effect S [23-26]

Thermal decomposition (degradation, destruction) is an important issue considering polyurethanes. Macromolecular compounds react, among others, with oxygen when subjected to heat and UV radiation, which can result in changes in chemical composition, degradation or destruction. Products capable of further reactions are created, which initiate depolymerization reactions, e.g., in reaction with oxygen, a hydrogen peroxide group, that can break into radicals capable of further reactions, is created in the polymer chain.

A process of polymer decomposition called pyrolysis can take place under the influence of high temperature in the atmosphere of neutral gas or in vacuum. As a result of this process, volatile gases and polymer fragments are produced. The temperature and the speed of polymer pyrolysis depends on its thermal stability and chemical reactions of the decomposition (i.e., depolymerization, destruction and degradation) happening in current conditions. For the material to be decomposed, an adequate energy needs to be supplied which will be sufficient to break the bonds between individual atoms, from which those compounds are built. The amount of polymer activation energy decides whether this reaction can take place.

Thermal decomposition of the materials is an endothermic and an irreversible process initially conducted under externally supplied heat and later on, under heat produced during the burning process. Due to the fact that the material surface warms up quicker than the lower layers, a gradient of temperatures is created and the pyrolysis speed is the largest on the material surface. When the material has a large specific surface area, the surface heating decides about the pyrolysis speed. In relation to that, foamed materials, spongy, in the form of foil, etc. that have large specific surface area are the largest fire hazard in the presence of air. During thermal decomposition, depending on the material type, the following products may be produced:

a) flammable gases or vapors that burn in the presence of air (methane, ethane, ethylene, formaldehyde, acetone, carbon oxide),

b) nonflammable gases or gases that does not burn in the presence of air-carbon dioxide, hydrogen chloride, hydrogen bromide, water steam,

c) liquids, usually partially decomposed polymer and organic compounds with large molecular weight

d) solid products-usually charred remains, charcoal or ash,

e) shattered solid particles or polymer fragments in the form of smoke [13,27,28].

The synthesis of new cheap polyol for rigid PUR-PIR foams was the aim of the research. The polyol was synthesized using the cheapest carboxyl acid available on the market. The synthesis of the new compound using the acid was possible because it contained the –OH functional groups. The availability of this acid enabled the synthesis of a cheap polyol. The properties of the obtained polyol were not much different from the industrial polyols, which makes is suitable for foam synthesis. The foam obtained using the polyol have properties similar to the foams obtained using industrial polyol (Rokopol RF551).

Experimental and Methods

Characteristics of raw materials

The Rokopol RF551 polyether (polyoxypropylenehexol with 400÷440 mgKOH/g hydroxyl number, molecular weight of 660) produced by Zakłady Chemiczne PCC Rokita S.A. in Brzeg Dolny (Table 1), and a technical diisocyanate Ongromat CR 30-20 produced in Hülls (Hungary), whose main component is diphenylmethane 4.4´-diisocyanate, were used to prepare rigid PUR-PIR foams. The density of Ongromat at 25°C was 1.23 g/cm³, viscosity was 200 mPa.s and the content of NCO groups was 31.0%. The polyether and diisocyanate were characterized according to ASTM D 2849-69 and ASTM D 1638–70 standards. The catalyst used to produce the foams was anhydrous potassium acetate (POCh Gliwice) applied in a form of 33% solution in diethylene glycol - DEG (catalyst 12), and "DABCO 33LV" (triethylenediamine, Hülls, Germany) applied in a form of 33% solution in DEG. The foam structure stabilizer was polyoxosilanepolyoxyalkene surfactant, Silicone L-6900" (Witco, Sweden). Carbon dioxide formed during a reaction of water with isocyanate groups acted as blowing agent. A liquid flame retardant Roflam P (TCPP, tri(2-chloro-1-methylethyl) phosphate), Albright and Wilson, Great Britain, was added to the foams. Foam synthesis was conducted by adding new compound E16.

For obtaining new E16 and the monohydrate N 1560 2-hydroxypropane-1,2,3-tricarboxylic acid produced by Brenntag Poland LLC company in Kędzierzyn Koźle (storage in Toruń) was used. The amount of water in the acid was 7.5-8.8%. The acid was dried in ventilated drier until the water volume reached 3.35%. The water amount was measured in a moisture analyzer according to the PN-A-79005-04/1997 Polish Standard. The 1.5-pentanediol (1.5-PD) - E16 was used in the synthesis (producer - POCh, Gliwice).

E16 compound was added to the foams in the amount of 0.1 to 0.5 R (equivalent) to the Rokopol RF551 quantity, which was reduced from 0.9 to 0.5 R.

The synthesis process of tris(5-hydroxypentyl)-2-hydroxypropane-1,2,3-tricarboxylate (E16) for PUR-PIR foams

New compound was synthesized by esterification with solvent method. In reaction 1, 1.5-pentanediol (1.5-PD) was used as alcoholic compound, producing E16 compound (equation 1). In this reaction the 2-hydroxypropane-1.2.3-tricarboxylic acid was the acidic compound.

In synthesis 1, 0.04% of the catalyst was used which was the Lewis acid by the name of tetraisopropyl titanate (market name-Tyzor TPT) produced by Du Pont. It is a transparent, yellowish liquid, very sensitive to moisture, with a freezing temperature of 19°C. It crystalizes in low temperatures (however after rising the temperature again, it is re-useable). It contains 28.1% TiO_2. Its molecular mass equals 284 g/mol, density in 20°C is 0.95 g/cm³, viscosity 3.5 mPa·s, boiling temperature 232°C. It mixes with most organic solvents.

Synthesis was conducted in three-neck glass flask (500 ml vol each) equipped with reflux condenser, thermometer, stirrer and Deana-Stark's head. The reaction took place under xylene. The generated water was collected in the head. Flask contained 1.5-PD in the amount of 156 g, which was mixed with 2-hydroxypropane-1.2.3-tricarboxylic acid (96 g). 0.096 g of catalyst was added at the end of synthesis 1. Flask was heated for about 20 minutes in an electric bath (until the acid dissolved, i.e., about 85°C). It was heated until the substances inside them boiled. From this moment the reaction time was measured. Synthesis 1 took

8.5 hours altogether. The temperature of the reaction was 161°C. As a result of the synthesis 1 (Scheme 1) was citrate (polyol).

Obtaining PUR-PIR foams

The foams were prepared on a laboratory scale using one-stage method from two-component system at equivalent ratio of NCO to OH groups equal to 3:1. Component A was obtained by thorough mixing (stirrer rotation speed - 1800 rpm, mixing time 10 s) of the appropriate amount of Rokopol RF-551, new compound E16, catalysts, flame retardant, surface-active agent and porophor (water). The amount of water was reduced by the content of water in E16. Component B was Ongromat 30-20 (Table 1). Both components were mixed (10 s mixing time) in an appropriate mass ratio and were poured into a metal open rectangular mould with dimensions of 195 × 195 × 240 mm (internal dimensions). Serie P16 (with E16) of foams was obtained by this method.

For raw materials containing reactive hydroxyl groups, a so-called R equivalent was calculated according to the equation:

$$R = \frac{56100}{L_{OH}}$$

in which L_{OH} – stands for the hydroxyl number.

Measurements

Citrate: Hydroxyl number of citrate was performed according to WT/06/07/PURINOVA formula, Purinova Bydgoszcz. The viscosity at the temperature of 25°C (EN ISO 12058-1) using a Hoeppler viscometer. The density of polyols in the pycnometer was measured at 25°C (298 K) according to PN-92/C-04504. Depends on the number of hydroxyl number of E16 used in the foam and the polyisocyanate

Compound	Unit	W	P16.1	P16.2	P16.3	P16.4	P16.5
Rokopol RF-55	R	1	0.9	0.8	0.7	0.6	0.5
	g	66.8	60.1	53.4	46.8	40.1	33.4
E16	R	0	0.1	0.2	0.3	0.4	0.5
	g	0	6.4	12.9	19.3	25.7	32.2
Silicon Tegostab 8460	g	4.7	4.7	4.7	4.7	4.7	4.7
	%mas.	1.5	1.5	1.5	1.5	1.5	1.5
DABCO	g	2.9	2.9	2.9	2.9	2.9	2.9
	%mas.	0.9	0.9	0.9	0.9	0.9	0.9
Catalyst 12	g	6.7	6.7	6.7	6.7	6.7	6.7
	%mas.	2.1	2.1	2.1	2.1	2.1	2.1
Roflam P	g	47.6	47.6	47.6	47.6	47.6	47.6
	%mas.	15	15	15	15	15	15
Water	R	0.7	0.7	0.7	0,7	0.7	0.7
	g	3.15	3.14	3.14	3,13	3.12	3.11
Ongromat 30-20	R	3.7	3.7	3.7	3.7	3.7	3.7
	g	250.7	250.7	250.7	250.7	250.7	250.7

Table 1: Formulation for P16 foam series.

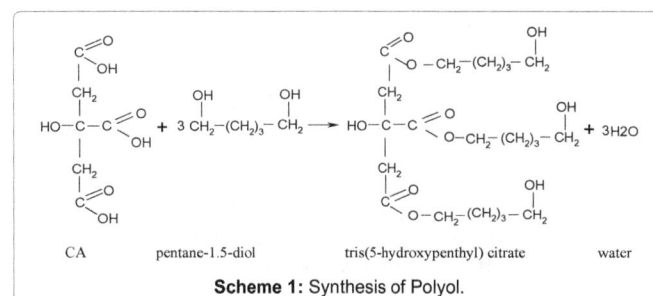

Scheme 1: Synthesis of Polyol.

necessary to produce foam. Density and viscosity are very important during the processing process. The water content was tested by Karl Fisher (PN-81/C-04959), in which the solvent used was a mixture of methanol and carbon tetrachloride in a 1:3 ratio, the titration reagent Combo. Marking is to dissolve the appropriate test portion of the product in Titraqual (Titrant for Titration) and potentiometric titration of the solution to the equivalence point.

Foams: Times of foam processing were measured during foaming i.e., start time, time of expansion and time of gelation (always counted from the moment of mixing of all components). Start time -to achieve a state of cream, which is the start of volume expansion foam). Time of expansion-increase the time until you have the maximum foam volume. Time of gelation-total gel time until the free surface of the foam stops to attach a clean glass rod.

During the synthesis and right after its end, the temperature of processes occurring inside the foam was measured for 1.5 h. The temperature changes were measured using thermocouple, whose sensor was placed inside the foam. The measurements were taken every minute.

Foams were thermostated for 4 h at the temperature of 120°C. They were then seasoned for 48 h at temperature of 20 ± 4°C and cut into pieces.

Their thermal properties as well as changes in linear, volume and mass dimensions were determined under increased temperature (120°C) in relation to initial sample length, volume and weight according to ISO 1923:1981 and PN-ISO 4590:1994.

Thermal resistance (thermostability) was measured with a derivatograph operating in Paulik-Paulik Erdey system produced by MOM-Budapest. Softening point was measured with Vicat apparatus in compliance with DIN 53424 standard, changes in linear dimensions, volume and weight after 48 hours of thermostating in ventilated drier at 120°C. Heating to 1000°C under oxygen or nitrogen with heating rate of 5°C/min.

Softening temperature was made by Vicat according to PN-EN ISO 306:2005. Determining the temperature in which a normalized needle immerses into the surface of the tested sample 1 mm deep, under appropriate weight with the temperature rising at steady speed.

The examination of melting point was conducted using camera Boetius.

The examination of changes in foams was conducted using DSC Q200 differential scanning calorimeter by TA Instruments. The apparatus range of work is from -90°C to +725°C (foams examinations were conducted in a range from 0°C to 400°C). The apparatus has built in Advanced Tzero technology. The DSC examination was performed under nitrogen and oxygen atmospheres. The weighted portion was 4 mg.

Thermal resistance (thermostability) was measured in two ways:

a. In air atmosphere with a derivatograph operating in Paulik-Paulik Erdey system produced by MOM-Budapest. Heating speed was 5°C/min, weighted portion was 100 mg. Heating to 1000°C

b. Under nitrogen atmosphere using thermogravimetric analyzer: TG Q500 by TA Instruments. Heating to 1000°C. The sample's weighted portions were 80 g. Heating speed was 10°C/min

FTIR examinations were performed using FTIR Spectrophotometer: Nicolet iS10 by Thermo Scientific, with spectrum range from 7800 to 350 cm⁻¹ and with maximum resolution capability <0.4 cm⁻¹ with DTGS detector.

FTIR analysis was marked in compliance with the standards in force. Laser beam (in an infrared spectrum of 4000 to 500 cm⁻¹) was passed through a sample in form of film, liquid substance (polyol) or a pill (foam sample in KBr). The presence of function groups in the compound was being determined.

Elemental analysis of E16 was performed using EA 1108 Carlo-Erba analyzer (Vario). Content of carbon and hydrogen where measured.

Number-average molecular weight (M_n), weight-average molecular weight (M_w), z-average molecular weight (M_z) and molecular weight distribution were determined using Viscotec T60A gel chromatograph equipped with three detectors' system: RI (refractive index detector), LS (light scattering detector) and DV (viscometer detector). Separation has been performed using two independent columns: PSS SDV (of 7.8 mm × 300 mm size with TSK bed-100 and 1000 Å pore diameter gel, using the following recording parameters: temperature 25 ± 0.1°C, volume flow of eluent 1 ml/min, the injection loop volume of 20 ml, concentration of polymer solution 4-5 mg/ml, analysis time 30 min. THF was used as eluent (distilled from over sodium prior to use). Calibration was based on common polystyrene references.

The remaining physico - mechanical properties (endurance, brittleness, absorptiveness, heat conductivity and others) will be the subject on other papers.

Results and Discussion

Citrate E16

In the laboratory the Department of Chemistry and Technology of Polyurethane resulted the product of synthesis of 2-hydroxypropane-1.2.3-tricarboxylic acid (citric acid) with pentane-1.5-diol with the use of 0.04% catalyst (Tyzor TPT) named tris(5-hydroxypentyl) citrate (E16). It is characterized by 496 mgKOH/g hydroxyl number. It is a straw-yellow liquid characterized by 496 mgKOH/g hydroxyl number. The hydroxyl number is the basic parameter of the oligomerol that is needed to calculate the polyurethane compound recipe. Based on that, the R equivalent is determined, which corresponds to the amount of compound containing hydroxyl groups (oligomerol) in relation to the amount of remaining foam components. The R equivalent is calculated according to the following equation (5):

$$R = \frac{56100}{HN}$$ (5); R-equivalent, HN – hydrogen number

It is very important to arrange the foam recipe and to determine the amount of water in the used polyurethane polyol. Water is added to the foam as a blowing agent. Excess water can cause cracking of the foam cells or even its fall. The value of hydroxyl number is within the range of HN of industrial polyols. As a result of that, the foaming process of PUR foams with the addition of the new E16 compound should progress similarly to the synthesis of foams with polyols available on the market.

The results of the examination of the new E16 compound show that density (equaling 1.053 g/cm³) does not limit the possibilities of using it in apparatuses for the synthesis of rigid PUR-PIR foams. The E16 viscosity (equaling 10364.1 mPa·s) is close to the viscosity that should be characteristic of raw materials used in standard appliances for PUR processing available on the market, i.e., 15000 mPa·s.

Also, the pH value of E16 (5.1) is close to the range defined by the pH of standard polyols [from 6.5 (Rokopol RF-55) to 8.7 (Rokopol 4845)]; that is why there is no need to expect great differences in the process of obtaining rigid PUR foams containing various amounts of E16, which was added instead of the commercial polyols.

The E16 compound contained 0.98% of water. The amount of water in the polyol is a significant parameter, especially in foams foamed with water. When the recipe is being determined, the amount of water in the oligomerol needs to be accounted for to choose the right water amount for foaming. The polyol added to polyurethane foams should not contain more than 1% of water.

As a result of the reaction using Dean Stark head, 27.7 cm³ of water was distilled which was in correspondence with the stoichiometric calculation of water amount (27.0 cm³).

The melting temperature was determined using Boetius apparatus. The method's principle was to observe the moment, when there would be the complete disappearance of crystallinity phase as a result of constantly increasing temperature. The T_t melting temperature of the acid component used in the synthesis of the new E16 compound (citric acid) was 158°C, and the T_r was 164°C. However, for the newly obtained E16 compound T_t=190°C, T_r=270°C.

The percentage value of carbon in the E16 compound was 54.31%, which is close to the calculated value (53.09%). The amount of hydrogen is 9.25% and it also did not deviate from the calculated value, which equals 7.96%. The examined molecular mass of the E16 compound is 310 g/mol and differs by 10% from the value calculated molecular weight.

Thermal properties of E16

Heat resistance plays a significant role in determining whether a specific polyol is useful for foam synthesis. Using a polyol with better thermal stability increases the temperature range of foam application. The E16 thermal stability was tested under neutral atmosphere (in nitrogen). The results were compared with the thermal stability of the industrial Rokopol RF551 (Tables 2 and 3). Based on TG curve (weight change) and DTG (Figure 1) curve (derivative weight change) the following characteristic quantities were measured (Table 3): the temperature of the foam's first weight loss (T_1,°C), the temperature of beginning of the decomposition (T_2,°C), and the temperature of the foam's highest speed of weight loss, so-called maximum of thermal effect (T_{max},°C, corresponding to the extreme on the DTG curve).

During the first stage of E16 decomposition (up to 135°C temp.) it can be observed that the compound lost circa 1.5% of its weight. This can be attributed to the water vaporizing after it was released due to the breaking of hydroxyl bonds. The quickest mass loss, which was circa 20%, can be observed at 292°C temperature, i.e., before reaching the square surface equal to the heat effect (T_{max}=366.1°C).

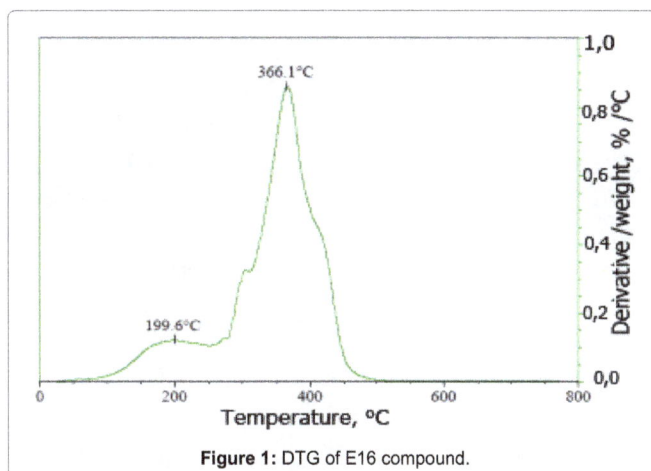

Figure 1: DTG of E16 compound.

The thermostability was defined as the temperature, at which the compound loses 5% or 10% of its mass. A so-called temperature of compounds half-live (Th) was also determined, at which the 50% mass loss occurred. The 5% mass loss under nitrogen atmosphere for the E16 (Table 4) occurs at 170°C temp., and the 10% at 212°C temp. The 20% mass loss for the E16 compound was also determined (292°C). Moreover, the Th was 358°C, and the T_{max} was equal 366°C. At 800°C temperature, only 1.5% of the E16 is left (as ash).

The heat resistance of the described E16 compound is lower by over a dozen percent from the industrial polyols used in rigid foam production. Namely, the quickest mass loss (of 70%) under nitrogen atmosphere in case of industrial used polyol-Rokopol RF-551-is reached at maximum thermal effect T_{max}=411°C. In the case of the E16, the highest mass loss (at T_{max}=366°C) is 55%.

When comparing the thermal stability (under nitrogen atmosphere) of the industrial polyol (Rokopol RF-551) with the applied E16 compound, a lower temperature of the beginning of mass change by 15.6% can be seen for the E16, lower temperature of the decomposition start by 17.5% and lower temperature of the maximum of thermal effect by 10.9%.

An examination of the E16 compound, obtained in the laboratory of the Department of Chemistry and Polyurethanes' Technology, was performed using DSC method under nitrogen atmosphere. Based on the DSC curve (Figure 2), two endothermic changes were observed. The beginning of E16 melting at 108°C. The proper melting temperature is 147°C and the enthalpy (H) at this point is 50.5 J/g. Second endothermic peak was also observed on the DSC curve. The beginning of melting happens at 304°C temperature, and the proper melting occurs at 354°C, enthalpy H=165 J/g. The first peak is connected to the degradation of -OH bonds and with the loss of water, which absorbed only 50.5 J/g of the heat. The second peak is related to the degradation of ester bonds, for which 165 J/g of the heat was needed.

Processing and thermostating parameters of PUR-PIR foams (P16)

The obtained E16 compound was used as a component in polyol premix for rigid PUR-PIR foams instead of the industrial Rokopol RF551. A series of 16.0-16.5 foams containing 0-0.5 R of the E16 was produced this way.

The addition of E16 had an effect on the processing parameters of the foams that increased lineally along with the increase of the E16 compound (Table 4). The increase in those parameters is caused by

Polyol	Temperature of the foam's first weight loss		Temperature of beginning of the decomposition		Maximum of thermal effect	
	T_1, °C	Mass loss, %	T_2, °C	Mass loss, %	T_{max}, °C	Mass loss, %
Rokopol RF551	160	1	205	5	411	70
E16	135	1.5	169	4	366	55

Table 2: Temperature of the beginning of mass change, decomposition start and the beginning of quickest mass loss in the E16 compound (in nitrogen atmosphere).

Polyol	$T_{5\%}$, °C	$T_{10\%}$, °C	$T_{20\%}$, °C	$T_{50\%}$ (T_h), °C	Pozostałość w temp. 800°C, %
Rokopol RF 551	205	284	378	404	0
E16	170	212	292	358	1,5

Table 3: Thermal stability of E16 compound (in nitrogen atmosphere).

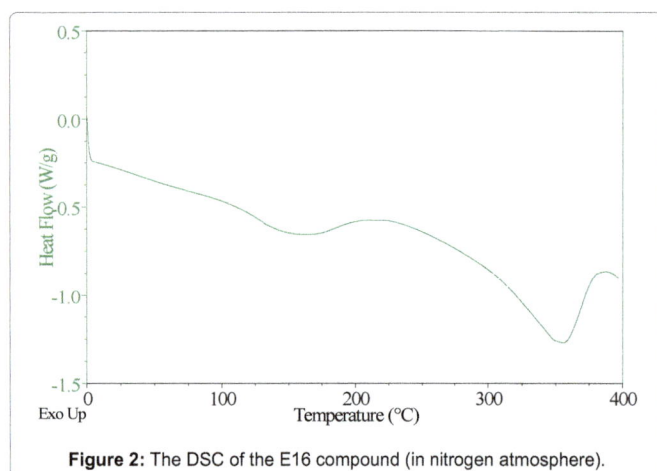

Figure 2: The DSC of the E16 compound (in nitrogen atmosphere).

Foam	Start time, s	Growth time, s	Gelation time, s
P16.0	10	30	27
P16.1	17	44	46
P16.2	19	48	44
P16.3	20	53	65
P16.4	21	60	82
P16.5	25	61	92

Table 4: Processing parameters of P16 foams.

the greater viscosity of polyol premix, whose value rises along with the amount of E16 in it. Therefore, the premix viscosity not only has an effect on the mixing of the components, but also on the elongation of processing times. The foaming speed influences the cellular shape. If the foaming reaction proceeds slowly, the cells become isotropic in shape which positively influences the application properties of the foamed materials, as they show identical values regardless of the examination direction.

The temperature changes during 90 minutes were determined in the foam (16.3 foam) after its production and taking it out of the mould (Figure 3). The time was measured from the moment of foam's rising start. In the first 29 minutes, the temperature outside of the foam increases and reaches the maximum value (132°C) in the 29ᵗʰ minute of the examination. This value stays unchanged for couple of minutes and then the temperature falls to 119°C, and in the 41ˢᵗ minute the cooling of the foams begins. A rapid decrease can be observed in the time from 41ˢᵗ and 53ʳᵈ minute. Later on, a more gentle decrease is seen for the next 71 minutes until the 67°C temperature is reached (after 91 minutes of examination). The changes in temperature during foam's seasoning in the first 1.5 hours, after it is produced, are caused by the changes occurring in the foam due to the occurring reactions. The increase of the reaction mixture temperature is caused by the exothermic reactions of isocyanate groups with water.

The amount of E16 compound in the foam does not have any effect on the mass change after 48 h of foam's thermostating in the 120°C temperature (Table 5). The values range from 1.1% (P16.1 foam containing 0,1R of the E16 compound) to 1.2% (foams P16.2 and P16.5) and to 1.4% (foams P16.3 and P16.4). A lack of visible linear dependency can be seen when it comes to changes in dimensions or volume of the foams after 48 h of thermostating, in relation to the amount of tris(5-hydroxypentyl) citrate (E16) in them. Those changes are very small and do not exceed 2.5% are probably are caused by the reaction in the presence of foams which occur during tempering foams.

Thermal and heat parameters of PUR-PIR foams (P16)

The TG foam analysis was conducted in neutral atmosphere in nitrogen (Figure 4) and in oxidizing atmosphere (mixture of nitrogen and oxygen). Based on TG curve (weight change) and DTG curve (derivative weight change) the following characteristic quantities were measured (Table 6): the temperature of the foam's first weight loss $(T_1, °C)$, the temperature of beginning of the decomposition $(T_2, °C)$, and the temperature of the foam's highest speed of weight loss, tzw. so-called maximum of thermal effect $(T_{max}, °C$, corresponding to the extreme on the DTG curve).

Table 6 shows the dependence of T_1, T_2, T_{max} temperatures on the amount of E16 in foam series 16 in nitrogen atmosphere. With the increasing amount of 0R (equivalent) to 0.5 R of the E16, the temperature of the first foam's weight loss T_1 (the beginning of weight change) slightly decreases from 77°C (16.0 foam containing 1 R of industrial Rokopol RF-551) to 75°C (16.5 foam containing 0.5 R of E16). The addition of more and more E16 also causes the decrease in beginning of the decomposition T_2, from 293°C (16.0 foam) to 219°C (16.5 foam). E16 additive does not affect the temperature of the beginning of the fastest decomposition (T_{max}) of the foam and as soon as the loss of weight in T_{max}. The decrease in temperature of the beginning of the decomposition from 293°C (16.1) to 217°C is caused by a slightly poorer thermal stability of the added E16, in comparison to the thermal stability of Rokopol RF-551.

The foams' thermostability is described as $T_{5\%}$ and $T_{10\%}$, and as $T_{20\%}$ (Table 7). A so-called "polymer's half-live temperature" was also determined-a temperature, in which the sample lost 50% of its initial weight (thermostability Th). The temperature $T_{5\%}$, $T_{10\%}$, $T_{15\%}$,

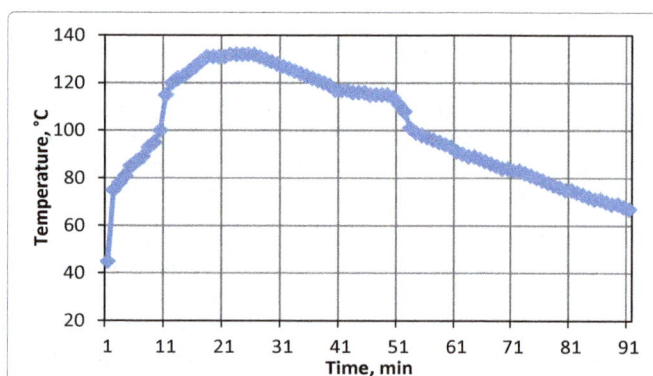

Figure 3: Temperature changes inside the foam in the first 90 minutes (16.3 foam).

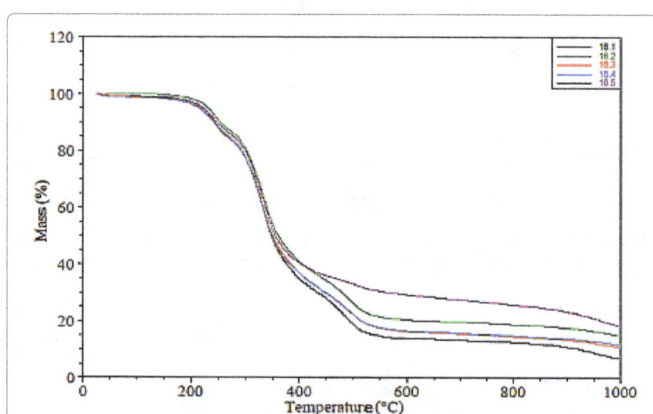

Figure 4: TG of foam P16.1-P16.5 (in nitrogen atmosphere).

Foam	Sample's dimensions changes according to the growth direction Δl_z, %	Sample's dimensions changes opposite to the growth direction Δl_z, %	Volume change, ΔV, %	Mass change Δm, %
P16.0	0.0	0.0	0.0	0.0
P16.1	-1.10	-1.10	-0.90	1.1
P16.2	-0.17	0.05	1.10	1.2
P16.3	0.02	2.10	1.08	1.4
P16.4	0.01	1.60	2.40	1.4
P16.5	-0.06	- 0.38	-1.20	1.2

Table 5: The stability of lineal dimensions, volume and mass of the foams after 48 hours of thermostating in 120°C temp.

Foam	First weight loss		Beginning of the decomposition		Maximum of thermal effect	
	Temp. T_1, °C	Weight loss, %	Temp. T_2, °C	Weight loss, %	Temp. T_{max}, °C	Weight loss, %
In nitrogen atmosphere						
16.0	80	0.5	205	4	336	30
16.1	77	0.5	293	5	334	55
16.2	76	0.5	225	3	336	53
16.3	75	0.5	214	5	334	56
16.4	75	0.5	210	4	334	59
16.5	75	0.5	217	4	336	52
In air atmosphere						
16.0	50	0.5	200	4	304	35
16.1	70	0.5	160	3	299	28
16.2	70	0.5	160	3	299	28
16.3	70	0.5	160	3	298	27
16.4	70	0.5	160	3	298	27
16.5	70	0.5	160	3	297	26

Table 6: The temperature of the foam's first weight loss (T_1), the temperature of beginning of the decomposition (T_2) and the temperature of beginning of the fastest decomposition of the (T_{max}) (in nitrogen and in air).

Foam	$T_{5\%}$, °C	$T_{10\%}$, °C	$T_{20\%}$, °C	$T_{50\%}$ (T_h), °C	The residue at 1000°C temp, %
In nitrogen atmosphere					
16.0	226	255	304	387	19
16.1	221	243	293	348	7
16.2	231	253	302	360	15
16.3	214	241	292	350	15
16.4	213	241	292	348	12
16.5	212	240	291	345	10
In air atmosphere					
16.0	200	270	303	480	0
16.1	174	248	286	500	0
16.2	174	249	286	500	0
16.3	174	249	286	499	0
16.4	174	250	287	499	0
16.5	174	250	287	499	0

Table 7: Thermal stability of PUR-PIR foams (in nitrogen and in air atmosphere).

$T_{20\%}$ decrease with increasing amounts of the new compound tris (5-hydroxypentyl)-2-hydroxypropane-1,2,3-tricarboxylate in the case of testing on TG under nitrogen. Can't be observed linear relationship between the quantities in foam of described above compound and $T_{50\%}$. The addition of the compound doesn't affect the $T_{5\%}$, $T_{10\%}$, $T_{15\%}$, $T_{20\%}$ for TG test under nitrogen, and they are suitably 174°C, 250°C, 287°C and 500°C. In comparison with the reference foam 16.0 observed decreasing T_5 of about 26%, $T_{10\%}$ and $T_{50\%}$ of about 20°C and $T_{15\%}$ of about 15°C.

Softening point decreases from 230°C (16.0 foam without E16) to 169°C (16.5 foam with 0.5 R of E16) (Figure 5). The lowering of softening temperature is caused by the addition of 0-0.5 R of the E16 compound to the P16 foam series.

Temperatures of transformation studied by DSC method taking place within the 16.3 foam in an oxygen atmosphere and under a nitrogen atmosphere show the Figure 6. Foam examination using DSC method under oxygen atmosphere only registers foam decomposition starting at around 170°C temperature.

In the DSC chart (16.3 foam sample, Figure 6) of the examination conducted under nitrogen for 16.0-16.5 foams, three endothermic peaks are visible at T_a, T_b and T_c temperatures (sample 16.3 foam, Figure 6). Depending on the amount of E16 in the foams, the peak shift in time is not large and reaches maximum of 2 minutes (for the 16.5 foam in relation to the 16.0 foam). The enthalpy in respective temperatures (T_a, T_b, T_c) is about: 34 J/g, 9 J/g, 32 J/g.

The first T_a peak (at around 75°C temp.) is related to the diffusion of carbon dioxide from the foams, which acts as the porophor. It is created as a result of the reaction of excess isocyanate groups with water. At this temperature, a diffusion of triethylenediamine (DABCO) from the foams can occur. DABCO, as a 33% solution in dipropylene glycol, is used as the catalyst for PUR synthesis reaction.

The T_b peak is related to the decomposition of the urea (at 250°C) (produced as a result of polyisocyanate reaction with water) as well as to the decomposition of urethane group (at 200°C). In 260°C temperature also ester bonds decompose. The T_b for the 16 foam series is about 275°C. The last endothermic peak in T_c temperature is connected to the thermal dissociation of isocyanurate bonds (whose decomposition starts at 300°C temperature). The T_c for foams from 16 series, is 330°C.

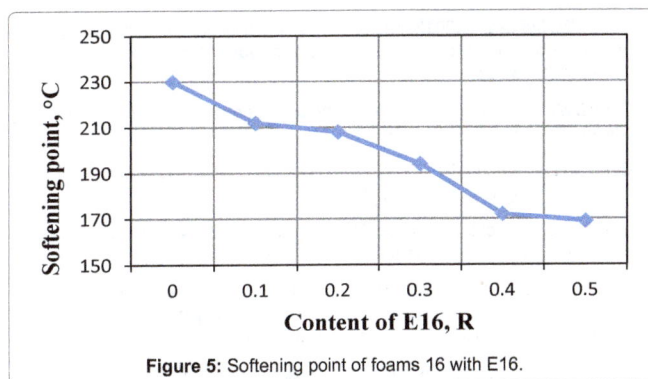

Figure 5: Softening point of foams 16 with E16.

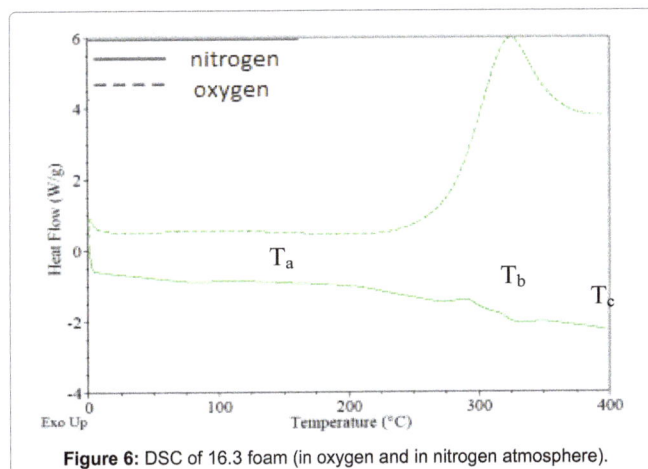

Figure 6: DSC of 16.3 foam (in oxygen and in nitrogen atmosphere).

Conclusion

New tris(5-hydroxypentyl) citrate as raw material for the production of PUR-PIR foams was obtained., the hydroxyl number of the polyols used in its synthesis must be determined. The hydroxyl number of the obtained compound is 496 mgKOH/g. The amount of water in the new compound was 0.98%. Due to this amount, the foam recipe does not need to be modified. The obtained foam series contain from 0.1 R do 0.5 R of the new compound. The foams, whose thermostability was examined, are characterized by higher T_1, T_2, T_{max} temperatures under nitrogen atmosphere, and by lower temperatures under oxygen. Moreover, under oxygen atmosphere the foams decomposed completely, and under nitrogen, 12-19% of the residue remained. The $T_{5\%}$ values were higher by about 50% for the foams examined under nitrogen. The foams showed similar $T_{10\%}$ values under both atmospheres. The $T_{20\%}$ were higher by around 15°C for foams heated in nitrogen. The difference in the $T_{50\%}$ was circa 150°C and foams heated in oxygen registered better value. The foams showed slight decrease in the $T_{5\%}$, $T_{10\%}$, $T_{20\%}$ and $T_{50\%}$ values in both atmospheres, as well as a decrease in the softening temperatures, along with the increasing amount of the tris(5-hydroxypentyl) citrate compound in them. The resulting tris(5-hydroksypentyl) citrate may be used in the synthesis of PUR-PIR rigid foam. The obtained foams can be successfully used as insulating boards (as a laminate), and as protective packaging during transport of the equipment, devices and glass.

References

1. Praca Z (1971) Analiza polimerów syntetycznych. WNT, Warszawa.
2. Saechtling M, Zebrowski W (1908) Tworzywa sztuczne. WNT, Warszawa.
3. Lisiak A, Weaver B (2012) Lowering the flammability of polyurethane foams. Chemical industry 91: 1912-1917.
4. Ozóg MM, Lubczak J (2012) The use oligoeterolu synthesized from melem and propylene carbonate for the preparation of polyurethane foams. Czasopismo Techniczne 9-M: 26.
5. Ozóg MM, Lubczak J (2009) Materialy IV Ogólnopolskiej Konferencji. Naukowej Nauka i Przemysl, Kraków 153.
6. Ozóg M, Lubczak J (2011) Materialy VI Ogólnopolskiej Konferencji Naukowej. Nauka i Przemysl, Kraków 122.
7. Broniewski T, Kapko J, Placzek W, Thomalla J (2000) Metody badan i ocena wlasciwosci tworzyw sztucznych. WNT, Warszawa.
8. Czuprynski B, Liszkowska J, Paciorek-Sadowska J, Lewandowski R (2010) Properties of rigid polyurethane-polyisocyanurate obtained with the addition of products of glycolysis. Chemical industry 89: 734-741.
9. Zielenkiewicz W (2000) Pomiary efektów cieplnych. WNT, Warszawa.
10. Czuprynski B, Liszkowska J, Paciorek-Sadowska J (2008) Inzynieria i Aparatura Chemiczna 4: 15-16.
11. Paciorek-Sadowska J (2011) Badania nad wplywem pochodnych kwasu borowego i N,N-(dihydroksymetylo) mocznika na wlasciwosci sztywnych pianek poliuretanowo-poliizocyjanurowych. Wydawnictwo Uniwersytetu Kazimierza Wielkiego, Bydgoszcz.
12. Dick C, Dominges-Rosado E, Eling B, Liggat JJ, Lindsay CI, et al. (2001) The flammability of urethane-modified polyisocyanurates and its relationship to thermal degradation chemistry. Polymer 42: 913-923.
13. Janik H (2010) Progress in the studies of the supermolecular structure of segmented polyurethanes. Polimery 55: 419-500.
14. Kuranska M, Prociak A, Mikelis K, Ugis C (2013) Porous polyurethane composites based on bio-components. Com Sci and Technol 75: 70-76.
15. Reich L, Stivala S (1971) Elements of polymer degradation. Mc Graw-Hill Inc.
16. Yick KL, Wu L, Yip J, Ng SP, Yu W (2010) Study of thermal–mechanical properties of polyurethane foam and the three-dimensional shape of molded bra cups. Jour of Mat Proc Technol 210: 116-121.
17. Tang Z, Maroto-Valer M, Andersen J, Miller J, Listemann L, et al. (2002) Thermal degradation behavior of rigid polyurethane foams prepared with different fire retardant concentrations and blowing agents. Polymer 43: 6471-6479.
18. Hakim AA, Nassar M, Emam M, Sultan M (2011) Preparation and characterization of rigid polyurethane foam prepared from sugar-cane bagasse polyol. Mat Chem and Physics 129: 301-307.
19. Lazarewicz T, Haponiuk JT, Balas A (2006) e-Polymers 51: 3-16.
20. Krasodomski M, Krasodomski W (2009) The use of thermal analysis methods in the study of petroleum products. The work of the Institute of Oil and Gas 159: 1-92.
21. Krasodomski M, Krasodomski W (2012) Research thermal stability additives dispersing cleaning motor fuels using thermal analysis techniques. Part I. State of the art. Nafta-Gaz 10: 684-692.
22. Chartoff RP, Sircar AK (2005) Thermal Analysis of Polymers. Encyclopedia of Polym Sci Technol, John Wiley & Sons Inc.
23. ASTM E 2550-07 (2007) Standard Test Method for Thermal Stability by Thermogravimetry, ASTM International.
24. Schultze D (1973) Termiczna analiza róznicowa. PWN, Warszawa.
25. Stoch S (1998) Przeglad metod analizy termicznej. Szkola Analizy Termicznej, Zakopane.
26. Pielichowski K (2002) Zastosowanie analizy termicznej w badaniu materialów organicznych. Szkola Analizy Termicznej, Zakopane.
27. Frish KC (1996) An overview of recent technical developments in polyurethanes. I. General introduction and substitutes for chlorofluorocarbons (CFCs) and their applications. Polimery 41: 193-197.
28. Czuprynski B, Liszkowska J, Paciorek-Sadowska J (2014) Modyfication of the Rigid Polyurethane-Polyisocyanurate Foams. Journal of Chem 1-12.

26

Study of the Selectivity of Methane over Carbon Dioxide Using Composite Inorganic Membranes for Natural Gas Processing

Habiba Shehu, Edidiong Okon, Ifeyinwa Orakwe and Edward Gobina*

Center for Process Integration and Membrane Technology, Robert Gordon University Aberdeen, United Kingdom

Abstract

Natural gas is an important fuel gas that can be used as a power generation fuel and as a basic raw material in petrochemical industries. Its composition varies extensively from one gas field to another. Despite this variation in the composition from source to source, the major component of natural gas is methane with inert gases and carbon dioxide. Hence, all natural gas must undergo some treatment with about 20% of total reserves requiring extensive treatment before transportation via pipelines. The question is can mesoporous membrane be highly selective for methane and be used for the treatment of natural gas? A methodology based on the use of dip-coated silica and zeolite membrane was developed. A single gas permeation test using a membrane reactor was carried out at a temperature of 293 K and a pressure range of 1×10^{-5} to 1×10^{-4} Pa. The permeance of CH_4 was in the range of 1.15×10^{-6} to 2.88×10^{-6} mols^{-1}m^{-2}Pa^{-1} and a CH_4/CO_2 selectivity of 1.27 at 293 K and 0.09 MPa was obtained. The pore size of the membrane was evaluated using nitrogen adsorption and was found to be 2.09 nm. The results obtained have shown that it is possible to use a mesoporous membrane to selectively remove carbon dioxide from methane to produce pipeline quality natural gas. There is a need for further study of the transport mechanism of methane through the membrane since this is essential for the separation of other hydrocarbons that could be present as impurities.

Keywords: Membrane; Natural gas; Permeation; Silica; Zeolite

Introduction

Natural gas is an important fuel gas that can be used as a power generation fuel and as a basic raw material in petrochemical industries. Its composition varies extensively from one gas field to another; a particular field might have about 95% methane, with small quantities of other hydrocarbons, nitrogen, carbon dioxide, hydrogen sulphide and water vapor, while another field may have about 10% of lower hydrocarbons like propane, butane or ethane as well as high carbon dioxide contents. Although there is variation in the composition from source to source, the major component of natural gas is methane with other hydrocarbons and unwanted impurities. Hence, all natural gas must undergo some form of treatment which might involve in some cases just the removal of water. However, about 20% of natural gas currently being produced worldwide requires more extensive treatment due to an increased amount of impurities before transportation via pipelines [1]. There are regulations in place to tightly regulate the composition of the natural gas transported to the pipelines. Membrane technology has only about 5% of the market for processing natural gas in the United States [2]. This percentage is expected to rise as better carbon dioxide selective membranes are developed [1,2]. High pressures in the range of 500-1500 psi are usually required to transport natural gas to a gas processing plant and for a membrane to be used to remove impurities and to minimize recompression cost; the membrane must selectively remove the impurities from the gas stream. This requirement determines the type of membrane that can be suitable [3].

Dehydration of natural gas

The current technology that is being used for the removal of water vapor from natural gas is glycol absorption [4]. Water is an easily condensable compound hence; there are many membranes with high water permeability as well as high water/methane selectivity. The use of glycol absorption is quite prominent and it has a low operational cost. For membrane technology to be competitive, it must cut down the rate of loss of methane with the permeate water. Offshore platforms glycol units are not suitable due to space; hence the use of membranes can be competitive [5].

Removal of nitrogen

The specification of inert gases in the natural gas pipelines is less than 4%. Gas reserves having higher contents are of low quality, although gas containing about 10% inert gases can be blended with low nitrogen content gas to achieve pipeline quality gas [5]. The economic importance of the content of nitrogen in natural gas is high. In the United States, the value of shut-in gas containing 10 to 15% nitrogen is about $30 billion [6], as a result there are numerous processes that have been evaluated for the removal of nitrogen. The current technology that is used now in large scale is the cryogenic plants. Membranes can be used to achieve these separations, the challenge being to develop membranes with high methane/nitrogen separation efficiency. The membrane system as compared to the cryogenic plant reduces the concentrations of water, hydrocarbons like propane and butane to a very low level as these components permeate preferentially to the membrane [5].

Carbon dioxide removal

A typical plant for the removal of carbon dioxide from natural gas uses absorption technology. This consists of two towers in which the first tower contains the feed gas at high pressure and an absorbent liquid flowing counter-current to the feed gas. The absorbent liquid that contains the absorbed carbon dioxide and heavy hydrocarbons is removed from the bottom of the tower [2]. Membrane technology is competitive against absorption for the removal of carbon dioxide from natural gas [7] as the high pressure absorber tower is an expensive,

*****Corresponding author:** Edward Gobina, Center for Process Integration and Membrane Technology, Robert Gordon University Aberdeen, United Kingdom
E-mail: e.gobina@rgu.ac.uk

large thick walled heavy vessel. The mass of the components absorbed is related to the size of the tower. Furthermore, these absorption units are quite difficult to maintain and corrosion is an important maintenance problem [2]. Membrane technology could offer a more competitive method for the removal of carbon dioxide from natural gas, although one of the disadvantages of using membranes is that current polymeric membrane could degrade and plastize due to the presence of components like water, carbon dioxide and C_{4+} hydrocarbons [1]. As inorganic membranes are more stable under harsh conditions than polymeric membranes, they offer a better choice of material although the cost implications of these inorganic membranes could be a limiting factor.

Membranes for hydrocarbons recovery

Membranes can be defined as selective barriers between two components through which selective transport can occur [8]. Gas separation membranes are used for numerous applications. Membranes used for gas separations can be generally classified into organic polymeric membranes and inorganic membranes. The organic/polymeric membranes that are used for gas separations are hollow asymmetric and nonporous. An important feature in the preparation of polymer membrane for gas separations is the process of spinning them into hollow fibre membranes which due to its large area is suitable for large scale industrial applications [9]. The major drawback for the use of these polymeric membranes is that they can't stand high temperatures and harsh chemical conditions. In petrochemical plants, natural gas treatment plants and refineries, feed gas streams of heavy hydrocarbons can be a problem as the polymer membranes can be plasticised or become swollen [10]. The development of inorganic membranes is riveting as they can withstand high temperatures and harsh chemical conditions. The major drawback for these membranes is their high cost, brittleness, low membrane area and low permeability in the case of highly selective dense membranes [10]. Inorganic membranes based on alumina, zeolites, carbon and silica have been used for the capture of CO_2 at elevated temperatures [11]. For the separation of hydrocarbons, zeolite membranes have shown interesting separation characteristics, although their separation efficiency depends on the operating conditions like temperature, composition and total pressure [12]. In a membrane separation unit, the temperature and pressure are usually kept constant; hence a study of the separation features of the membrane is needed to get the optimal separation conditions [13-15].

Gas transport through inorganic membranes

The separation of gases in membranes is possible due to the difference in the rate of movement of the different species through the membrane. For membranes having large pore sizes of 0.1 to 10 μm, the gases permeate via convective flow and there is not much separation of the gases observed because flow depends on the viscosity of the gases. For mesoporous membranes, separation is based on the collision between the gas molecule and the membrane pore wall and hence the mean free path of the gas molecules is greater than the pore size. The diffusion here is governed by Knudsen mechanism and the rate of transport of any gas is inversely proportional to the square root of its molecular weight. However, for a microporous membrane with pore size less than 2 nm, separation of gases is based mostly on molecular sieving. The transport mechanism in these membranes is often complex and involves surface diffusion that occurs when the permeating species exhibit a strong affinity for the membrane surface and thus adsorbed on the walls of the pores [8].

The permeation of gases through a membrane is dependent on both the diffusion and the concentration gradient of the species along the membrane. The selective transport of a gas molecule through a membrane is often associated with the pressure, temperature, electric potential and concentration gradient. The permeability and selectivity are some of the parameters that are used to determine a membrane's performance. The permeance P (molm^{-2}s^{-1}Pa^{-1}) represents the proportionality coefficient with the flux at steady state of a particular gas through a membrane:

$$P = \frac{Q}{A \times \Delta p} \tag{1}$$

Where Q is the molar gas flow rate through the membrane (mol s^{-1}), A is the membrane surface area (m^2) and Δp is the pressure difference across the membrane (Pa). The permeance is therefore a measure of the quantity of a component that permeates through the membrane.

The calculated gas selectivity is the ratio of the permeability coefficients of two different gases as they permeate independently through the membrane is given by:

$$\alpha_{ij} = \frac{P_i}{P_j} \tag{2}$$

where P_i and P_j is the permeance of the single gases through the membrane (mol m^{-2} s^{-1} Pa).

The selectivity is the measure of the ability of a membrane to separate two gases and it is used to determine the purity of the permeate gas as well as determine the quantity of product that is lost.

Experimental

Membrane modification

Two types of membranes (silica and zeolite) were prepared on α-alumina support and tested for the flux of methane, propane, nitrogen, helium, argon and carbon dioxide at varying feed pressures and in the case of the silica membrane the temperature was varied. The membranes were fabricated using the dip coating method as illustrated in Figure 1.

Silica membrane preparation: As depicted in Figure 1, the membrane was prepared by the dip-coating method. The support outside surface was exposed into a solution that comprises of silicone elastomer, curing agent and isopentane in the ratio 10:1:100 respectively and the volumes used are given in Table 1. The mixture was first homogenised with magnetic stirring for 2 hours before the support was dipped for 1 hour with constant stirring to prevent the

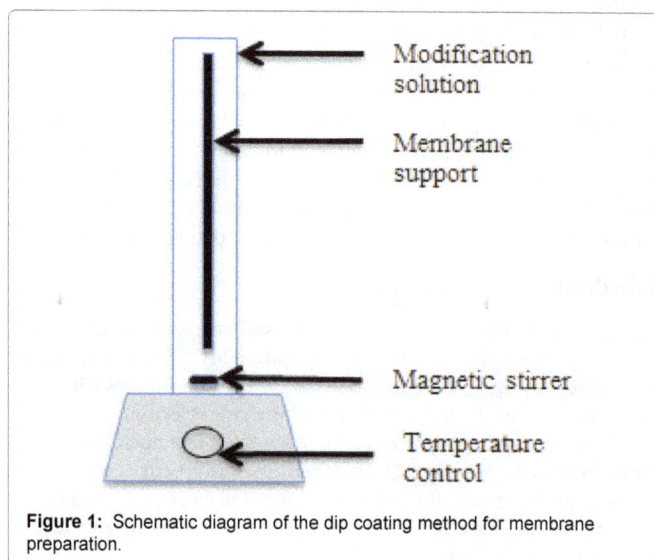

Figure 1: Schematic diagram of the dip coating method for membrane preparation.

mixture from coagulating. The membrane was air dried for 30 minutes and thermally treated at 333 K for 2 hours prior to permeation test.

Zeolite membrane preparation: The zeolite membrane was prepared also by the dip-coating method. Here, however a solution containing silicone oxide, aluminium oxide, sodium oxide and deionised water was prepared and homogenised at room temperature for 20 hours, the amount of each substance used is given in Table 2. Zeolite crystals were deposited on the outside surface of the alumina support which was then dipped into the solution and maintained for 20 hours at 343 K. The membrane was withdrawn and washed with deionised water and the pH of the rinse water was monitored. When the rinse water pH was neutral the membrane was air dried for 20 minutes and thermally treated in the oven at 338 K for 2 hours prior to permeation tests.

Characterization

The membranes were characterized using nitrogen physisorption measurements carried out by a nitrogen pysisorption Quantachrome gas analyzer.

Permeation set up

The permeance of carbon dioxide, oxygen, methane nitrogen and propane were each determined at various pressures ranging from 1×10^{-5} to 1×10^{-4} Pa using the gas permeation set up in Figure 2. This pressure range was selected as a scale down experimental value to determine the effect of pressure on gas permeance. The single gases used for this work were obtained from BOC, UK and have a purity assay of 99.9%. The permeate side was maintained at atmospheric pressure. The flux of the permeate gas was measured with a Cole-Palmer volumetric digital flow meter (L min^{-1}).

Results and Discussion

Characterisation using nitrogen physisorption

One of the most important techniques for the characterisation of nano- sized porous materials in terms of surface area, pore volume and pore size distribution is the physical adsorption of gas on the surface of the material. Different types of physisorption isotherms (Figure 3) are observed for different materials. Type I: microporous, type II: non-porous or macro-porous, type III: non-porous or macro-porous with week interaction, type IV: mesoporous, type V: mesoporous with weak interaction and type VI: layer-by-layer adsorption.

The specific surface area of the silica and zeolite membrane was determined from the adsorption of nitrogen on the external and internal surface of the membranes at 77.35 K using a quantachrome adsorption gas analyser. The operating conditions of the instrument in given in Table 3.

The adsorption and desorption isotherm of the zeolite membrane is presented in Figure 4 and it corresponds to type III isotherm from Figure 2. This indicates that the zeolite may be macro porous or non-porous adsorbent with weak adsorbent-adsorbate interaction (1). In theory, zeolites and silica are highly porous and have very large surface area.

The physisorption isotherm for the silica membrane is presented in Figure 5 and it shows the adsorption and desorption isotherms which corresponds to type IV or V which indicates the membrane is a mesoporous adsorbent, the pore sizes and specific surface area of the membrane is given in Table 4.

Substance	Amount (ml)
Curing agent	5
Isopentane	500
Silicone elastomer	50

Table 1: Composition of the modification solution for silica membrane.

The adsorption behaviour of mesoporous materials is determined by the adsorbent-adsorbate interactions. Hence the Kelvin equation (equation 3) which is based on cylindrical pores is used for the evaluation of the pore size distribution of the membrane layer by the

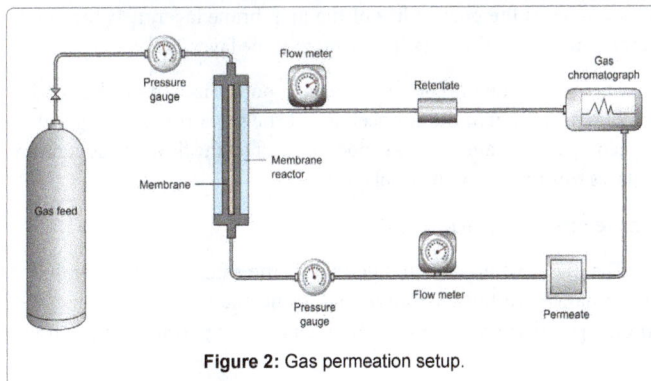

Figure 2: Gas permeation setup.

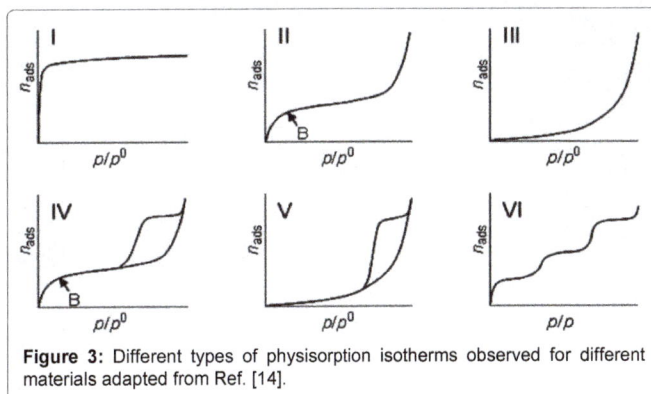

Figure 3: Different types of physisorption isotherms observed for different materials adapted from Ref. [14].

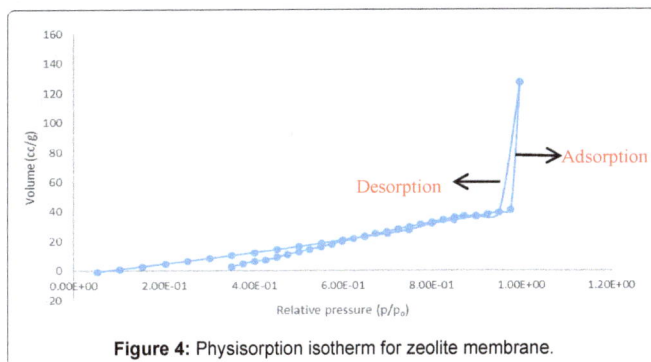

Figure 4: Physisorption isotherm for zeolite membrane.

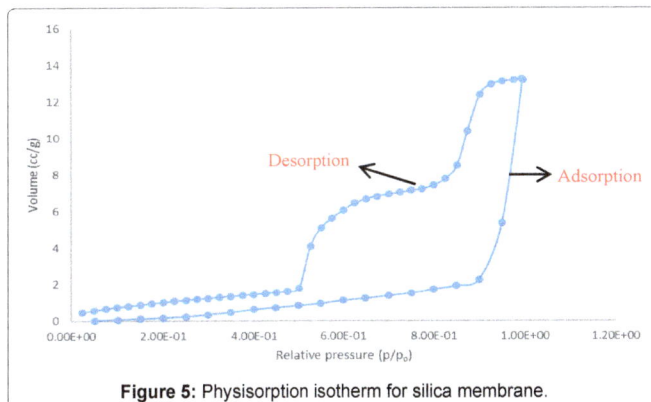

Figure 5: Physisorption isotherm for silica membrane.

Barrett-Joyner-Halenda (BJH) method. The BJH graph (Figure 6 (a) and (b)) shows the adsorption and desorption branches which are used to determine the pore sizes of the membrane.

$$r_p = r_k + t \qquad (3)$$

where r_p is the pore radius of the membrane layer, r_k is the kelvin radius and t is the thickness of the membrane layer.

The silica membrane has smaller pore size than the zeolite membrane, although the surface area of the silica is much larger than the zeolite membrane. This is supposed to affect the flow and separation of gases through these materials.

Single gas permeation tests

The main parameters that determine the efficiency of a membrane performance are the separation factor and the permeation flux [15]. At only gas phase conditions in the feed and the permeate sides, single

gases CH_4 and CO_2 where fed into the membrane reactor at a pressure range of 0.1 to 1×10^{-5} Pa and a temperature of 298 K.

Carbon dioxide has a much higher permeation flux through the silica membrane than the zeolite membrane (Figure 7 (a)). The maximum flux of CO_2 through the zeolite membrane at 1 bar was 9.9×10^{-2} mol s^{-1} m^{-2} while through the silica membrane it was 2.3×10^{-1} mol s^{-1} m^{-2}. This could be as the result of the diffusion of CO_2 through the silica layer by adsorptive surface flow. The pore size of the silica membrane (4.183×10^{-9} m) is smaller than the pore size of the zeolite membrane (11.394×10^{-9} m) as observed in Table 4. In the case of methane, the permeation flux is higher at 2.0×10^{-1} mol s^{-1} m^{-2} at 1 bar through the zeolite membrane (Figure 7 (b)) than 1.2×10^{-1} mol s^{-1} m^{-2} for the silica membrane at 1×10^{-5} Pa. The molecular sieving abilities as well as selective sorption properties of zeolites that make some specie to absorb on the surface of the membrane at a greater rate than another. The selectivity α of CH_4/CO_2 as determined from equation (2) is expressed in Figure 8.

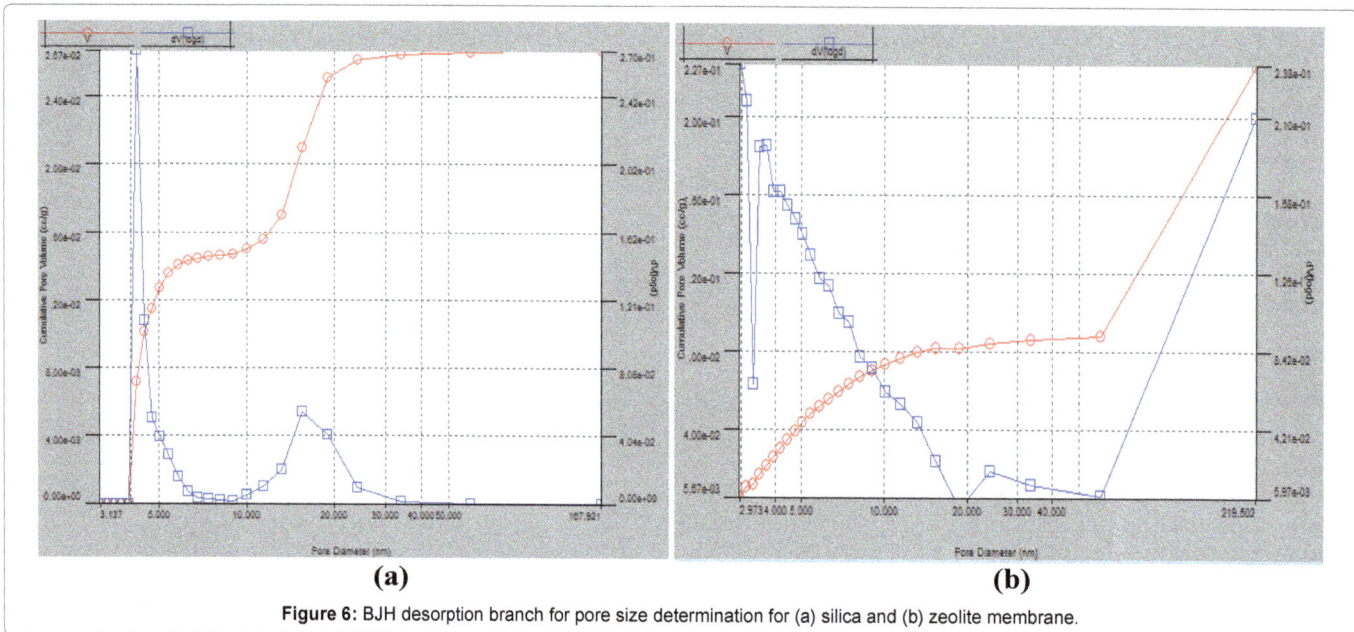

(a)   **(b)**

Figure 6: BJH desorption branch for pore size determination for (a) silica and (b) zeolite membrane.

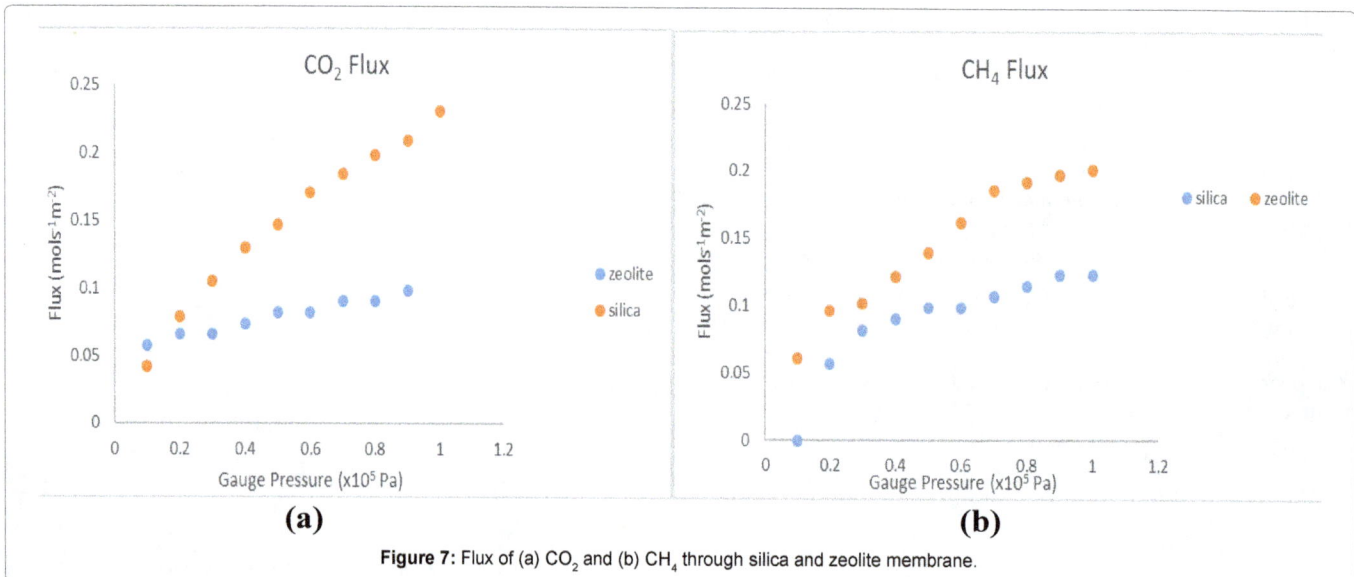

(a)   **(b)**

Figure 7: Flux of (a) CO_2 and (b) CH_4 through silica and zeolite membrane.

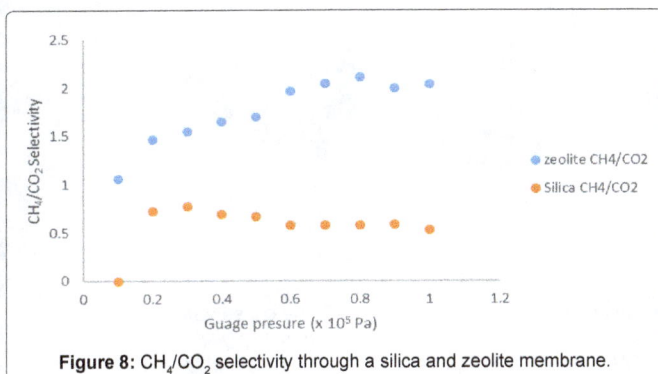

Figure 8: CH_4/CO_2 selectivity through a silica and zeolite membrane.

Chemical	Amount (ml)
Aluminium oxide	10
Sodium hydroxide	14
Deionised water	798
Silicone oxide	1

Table 2: Composition of the modification solution for zeolite membrane.

Parameter	Value
Area (A2 mol^{-1})	16.2
Non-Ideality (1/mm Hg)	6.58×10^{-5}
Sample cell type (mm)	12
Analysis time (mins)	237
Mol weight (g mol^{-1})	28.0134
Ambient temperature (K)	300
Bath temperature (K)	77

Table 3: Optimum operating conditions of the Quantachrome Gas Analyser.

	Silica membrane	Zeolite membrane
Pore size (× 10^{-9} m)	4.183	11.394
Specific surface area (m^2/g)	10.692	0.619

Table 4: Pore size and surface area of silica and zeolite membranes.

It is obvious from Figure 8 that there is a linear increase in the separation factor of CH_4/CO_2 with the zeolite membrane with the increase in the pressure drop, it can be assumed that the increase in permeating flux is also an influencing parameter for the selectivity. The silica membrane showed a decrease of selectivity with increase in the pressure drop. At 0.8 bar, zeolite membrane had the highest separation factor of 2.1 while silica membrane had a separation factor of 0.6. Hence it can be assumed that the pressure drop or permeating flux has little or no influence on the selectivity of CH_4/CO_2.

Conclusion

With this work a general impact of pressure drop and pore size on the separation performance of mesoporous membranes for the separation of carbon dioxide and methane was demonstrated. Two different types of membranes were studied and characterized (silica and zeolite). It was found that despite the lower pore of the silica membrane and the higher flux of carbon dioxide through the membrane, the zeolite membrane had a much higher selectivity, which increased linearly with pressure drop across the membrane. This has shown that zeolite membrane could be used for the removal of carbon dioxide from natural gas. Further studies are planned to demonstrate to how the membrane could be used to separate the heavier components of natural gas mixtures that arise during dew point adjustments, thermal

problems during transportation as well as when expanding highly compressed natural gas components.

Acknowledgements

The authors thank the Center for Excellence in Process Integration and Membrane Technology and the School of Engineering at the Robert Gordon University Aberdeen for the financial support.

References

1. Baker RW, Lokhandwala K (2008) Natural gas processing with membranes: an overview. Industrial & Engineering Chemistry Research 47: 2109-2121.

2. White LS (2010) Evolution of natural gas treatment with membrane systems. Membrane Gas Separation, p: 313.

3. Baker RW (2012) Membrane Technology and Applications. John Wiley and Sons Ltd., United Kingdom.

4. Graham J, Krenek M, Maxon D, Peirson J, Thompson J (1994) Natural Gas Dehydration: Status and Trends. Gas Research Institute Report GRI-94-099.

5. Baker RW (2002) Future directions of membrane gas separation technology. Industrial & Engineering Chemistry Research 41: 1393-1411.

6. Tannehill CC, Raven M, Brown K (1999) Nitrogen Removal Requirements from Natural Gas: Topical Report. Gas Research Institute.

7. Kohl A, Nielsen R (1997) Gas purification. Gulf Professional Publishing, Houston, TX, USA.

8. Abedini R, Nezhadmoghadam A (2010) Application of Membrane in Gas Separation Processes: Its Suitability and Mechanisms. Petroleum & Coal 52: 69-80.

9. Strathmann H (2001) Membrane separation processes: current relevance and future opportunities. AIChE Journal 47: 1077-1087.

10. Bernardo P, Drioli E, Golemme G (2009) Membrane gas separation: a review/state of the art. Industrial & Engineering Chemistry Research 48: 4638-4663.

11. Lin YS (2001) Microporous and dense inorganic membranes: current status and prospective. Separation and Purification Technology 25: 39-55.

12. van de G, Jolinde M, van der Bijl E, Stol A, Kapteijn F, et al. (1998) Effect of operating conditions and membrane quality on the separation performance of composite silicalite-1 membranes. Industrial & Engineering Chemistry Research 37: 4071-4083.

13. van de G, Jolinde M, Kapteijn F, Moulijn JA (1999) Modeling permeation of binary mixtures through zeolite membranes. AIChE Journal 45: 497-511.

14. Weidenthaler C (2011) Pitfalls in the characterization of nanoporous and nanosized materials. Nanoscale 3: 792-810.

15. Neubauer K, Dragomirova R, Stöhr M, Mothes R, Lubenau U, et al. (2014) Combination of membrane separation and gas condensation for advanced natural gas conditioning. Journal of Membrane Science 453: 100-107.

27

Spectroscopic Study of Poly(Vinylidene Fluoride)/Poly(Methyl Methacrylate) (PVDF/PMMA) Blend

Benabid FZ[1]*, Zouai F[2], Douibi A[1] and Benachour D[1]

[1]LMPMP, Faculty of Technology, Université Sétif-1, Algeria
[2]Unité de Recherche Matériaux Emergents, Université Sétif-1, Algeria

Abstract

Poly(vinylidene fluoride)/poly(methyl methacrylate) blends casted in the DMF could be used in the conservation of historic structures (monuments) exposed to atmospheric agents or as a coating to replace and maintain parts or missing pieces. This study deals with the effect of blending of PVDF to PMMA to enhance the properties and their properties were studied using the FTIR and UV-visible spectroscopy. In FTIR spectra, it was found that PVDF/PMMA blend casted in the Dimethylformamide (DMF) showed the superposition of the spectra of all compositions, with the exclusion of any chemical reaction between two polymers or the presence of the double bonds characteristic of PVDF dehydrofluoration. The UV-visible spectroscopy before and after exposure to artificial weathering, showed that the PVDF is very stable (the invariant absorbance values at 200 nm wavelength after the equivalent of two years of aging). In contrast, the absorbance of PMMA has changed at the same wavelength explaining its tendency of degradation.

Keywords: Blend; PVDF; PMMA; Monuments

Introduction

The conservation and protection of historic monuments or culturally significant structures have recently attracted much attention from material scientists [1]. A few years ago, various synthetic polymers have been widely used in the treatment of construction materials of historical monuments for consolidation and conservation of such structures [2]. Using polymeric coatings for this area has created serious challenges for the surface science and technology. Some of the challenges are as follows [3,4].

Van Hees and Brocken [5] evaluated the salt growth in brick masonry specimens, coated with a water repellent, during a salt crystallization test. They demonstrated that the behavior of different salts on development of salt damages is completely different. However, it is demonstrated that the adsorption of dusts suspensions and water-soluble air pollutants decreases with increasing hydrophobicity of the surface of building materials [1]. The fluorine substitution of the hydrogen atoms present in a macromolecular chain improves the heat resistance and chemical resistance, delays or inhibits flame propagation, lowering the critical surface tension and exalts the dielectric characteristics [6].

Acrylic resins undergo deterioration face conditions under UV radiation and their climatic exposure causes degradation of their structure. However, their physical characteristics and low price always consider important research topics. In this research work, the focus was on the development of films of PVDF and PMMA blends and their spectroscopic analysis (FTIR and UV-visible spectroscopy).

Materials and Methods

PVDF (Hylar 5000), special coating as a white powder, manufactured by Ausimont, Italy [7] and PMMA (Vedril Spa-Resina Metacrilica) manufactured by Mont Edison, Italy were used [8]. The PVDF/PMMA films were obtained by casting each polymer separately in with 1% concentration of the polymer in the solvent (DMF) at temperature, 70°C (Table 1).

Moisture, temperature and ultraviolet radiation contribute to material degradation. Accelerated weathering test (Xeno Test) is the simulation of these conditions using special environmental chambers and instruments in order to speed up the weathering process and measure its effects on parts, components, products, and materials. On the other hand, the Salt spray testing is a test method for evaluating a product or a coating resistance to corrosion in the face of extended exposure to a saline, or salted, spray.

FTIR spectroscopy

Fourier-transform infrared spectroscopy (FTIR) spectra were recorded by means of a Perkin Elmer Spectrum 1000 spectrometer with a wavenumber resolution of 4 cm^{-1} in the range from 450 to 4400 cm^{-1}, using attenuated total reflection.

Each spectrum results from an average of 200 scans to detect any changes in the chemical structure of various compositions of PVDF/PMMA blends before and after exposure to artificial weathering (Xeno Test and Salt Spray) according to ISO 11507:1997(F) and ISO 7253:1996(F) respectively.

UV-visible spectroscopy

The UV-visible spectroscopy was performed using a Unicam UV 300 spectrophotometer to determine the evolution of the absorbance of various compositions of PVDF/PMMA blends before and after exposure to artificial weathering (XENO Test and Salt Spray).

PVDF (%)	100	90	80	70	60	50	40	30	20	10	0
PMMA (%)	0	10	20	30	40	50	60	70	80	90	100

Table 1: Compositions in volume percentage of different compositions proposed in this study.

*Corresponding author: Benabid FZ, LMPMP, Faculty of Technology, Université Sétif-1, Algeria, E-mail: fzbenabid@yahoo.fr

Results and Discussion

FTIR spectroscopy

FTIR spectra of PVDF/PMMA blends showed the spectral superposition of all the compositions where the exclusion of any chemical reactions between the two polymers.

The results of the FTIR of the PVDF/PMMA blend solutions in the DMF (Figures 1, 2 and 3) showed the appearance of a band at 3470 cm^{-1} for the PVDF and at 3538 cm^{-1} for the PMMA corresponding to the hydrogen bonds of the hydroxyl group initiated by the solvent. The DMF is also the source of the occurrence of the stretching vibration of C=O amide 1673 m^{-1} and for the PVDF to 1675 cm^{-1} for the PMMA.

In the films of PVDF/PMMA blend before and after exposure to artificial weathering and to the salt spray it is noted that the bands at 510, 839, 880 and 1406 cm^{-1} are those of the β crystallinity phase of the PVDF [9,10]. The α-phase is identified by the presence of peaks at 763 and 948 cm^{-1} [11,12].

It should be noted that according to the literature [12,13], the increase in the temperature reduces the presence of the crystallization of the β phase and it is clear that for a high proportion of this phase must perform complete drying films at 75°C. The strong band at 1233 cm^{-1} is due to CF$_2$ and while the CH bonds showed peaks at 2980 and 3022 cm^{-1}. The deformation of CH$_2$ bonds is located at 1429 cm^{-1}.

Concerning the PMMA, a strong peak at 750 cm^{-1} present identifying the ρ (C-H$_2$) rocking; The C-O-C bond (ether ester) is indicated by the presence of peaks at 1151 and 1256 cm^{-1}. The peaks at 1438 and 1458 cm^{-1} are characteristics of C-H bonds(bending) and the one at 1720 cm^{-1} is due to presence of carbonyl group (C=O). Peaks characterizing the vibration of symmetric and asymmetric -CH stretching are at 2859 and 2930 cm^{-1} respectively.

UV-visible spectroscopy

In the Table 2 and Figure 4 showing the absorbance variations a function of the different compositions of the PVDF/ PMMA blends in solution in DMF before and after exposure to artificial weathering and salt spray.

The Table 2 concerning films, before and after exposure to artificial weathering, it has been observed that for the PVDF, the absorbance value is the same compared to the reference film (before aging) (0.914) and that after exposure to artificial aging (0.927) which explains the stability of PVDF even after the equivalent of two years of aging.

However for the PMMA, the absorbance values changed compared to the reference film (before aging) (1.690), that after exposure to artificial aging (0.565) which explains the tendency of PMMA degradation.

The Table 3 concerning films, before and after exposure to salt spray, it has been observed that for the PVDF, the absorbance value is the same compared to the reference film (before exposure) (0.914) and that after exposure to salt spray (0.905) which explains the stability of PVDF.

However for the PMMA, the absorbance values changed compared to the reference film (before exposure) (1.690), that after exposure (0.661) which explains the tendency of PMMA degradation. The Figure 4 which represents the absorbance variation as a function of the different compositions of the PVDF/PMMA blends dissolved in the DMF showed that there is a synergism with an optimum at 30/70.

Conclusion

From the different results obtained, it can be concluded that the infrared spectroscopy (FTIR) of PVDF/PMMA blends casted in the

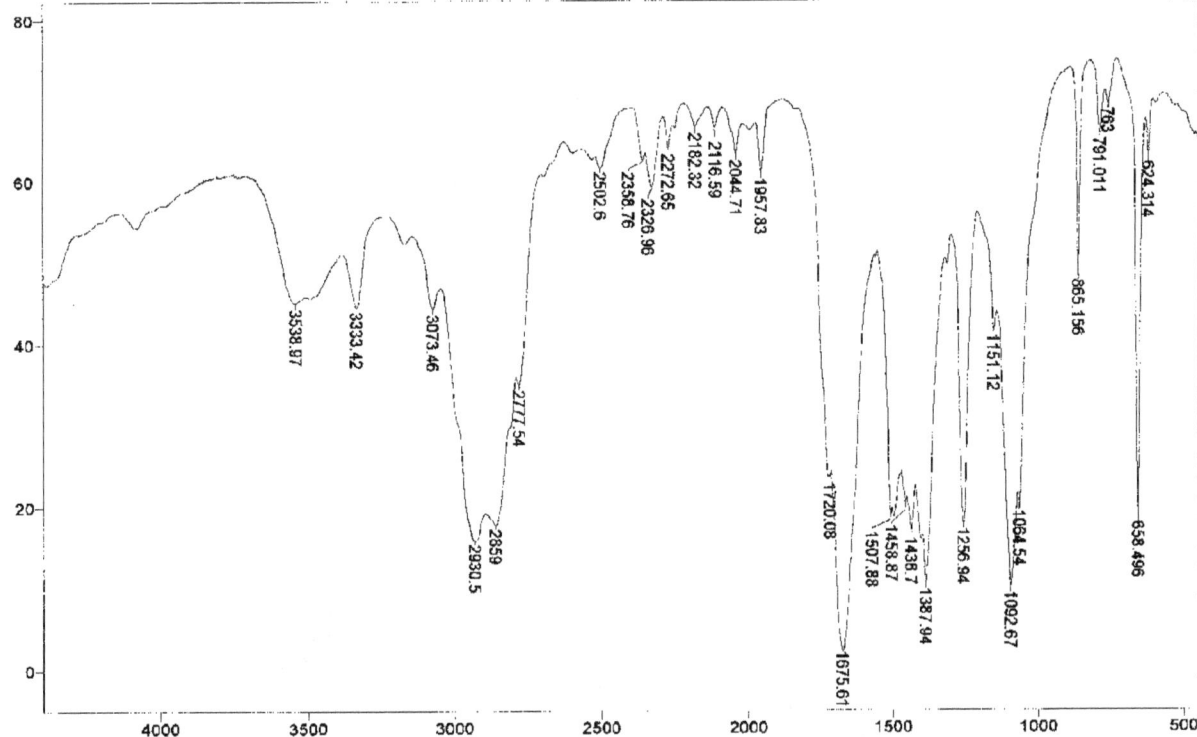

Figure 1: Evolution of IR transmittance (%) of the different compositions of the PMMA casted in the DMF as a function of the wave number (cm^{-1}).

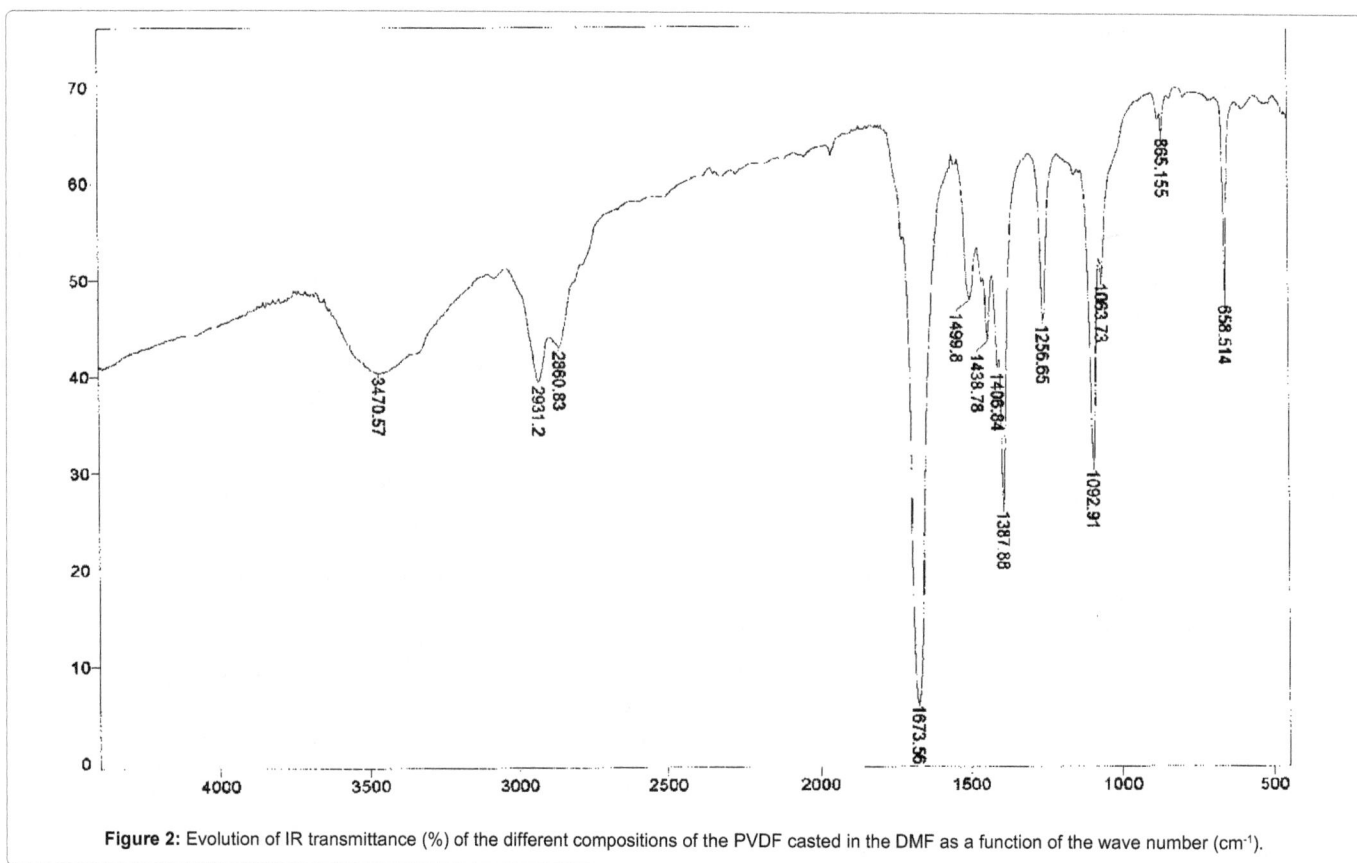

Figure 2: Evolution of IR transmittance (%) of the different compositions of the PVDF casted in the DMF as a function of the wave number (cm^{-1}).

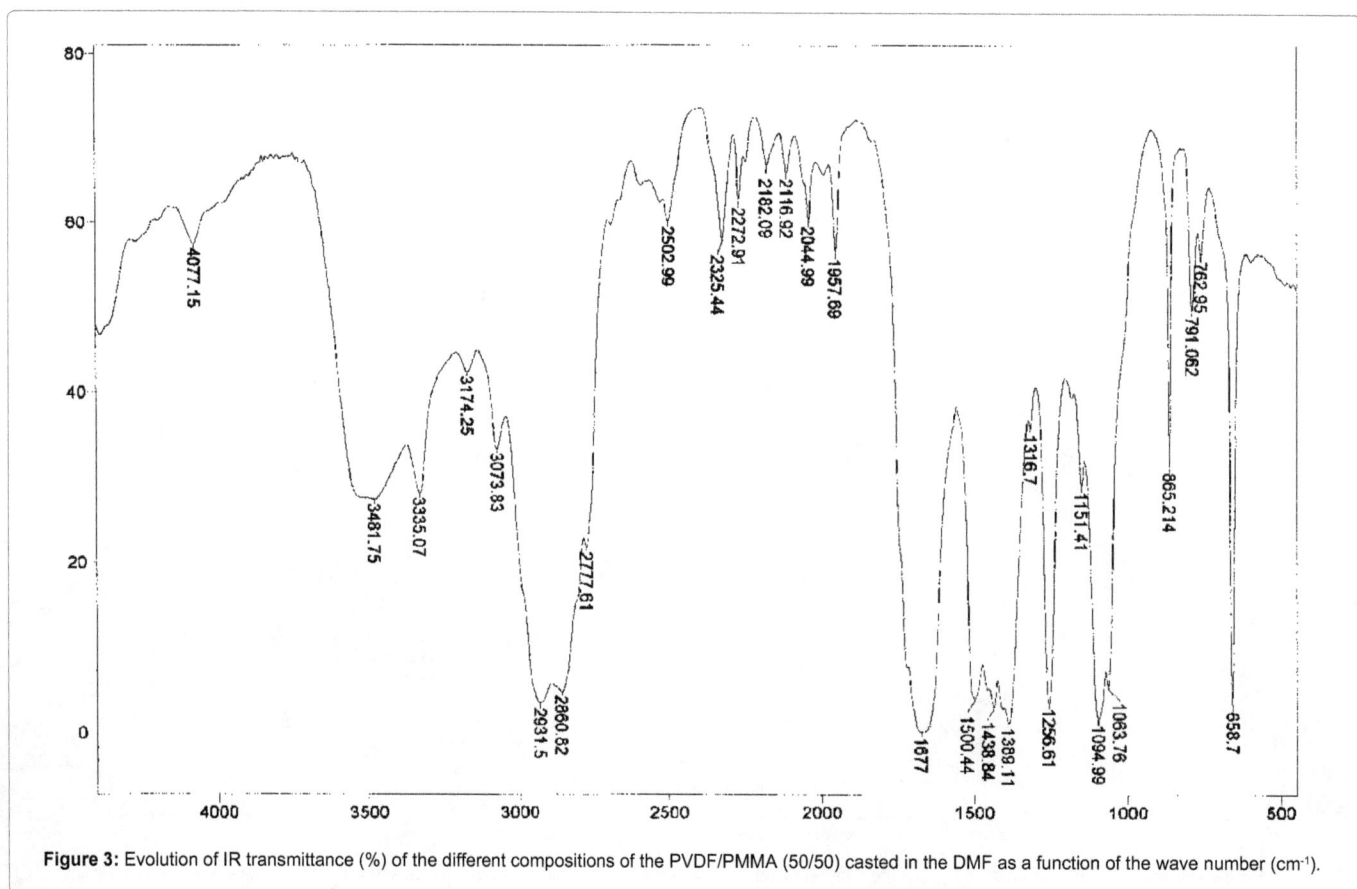

Figure 3: Evolution of IR transmittance (%) of the different compositions of the PVDF/PMMA (50/50) casted in the DMF as a function of the wave number (cm^{-1}).

Absorbance	PVDF	PMMA
Before exposure	0.914	1.690
After exposure	0.927	0.565

Table 2: Absorbance variation versus different compositions of PVDF/PMMA blends at before and after exposure to artificial weathering at wavelength of 200 nm.

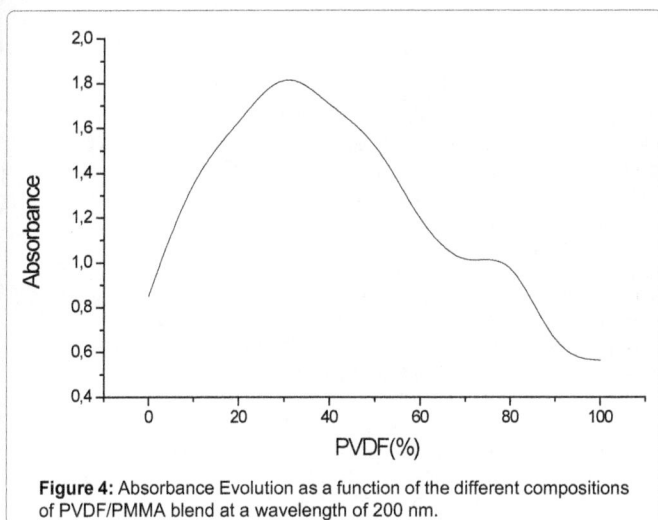

Figure 4: Absorbance Evolution as a function of the different compositions of PVDF/PMMA blend at a wavelength of 200 nm.

Absorbance	PVDF	PMMA
Before exposure	0.914	1.690
After exposure	0.905	0.661

Table 3: Absorbance variation versus different compositions of PVDF/PMMA blends at before and after exposure to salt spray a wavelength of 200 nm.

DMF showed the superposition of the spectra of all compositions where the exclusion of any chemical reactions between two polymers or the presence of the double bonds characteristic of deshydrofluoration. On the other hand, the PVDF/PMMA studied blend showed the presence of two crystallinity phases α and β. The UV visible spectroscopy showed that the PVDF is very stable (the invariant absorbance values at 200 nm wavelength) after the equivalent of two years of aging. In contrast,

the absorbance of PMMA has changed at the same wavelength (200 nm), which explains its tendency to degradation. The UV-visible spectroscopy showed that for the various compositions of the PVDF/PMMA blends dissolved in the DMF, the absorbance increased and therefore the solubility of the blend.

References

1. Sadat-Shojai M, Ershad-Langroudi A (2009) Polymeric Coatings for Protection of Historic Monuments: Opportunities and Challenges. J Appl Polym Sci 112: 2535-2551.

2. Favaro M, Mendichi R, Ossola F, Russo U, Simon S, et al. (2006) Evaluation of polymers for conservation treatments of outdoor exposed stone monuments. Part I: Photo-oxidative weathering. Polym Degrad Stab 91: 3083-3096.

3. Michoinova D - New Materials for the Protection of Cultural Heritage.

4. Han JT, Zheng Y, Cho JH, Xu X, Cho K (2005) Stable superhydrophobic organic-inorganic hybrid films by electrostatic self-assembly. J Phys Chem B 109: 20773-20778.

5. Van Hees RPJ, Brocken HJP (2004) Damage development to treated brick masonry in a long-term salt crystallisation test. Constr Build Mater 18: 331-338.

6. Teng H (2012) Overview of the Development of the Fluoropolymer Industry. Appl Sci 2: 496-512.

7. http://www.Solvaysolexis.com/pdf/Hylar5000_lg.pdf

8. http://www.Plasticstechnology.com/articles/200210bib1.html

9. Davis BL, Perry JE, Neth DC, Waters KC (1998) A Device for Simultaneous Measurement of Pressure and Shear Force Distribution on the Plantar Surface of the Foot. J Applied Biomechanics 14: 93-104.

10. Gregorio R, de S Nociti NCP (1995) Effect of PMMA addition on the solution crystallization of the α and ß phase of poly(vinylidene fluoride) (PVDF). J Phys D: Appl Phys 28: 432-436.

11. Lu FJ, Hsu SL (1986) Study of the crystallization behavior of poly(vinylidene fluoride) from melt under the effect of an electric field. Macromolecules 19: 326-329.

12. Suess M, Kressler J, Kammer HW (1987) The miscibility window of poly(methylmethacrylate)/poly(styrene-co-acrylonitrile) blends. Polymer 28: 957-960.

13. Gregorio R, Cestari M (1994) Effect of crystallization temperature on the crystalline phase content and morphology of poly(vinylidene fluoride). Polymer Physics 32: 859-870.

28

Using Slow Sand Filtration System with Activated Charcoal Layer to Treat Salon Waste Water in a Selected Community in Cape Coast, Ghana

Isaac Mbir Bryant* and Roberta Tetteh-Narh

Department of Environmental Sciences, University of Cape Coast, Ghana

Abstract

Discharge of untreated salon waste water into the surrounding environment in Ghana remains so probably because of poor knowledge of Ghanaians about treated waste water and its reuse as well as ignorance of waste water to ground water pollution. In Ghana, there is little or no knowledge on waste water treatment technology for salon waste water. In addition, a greater proportion of Ghanaians have no knowledge regarding reusability potentials of treated waste water. Thus, this study assessed the efficiency of a simple slow sand filtration system integrated with activated charcoal layer for salon waste water treatment. The study also assessed the perception of some selected Ghanaians in Cape Coast on reuse of treated salon waste water. For sixteen weeks, salon waste water collected from five different beauty salons in Amamoma was homogenized and treated. Selected parameters of both influent and effluent were analyzed. The percentage removals of some selected heavy metals present in the treated waste water (Effluent) show Copper $32.836 \pm 7.013\%$, Cadmium $59.259 \pm 8.006\%$, Zinc $83.333 \pm 6.881\%$, Iron $38.095 \pm 2.002\%$, Lead $100.000 \pm 12.939\%$ and Arsenic $100.000 \pm 11.573\%$. pH $9.877 \pm 1.107\%$, Conductivity $6.250 \pm 0.819\%$, Total Dissolved Solids $5.810 \pm 0.629\%$, Biological Oxygen Demand $21.780 \pm 1.578\%$, Turbidity $93.798 \pm 6.073\%$, Nitrates-nitrogen $67.727 \pm 5.759\%$, Phosphate-phosphorus $67.614 \pm 3.264\%$, Ammonia-Nitrogen $79.249 \pm 8.311\%$, Total Suspended Solids $94.043 \pm 0.948\%$ and Chemical Oxygen Demand $84.487 \pm 2.823\%$. All effluent parameters conformed to EPA Ghana standards for effluent discharge except turbidity, $N\text{-}NO_3$, conductivity, TSS, COD and $N\text{-}NH_3$. The results proved that treatment of salon waste water using an integration of slow sand filtration system and activated charcoal layer could be adopted as domestic waste water treatment technology especially in developing countries like Ghana since the percentage removal for four of the treated heavy metals (Cadmium, Zinc, Iron and Arsenic) were around 60% and above except for Iron and Copper which were below 40%.

Keywords: Slow sand filtration system; Activated charcoal layer; Salon waste water

Introduction

With the rising population in recent times, many countries worldwide battle with the issue of waste management, especially the efficient treatment of waste water as well as its disposal. This has given rise to various forms of pollution. Waste water can be described as water that is not whole to be used as a result of influence by humans [1]. This includes liquid waste discharges from domestic residences, commercial properties, small-scale industries and institutions. This study centers on salon waste water. To satisfy the demands of the increasing population of people who patronize salon services, proliferation of salons is on the ascendancy and the result is that too much waste is generated in providing more of the beauty care services. Salons offer a wide range of services such as hair styling and treatment, make-up application and the like. The waste water contains lots of organic and inorganic compounds. These components usually end up in the sanitary sewer system, where it can affect environmental health adversely [1].

In view of the issues associated with waste water and the increasing demand in water and environmental needs, efforts have been made to treat waste water for the sake of environmental health and conservation. Water treatment plays a greater role in the water supply chain. It helps in the sustainable management of vital water resources alongside water conservation and efficiency. This project therefore aims to determine the efficiency of slow sand filtration system with activated charcoal layer in pollutant removal. The slow sand filtration system is composed of shallow layers of stones and gravels beneath a deep layer of sand, and schmutzdecke on the topmost layer of the system where most of the removal and biologic activity occurs [2]. Apart from desalination and reverse osmosis, the slow sand filtration is said to the most effective single treatment for drinking water purification. It is used for water supply to large cities, small villages and even for use in individual

households [3]. Activated charcoal refers to a form of processed carbon with very high porosity and a very large surface area for adsorption. It can effectively reduce certain organic and inorganic compounds such as micro pollutants, lead, chlorine, fluorine, dissolved radon, dissolved oxygen, color, harmless taste and odor causing compounds, which may not be removed in slow sand filtration [4-6]. In Ghana, there is little or no knowledge on waste water treatment technology for salon wastewater. In addition, a greater proportion of Ghanaians have no knowledge regarding reusability potentials of treated wastewater. Thus, this study assessed the efficiency of a simple slow sand filtration system integrated with activated charcoal layer for salon waste water treatment. The study also assessed the perception of some selected Ghanaians on reuse of treated salon wastewater.

Materials and Methods

Study area

Amamoma is a community in the Cape Coast Metropolis located in the Central Region of Ghana (Figure 1). Cape Coast is located South to the Gulf of Guinea (5° 6' 0" North, 1° 15' 0" West).

The major rainy season occurs between May to July and minor rainy

---

*Corresponding author: Isaac Mbir Bryant, Department of Environmental Sciences, University of Cape Coast, Ghana, E-mail: ibryant@ucc.edu.gh

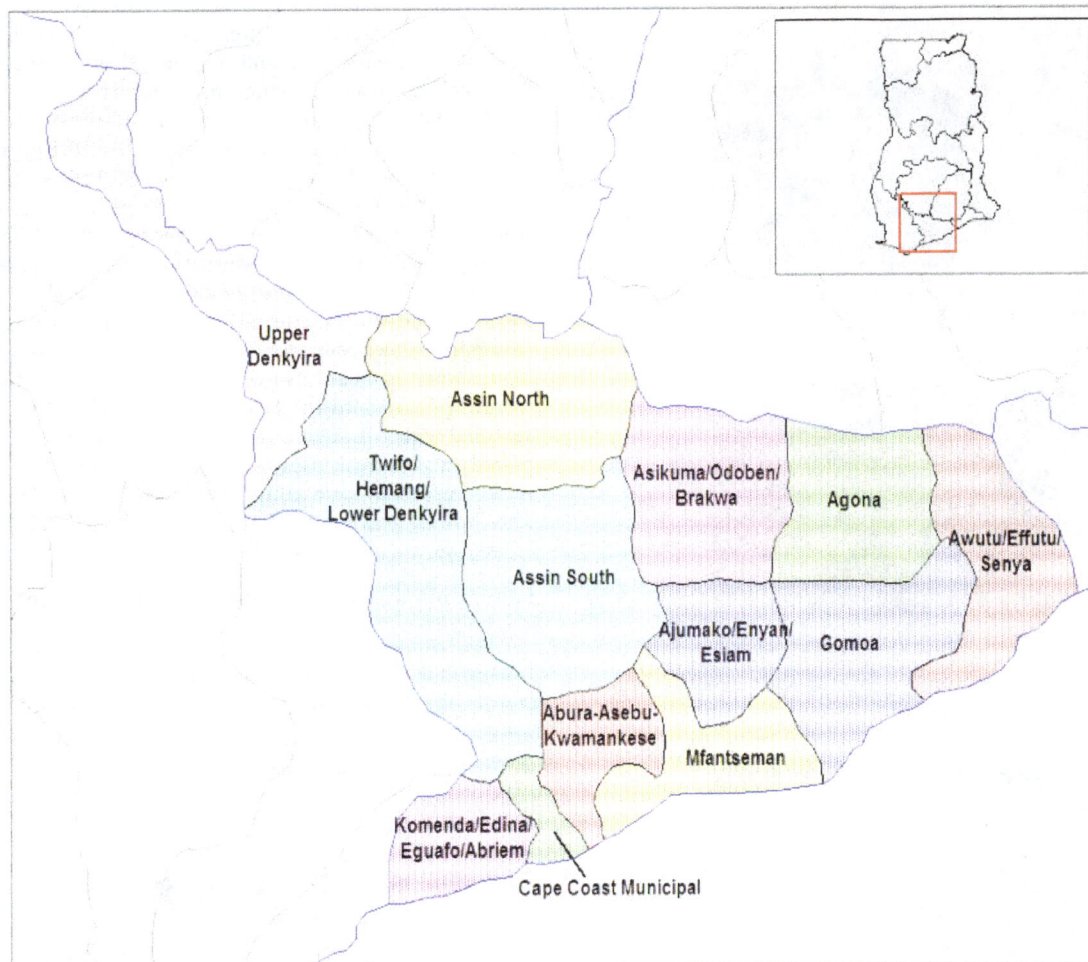

Figure 1: Geographical location of the Central Region showing Cape Coast Municipal.

season falls within November to January [7]. The majority of students of the university are also found there and this has made salon services a booming business in Amamoma. Ground water is the primary water source due to irregular supply of tap water, especially during the dry season. During this season, supply of potable water through pipes reduces thus attention is shifted to underground water.

Sample collection and preparation

A total of about 800 L of waste water was collected from five selected salons in Amamoma, via random sampling. The collection was carried out for sixteen weeks, between the period of January and April, 2015. About 10 L of waste water was collected from each salon once a week during this period. Prior to treatment, the waste water from all five salons was homogenized. A disinfected 750 ml water bottle was filled with the homogenized waste water and kept in a refrigerator at a temperature of 0-4°C, before transportation to the laboratory for analysis. The rest of the waste water was introduced into the slow sand filtration system.

Design of the slow sand filtration system

The slow sand filtration system was designed, using a 100 L plastic container. A white thread socket covered with a mesh was fitted through a hole created at the base of the container. The mesh was used to prevent particles (e.g., fine sand) from clogging the effluent outlet.

Connected to the socket was a PVC tube which served as effluent outlet. The lid of the container was perforated to serve as diffusion plate. This is to enable even distribution of the influent into the filter bed and to trap large solid materials in the influent such as hair strands. The filter bed was made of a bottom layer of stones (average of 6 cm in diameter) to a depth of 8 cm, followed by 11 cm depth of gravels (average of 3 cm in diameter). The strata were covered with a mesh to prevent mixing of the gravels and the coarse sand (which occupied the next stratum). The coarse sand with 2 mm diameter and depth of 8.5 cm occupied the next layer. This layer was followed by fine sand of about 0.125 mm diameter to a depth of 31 cm. The high depth of fine sand was so to increase retention time in order to ensure efficient treatment. The final layer of the filter bed was activated charcoal (from Zeal Technology-Takoradi, Ghana) of a depth of about 5 cm. The activated charcoal has the following properties: 0.5 mm square surface area, moisture content=5.0% w/w, ash content=6% w/w, adjustable pH value (7), bulk density=0.5 g/cm^3, hardness number=90, density=21 g/cm^3, appearance=black and insoluble matter=3% w/w. In effect, the whole filter bed occupied 63.5 cm of the plastic container. The perforated lid was used to cover the set-up during treatment to prevent foreign particles from dropping into the treatment system.

The detailed layers of the slow sand filtration bed used in this study are shown in the Figures 2-5.

Figure 2a: Picture of stones (Average 6 cm in diameter) used in slow sand filtration system design (8 cm depth).

Figure 2b: Picture of gravels (Average 3 cm in diameter) used in slow sand filtration system design (11 cm depth).

Figure 3a: Picture of coarse sand (2 mm in diameter) used in slow sand filtration system design (8.5 cm depth).

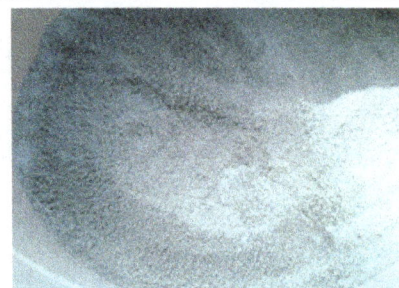

Figure 3b: Picture of fine sand (0.125 mm in diameter) used in slow sand filtration system design (31cm depth).

The flow rate for the effluent was calculated by recording the volume of effluent collected and the difference in the time the effluent started flowing out of the system and the time it stopped flowing in drops. After filtration, a disinfected 750 ml water bottle was also filled with the treated water for analysis. On a weekly basis, two well labeled samples were kept in a cooler containing ice blocks and transported immediately to the Ecology Laboratory of the University of Ghana in Accra.

Heavy metals analysis

The heavy metals that were determined included Cadmium, Lead, Copper, Zinc, Iron and Arsenic. These were analyzed using the Graphite Furnace Atomic Absorption Spectrophotometer 900T (GFAAS 900T) by Perkin Elmer with model pinAAcle 900T. The GFAAS 900T measured the heavy metals based on the following wavelengths and slit respectively: Cadmium (326.1 nm, 0.7), Lead (405.8 nm, 0.2), Copper (324.7 nm, 0.7), Zinc (213.8 nm, 0.2), Iron (372.0 nm, 0.2) and Arsenic (235 nm, 0.2). The GFAAS is a technique which employs the use of a graphite- coated furnace to vaporize and it is capable of producing very high temperatures as 3000°C. It measures samples in parts per billion hence detects even very minimal concentrations of heavy metals in samples. The set-up was warmed up and calibrated three times after which an aliquot was introduced into the heated graphite tube. The samples were acidified with Nitric acid (a drop) and deposited in the graphite-coated tube to heat, vaporize, ash and atomize the samples. The results measured for a sample is the average of triplicate analysis. This process was repeated for both samples of the influent and effluent for each heavy metal.

Laboratory analysis of physico-chemical parameters

The methods outlined in the Standard Methods for the Examination of Water and Waste water [8] and HACH Company Ltd. (1996) DR/2010 Spectrophotometer Procedure Manual was followed for the analyses of all the physico-chemical parameters. The process was repeated for both samples of the influent and effluent for each physicochemical parameter.

Figure 4a: Picture of the overall set-up of the slow-sand filtration system used for the treatment of salon waste water in Cape Coast.

Figure 4b: Picture of Activated Charcoal layer used for the treatment of salon waste water in Cape Coast.

Results

Table 1 shows mean values obtained from heavy metals analysis of the homogenized salon water before and after treatment over sixteen weeks. Lead and Arsenic were Below Detectable Limits (BDL). Iron recorded the highest value both in the treated (0.039 mg/L) and untreated water (0.06). Highest percentage removal of heavy metals was recorded in Cadmium (80.135 ± 32.024%) while copper recorded the least (31.876 ± 108.052%). All the parameters were within the acceptable limits of EPA Ghana.

Table 2 shows mean results obtained from analysis of physicochemical parameters in the homogenized salon waste water before and after treatment over sixteen weeks. pH, nitrates, BOD and COD were within the limits of EPA Ghana while turbidity, $N-NO_3$, conductivity, TSS, COD and $N-NH_3$ exceeded the limits.

Figures 6 to 11 show the relationship between percentage removal and flow rate, as well as percentage removal and HRT for the heavy metals. The graphs reveal an inverse relationship between flow rate and HRT and percentage removal.

Perception of some residents of Cape Coast on reuse of treated salon wastewater

A total of 50 individuals were randomly selected and issued with the questionnaire to find out their perception on the reuse of treated salon waste water. The results were compiled into the Figures 12-23.

Discussion

Turbidity, electrical conductivity and pH

Turbidity values exceeded EPA Ghana standard limit of 75 mg/L. This could be because of the high levels of the suspended and dissolved particles in the homogenized salon waste water as a result of high variability in the salon waste water characteristics from each salon such as source of water used in washing of hair, products used in the various salons (detergents, shampoos, soaps, dyes, etc.). Raw water turbidity recorded in this study was high compared to the study by ref. [9] which recorded a raw water turbidity value of 950 NTU. This is probably because the sampling was from a particular salon, unlike the homogenized samples from five salons for this study. Conductivity however, was beyond EPA's acceptable limit of 1500 μS/m (Table 2). The high level of conductivity in the influent could be attributed to the high concentration of dissolved ions present the wastewater. On the other hand, the conformity of the mean pH value to Ghana Environmental Protection Agency's acceptable limits of 6.5-9 could be attributed to the large surface area of both the activated charcoal and the fine sand. In treating salon waste water using activated charcoal only, Egbon et al. [9] recorded pH of 7.10 ± 0.10. The results of this study (pH of 7.0), however, is contrary to the findings of Egbon et al. [9]. This may be due to the integration of slow sand filtration with activated charcoal in the study. The alkalinity may also be traced to input from the chemical composition of the hair products used at the salons.

Phosphate, nitrates-nitrogen and ammonia-nitrogen

Phosphate concentrations of 0.1-5.6 μl/l over a long period of time are enough to cause eutrophication in a water body, according to Browman et al. [10]. In this study however, the levels of phosphate were 2.4 mg/L and 0.8 mg/L for untreated and treated waste water respectively (Table 2). Phosphate concentration conformed to the guidelines of the Environmental Protection Agency's permissible limit of 2 mg/L (Table 2). The level is not enough to cause much eutrophication to cause adverse effects in water bodies if discharged

into any, except over a long period of time, as asserted by Browman et al. [10]. The Nitrate concentration from this study conformed to the Ghana EPA acceptable limit of 50 mg/L (Table 2). This may be due to the fact that less organic matter was broken down to Oxides and Nitrate [11] Egbon et al. [9], recorded 0.05 mg/L Nitrate. The difference between the two studies may be attributed to the homogenized effect of nitrates from the five salons unlike from a single salon in the case of Egbon et al. [9]. The ammonia nitrogen concentration (2.885 mg/L)

Figure 5a: Picture of untreated salon waste water.

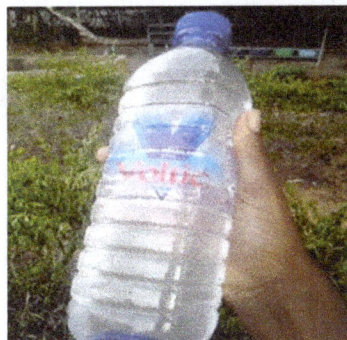

Figure 5b: Picture of the treated salon waste water.

Parameter	Influent	Effluent	% Removal STD	EPA GHANA
Cu (mg/L)	0.067 ± 0.045	0.045 ± 0.022	2.836 ± 27.013	0
Cd (mg/L)	0.027 ± 0.004	0.011 ± 0.005	59.259 ± 8.006	0.1
Zn (mg/L)	0.054 ± 0.009	0.009 ± 0.003	83.333 ± 6.8805	0.5
Fe (mg/L)	0.063 ± 0.010	0.039 ± 0.022	38.095 ± 2.002	0.3
Pb (mg/L)	0.010 ± 0.004	0.000 ± 0.000	100.000 ± 12.939	0.1
As (mg/L)	0.002 ± 0.000	0.000 ± 0.000	100.000 ± 11.573	0.1

Table 1: Mean results for analyzed heavy metals of homogenized Salon waste water samples collected for sixteen weeks.

Parameter	Influent	Effluent	% Removal	EPA STD
pH	7.776 ± 0.497	7.008 ± 0.166	9.877 ± 1.107	6.0-9.0
EC (μs/m)	1780.125 ± 178.274	1668.875 ± 111.286	6.25 ± 0.819	1500
TDS (mg/L)	891.375 ± 88.868	839.625 ± 55.659	5.81 ± 0.629	1500
DO (mg/L)	3.725 ± 0.394	4.038 ± 0.551	n.a.	1000
BOD (mg/L)	29.325 ± 5.224	22.938 ± 5.294	21.780 ± 1.578	50
Turbidity (NTU)	1556.675 ± 253.136	96.538 ± 36.466	93.798 ± 6.073	75
N-NO₃ (mg/L)	24.025 ± 7.710	7.863 ± 2.471	67.727 ± 5.759	50
PO₄ (mg/L)	2.393 ± 0.939	0.775 ± 0.324	67.614 ± 3.264	2
N-NH₃ (mg/L)	13.903 ± 1.503	2.885 ± 0.79125	79.249 ± 8.311	1
TSS (mg/L)	1662.000 ± 273.257	99.000 ± 27.310	94.043 ± 0.948	50
COD (mg/L)	2925.000 ± 300.223	453.750 ± 50.959	84.487 ± 2.823	250

Table 2: Mean results for analyzed physico-chemical parameters of homogenized Salon waste water samples collected for sixteen weeks.

6a)

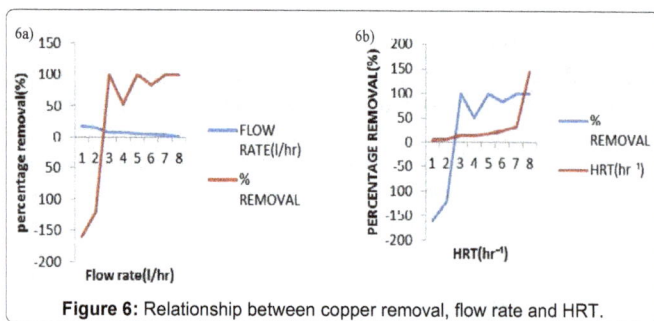

Figure 6: Relationship between copper removal, flow rate and HRT.

7a)

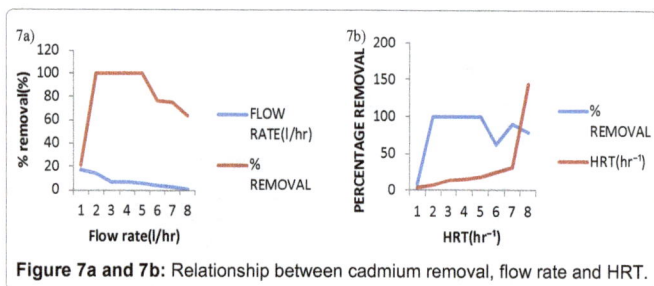

Figure 7a and 7b: Relationship between cadmium removal, flow rate and HRT.

8a)

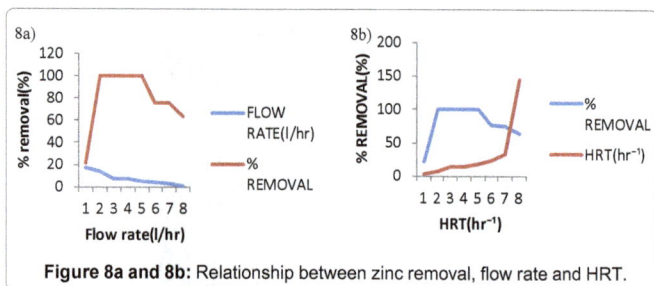

Figure 8a and 8b: Relationship between zinc removal, flow rate and HRT.

9a)

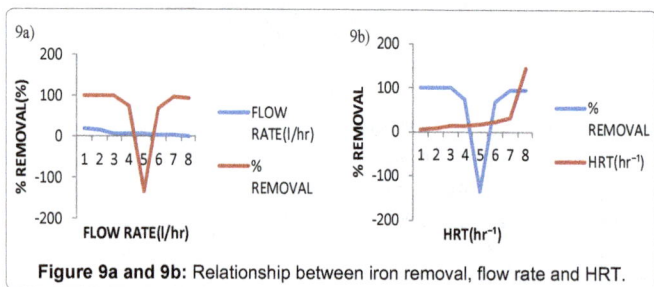

Figure 9a and 9b: Relationship between iron removal, flow rate and HRT.

10a)

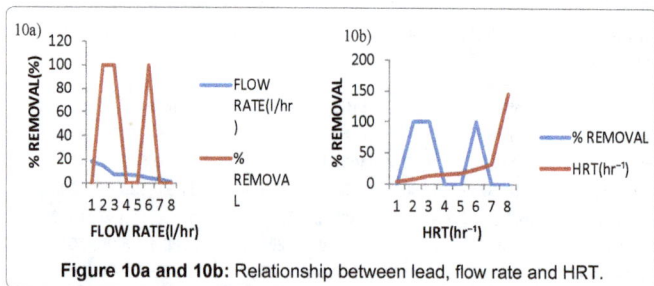

Figure 10a and 10b: Relationship between lead, flow rate and HRT.

however, did not conform to Ghana EPA permissible limit of 1.0 mg/L (Table 2). This could be attributed to the high levels of Nitrogen present in the wastewater. However, it recorded a percentage removal of 79.2%, which is appreciable. This high removal is attributed to the low effluent flow rates.

Biochemical oxygen demand (BOD), Dissolved oxygen (DO) and Chemical oxygen demand (COD)

BOD is the most important variable in water pollution control since it indicates the actual level of biodegradable pollutants in the water [12]. The level of BOD_5 in the effluent (22.9 mg/L) was within the acceptable limit (50 mg/L) of EPA Ghana. This suggests that it will cause no harm to aquatic life since the oxygen content is not high enough to cause anaerobic conditions. From the study of Egbon et al. [9], BOD of the effluent was 5.19 mg/L and removal was 74%. In contrast, several inconsistencies were recorded in the BOD removal over the sixteen weeks in this study. This could be attributed to the fact that the suspected increase in BOD is due to input from the activated charcoal since it is also organic. The inconsistency is also suspected to be due to the fact that the schmutzdecke may not be fully developed in the initial stages to remove the BOD. The non-conformity of dissolved oxygen (Table 2) could be due to lower levels of suspended and dissolved solids in the water. The degree and extent of DO increase depends on the BOD of the effluent in that, the lower the BOD, the higher the DO and vice versa. The overly high concentration of COD (Table 2) is possibly due to the high levels of dissolved and suspended solids which were not efficiently removed. The study however, recorded satisfactory average removal percentage of 84.5 over the sixteen weeks. Egbon et al. [9], recorded COD removal of 60.5% which was lesser compared

11a)

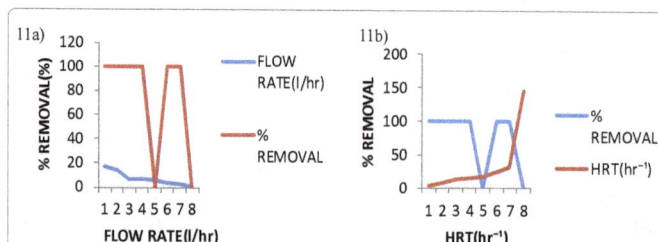

Figure 11a and 11b: Relationship between arsenic removal, flow rate and HRT.

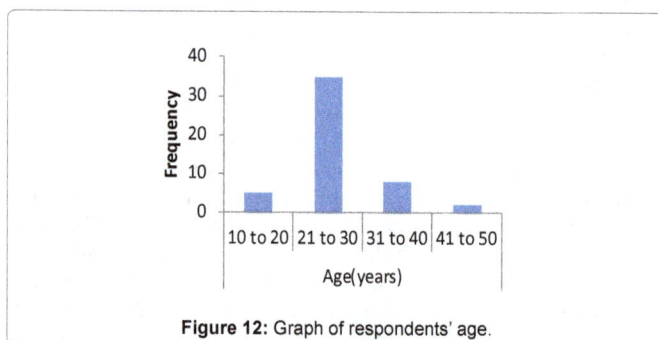

Figure 12: Graph of respondents' age.

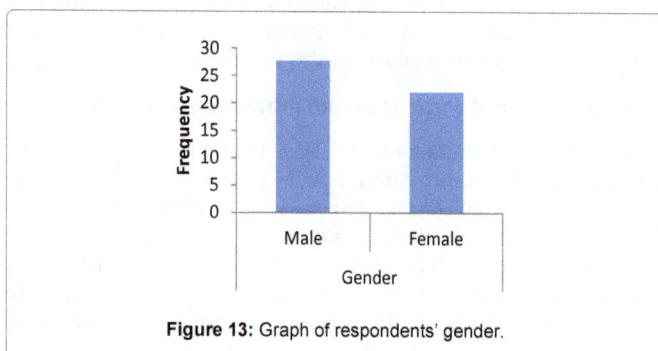

Figure 13: Graph of respondents' gender.

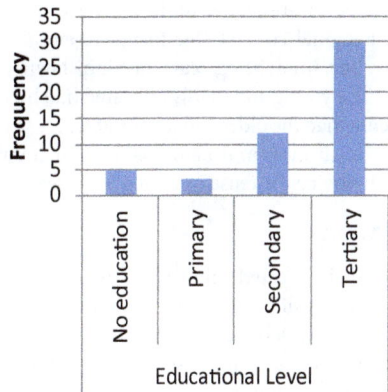

Figure 14: Graph of respondents' educational level.

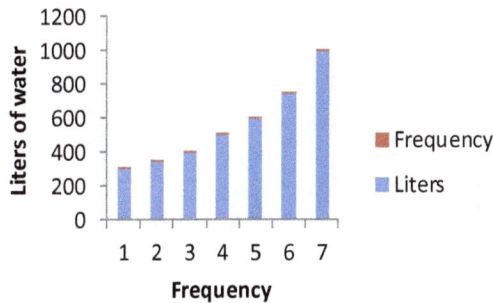

Figure 15: Graph of liters of water used by the salons.

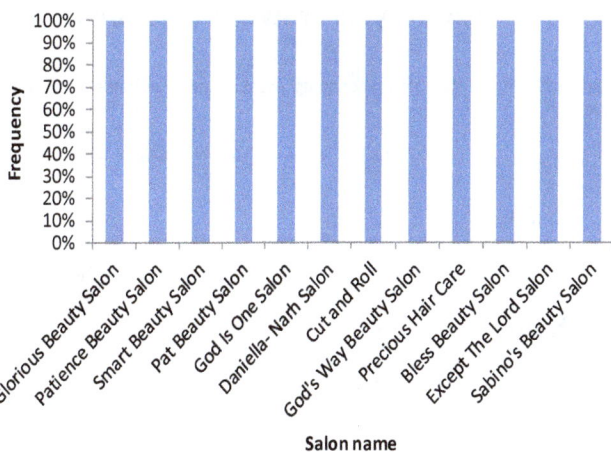

Figure 16: Graph of names of hairdressers salons.

to the large porosity and surface area of the activated charcoal and fine sand which was able to remove the metals onto its surface. Even in trace amounts heavy metals pose threats to the human body if exposed to the body continuously. This implies that there may be no safe level of exposure to lead [13]. For instance, cadmium exposure even in lower amounts could cause detrimental health problems [14]. Beyond the recommended limits, zinc exposure could result in significant bioaccumulation with possible toxic effects for aquatic organisms. Lead exposure could be said to be a less significant issue when it comes to salon waste water. This may be because the chemical composition of products used at the five salons contained little or no Lead. However, heavy metals are very toxic components in the shampoos and other cosmetic products used at the salon even at low levels. The large surface area of the sand and the activated charcoal which was able to remove all the Arsenic in the waste water although the Arsenic composition may be insignificant in the hair products used in the salons. However, the continued use of products contaminated with Arsenic, may cause slow release into the human body and cause adverse effects [15].

Discussion of Results

Although majority of the respondents were aware of waste water reuse, the idea of practicing it is not accepted by most. One participant (tertiary student) made an input that the mother who owned a salon always supplied raw salon waste water to many households in Kumasi in the Ashanti Region of Ghana for flushing toilet and that the use of raw waste water was very common in Kumasi. This indicates that the concept of waste water reuse (even in the untreated form) was accepted and employed in Kumasi but not in Amamoma. The reason for the difference could be that the people of Amamoma are not enlightened on

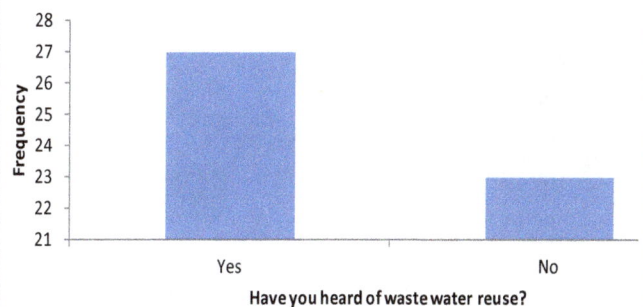

Figure 17: Graph of respondents' knowledge on waste water reuse.

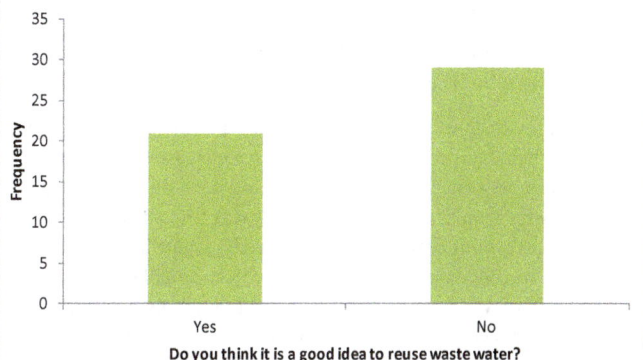

Figure 18: Graph of respondents' views on waste water reuse.

to the percentage removal recorded by this study which employed slow sand filtration system and activated charcoal.

Heavy metals

The values recorded for the heavy metals were within acceptable limits of effluent discharge of the Ghana EPA. This could be attributed

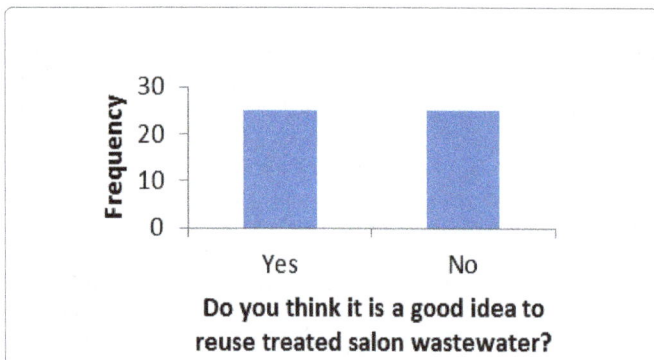

Figure 19: Graph of respondents' views on treated salon waste water reuse.

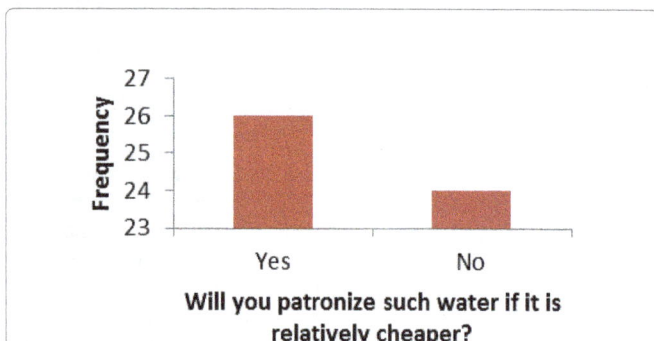

Figure 20: Graph showing willingness of respondents to buying cheap treated waste water.

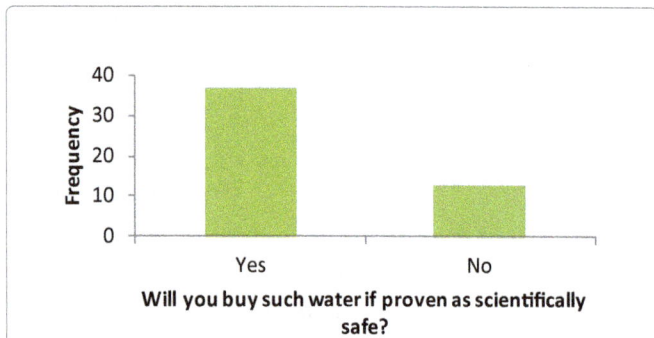

Figure 21: Graph showing respondents' patronization of treated waste water scientifically proven as safe.

A similar work was done by Robinson and Hawkins (2005) [16], in USA to assess public perception on waste water reuse. The study revealed that the people were against the idea of reusing it for application involving close, personal contact (for instance, laundry) but had no problem using it for firefighting, car washing, lawn irrigation and agricultural uses. The young and highly educated had higher knowledge on waste water reuse than the older and uneducated. Close resemblance of this study to the present study in Amamoma shows that public education is imperative in employing the reuse of treated waste water.

Recommendations

In the reuse of the treated salon waste water, public health is a major consideration. As such, it is recommended that further research should be carried on the microbial parameters of the waste water before and after treatment to assess the risk involved in potential disease transmission. It is again recommended that pre-treatment of the waste water should be carried out in further studies, where all the particles in the waste water are allowed to settle for some time before introducing it into the slow sand filtration system. In addition, it is recommended that

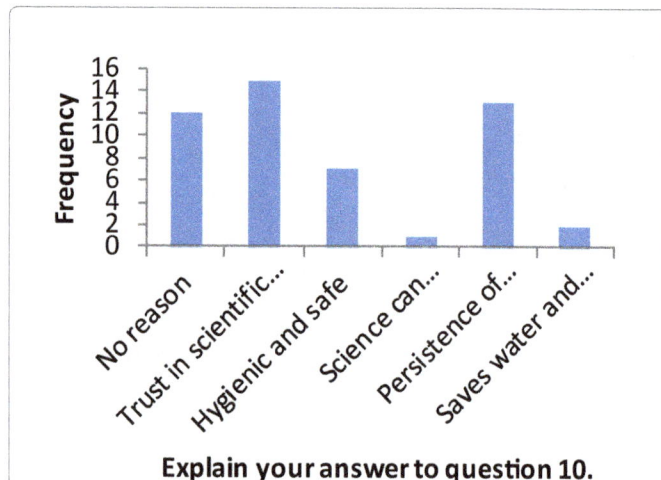

Figure 22: Graph showing respondents' explanation to response to previous question.

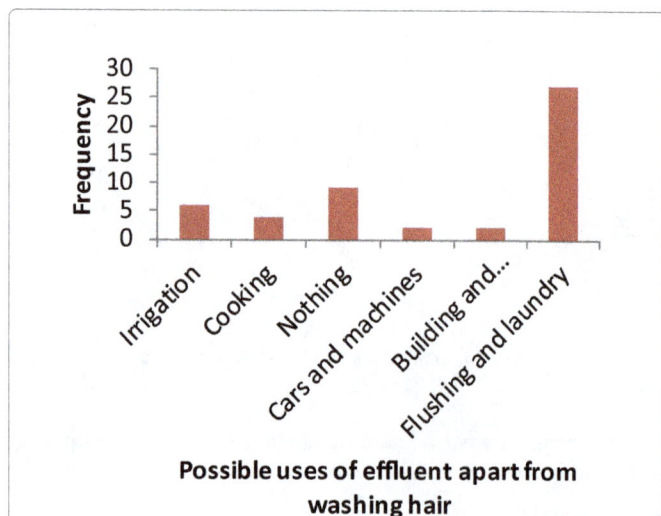

Figure 23: Graph on possible uses of effluents.

water conservation for sustainability. Their belief (people using waste water generated by others for rituals) may also inform their decision.

That notwithstanding, they are very price-sensitive; cheap prices influenced their decision to patronize the treated waste water (although the difference between those who would patronize it due to cheap prices and those who would not patronize at all was 2%). If the people are educated on recycling of waste water, backed by scientific confirmation, they would readily employ it, but only for non-potable domestic use. The amount of water generated daily by the salons averagely, also show that treating the salon waste water for non- potable domestic use or even for washing hair a second time before finally treating it for discharge may be of significant value.

this treatment technology is used for the treatment of other waste water such as industrial and sewage. Furthermore, public education should be carried out to enlighten the public on the need to embrace reuse of treated waste water. Finally, utility of the treated water may be further enhanced and maximized with the addition of Chlorine and Alum.

Conclusions

In effect, the slow sand filtration system with the activated charcoal layer could be adopted as domestic waste water treatment technology especially in developing and low-income countries since the percentage removal for four of the treated heavy metals (Cadmium, Zinc, Iron and Arsenic) were around 60% and above except for Iron and Copper which were below 40%. It was established that the flow rate was inversely proportional to the hydraulic retention times and retention time directly proportional to percentage removal. Finally, it was established that it is possible to use the treated waste water for purposes such as non-potable domestic reuse like flushing toilets, cleaning and gardening, irrigation of agricultural areas (for crops that cannot be eaten unwashed), washing vehicles, industrial use for cooling, or discharge into nearby streams, lakes or other water bodies.

Acknowledgements

We express our profound appreciation to the staff of Ecology Laboratory-University of Ghana and all hairdressers and respondents who participated in this study.

References

1. Bowers F, Cole K, Hoffman J (2002) Characterizing Beauty Salon Waste water for the Purpose of Regulating Onsite Disposal Systems. pp.1-3.

2. Lahlou M (2000) Slow Sand Filtration. Techbriefs.

3. ITACANET (2005) An Introduction to Slow Sand Filtration.

4. Amirault R, Chobanian G, McCants D, McCann A, Burdett H, et al. (2003) Activated Carbon Treatment of Drinking Water Supplies. Healthy Drinking Waters for Rhode Islanders. University of Rhode island, USA.

5. Adegoke R, Adekola FA (2010) Removal of Phenol from Aqueous Solution by Activated Carbon Prepared from Some Agricultural Materials. Advances in Natural and Applied Sciences 4: 293-298.

6. Dvorak IB, Skipton OS (2013) Drinking Water Treatment: Activated Carbon Filtration. University of Nebraska-Lincoln, University of Nebraska-Lincoln Extension Publications, USA.

7. Faanu A, Adukpo OK, Okoto RJS, Diabor E, Darko EO, et al. (2011) Determination of radionuclides in underground water sources within the environments of university of coast. Res J Env Earth Sci 3: 269-274.

8. APHA/AWWA (1998) Standard Methods for the Examination of Water and Wastewater. Australia and New Zealand (ARMCANZ). pp.2-8.

9. Egbon EE, Idode OV, Egbon IE, Chukwuma AP (2013) Treatment of Saloon Wastewater Using Activated Carbon. Chemical and Process Engineering Research 17: 24-28.

10. Browman MG, Harris RF, Ryden JC, Syers JK (1979) Phosphorus loading from urban storm water runoff as a factor in lake eutrophication-Theoretical considerations and qualitative aspects. Journal of Environmental Quality 8: 561-566.

11. Nkegbe E, Emongor V, Koorapetsi I (2005) Assessment of Effluent Quality at Glen Waste water Treatment Plant. Journal of Applied Sciences 5: 647-650.

12. Hammer MJ, Hammer MJJ (1996) Water and waste water technology. Prentice-Hall Inc. p.519.

13. Canfield RL, Henderson CR Jr, Cory-Slechta DA, Cox C, Jusko TA, et al. (2010) Intellectual Impairment in Children with Blood Lead Concentrations Below 10 microg per deciliter. N Engl J Med 348: 1517-1526.

14. Young R (2005) Toxicity profile of cadmium.

15. Gondal MA, Seddigi ZS, Nasr MM, Gondal B (2009) Spectroscopic detection of health hazardous contaminants in lipstick using Laser Induced Breakdown Spectroscopy. J Hazard Mater 175: 726-732.

16. Robinson KG, Robinson CH, Hawkins SA (2005) Assessment of Public Perception Regarding Waste water Reuse: Water Science and Technology. Water Supply 5: 59-65.

29

Study on the Effect of Nanoparticle Loadings in Base Fluids for Improvement of Drilling Fluid Properties

Chai YH[1], Suzana Yusup[2]* and Chok VS[3]

[1]Chemical Engineering Department, Universiti Teknologi PETRONAS, Bandar Seri Iskandar, 31750 Tronoh, Perak, Malaysia

[2]Associate Professor, Chemical Engineering Department, Universiti Teknologi PETRONAS, Bandar Seri Iskandar, 31750 Tronoh, Perak, Malaysia

[3]Director of Technology, Platinum Nanochem Sdn. Bhd., Lot 15-19 & PT 1409, Senawang Industrial Estate, Batu 4 Jalan Tampin, 70450 Seremban, Negeri Sembilan Darul Khusus, Malaysia

Abstract

Nanotechnology is increasingly capturing the attention of material researchers as this technology pushes the limits and boundaries of the pure material itself. Liquids dispersed with nanoparticles generally have higher physical properties enhancements. In the current study, ball-milled functionalized –COOH carbon nanoparticles were introduced into the targeted base fluid of drilling mud. The investigating parameter involved in this study is carbon nanoparticle loadings, ranging from 0 wt% to 1.0 wt%, which is readily dispersed in the base fluid. The method of dispersion chosen is through indirect dispersion in ultrasonic bath. The effect and significance of the investigating parameters were studied based on the desired physical properties of ideal base fluids for drilling muds, mainly thermal conductivities and viscosity of the fluid. The conditions for dispersion in this study were ball-milled functionalized –COOH carbon nanoparticle size at an average size of 10 µm with 90 minutes of indirect ultrasonic dispersion. Result shows that addition of functionalized nanoparticles into base fluids yields as much as 6% of thermal conductivity enhancement while approaching viscosity of pure base fluid at higher shear rate.

Keywords: Nanoparticle; Ultrasonic dispersion; Drilling fluid; Thermal conductivity; Viscosity

Introduction

Drilling fluid, commonly known as drilling mud, is one of the most important aspects in any drilling process. It primarily aids to cool and lubricate the drilling bit, remove solid fragments from drilling area to the surface and counterbalancing the formation pressure within the wellbore.

However, the exploration of oil and gas gradually ventures deeper and further into deep sea drilling when high volume of oil and natural gas can be found. Deep sea drilling is often conducted in a pre-set environment at high temperature high pressure (HTHP) due to frictional forces and pressure due to extreme depth of the wellbore. Commonly, this poses mechanical malfunctions to drilling equipment used. Such problems include overheating of equipment as well as lost circulation is common in any drilling operations due to the limitations posed by drilling mud.

In recent years, nanotechnology is starting to make its impact globally by expanding the current limitations of current available materials and fluids. The properties of materials are differentiated from the conventional material used. Nanomaterials have a relatively larger surface area. Thus, this enhances the material as they are chemically more active and enhancement in terms of strength, electrical and thermal properties can be seen [1]. The integration of nanotechnology into oil and gas exploration sector opens up new opportunities to further explore the capabilities of oil and gas exploration in deep water.

Dispersion of nanoparticles is commonly carried out using high intensity ultrasound. However, various parameters generally affect the dispersion of dry powdered nanoparticles, mainly particle adhesion/cohesion forces [2], surface tension of fluids [3], and nanoparticle size [4,5].

Nanofluids generally produce higher thermal properties compared to its counterpart base fluid [3,4]. However, the viscosity parameter has significant fluctuating behaviours as there is sharp increment of

nanofluid viscosity before approaching base fluid's viscosity [4].

In this study, the experimental design setup for dispersion of nanoparticles involves preparation of carbon nanoparticles beforehand. Ball-milling of carbon nanotubes is carried out to ensure reduction of size of tubular carbon tubes at nanoscale. Nanoparticle loadings considered in this research study are in accordance to the range carried out by Sedaghatzadeh, et al. [5] at 0.2 wt%, 0.4 wt%, 0.6 wt%, 0.8 wt% and 1.0 wt% ratio of nanoparticle mass to mass of base fluid. Functionalised carbon nanoparticles are dispersed into oil-based mud, which is a non-polar liquid. The dispersion of hydrophobic nanoparticles into a non-polar fluid is suitable as it more stable and more soluble due to its counterparts. Previous studies have been conducted on hydrophobic nanoparticles dispersed in oil-based fluid are able to obtain excellent dispersion process through additional adjustments in pH and surface-capping of nanoparticles in the nanofluid suspension [6]. Functionalized nanoparticles are able to stabilize itself well with stability up to six (6) months without sedimentations [7]. Besides that, influences of ball-milling, temperature and ultra-sonication intensity are considered in this study.

This article aims to portray that the addition of minimum amount of functionalized carbon nanoparticles are able to yield enhancement in thermal conductivity as well as retaining the viscosity of the nanofluid

*Corresponding author: Suzana Yusup, Associate Professor, Chemical Engineering Department, Universiti Teknologi PETRONAS, Bandar Seri Iskandar, 31750 Tronoh, Perak, Malaysia, E-mail: drsuzana_yusuf@petronas.com.my

suspension as it approaches higher shear rate. This effect is particularly important in any drilling operations as higher thermal conductivity of drilling fluids are required to channel heat load away from the drilling bits to avoid equipment overheating. Besides that, viscosity is an important parameter as addition of functionalized carbon nanoparticles gives higher viscosity at low shear rate while it approaches viscosity of the base fluid at higher shear rate. This effect is beneficial to any drilling operations as it shows that less power consumption is actually needed for drilling operations which utilizes drilling nano-fluids.

Materials and Methods

Materials

Functionalized carbon nanotubes are selected in this research study. The carbon nanotubes have an average size of 24 μm. A particle size distribution (PSD) analysis is provided in Figures 1 and 2.

Nanoparticles preparation

Planetary Mono Mill PULVERISITTE 6 *classic line* series is used in the ball milling process. The cylindrical bowl with zirconium oxide material of construction with 250 mL volume containing 25 zirconium oxide balls with mean diameter of 5 mm are filled with 20 g of carbon nanotubes. The functionalized carbon nanotubes are ball-milled into carbon nanoparticles through ball-milling operation. The approach taken in this preparation follows Tucho, *et al.* [8]. The total ball-milling duration is three (3) hours with intensity of 500 rpm. In this study, each hour of ball milling duration has an allowance of 10 minutes stoppage time and is repeated until total of 3 hours of ball milling duration is achieved.

Nanoparticles characterization

Two traits are characterised in this study approach, mainly the particle size distribution of the nanoparticles and the chemical bonds and functional groups present in the carbon nanoparticles.

The particle size distribution (PSD) analysis is conducted using Malvern Mastersizer 2000 Particle Size Analyzer, where it is used to measure the particle size where the scattered light intensity from the passing laser beam through the dispersed particulate sample is measured.

Fourier Transform Infrared (FTIR) spectroscopy analysis is carried out using Perkin Elmer Transform Infrared Spectrometer where transmittance of light at specific wavelengths is measured. The characteristics of the peaks recorded are classified into type of peaks, chemical bonds and functional groups present.

Base fluid and nanofluid characterization

The thermal conductivity and density are measured in this study.

For density measurement, Anton Paar Density Meter 4500 M is used. A 10 mL of pure base fluid is injected into the side port of the device. The method of measurement is through the differences of fluid in U-tube located within the device. The reference temperature of the measurement is taken at 25°C.

For thermal conductivity analysis, P.A. Hilton Thermal Conductivity of Liquids and Gases Unit H471 is used. The heater is jacketed by a plug with two thermocouples positioned in two different locations: at the plug and at the jacket. The thermal conductivity of base fluid is measured by determining the temperature difference between two thermocouples across a fixed heat transfer area. The radial clearance used in this research is 0.3 mm with 0.0134 m² heat transfer area. The voltage applied to the heater is constant at 60 V for liquid while 40 V is applied for gas. The formula of conductivity used in this analysis is shown below:

$$\text{Thermal conductivity, } k = \frac{Q\Delta r}{A\Delta T} \quad (1)$$

where Q is the heat transfer by conduction, W; Δr is the radial clearance, m; A is the surface area of heat conduction, m²; and ΔT is the temperature difference between two thermocouple points, °C.

The incidental heat transfer present must not be neglected when performing calculations of thermal conductivity of the base fluid. The experiment was conducted twice and the average mean values were taken and plotted.

The viscosity analysis of the base fluid and nano-based fluid is carried out using Brookfield Cap 2000 + L series which has a supported temperature range from 5°C to 75°C. In this experimental study, the shear rate, in rpm, is carried out at 100, 150 and 200 respectively at temperature of 30°C. The spindle used in this study is spindle 3, with a running time of 60 s and holding time of 10 s. The experiment was repeated twice with the same spindle number and the average mean value was taken and plotted.

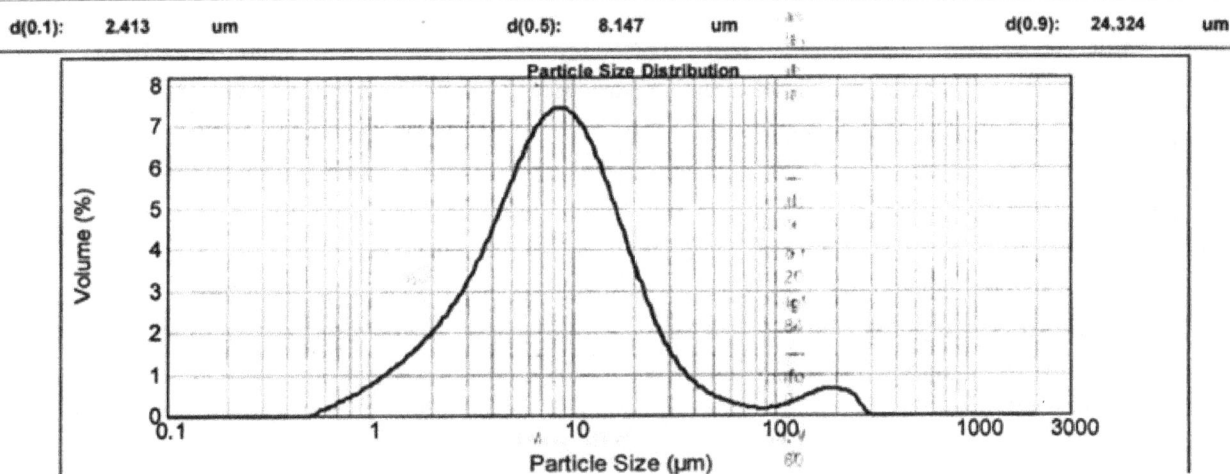

Figure 1: Particle size distribution of carbon nanotubes (CNTs) functionalised group –COOH

Nanoparticles dispersion

In this research study, indirect sonication dispersion is carried out to disperse carbon nanoparticles into the base fluids. Bath Ultrasonic Branson Model 8510E-DTH equipment is used to suspend samples within the tank. The tank has dimensions of 495×290×150 which allows multiple samples to be sonicated together at the same time. The delivering power of the equipment is 320W with a frequency of 40 kHz. Five (5) samples of base fluid measuring 100 mL is mixed crudely with 0.2wt%, 0.4wt%, 0.6wt%, 0.8wt% and 1.0wt% of carbon nanoparticles respectively in 150 mL beakers. The carbon nanoparticle mass loadings are calculated beforehand based on the density of pure base fluid. The correlation used to estimate ultrasonic dispersion follows Yang et al. [9]. The correlation is as stated below:

$$E = \frac{P \times t}{V} \qquad (2)$$

where P is the output power of the sonicator, W; t is the sonication time, in s; V is the total volume of liquid sonicated, in mL. The optimum sonication energy estimated by Yang et al. is averagely 2250 ± 250 J/mL. From Equation 2, the time taken required to disperse nanoparticles with 100 mL of base fluid with constant power deliverance of 320 W is roughly 13 minutes. However, the dispersion of samples was also carried out at 15 minutes, 30 minutes and 90 minutes respectively.

Results and Discussions

Particle size distribution (PSD) analysis

The analysis of the particle size distribution is provided in the figures below. Figure 1 and Figure 2 display the results of CNTs before ball-milling and after ball-milling respectively. In Figure 1, 10% of the total population measured is averagely 2.413 μm and 90% of the total population giving an average measurement of 24.324 μm. In Figure 2, 10% of the total population gives a reduction in measurement length at 1.236 μm while 90% of the total population also gives a reduction in length at 12.273 μm.

Overall, the comparison in size of carbon nanoparticles before and after ball milling shows a total reduction in size by 48.8% with more or less reduction size of half of the length of the carbon nanotubes.

For the preparation of ultrasonic dispersion of nanoparticles, the pre-requisite for a good dispersion and a stable suspension requires nanoparticles at 200 nm or shorter. This is further supported by literature reviews of previous experimental works which dispersed similar sizes of nanoparticles. In this experiment, the ball-milled samples do not fully fit the requirement as 90% of the total population distribution yields sizes larger than 200 nm. The nature of the size of the carbon nanoparticles may give rise to the tendency of the carbon nanoparticles to agglomerate or tangled each other when dispersed under ultrasonic dispersion.

Ball milling of carbon nanotubes experiment is carried in accordance to the study conducted by Tucho et al. (2010) [8], where key parameters involved in this experiment are the speed factor and time duration for the ball milling process. 500 rpm and 3 hours of total duration of intense ball milling experiment were carried out. However, from previous experience the high intensity of the impact between the grinding balls and the walls of the grinding bowl might scrape off trace amounts of zirconium oxide into the samples.

Therefore, to prevent any contamination of zirconium oxide into the carbon nanoparticle samples, at every hour of ball milling operations, there will be an hour of interval stoppage before the next hour of ball milling. At the same time, this could also prevent deformation of the morphology of the carbon nanoparticles due to high temperature as a result of extreme intensity from the collision of the grinding bowls.

The collision intensity of the grinding balls might not build up sufficiently over the total 3 hours of ball milling duration as suggested by Tucho et al. (2010) [8] as there is an interval between each hour of ball milling operations.

Fourier transform infrared (FTIR) spectrometry

Figure 3 below shows the Fourier Transform Infrared Spectroscopy (FTIR) analysis result of ball milled carbon nanoparticles functionalised group –COOH. A table of characteristic IR is used to determine the presence of the bonds and functional group in the samples.

In Figure 3, a broad peak with wavelength 3428.7 cm^{-1} can be seen which shows presence of O-H stretch bond or H-bonds which shows the presence of possible alcohol or phenol groups. A small peak can be seen at wavelength 2917.58 cm^{-1} showing the presence of O-H stretches with carboxylic acids as its main functional group. The analysis shows medium peak at 1635.13 cm^{-1} containing carboxylic acid functional group as well. Another smaller broader peak at 1130.35 cm^{-1} also shows C-O stretch bonds containing carboxylic acid group. From the FTIR results obtained it can be concluded that no external organic

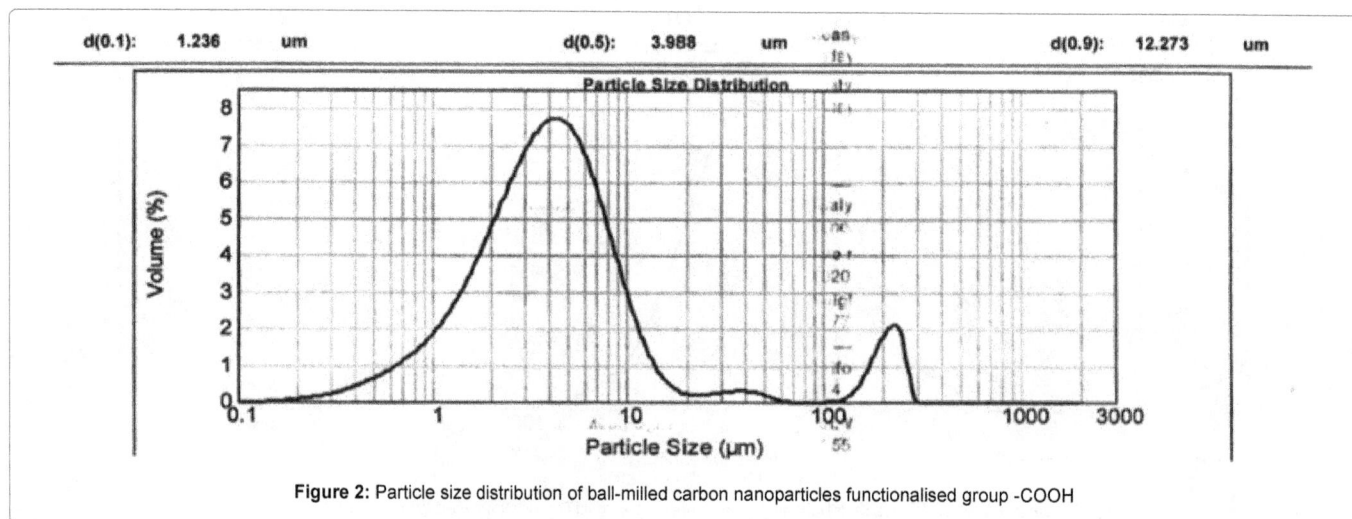

Figure 2: Particle size distribution of ball-milled carbon nanoparticles functionalised group -COOH

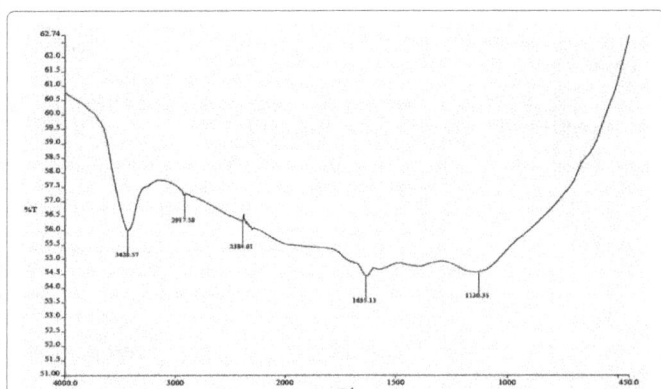

Figure 3: IR spectroscopy graph of ball milled carbon nanoparticle functionalised group –COOH

contaminants can be traced in this carbon nanoparticle sample.

Thermal conductivity analysis

In this experiment, the thermal conductivity of nano-basefluid is measured at temperature of 60 V settings. The temperature at the maximum voltage delivered is at mean temperature of 43°C. The thermal conductivity enhancement of the nanofluid can be summarised in Figure 4.

From Figure 4, a fluctuating trend can be seen when nanoparticle loadings are increased from 0.2 wt% to 0.6 wt%. Thermal conductivity of nanofluid increases nonlinearly with increasing mass fraction of nanoparticles [8]. This is true from the results obtained where the profile of relative thermal conductivity enhancement against nanoparticle loadings (wt%) increased nonlinearly. However, increasing nanoparticles concentration decreases the distance between particles while random collision due to Brownian motion from the heating process accelerates agglomerations of nanoparticles. The size of nanoparticles plays an important factor in determining the nanofluid's thermal conductivity. The ball milled nanoparticles in this study yield nanoparticle sizes of 4 μm averagely. However, large particles do not possess Brownian motion anymore as the particles approach micrometre size [9], thus leading to lower thermal conductivity enhancements. From the results obtained, 0.2 wt% has a considerable high relative thermal conductivity than 0.4 wt% and 0.6 wt% respectively due to increase distance between particles. However, higher nanoparticle loadings gives greater thermal conductivity enhancement but induced higher settlement of nanoparticle cluster sizes in the end.

Another possible cause for the fluctuating trends can be attributed to the inconsistency of water flow around the jacketed area of the equipment. Higher water flow rate cools down the shell side's temperature, resulting in higher difference of temperature gradient between both thermocouple points at the heater and shell side respectively.

Viscosity analysis

The analysis is conducted at a fixed temperature of 30°C to compare the viscosity of the pure base fluid with nanofluid samples. From the plot above, the pure base fluid has the lowest viscosity while 0.4 wt% of dispersed nanoparticle in base fluid has the highest viscosity at shear rate of 200 rpm.

The viscosity of the nanofluid suspensions has been widely debated where numerous studies found that there had been fluctuating results

from various researchers which can hardly be concluded [4]. The result shown by the plot shows that although it is true all nanofluids of various nanoparticle loadings have higher viscosity compared to its base fluid, the trendline shows that at higher shear rate, the viscosity of the nanofluid approaches that of the base fluids'. From Figure 5, the nanoparticle concentrations of 0.4 wt% are just slightly higher than that of 1.0 wt%. This can be attributed to the uneven dispersion of nanoparticles. A larger nanoparticle size will coagulate more easily to form larger agglomerates as larger size of nanoparticles cause decrement in distance between particles and increases the attractive van der Waals forces. Thus, nanoparticles at the point of extraction for sample 0.4 wt% are generally larger in size which attributes to the higher viscosity of nanofluid. As comparison, 0.4 wt% nanoparticle loadings also yield the lowest thermal conductivity as compared to the remaining loadings. It can be concluded that 0.4 wt% nanofluid sample has agglomerated sufficiently to yield lower thermal conductivity but higher viscosity [10].

Conclusions

The study on effects of nanoparticle dispersion loadings in base fluids is relative new research discoveries, in which limited research experimental data are available. The addition of nanoparticles into base fluids shows thermal conductivity enhancements at higher loadings. However, the size of nanoparticles plays a crucial role in affecting the physical properties of the base fluids.

Recommendations

Use of surfactants to suspend nanoparticles within the base fluid is recommended in future study. Study shows that usage of surfactants gives longer suspension duration of nanoparticles, resulting in better dispersion of the nanoparticles.

Acknowledgements

The authors are grateful for the contribution of materials from Platinum NanoChem Sdn. Bhd. for making this research study possible.

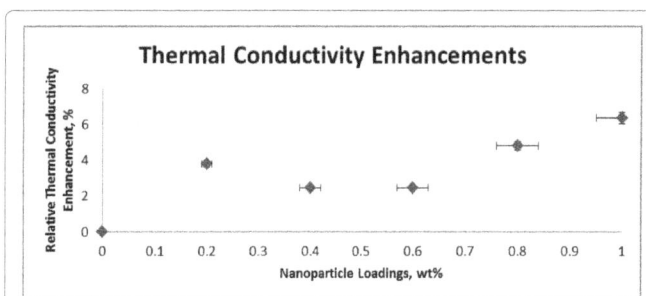

Figure 4: Thermal conductivity enhancement as a function of nanoparticle loadings

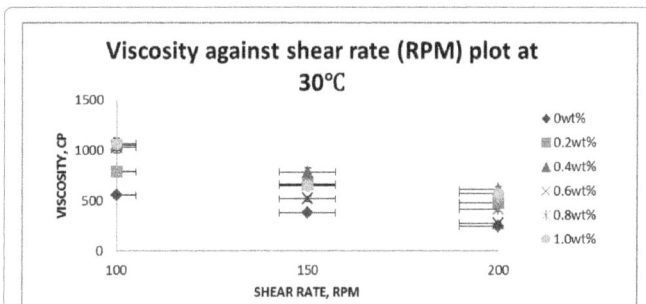

Figure 5: Viscosity against shear rate, rpm, and plot at 30°C

References

1. Hashim U, Salleh S, Nadia E (2009) Nanotechnology development status in Malaysia: industrialization strategy and practices. International Journal Nanoelectronics and Materials 2: 119-134.

2. Tinke AP, Govoreanu R, Weuts I, Vanhoutte K, De Smaele D (2009) A review of underlying fundamentals in a wet dispersion size analysis of powders. Powder Technology 196: 102-114.

3. Tanvir S, Qiao L (2012) Surface tension of Nanofluid-type fuels containing suspended nanomaterials. Nanoscale Res Lett 7: 226.

4. Ruan B, Jacobi AM (2012) Ultrasonication effects on thermal and rheological properties of carbon nanotube suspensions. Nanoscale Res Lett 7: 127.

5. Sedaghatzadeh M, Khodadadi A, Tahmasebi MB (2012) An Improvement in Thermal and Rheological Properties of Water-based Drilling Fluids Using Multiwall Carbon Nanotube (MWCNT). Iranian Journal of Oil & Gas Science and Technology 1: 55-65.

6. Dan L, Biyuan H, Wenjun F, Yongsheng G, Ruisen L (2010) Preparation of Well-Dispersed Silver Nanoparticles for Oil-Based Nanofluids. Industrial & Engineering Chemistry Research 49: 1697-1702.

7. Yang X, Liu ZH (2010) A kind of nanofluid consisting of surface-functionalized nanoparticles. Nanoscale Res Lett 5: 1324-1328.

8. Tucho WM, Mauroy H, Walmsley JC, Deledda S, Holmestad R, et al. (2010) The effects of ball milling intensity on morphology of multiwall carbon nanotubes. Scripta Materialia 63: 637-640.

9. Yang K, Yi Z, Jing Q, Yue R, Jiang W, et al. (2013) Sonication-assisted dispersion of carbon nanotubes in aqueous solutions of the anionic surfactant SDBS: The role of sonication energy. Chinese Science Bullentin 58: 2082-2090.

10. Duan F (2012) Thermal Property Measurement of Al2O3-Water Nanofluids. Smart Nanoparticles Techonology InTech 576.

30

Studies on the γ-Irradiated Polyvinyl Alcohol (PVA) Blended Gelatin Films

AM Sarwaruddin Chowdhury[1*], Moshfiqur Rahman[1], Pinku Poddar[1], Syed Rashedul Alam[2], Kamol Dey[1], Noor Md Shahriar Khan[1], Md Ali Akbar[1], Ruhul A Khan[3] and Zinia Nasreen[4]

[1]Department of Applied Chemistry and Chemical Engineering, Faculty of Engineering and Technology, University of Dhaka, Dhaka, Bangladesh
[2]Research and Development (R&D), Berger Paints Limited, Bangladesh
[3]Nuclear and Radiation Chemistry Division, Institute of Nuclear Science and Technology, Bangladesh Atomic Energy Commission, Dhaka, Bangladesh
[4]Department of Chemistry, Dhaka University of Engineering Technology (DUET), Gazipur, Bangladesh

Abstract

Different concentrations of PVA blended Gelatin were taken for grafting under γ-radiation. Better grafting means better cross-linking of polymer. We observed the effect of concentration of polymers and the effect of radiation doses. Various physico-mechanical and morphological properties like tensile strength (TS), elongation at break (Eb), FTIR, Scanning Electron Microscopy (SEM) of films were observed. For untreated (virgin) Film, radiation doses were optimized for such types of properties. By casting process Gelatin-based polyvinyl alcohol (PVA) films were prepared in different ratios. These films were irradiated under gamma radiation (^{60}Co) at different doses (0.5-5 kGy). The mechanical properties of these films were evaluated. It was found that 95% gelatin +5% PVA film exhibited the highest tensile strength (TS) value at 0.5 kGy gamma radiation (51 MPa), which was 46% higher than that of non-irradiated films.

Keywords: Radiation; Cross-linking; Polymer; Polyvinyl alcohol; Mechanical

Introduction

Polymer science and technology is one of the most active and promising fields in embracing a multitude of topics from natural polymers such as cellulose, wool, silk, jute etc. which are of utmost importance for living systems, to the synthetic high polymers [1,2]. Gelatin is one of the most versatile natural products known. Gelatin represents a typical renewable material from natural resources of animal origin. Gelatin was one of the first macromolecules employed in the production of biomaterials [3]. This biopolymer still attracts the attention of researchers because it is produced abundantly practically worldwide, has a relatively low cost and possesses excellent functional and filmogenic properties [3,4]. For this reason gelatin has been studied in film technology both alone [5-10] and in blends with other biopolymers [11]. Animal glue and gelatins normally contain about 15% of water and 1-4% of inorganic salts. They may also contain a small amount of grease. The main high molecular weight impurity that has been identified is a heat-coagulable mucoprotein complex.

Bone glues may contain as much as 6% of these materials. These impurities are of little or no importance in the majority of glue uses. The properties depend on the major protein constituent derived from the breakdown of collagen. This pure, and to some extent hypothetical, material is described here as gelatin (without an "e"). Gelatin then is regarded as the major protein constituent of gelatin and glue. The purest gelatins contains largely of gelatin and water. Gelatin, like its precursor collagen, contains carbon 15%, hydrogen 26%, nitrogen 18%, oxygen 25%, sulfur 0.1% and phosphorus traces (Figure 1).

Gelatin is a biopolymer, which possesses a number of properties. It has versatile application owing of its inherent properties. As a natural polymer it always gets priority for various applications. It degrades very quickly to the environment and this property makes some advantages and disadvantages for its application. Its water aging nature makes it alluring for bio medical implants. Therefore it is necessary to modify its quality and characteristic properties and improve its end products

for diversified application like biomedical structural material. PVA is recognized as one of the few synthetic polymers truly biodegradable under both aerobic and anaerobic conditions. The irradiated blend films exhibit higher mechanical properties compared with the non-irradiated films [12]. Grafting of gelatin by various polymers has been studied with the objective of improving or modifying the properties of gelatin and in order to develop new materials combining the desirable properties of both natural and synthetic polymer. Many attempts, such as physical and chemical treatments, have lead to changes in the surface structure and surface energy of the films. Among them, physical treatments, such as ionizing or non-ionizing radiation, can introduce better surface cross-linking between natural and synthetic polymers, and reduce the hydrophilic nature of the film. Surface modification of the films can be carried out by the monomer treatment. The acrylate monomer 1,4-butanediol diacrylate (BDDA) induced cross-linking using their double bonds [13-16].

The present work was under taken to prepare a bio-synthetic blend of Gelatin with PVA, to modify its preparation and also modifying its quality and characteristic properties. For this reason, γ-radiation technique was applied. Gelatin and PVA are cheap, eco-friendly and available in Bangladesh. Considering all the factors, the ultimate goal of this research work is to improve the property of Gelatin through enhancement of tensile strength, elongation at break, stability etc. and finding its possibilities as bio medical material for implementation.

*Corresponding author: AM Sarwaruddin Chowdhury, Department of Applied Chemistry and Chemical Engineering, Faculty of Engineering and Technology, University of Dhaka, Curzon Hall, Dhaka-1000, Bangladesh
E-mail: profdrsarwar@gmail.com/sarwar@du.ac.bd

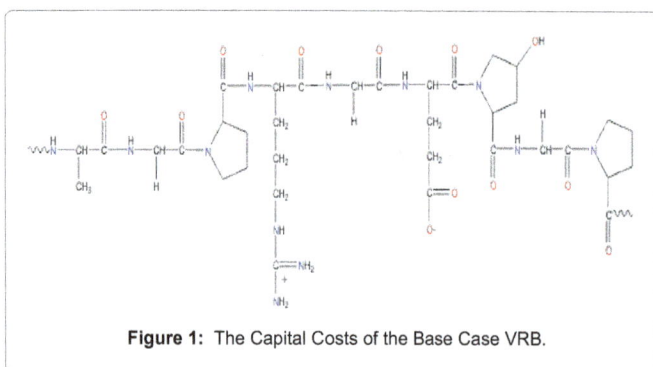

Figure 1: The Capital Costs of the Base Case VRB.

Experimental

Materials

Gelatin (185 Bloom; Type A, pharmaceutical grade) was collected from the Opsonin Pharma Limited, Barishal, Bangladesh. The monomer BDDA was purchased from E. Merck, Germany, and acetone from BDH Chemicals Limited, England. Methanol was purchased from E. Merck, Germany. The synthetic polymer PVA (Molecular Weight: 72000) was purchased from Fluka Chemie AGCH-9470 Buchs.

Methods

Preparation of film: Gelatin was dissolved in hot water to form solution. Poly vinyl alcohol (PVA) was also dissolved in hot water with constant stirring. Then these two solutions were mixed. The mixture was continuously stirred with the help of magnetic stirrer to form homogeneous mixture and heated for about one and half an hour. The solution was then cast on to the silicon paper covered glass plate to form film. The solution layer was maintained into a thickness of 4 mm on the glass plate. The solution was dried into films at room temperature for 48 hours.

The dried films were then peeled from the silicon cloth and cut with a scissor into small pieces of length 70 mm and width 10 mm. Thickness was measured by slide calipers and that was 1.5 mm on an average. These samples were stored in laminated polythene bag and kept in desiccators at room temperature prior testing. These samples were irradiated under γ (gamma)-radiation with different doses (total dose) like 0.5, 1.0, 1.5, 2.5 and 5.0 kGy.

Preparation of soaking formulation using 1,4-Butanediol diacrylate (BDDA): One soaking formulation was prepared with BDDA (3%) in methanol.

Treatment of films: Gelatin and Gelatin/PVA bio blended Films were irradiated under γ-radiation at different doses of gamma radiation like 0.5, 1.0, 1.5, 2.5 and 5.0 kGy using γ-ray from ^{60}Co. The Films were then subjected to various characterizations.

Characterization of Gelatin/PVA bioblended film: The films were exposed at room temperature for well grafting for about 24 hrs. After this tensile strength (TS), Elongation at break (Eb), Scanning Electron Microscopy (SEM), and FTIR were measured.

Treatment of films with soaking formulation: Gelatin and Gelatin/PVA bioblend film of the composition (95:5) was soaked in 3% BDDA at 3 minutes soaking time. After soaking, these films were irradiated under γ- radiation at different doses of gamma radiation like 0.5, 1.0, 1.5, 2.5 and 5.0 kGy using γ-ray from ^{60}Co. The films were then subjected to various characterizations.

An electromagnetic wave, a gamma ray is similar to ordinary visible light but differs in energy or wavelength. Sunlight consists of a mixture of electromagnetic rays of various wavelengths, from the longest, infrared, through red, orange, yellow, green, blue, indigo, and violet, to the shortest in wavelength, ultraviolet. A gamma rays wavelength is far shorter than ultraviolet (i.e., it is far higher in energy). Gamma rays are produced following spontaneous decay of radioactive materials, such as cobalt-60 and cesium-137. A cobalt-60 gamma ray can penetrate deeply into the human body, so it has been widely used for cancer radiotherapy.

Gamma radiation has more energy and therefore it is possible for the electrons to be discharged. This results in the creation of electrically charged particles, which are called ions. The amount of energy deposited in the product is referred to as the "absorbed dose" (1 kilo Gray=1 kilojoules/kg) (Figure 2).

If we relate it to heat, 10 kGy is equal to the amount of energy required to raise the temperature of 1 kg of water by 2.4°C.

Property measurement

Tensile properties: Tensile properties; tensile strength (TS) and elongation at break (Eb) of the cured films are measured with Universal Testing Machine (INSTRON, model 1011, UK). The load capacity is 500 N, efficiency is within ± 1%. The crosshead speed is 10 mm/min. Gauze length is 30 mm. Following equations are used to measure the tensile properties.

$$\text{Tensile strength, TS (MPa)} = \frac{\text{Load (N)}}{\text{Thickness (mm)} \times \text{Width (mm)}}$$

$$\text{Elongation at break, Eb (\%)} = \frac{\text{Displacement at break}}{\text{Gauze length}} \times 100$$

FTIR analysis: Pure gelatin film, pure PVA film and a blend (95% gelatin+5% PVA) was characterized by FT-IR.

Morphological characteristics: The surface morphology of gelatin, PVA and their blends were determined by Scanning Electron Microscopy (SEM).

Results and Discussion

Generally Gelatin is insoluble in cold water. So, Gelatin was dissolved in hot water. Gelatin was blended with poly (vinyl alcohol), a biodegradable synthetic polymer, in order to improve physico-mechanical and thermal properties in the films.

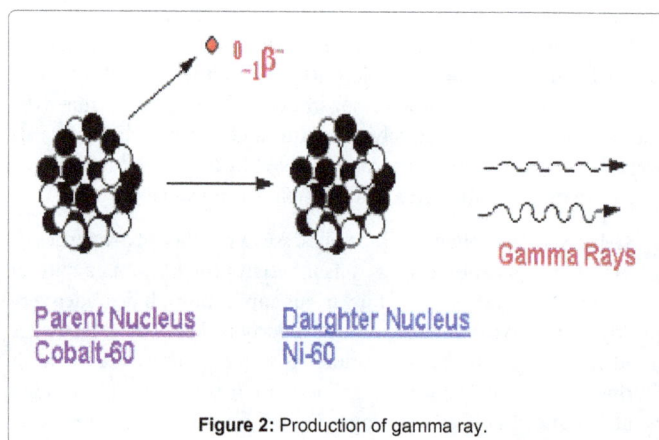

Figure 2: Production of gamma ray.

Preparation and chractarization of Gelatine/PVA blend

Gelatin containing 5, 10 and 15% PVA solution and PVA containing 5, 10, 15 and 20% gelatin solution were blended for preparing Film. These solutions were blended in hot water for about one and half an hour to produce homogenous solution at the end of the process. During cooling, the solution was transformed into semi gel and then was casted on to silicon paper covered glass plate for Film formation.

γ-Radiation process and optimization of γ-radiation dose

Different concentrations of PVA blended Gelatin were taken for grafting under γ-radiation. Better grafting means better cross-linking of polymer. We observe the effect of concentration of polymers and the effect of radiation doses. Various physico-mechanical and thermal properties like tensile strength (TS), elongation at break (Eb), FTIR, and SEM (Scanning Electron Microscope) of films were observed. For untreated (virgin) Film, radiation doses were optimized for such types of properties.

Mechanical properties of untreated and treated film

Four different blends were prepared from various concentrations of gelatin/PVA solutions. When blends were prepared by mixing gelatin and PVA water solution, a homogeneous water solution was produced, thus showing compatibility of the two components in the solvent, whereas cast films appeared homogeneous only for a limited amount (20%) of one component into the other. Thus, in blends with the same amount of PVA and gelatin, phase separation and opacity were evident. These were impossible to handle. So we investigated the physico-mechanical properties of films of above mentioned compositions.

Optimization of grafting condition with extent of physic-mechanical properties

Tensile strength (TS): Tensile strength (TS) is very important in selecting diverse application of polymer. The results of TS values of the not radiated Films (gelatin and PVA based) were plotted in Figure 3 for 0%, 5%, 10% and 15% (GP1, GP2, GP3 and GP4) PVA containing gelatin Films. The TS values of irradiated Films were plotted in Figure 4 against total dose (γ-radiation dose). From the Figure 3 it was observed that with the loading of PVA into gelatin the TS (tensile strength) of the base polymer were significantly decreased. But, from Figure 4 it was seen that, due to incorporating radiation the TS values were improved up to some radiation dose and then again decreased. The highest TS for blends were observed for 5% PVA containing Gelatin Film at 0.5 kGy dose (51 MPa). In case of pure Gelatin Film TS values was also increased with the increase of radiation doses and TS value attained maximum at 0.5 kGy (total dose) and then decreased with increasing radiation doses. Higher γ-radiation dose might have caused degradation of the polymer and the Film became hard and brittle whilst at lower doses cross linking might have dominated over chain seasoning.

In case of irradiated Film, TS values were reached a maximum up to approximately 0.5, 1.5 kGy (radiation dose) and then decreased for further increasing of radiation intensity as well as Gelatin concentration. When the Gelatin Film subjected to the radiation, hydroxyl group from Gelatin radicals were initiated to form cross linked network. So, TS value increases with radiation, but higher radiation doses caused degradation due to the breaking of the polymer chains. So, at higher radiation doses TS decreased. From the figure it is clear that TS value of gamma treated Film is higher than that of untreated Film.

Tensile Strength (TS) of different formulations change with γ-irradiation dose. At 0.5 kGy tensile strength of GP1, GP2, GP3, and GP4 is 56.65, 51.67, 32.11, and 28.45 respectively. But tensile strength of non-irradiated GP1, GP2, GP3, and GP4 is 41.34, 34.94, 29.19, and 11.54 respectively.

Elongation at break (Eb): Elongation is an important mechanical property in the application of polymer. The results of elongation at break (%) of the not radiated Films (gelatin and PVA based) were plotted in Figure 5 for 0%, 5%, 10% and 15% (GP1, GP2, GP3 and GP4)

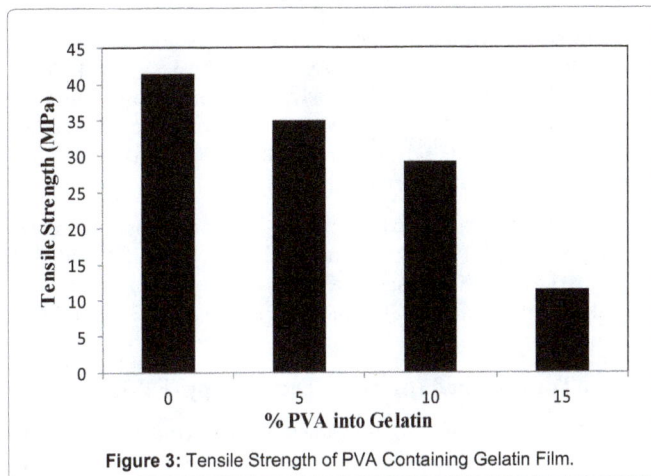

Figure 3: Tensile Strength of PVA Containing Gelatin Film.

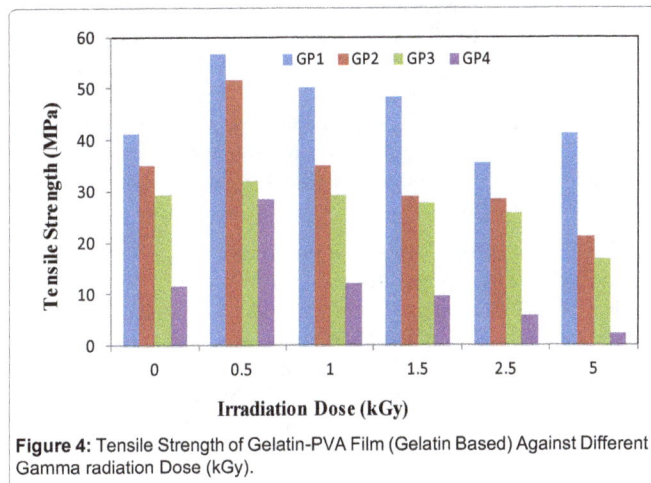

Figure 4: Tensile Strength of Gelatin-PVA Film (Gelatin Based) Against Different Gamma radiation Dose (kGy).

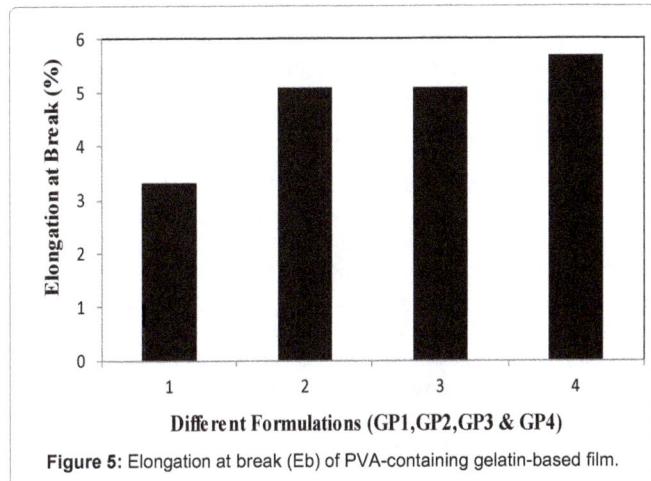

Figure 5: Elongation at break (Eb) of PVA-containing gelatin-based film.

PVA containing gelatin films. The Eb (%) values of irradiated films were plotted against total dose (γ-radiation dose) in Figure 6. It is observed that Eb value increases for blend films drastically due to incorporating PVA. This is due to the increasing concentration of highly flexible PVA into the blend films. The highest Eb is found to be 5.6% for the film obtained from GP4 formulation (85% gelatin+15% PVA). The Eb values of irradiated films are plotted (Figure 6) against gamma irradiation dose as a function of formulation. It is found that Eb values decrease with the increase of irradiation dose for all formulations.

In the case of blend films, the maximum Eb of 4.8% is observed for the GP2 formulation at 0.5 kGy dose, followed by 3.8% and 4.2% for GP3 and GP4 formulation, respectively, at the same irradiation dose. Elongation at break (Eb) changes with gamma irradiation dose at 0.5 kGy are shown in Table 1. It is clear that the Eb of gamma treated film is lower than that of untreated film [12].

Elongation at break (%) changes with γ-irradiation dose. At 0.5 kGy elongation at break (%) of non-irradiated GP1, GP2, GP3, and GP4 is 3.2, 5.1, 5.09, and 5.67 respectively. But after irradiation elongation at break (%) of GP1, GP2, GP3, and GP4 is 3.16, 4.79, 3.83, and 4.23 respectively.

Fourier Transformed Infrared Spectroscopy (FTIR) analysis

FTIR is of importance to study the molecular structure. The width and intensity of spectral bands as well as position of peaks are all sensitive to environmental changes and to conformations of macromolecule on molecular level. Intermolecular interactions occur when two polymers are compatible. So the FTIR spectra of the blends are different from those of the pure polymer, which is advantageous to study the extent of compatibility of the blend polymers. FTIR spectrophotometer has also been found to be a valuable tool in studying graft copolymerization reactions. Figures 7, 8 and 9 shows the infrared spectra for the films of pure gelatin, pure PVA and the irradiated blend in the wave number range of 2400-500 cm^{-1}.

The FTIR spectra of untreated pure gelatin shown in Figure 7. The most distinctive spectral features for the protein were the strong amide I and II bands centered at approximately at 1640 and 1550 cm^{-1}, respectively. The amide I absorption was primarily due to the stretching vibration of the C=O bond and the amide II band was due to the coupling of the bending of the -NH bond and the stretching of the C-N bond. The FTIR spectra of untreated pure PVA shown in Figure 8. The peaks at 1088 cm^{-1} indicated the C-O stretch of secondary alcoholic groups.

The FTIR spectrum of irradiated blend (95% Gelatin+5% PVA) was given in Figure 9. The grafted product does not show any characteristic peak corresponding to carbonyl group and amino group indicating the crosslinking through these groups.

Scanning Electron Microscopy (SEM)

Scanning electron microscopic (SEM) image of untreated pure gelatin, pure PVA and an irradiated blend (95% Gelatine+5% PVA) were shown in Figures 10, 11 and 12. In order to study surface morphology SEM study was undertaken. In Figure 10 some unbound micro granules was observed. The SEM of PVA showed better film forming property than gelatin. Figure 12 indicating some type of interaction between gelatin and PVA due to irradiation. Crosslinking and chain scission occurred when polymers were exposed to gamma irradiation [17]. Polysaccharides and other natural polymers generally degrade by breaking the glycosidic linkage under gamma radiation [17].

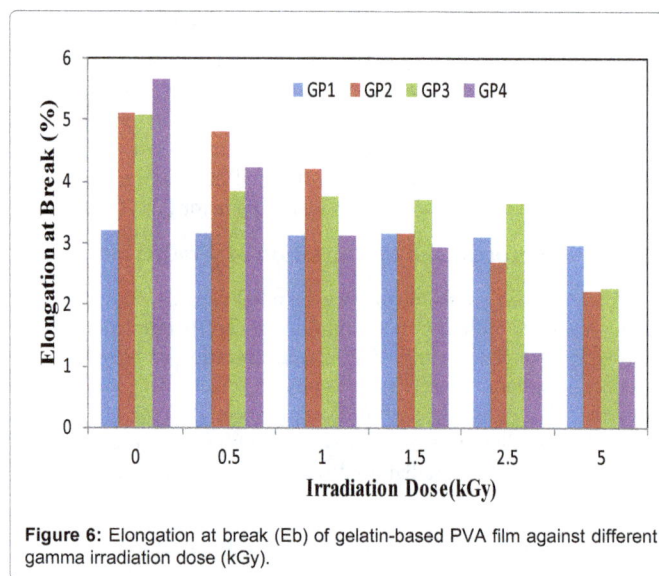

Figure 6: Elongation at break (Eb) of gelatin-based PVA film against different gamma irradiation dose (kGy).

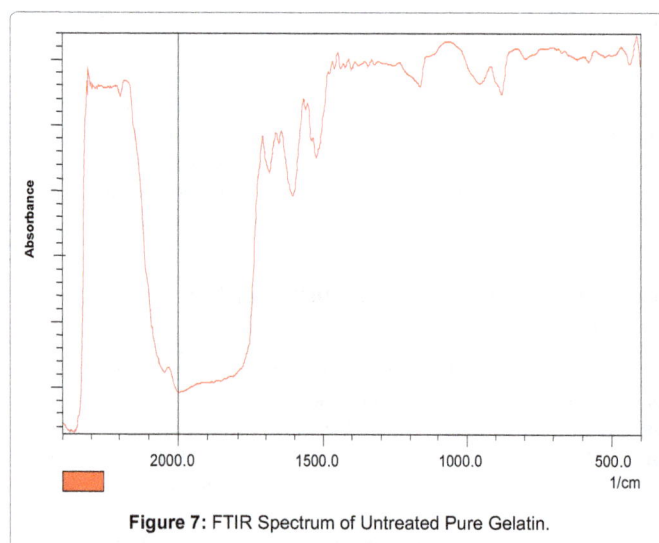

Figure 7: FTIR Spectrum of Untreated Pure Gelatin.

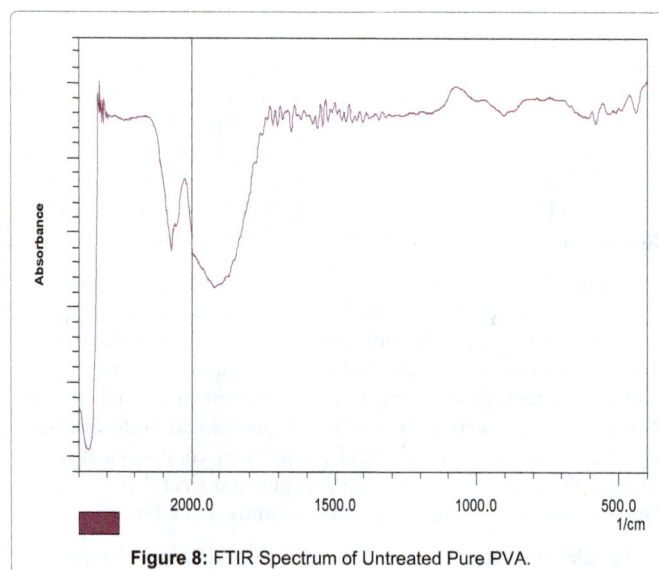

Figure 8: FTIR Spectrum of Untreated Pure PVA.

Composition (% w/w)		
Formulation	Gelatin (%)	PVA (%)
GP1	100	00
GP2	95	05
GP3	90	10
GP4	85	15

Table 1: Composition of different blending formulations (% w/w).

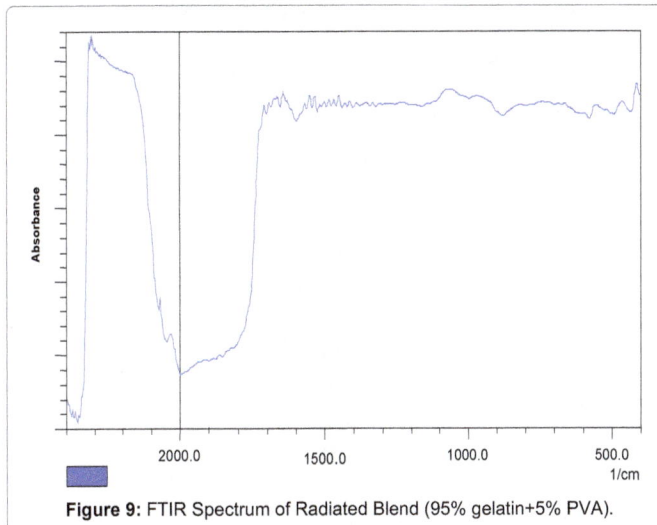

Figure 9: FTIR Spectrum of Radiated Blend (95% gelatin+5% PVA).

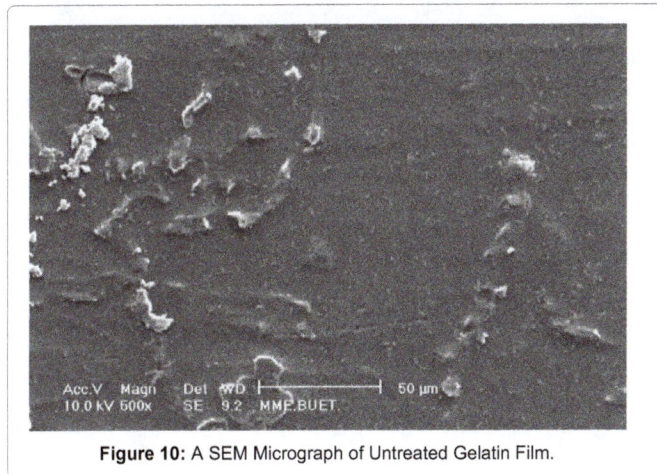

Figure 10: A SEM Micrograph of Untreated Gelatin Film.

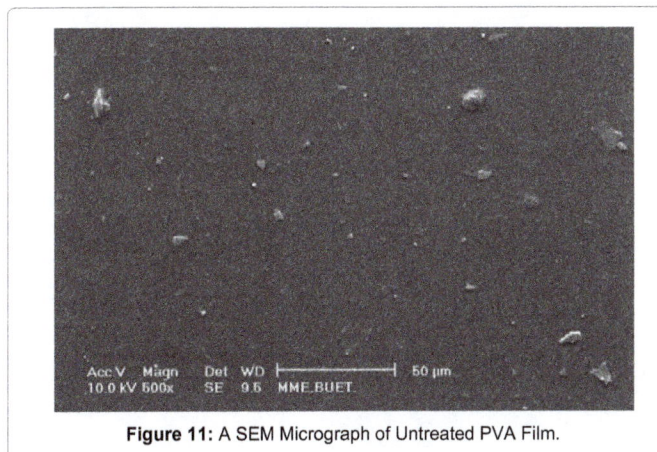

Figure 11: A SEM Micrograph of Untreated PVA Film.

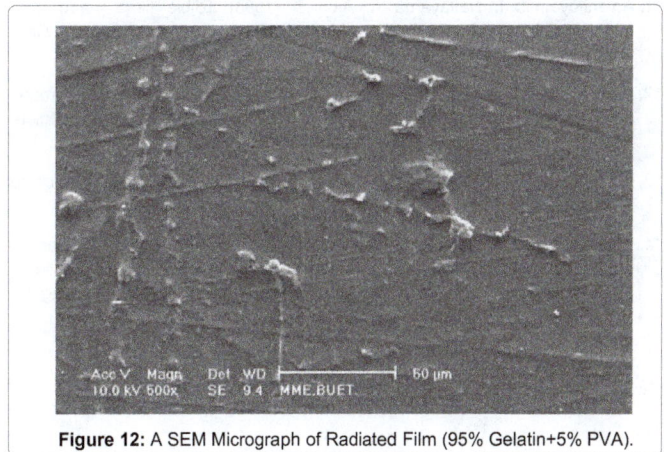

Figure 12: A SEM Micrograph of Radiated Film (95% Gelatin+5% PVA).

Conclusion

The mechanical and morphological properties of irradiated PVA mixed gelatin films indicated higher compared with the non-irradiated films. It was found that 95% gelatin+5% PVA film exhibited the highest tensile strength (TS) value at 0.5 kGy gamma radiation (51 MPa), which was 46% higher than that of non-irradiated films. Elongation at break (Eb) changes with gamma irradiation dose. The Eb of gamma treated film is lower than that of untreated film. The FTIR Spectrum of Radiated Blend (95% gelatin+5% PVA) blend shows that the grafted product does not show any characteristic peak. The SEM (Scanning Electron Microscopy) images indicate some sorts of interaction between gelatin and PVA due to irradiation. The ultimate target concluded that the enhanced property of PVA mixed gelatin films is obtained, which may be suitable for bio-medical applications.

References

1. Choudhury TR, Mollah MZI, Khan MA, Ali P, Chowdhury AMS, et al. (2014) Mechanical Properties Characterization of Jute Yarn Treated by Photo-curing with EG (Ethylene Glycol): Surface Treatment (KMnO₄). Journal of Composites and Biodegradable Polymers 2: 10-21.

2. Poddar P, Arafat Y, Dey K, Khan RA, Chowdhury AMS (2014) Effect of γ radiation on the performance of jute fabrics-reinforced urethane-based thermoset composites. Journal of Thermoplastic Composite Materials 1-11.

3. Gennadios A, McHugh TH, Weller CL, Krochta JM (1994) In: Krochta JM, Baldwin EA, Nisperos-Carriedo M (eds.) Edible coatings and films to improve food quality. Technomic Publishing Company, pp. 201-277.

4. Arvanitoyannis IS (2002) In: Gennadios A (ed.) Protein-based films and coatings. CRC Press, Boca Raton, pp. 275-304.

5. Achet D, He XW (1995) Determination of the renaturation level in gelatin films. Polymer 36: 787-791.

6. Sobral PJA, Menegalli FC, Hubinger MD, Roques MA (2001) Mechanical, water vapor barrier and thermal properties of gelatin based edible films. Food Hydrocolloid 15: 423-432.

7. Carvalho RA, Grosso CRF (2004) Characterization of gelatin based films modified with transglutaminase, glyoxal and formaldehyde. Food Hydrocolloid 18: 717-726.

8. Thomazine M, Carvalho RA, Sobral PJA (2005) Physical Properties of Gelatin Films Plasticized by Blends of Glycerol and Sorbitol. J Food Sci 70: E172-E176.

9. Vanin FM, Sobral PJA, Menegalli FM, Carvalho RA, Habitante AMQB (2005) Effects of plasticizers and their concentrations on thermal and functional properties of gelatin-based films. Food Hydrocolloid 19: 899-907.

10. Bergo PV, Sobral PJA (2007) Effects of plasticizer on physical properties of pigskin gelatin films. Food Hydrocolloid 21: 1285-1289.

11. Arvanitoyannis I, Nakayama A, Aiba S (1998) Edible films made from hydroxypropyl starch and gelatin and plasticized by polyols and water. Carb Polym 36: 105-119.

12. Rahman M, Dey K, Parvin F, Sharmin N, Khan RA, et al. (2011) Preparation and Characterization of Gelatin-Based PVA Film: Effect of Gamma Irradiation. International Journal of Polymeric Materials 60: 1056-1069.

13. Khan MA, Hasan MM (2004) Polymer Surface Modification: Relevance to Adhesion. CRC Press, Boca Raton 3: 263-283.

14. Khan M, Khan R, Noor F, Rahman MM, Noor-A-Alam M (2009) Studies on the Mechanical Properties of Gelatin and Its Blends with Vinyltrimethoxysilane: Effect of Gamma Radiation. Polymerplastics Technology and Engineering 48: 808-813.

15. Zaman H, Khan AH, Hossain MA, Khan M, Khan R (2009) Mechanical and Electrical Properties of Jute Fabrics Reinforced Polyethylene/Polypropylene Composites: Role of Gamma Radiation. Polymerplastics Technology and Engineering 48: 760-766.

16. Czvikovszky T (1995) Reactive recycling of multiphase polymer systems through electron beam. Elsevier 105: 233-237.

17. Song CL, Yoshii F, Kume TJ (2001) Macromolecular Science, Pure and Applied Chemistry Part A 38: 961.

31

Temperature-Dependence of Electrical Conductivity for Some Natural Coordination Polymeric Biomaterials Especially Some Cross-Linked Trivalent Metal-Alginate Complexes with Correlation between the Coordination Geometry and Complex Stability

Ishaq A Zaafarany*

Chemistry Department, Faculty of Applied Sciences, Umm Al-Qura University, Makkah Al-Mukarramah 13401, Saudi Arabia

Abstract

The electrical conductivity (σ) of cross-linked arsenic (III) - alginate complex as a coordination polymeric biomaterial in the form of circular discs have been measured as a function of temperature. The measured values of the electrical conductivity were found to be in the range of semiconductors. The change of electrical conductivity as a function of temperature was found to be of considerable complexity (Arrhenius plot of ln σ vs. 1/T). The appearance of a parabola zone at the early stages was explained by the release of waters of crystalline, whereas the sharp increase in σ values observed at the elevated temperatures was interpreted by the degradation process of the complex to give rise to the metal oxide as final product. The X-ray diffraction patterns indicated that the metal-alginate complex is amorphous in nature. Infrared absorption spectra revealed a sort of complexation between the trivalent metal cations and the functional carboxylate and hydroxyl groups of alginate macromolecule. A suitable conduction mechanism in terms of the complex stability in relation to the coordination geometry is suggested and discussed.

Keywords: Polymeric biomaterials; Alginate, Arsenic (III); Complexes; Semiconductors; Electrical conductivity.

Introduction

Alginate is a binary linear heteropolymer consisting of 1,4- linked β-D-mannuronic and α-L- guluronic acids units. The monomers are arranged in clockwise fashion along the macromolecular chains [1].

Conducting polymers offer the promise of achieving a new generation of polymers and materials that exhibit a high potential application in industrial technology [2-6]. Conductance measurements of polymers that sandwiched between two metal electrodes have attracted much attention owing to its potential applications in electronic devices [7] as well as to gain more information on the structure of materials that may be achieved from such measurements.

Although, conducting polymers in particularly that prepared from synthetic polymeric materials have been the subject of continuous research and development in recent years owing to their unique electrical, optical and chemical properties [8], a little attention has been focused to that of natural polymers such as the coordination polymeric biomaterials such as cross-linked metal-alginate complexes. This fact may be attributed to a lack of information on the electrical properties of such natural polysaccharide complexes. Even though, Hassan and co-workers studied the behavior of electrical conductivity for some coordination biopolymeric polysaccharides such as cross-linked metal-alginate complexes. For example the electrical conductivity of such complexes in either granule [8,9] or the gel forms[10,11] was investigated as a function of frequency earlier. Again, the electrical conductivity as a function of temperature for some metal-alginate complexes such as monovalent [12], divalent [13], trivalent [14,15], tetravalent [16-18] and hexavalent [19] metal cations in the granule forms have been studied elsewhere. Unfortunately, the suggested conduction mechanisms in these complexes are still incomplete owing to the discrepancies and variety observed in the behavior of electrical conductivity of studied complexes. It was found that no change in the electrical conductance with raising temperature for complexes involving chelated divalent [13] and trivalent [14] metal ions of one-

equivalent nature and the values of the electrical conductivity obtained (σ) lies in the range of insulators. On the other hand, metal alginate complexes involving trivalent [15], tetravalent [16-18] and hexavalent [19] metal ions of multi-equivalent nature showed an increase in the electrical conductivity with raising the temperature through parabolic shapes depending on the possible number of oxidation states formed during the reduction of chelated metal ion in that complex.

Preliminary experiments of the cited As (III)-alginate complex showed complicated behavior with respect to the trend of electrical conductance change with raising the temperature. However, As (III) ion cannot be reduced to another form, a parabola zone was appeared at the early stages, whereas a sharp increase in (σ) values followed by a sharp decrease in these values at higher temperatures was observed on contrary to that observed previously in similar complexes [13,14].

In view of the above discrepancies and our interest in physicochemial studies of such natural coordination biopolymeric complexes, the electrical conductivity of the cited complex seems to be of great significant and merits an investigation with the aims of shedding some light on the conduction mechanism in these metal-alginate complexes as well as to gain further information on the correlation between the electrical properties, geometrical structure and the stability in these

*Corresponding author: Ishaq A Zaafarany, Chemistry Department, Faculty of Applied Sciences, Umm Al-Qura University, Makkah Al-Mukarramah 13401, Saudi Arabia, E-mail: ishaq_zaafarany@yahoo.com

coordination polymeric biomaterial complexes as well as to compare the results obtained with those reported previously for other metal-alginate complexes.

Materials and Methods

Reagents

Sodium alginate used was Cica-Reagent (Kanto Chem. Co.). Doubly distilled conductively water was used in all preparations. All other materials used were of analytical grade.

All other materials were of BDH grade (with purity 99.99%) and were used without further purification. Doubly-distilled water was used in all preparations.

Preparation of As (III)-alginate granule complex

Generally, cross-linked metal-alginate granule complexes are prepared by the replacement of Na+ counter ions in alginate macromolecule by the corresponding polyvalent metal cations as described elsewhere [12-19]. This process was performed by stepwise addition of the alginate powder to the electrolyte solution of the As (III) ions while rapidly stirring the solution to avoid the formation of lumps, which swells with difficulty. After completion of the exchange process, the grains formed were washed with deionized water until the resultant water became free from non-chelated metal ions and then dried under vacuum over anhydrous $CaCl_2$ or P_2O_5.

Samples in the form of circular discs with diameter 13 mm and thickness 1-2 mm were obtained using an IR disc press at a constant pressure of 1500 psi.

X-ray diffraction

The X-ray diffraction patterns were obtained using a Philip 1710 diffractometer as described elsewhere [12-19].

IR spectra

The IR spectra were scanned on a Pye Unicam Sp 3100 spectrophotometer using the KBr disc technique (4000-400 cm^{-1}).

Conductance measurements

The specific mechanical properties of metal-alginate complexes in gel form make the measurements of their electrical properties by the usual methods very difficult. Therefore, a new technique of measurements involving a new cell device has been devised by us in order to overcome such difficulty [10,11].

The AC conductance was measured over the temperature range 290-560 K using a Keithely 610 C electrometer. The electrodes used were of circular shape forming M-S-M sandwich, where M represents the electrodes and S is the specimen. The metal-alginate complex discs were sandwiched between the two standard electrodes (graphite, copper or silver paste) mounted into a specially designed temperature-controlled electric furnace provided with a standard copper-constantan thermocouple. The samples were thermostated by kept short circuited for 3-5 h in air-thermostat within $\pm 1.0 \degree C$ to make it ready for experiments. The electrical resistance of the samples was measured, and from this the electrical conductivities (σ) were calculated as follows

$$\sigma = (1/R)(L/a) \qquad (1)$$

where R is the resistance (Ω), a is the area of the sample surface (cm^2) and L is the thickness of the specimen (cm).

Results and Discussion

Cross-linked trivalent metal-alginate complexes are formed when the Na+ counter ions of alginate macromolecule are replaced by an equivalent amount of quadrivalent metal cations. The interdiffused metal cations chelate the carboxylate and hydroxyl functional groups of alginate macromolecular chains through formation of partially ionic and partially coordinate bonds [20], respectively, and mediated through blocks of alginate macromolecular chains in an egg-carton like structure [20-23]. This exchange process occurs inherently and stoichiometrically even the valences and mobilities of the two exchanging counter ions are quite different [24].

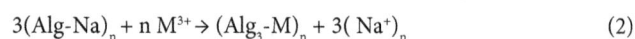

$$3(Alg\text{-}Na)_n + n\, M^{3+} \rightarrow (Alg_3\text{-}M)_n + 3(\,Na^+)_n \qquad (2)$$

solid aqueous solid complex aqueous

where (Alg – Na), $_(Alg_3$ – M) and M are sodium alginate, trivalent metal alginate complex and the interdiffused metal ion (As(III)), respectively.

The X–ray diffraction patterns indicate that the investigated alginate complex is amorphous in nature. Relevant infrared bands which provide considerable structural evidence for the mode of attachment between these trivalent metal cations and alginate functional groups are shown in Figure 1. The vibrational assignments of the bands showed that the bands of $\upsilon_s CO_2^-$ and $\upsilon_{as} CO_2^-$ are shifted from 1600 and 1400 cm^{-1} in alginate to higher frequencies in the complexes, indicating the complexation between the interdiffused metal cations and the carboxylate and hydroxyl functional groups of alginate macromolecular chains [25,26].

The broad band observed at around 3500 cm^{-1} which belongs to υ_{OH} of water (or OH-free functional groups) is also shifted to lower frequencies as shown in Table 1. The free-ligand has a strong band at 1735 cm^{-1} which can be assigned to the carbonyl stretching vibration of the carbonyl group [25]. The displacement of this band at around 1742-1747 cm^{-1} may also indicate the coordination of the carboxylate groups with the appearance of both symmetric (υ_s) and asymmetric (υ_{as}) vibrations of COO^- group. Again, the location of $\upsilon_s\, COO^-$ is a diagnostic of the bridging carboxylate groups.

The electrical conductivity of As (III)- alginate complex in the form

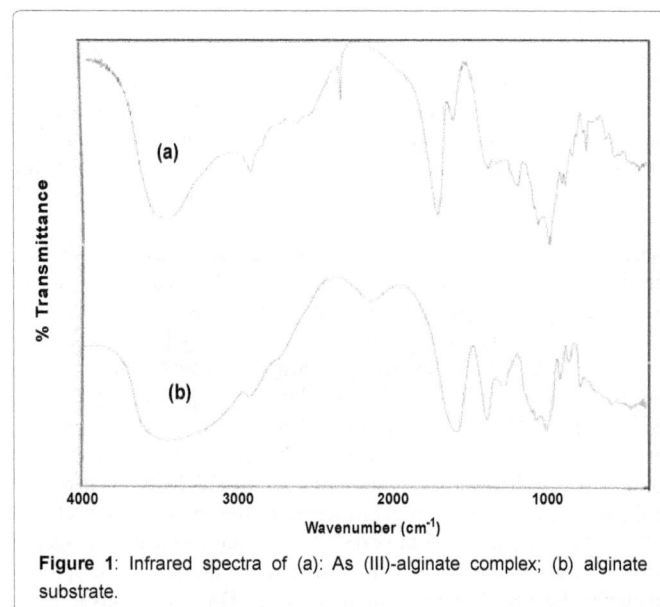

Figure 1: Infrared spectra of (a): As (III)-alginate complex; (b) alginate substrate.

Complex	ν_sOCO	ν_{as}OCO	ν_{OH}	ν_{M-O}	Reference
Na-alginate	1400	1600	3500	850	24,30
Ag^I-alginate	1400	1615	3444	830	12
As^{III}-alginate	1408	1634	3467	818	This work
La^{III}-alginate	1419	1598	3426	820	14
Ce^{III}-alginate	1421	1618	3441	815	14
Cr^{III}- alginate	1420	1637	3463	810	15
Fe^{III}-alginate	1418	1633	3448	817	15
Sn^{IV}-alginate	1414	1634	3457	811	16
Zr^{IV}-alginate	1408	1629	3447	813	16
Se^{IV}-alginate	1408	1629	3387	806	17
Th^{IV}-alginate	1419	1635	3461	890	18
U^{VI}-alginate	1410	1591	3410	817	19
Ce^{IV}-alginate	1424	1603	3445	809	37

Notes: ν_s Symmetry stretching vibrations of OCO, ν_{as}: Assymetry stretching vibrations of OCO, ν_{OH}: Stretching vibrations of H bond of OH, ν_{M-O}: Stretching vibrations of metal-oxygen bond.
Table 1: FTIR frequencies (cm^{-1}) for Sodium Alginate and Arsenic (III) cross linked Metal-Alginate complexes.

of circular disc was measured as a function of temperature. As is shown in Figure 2, a plot of ln σ against 1/T of Arrhenius equation displayed a complicated behavior where a small parabola zone was appeared at the early stages, but a sharp increase in (σ) values followed by a sharp decrease in these values was observed at the higher temperatures.

The formation of such parabolic zones was observed in analogous studies on the temperature-dependence of electrical conductivity for some other cross-linked metal-alginate complexes of multi-equivalent nature [15-19]. In those complexes, the observed parabolic behavior for the increase of the electrical conductivity with increasing the temperature at the early stages was explained by the formation of free-radical intermediates as a result of reduction of the chelated multi-equivalent metal cations to lower oxidation states. This suggestion was based on the possibility of transfer of electrons from the alginate macromolecule to the cross-linked metal cations. The number of transition zones is mainly dependent on the possible oxidation states obtained from the reduction of metal ion. For example, chromium (III) and iron (III) are well-known to be of two-equivalent oxidant nature [27], i.e. $M^{3+} \rightarrow M^{2+}$ on reduction. Therefore, the number of expected transition zones should be only one as was experimentally observed [15]. Again, uranium (VI) is a multi-equivalent oxidant [28] which gives various oxidation states on reduction, U^V, U^{IV} and U^{III}, through successive one-electron-transfer mechanism in a sequence by. So, three parabolic zones should be existed as was reported earlier [19]. On the other hand, the reduction of divalent- and some trivalent-metal cations of one-equivalent nature such as lanthanum (III) and cerium (III) cannot be reduced to lower oxidation states [27]. Hence, no parabolic zones should be expected as was confirmed experimentally [13,14].

In view of the above interpretations and the experimental observations, the formation of small parabolic zone at the early stage with respect the behavior of electrical conductivity with increasing temperature (Figure 2) in As(III)-alginate complex cannot be attributed to reduction of chelated As(III) metal ion since no further lower oxidation states are known for arsenic ion are known [27]. The most probable interpretation of such observed parabola may be explained by the evolution of water of crystallization as was reported for analogous cases [12].

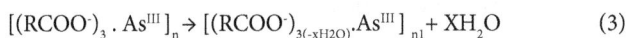

$$[(RCOO^-)_3 . As^{III}]_n \rightarrow [(RCOO^-)_{3(-xH2O)}.As^{III}]_{n1} + XH_2O \quad (3)$$

where RCOO$^-$ represents to the alginate monomers $(C_6H_7O_6^-)_n$

Again, the sharp increase in (σ) values observed at higher temperatures over 200°C (I) can be explained by the degradation of the complex to form its corresponding arsenic carbonate radical [29] with evolution of carbon dioxide and water vapor as follows

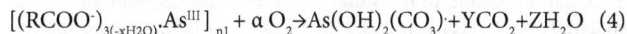

$$[(RCOO^-)_{3(-xH2O)}.As^{III}]_{n1} + \alpha\, O_2 \rightarrow As(OH)_2(CO_3)^{\cdot}+YCO_2+ZH_2O \quad (4)$$

Of course, the dimerization process of such free-radicals formed will lead to a decrease in the charge carriers and, hence, a decrease in the electrical conductivity should be expected as was experimentally observed (II). This dimerization can be expressed by the following equations.

$$2\,(As(OH)_2(CO_3)^{\cdot}) \rightarrow [As(OH)_2(CO_3)]_2 \quad (5)$$

On other hand, the sharp increase in σ values observed at elevated temperatures (III) can be explained by the decomposition of the arsenic carbonate to give rise to its corresponding more stable metal oxide as final degradation products [14,15,30,31]

$$[As(OH)_2(CO_3)]_2 + \frac{1}{2} O_2 \rightarrow As_2O_5 + 2H_2O + 2\,CO_2 \quad (6)$$

The values of electrical conductivities obtained using different electrodes were found to be in good agreement with each other confirming the reproducibility of the conductance measurements (within experimental errors of ± 3%). The correlation coefficient of the data obtained was (r^2>0.99). In other words, it means that the nature of electrodes has no influence on the behavior of electrical conductivity.

On the other hand, the kinetics of non-isothermal decomposition of some coordination biopolymer trivalent-metal alginate cross-linked complexes such as Al (III), Cr (III), Fe (III), Ce (III) and La (III) has been investigated using a Mettler TA 3000 thermal analyzer in static air and were reported elsewhere [32,33]. The TG curves showed three stages of weight-loss, whereas the DTG curves indicated the presence of a series of thermal peaks associated with the TG curves. The first stage was found to correspond to the thermal dehydration of coordinated water molecules in the first stage. The second stage was corresponding to the formation of intermediate complexes resulted from the decomposition of the dehydrated complexes, whereas the third stage represented to the degradation of the formed intermediates to give metal oxides as final products. These results may be considered as indirect evidence to support our suggested conductance mechanism of the cited work. The thermal decomposition of As (III)-alginate complex stills in progress in our laboratory.

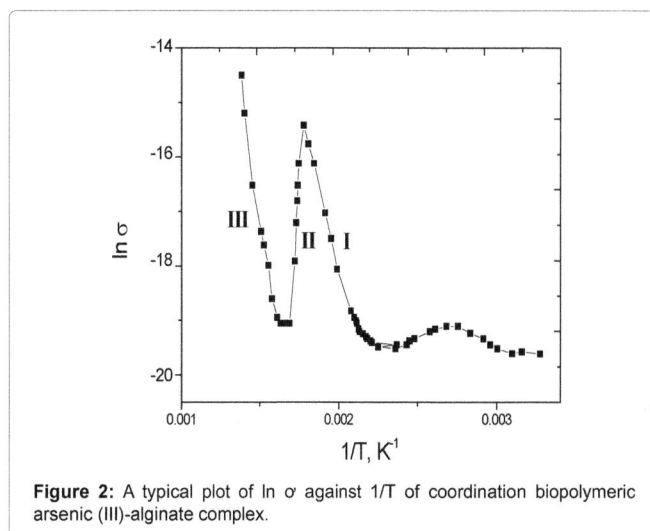

Figure 2: A typical plot of ln σ against 1/T of coordination biopolymeric arsenic (III)-alginate complex.

It was reported that the interdiffused metal ions chelate the functional groups of alginate macromolecular chains through two geometrical structures [34]. The first type corresponds to an intramolecular association in which the functional groups involved in chelation belong to the same chain. Hence, the plane involving the chelated metal ions is parallel with the plane of alginate macromolecular chains. This configuration may be called a planar geometry. The second structure corresponds to the intermolecular association in which the plane containing the metal ions is perpendicular to the plane of alginate chains. Here, the involved functional groups are related to different chains and the configuration obtained is termed a non-planar geometry. The type and nature of geometry depends on both valency and coordination number of the chelated metal ion in the complex.

Divalent [26] and hexavalent [27] metal cations are known to be of octahedral six coordination geometry in their complexes and, hence, tend to chelate the functional groups via either inter- or intramolecular association in order to attain the octahedral geometry in their complexes as shown in Schemes I and II. The priority of these two geometrical structures depends on the stability of the complexes formed. Again, trivalent- and tetravalent-metal ions are restricted to cross-link the functional groups of alginate macromolecule via only intermolecular association in their alginate complexes for geometrical reasons (Scheme III). The preferability of intermolecular association mechanism for chelation in case of the latter metal- alginate complexes can be explained by their tendency to decrease the bond stretching resulting from metal-oxygen bond elongation occurs in case of intramolecular association [34].

In view of the above interpretation, the geometrical structure in those metal alginate complexes can be achieved through chelation of the interdiffused metal ions with the carboxylate groups as principal units in addition to the hydroxyl and/or the water molecules of the amorphous phase. This is to complete the coordination number of the chelated metal ion and reach with the complex to a more stable geometry.

It is well known that the charge carriers drift their velocity to maximum in case of intermolecular association complexes owing to the presence of channels around the planes containing the metal ions rather than in case of intramolecular association [34-36]. Hence, the electrical conductivity is expected to be higher in non-planar than that in planar geometries [34]. As is shown in Table 2, the value of electrical conductivity lies in the range of semiconductors for monovalent, trivalent and tetravalent metal-alginate complexes indicating that the chelation in these metal-alginate complexes should be of intermolecular association geometry which may confirm our suggestion (Scheme III).

On the other hand, the values of the electrical conductivities for uranium (VI)- and silver(I)-alginate complexes listed in Table 2 may indicate that the uranyl ion prefers to coordinate with the functional group of alginate macromolecule by intramolecular association mechanism (Scheme IV), whereas silver(I) prefers to chelate via intermolecular association ones in its alginate complex (Scheme I) in a good accord with that reported earlier [12,34].

Generally, the conductance in polymeric compounds occurs via two conduction mechanisms named ionic and electronic conduction, respectively, depending on the nature of the charge carriers existing within the network of the macromolecular chains [6]. The value of activation energy may be considered as indirect evidence for the conduction mechanism. When the value of the activation energy, E_a, is larger than unity, it means that the cationic conduction mechanism tends to be the more prominent one, whereas that value is usually lower than unity in case of electronic conduction mechanism. The activation

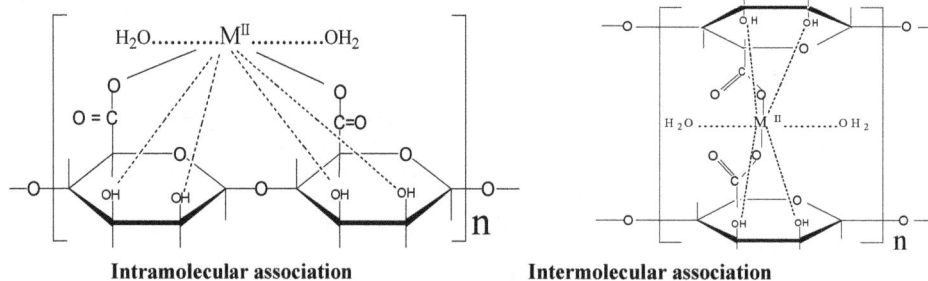

Scheme I: Chelation in Divalent Metal Ion Complexeas

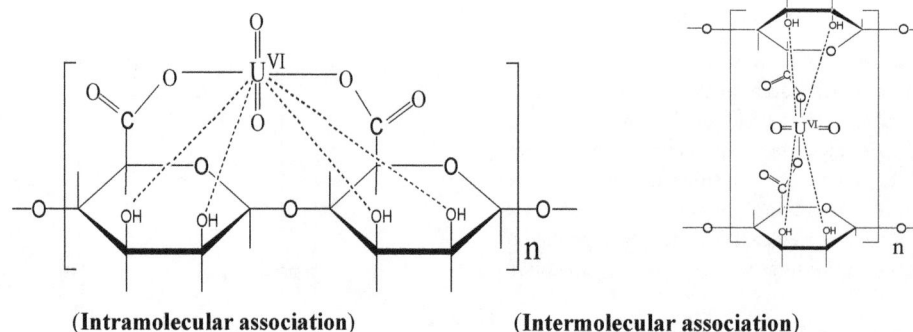

Scheme II: Chelation in Uranyl Ion Complex

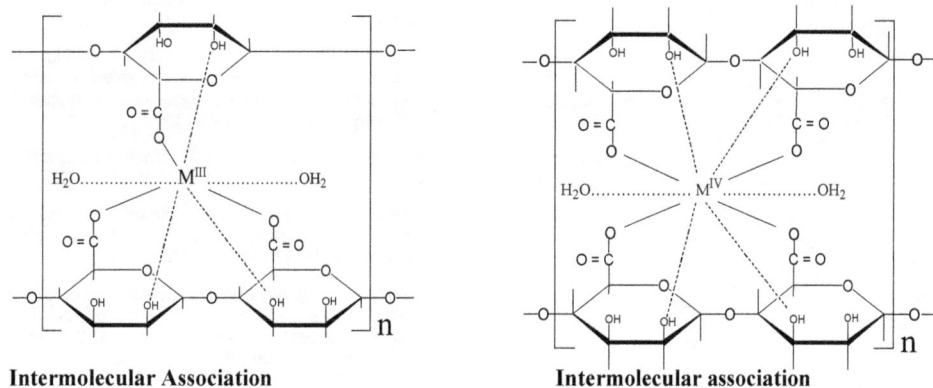

Scheme III: Chelation in Trivalent Metal Ion Complexes

Complex	E_a (Initial stage)	E_a (Final stage)	σ (Ω^{-1} cm^{-1})	Reference
Ag^I-alginate	0.21	3.15	1.8×10^{-8}	12
As^{III}-alginate	0.13	1.64	2.6×10^{-9}	This work
La^{III}-alginate	-	1.47	1.3×10^{-9}	14
Ce^{III}-alginate	-	1.19	1.0×10^{-9}	14
Cr^{III}-alginate	0.16	2.74	10.0×10^{-9}	15
Fe^{III}-alginate	022	1.41	2.0×10^{-9}	15
Sn^{IV}-alginate	0.33	0.98	1.1×10^{-9}	16
Zr^{IV}-alginate	0.36	0.85	1.2×10^{-9}	16
Se^{IV}-alginate	0.25	1.17	6.0×10^{-9}	17
Th^{IV}-alginate	0.86	1.20	2.01×10^{-9}	18
U^{VI}-alginate	0.37	0.83	1.7×10^{-12}	19
Ce^{IV}-alginate	0.28	0.84	1.04×10^{-9}	37

Table 2: The activation energies in eV and the electrical conductivity in Ω^{-1} cm^{-1} (290 K) for some cross linked Metal-Alginate complexes.

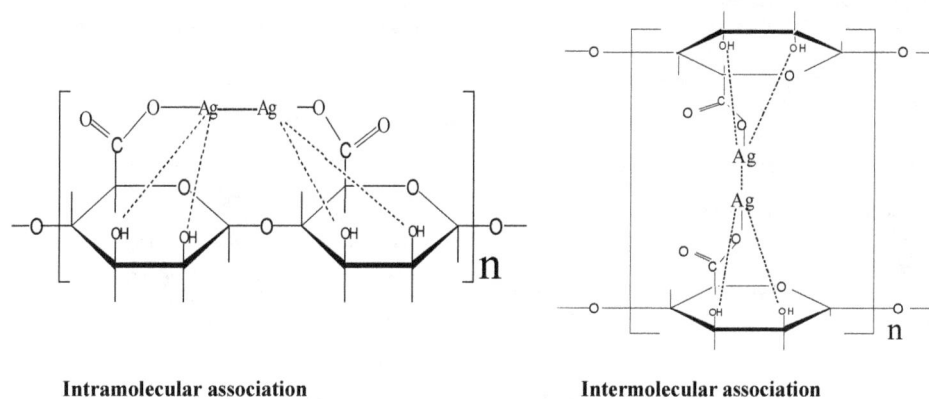

Scheme IV: Chelation in Silver (I) Ion Complex

energies in these cross-linked metal alginates can be evaluated from the slopes of ln σ –1/T plots using the Arrhenius equation

$$\sigma = \sigma^o \exp(-E_a/RT) \qquad (7)$$

where σ is the electrical conductivity, σ^o is a constant and E_a is the activation energy of the charge carriers. These values were calculated by using the least-squares method and are summarized along with that of other cross-linked metal alginate complexes in Table 2. The calculated values of E_a were found to be 0.13 ± 0.005 and 1.64 ± 0.04 for the initial and final stages, respectively. The lower values of activation energy <1.0 eV observed for studied tetravalent metal alginate complexes of the present work may reflect the prominent of the electronic conduction

mechanism. The formation of free-radicals through the treatment of the studied complexes can support this suggestion.

Conclusion

The electrical conductivity of coordination biopolymeric As(III)-alginate complex has been measured as a function of temperature. The Arrhenius plot of ln σ against 1/T showed a complicated behavior. The magnitude of the σ values indicated that the electrical conductivity of such complex lies in the range of semiconductors. It was found that the geometrical configuration between the chelated metal ion and the functional carboxylate and hydroxyl groups in the determining factor for the electrical conductivity behavior and play an important role in the stability of the formed complex.

References

1. Chanda SK, Hirst EL, Percival EG, Ross AG (1952) Structure of alginic acid. J Chem Soc 1833-1837.

2. Nandapure BI, Kondawar SB, Salunkhe MY, Nandapure AI (2013) Magnetic and transport properties of conducting polyaniline / nickel oxide nanocomposites. Adv Mat Lett 4: 134-140.

3. Nhivekar GS, Mudholkar RR (2011) Microcontroller based IR remote control signal decoder for home application . Adv Appl Sci Res 2: 410.

4. Jing X, Wang Y, Zhang B, (2005) Electrical conductivity and electromagnetic interference shielding of polyaniline/polyacrylate composite coatings. J Appl Polym Sci 98: 2149-2156.

5. Mead CA (1961) Operation of tunnel-emission devices. J Appl Phys 32: 646.

6. Sulaneck WR, Clark DT, Samuelson EJ (1990) Science and Application of Conducting Polymers. IOP Publishing Ltd., UK, USA.

7. El-Dossouky M, Hegazy EA, Dossuki AM, El-Sawy NM (1986) Electrical conductivity of anionic graft copolymers obtained by radiation grafting of 4-vinylpyridine onto poly(vinyl chloride). Radiat Phys Chem 27: 443-446.

8. Abdel-Wahab SA, Ahmed MA, Radwan FA, Hassan RM, El-Refae AM, et al. (1997) Relative permittivity and electrical conductivity of some divalent metal alginate complexes. Mater Lett 20:183-188.

9. Ahmed MA, Radwan FA, El-Refae AM, Abdel-Wahab SA, Hassan RM (1997) Temperature and frequency dependence of the electrical properties of metal alginate complexes. Ind J Phys 71: 39.

10. Hassan RM, Makhlouf M Th, Summan AM, Awad A (1989) Influence of frequency on specific conductance of polyelectrolyte gels with special correlation between strength of chelation and stability of divalent metal alginate ionotropic gels. Eur Polym J 25: 993-996.

11. Hassan RM (1989) Influence of frequency on electrical properties of acid and trivalent metal alginate ionotropic gels. A correlation between strength of chelation and stability of polye1ectrolyre gels. High Perform Polym 1: 275-284.

12. Hassan RM (1991) Alginate polyelectrolyte ionotropic gels. VII. Physicochemical studies on silver(I) alginate complex with special correlation to the electrical properties and geometrical structure. Coll Surf 60: 203-212.

13. Khairou KS, Hassan RM (2002) Temperature-dependence of electrical conductivity for cross-linked mono- and divalent metal alginate complexes. High Perform Polym 14: 93-99.

14. Zaafarany IA, Khairou KS, Hassan RM (2010) Physicochemical studies on some cross-linked trivalent metal-alginate complexes especially the electrical conductivity and chemical equilibrium related to the coordination geometry. High Perform Polym 22:69-81.

15. Hassan RM, El-Shatoury SA, Mahfouz RM, Osman MA, El-Korashy A, et al. (1995) Alginate polyelectrolyte ionotropic gels. VIII. Electrical properties of di- and trivalent metal alginate complexes specially iron(III) and chromium(III) alginate resins. Bull Fac Sci Assiut Univ Egypt 24 : 141.

16. Hassan RM, Zaafarany IA, Gobouri AA (2013) Temperature-dependence of electrical conductivity of some natural coordination polymeric biomaterials especially cross-linked tetravalent metal-alginate complexes with correlation between the coordination geometry and complex stability. Advan Biosen Bioelectro 2: 16.

17. Zaafarany IA, Khairou KS, Hassan RM, Ikeda Y (2009) Physicochemical studies on cross-linked thorium (IV)-alginate complex especially the electrical conductivity and chemical equilibrium related to the coordination geometry. Arabian J Chem 2: 1-10.

18. Ishaq IA, Khairou KS, Hassan RM (2009) Physicochemical studies on some natural polymeric complexes of quadrivalent metal cations. Electrical conductivity and chemical equilibrium of cross-linked selenium (IV)-alginate complex with correlation between the complex stability and geometrical structure. J Saudi Chem 13: 49-60.

19. Hassan RM, Ikeda Y, Tomiyasu H (1993) Alginate polyelectrolyte ionotropic gels. Part XV. Physicochemical properties of uranyl alginate complex especially the chemical equilibrium and electrical conductivity related to the coordination geometry. J Mater Sci 28: 5143-5147.

20. Awad A, El-Cheikh F (1981) Electrical resistance and anisotropic properties of some metal alginate gels. J Coll Interf 80: 107-110.

21. Hirst E, Rees DA (1965) The structure of alginic acid. Part V. Isolation and unambiguous characterization of some hydrolysis products of the methylated polysaccharide. J Chem Soc 1182-1187.

22. Rees DA, Scott WE (1971) Polysaccharide conformation. Part VI. Computer model-building for linear and branched pyranoglycans. Correlations with biological function. Preliminary assessment of inter-residue forces in aqueous solution. Further interpretation of optical rotation in terms of chain conformation. J Chem Soc B 469-479.

23. Schweiger RG (1962) Acetylation of alginic acid. II. Reaction of algin acetates with calcium and other divalent ions. J Org Chem 27: 1786-1789.

24. Hellferich H (1962) Ion Exchange, McGraw-Hill, New York.

25. Bellamy LJ (1966) The Infrared Spectra of Complex Molecules. (2nd edn), Chapman and Hall, London.

26. Cozzi D, Desider PG, Leppri L, Cinatelli G (1968) Alginic acid, a new thin layer material. J Chromatogr 35: 369-404.

27. Cotton AF, Wilkinson G (1972) Advanced Inorganic Chemistry. (3rd edn), John Wiley, New York.

28. Hassan RM (1992) A review on oxidation of uranium (IV) by polyvalent metal ions. A linear free-energy correlation. J Coord Chem Rev 27: 255-266.

29. Neuberger CS, Helz GR (2006) Thermodynamic assessment and chemical interpretations. Appl Geochem 20: 1218-1226.

30. Said AA, Abd El-Wahab MMM, Hassan RM (1994) Thermal and electrical studies on some metal alginate compounds. Thermochem Acta 233: 13-24.

31. Zaafarany IA, Khairou KS, Terkistani FA, Iqbal S, Khairy M, Hassan RM, et al. (2012) Kinetics and mechanisms of non-isothermal decomposition of Ca(II)-, Sr(II)- and Ba(II)- cross-linked divalent metal-alginate complexes. Int J Chem 4: 7-14;Said AA, Hassan RM (1993) Thermal decomposition of some divalent metal alginate gel compounds. Polym Degrad Stab 39: 393-397.

32. Zaafarany IA (2010) Non-isothermal decomposition of Al, Cr and Fe cross-linked trivalent metal-alginate complexes. JKAU 22: 193-202.

33. Terkistani FA, Hassan RM (2012) Kinetics and mechanisms of non-isothermal decomposition of some cross-linked metal-alginate complexes especially trivalent-metal-alginate complexes. Orien J Chem 28: 913-920.

34. Hassan RM (1993) Alginate polyelectrolyte ionotropic gels. XIII. Geometrical aspects for chelation in metal alginate complexes related to their physicochemical properties. Polm Inter 31: 81-86.

35. Thomas DP, Randal TC, Ralph M (2006) Molecular models of alginic acid: Interactions with calcium ions and calcite surfaces. Geochimica et Cosmochimica Acta 70: 3508-3532.

36. Braccini I, Grasso RP, Perez S (1999) Conformational and configurational features of acidic polysaccharides and their interactions with calcium ions: a molecular modeling investigation. Carbohyd Res 317: 119-130.

Permissions

List of Contributors

Mark Moore, Robert Counce, Jack Watson and Thomas Zawodzinski
Department of Chemical Engineering, University of Tennessee, USA

Marek Zieliński
Department of Inorganic and Analytical Chemistry, Faculty of Chemistry, University of Lodz, Poland

Kazusa Terasaka, Hiroyuki Imai and Xiaohong Li
Faculty of Environmental Engineering, The University of Kitakyushu, 1-1 Hibikino, Wakamatsu, Kitakyushu, Fukuoka, Japan

Eugeniusz Orszulik
Central Mining Institute, Pl. Gwarków 1, 40–116 Katowice, Poland

Maedeh Asari
Faculty of Engineering- Department of Chemical Engineering, Islamic Azad University, Shahrood Branch, Shahrood, Iran

Faramarz Hormozi
Faculty of Chemical, Gas and Petroleum Engineering, Semnan University, Semnan, Iran

Aashit Kumar Jaiswal and RR Yadav
Department of Physics, University of Allahabad, Allahabad, Uttar Pradesh, India

Gopal Nath
Department of Microbiology, Institute of Medical Sciences, Banaras Hindu University, Varanasi, Uttar Pradesh, India

Mayank Gangwar
Department of Microbiology, Institute of Medical Sciences, Banaras Hindu University, Varanasi, Uttar Pradesh, India

Department of Pharmacology, Institute of Medical Sciences, Banaras Hindu University, Varanasi, Uttar Pradesh, India

Chowdhury Kaiser Mahmud, Md. Asanul Haque and AM Sarwaruddin Chowdhury
Department of Applied Chemistry and Chemical Engineering, Faculty of Engineering and Technology, University of Dhaka, Bangladesh

Mohammad Abdul Ahad
Department of Chemistry, Faculty of Science, University of Chittagong, Chittagong, Bangladesh

Md. Abdul Gafur
Department of Pilot Plant and Product Development Center, Bangladesh council of Scientific and Industrial Research, Dhaka, Bangladesh

Montalbán MG, Trigo R and Víllora G
Department of Chemical Engineering, University of Murcia, Murcia, Spain

Collado-González M and Díaz Baños FG
Department of Physical Chemistry, University of Murcia, Murcia, Spain

Tushar Kumar Sheel, Pinku Poddar and AM Sarwaruddin Chowdhury
Department of Applied Chemistry and Chemical Engineering, Faculty of Engineering and Technology, University of Dhaka, Dhaka,

AJM Tahuran Neger
Institute of Glass and Ceramic Research and Testing (IGCRT), BCSIR Dhaka, Dhaka, Bangladesh

ABM Wahid Murad
Institute of Leather Engineering and Technology, University of Dhaka, Dhaka, Bangladesh

H Amani, Ahmad Zuhairi Abdullah and Abdul Rahman Mohamed
School of Chemical Engineering, Universiti Sains Malaysia, 14300 Nibong Tebal, Penang, Malaysia

Lilis Hermida
School of Chemical Engineering, Universiti Sains Malaysia, 14300 Nibong Tebal, Penang, Malaysia

Department of Chemical Engineering, Universitas Lampung, Bandar Lampung 35145, Indonesia

Mark Moore, Bonnie Herrell, Robert Counce and Jack Watson
Department of Chemical Engineering, University of Tennessee, USA

Eric M Adetutu and Andrew S Ball
School of Applied Sciences, RMIT University, Bundoora 3083, Australia

Abdulatif A Mansur
School of Applied Sciences, RMIT University, Bundoora 3083, Australia
Environmental and Natural Resources Engineering, Faculty of Engineering, Azawia University, Libya

Muthu Pannirselvam
School of Civil, Environmental and Chemical Engineering, RMIT University, Melbourne 3000, Australia

Khalid A Al-Hothaly
School of Applied Sciences, RMIT University, Bundoora 3083, Australia
Department of Biotechnology, Faculty of Science, Taif University, Kingdom of Saudi Arabia

Edidiong Okon, Habiba Shehu and Edward Gobina
Center for Process Integration and Membrane Technology (CPIMT), School of Engineering, The Robert Gordon University, Aberdeen, AB10 7GJ, United Kingdom

Eamor M Woo, Graecia Lugito, Cheng-En Yang and Shi-Ming Chang
Department of Chemical Engineering, National Cheng Kung University, Taiwan

Li-Ting Lee
Department of Materials Science and Engineering, Feng Chia University, Taiwan

Mahendra Kumar Trivedi, Rama Mohan Tallapragada, Alice Branton, Dahryn Trivedi, and Gopal Nayak
Trivedi Global Inc., 10624 S Eastern Avenue Suite A-969, Henderson, NV 89052, USA

Omprakash Latiyal and Snehasis Jana
Trivedi Science Research Laboratory Pvt Ltd, Hall-A, Chinar Mega Mall, Chinar Fortune City, Hoshangabad Rd, Bhopal, Madhya Pradesh, India

Venkat K. Rajendran, Andreas Menne and Axel Kraft
Fraunhofer Institute for Environmental, Safety and Energy Technology UMSICHT, Germany

Mohammed Nasir Kajama, Habiba Shehu, Edidiong Okon and Ify Orakwe
Centre for Process Integration and Membrane Technology (CPIMT), School of Engineering, The Robert Gordon University, Aberdeen, AB10 7GJ, United Kingdom

Haitham Mohammad Abdelaal
Ceramics Department, The National Research Centre, Al-Buhouth St. Dokki, Cairo, Egypt

Bernd Harbrecht
Department of Chemistry and Centre of Materials Science Philipps University, 35032 Marburg, Germany

Pinku Poddar, Noor MD Shahriar Khan and AM Sarwaruddin Chowdhury
Department of Applied Chemistry and Chemical Engineering, Faculty of Engineering and Technology, University of Dhaka, Dhaka, Bangladesh

MD Tariqul Islam
Department of Applied Chemistry and Chemical Engineering, Faculty of Engineering and Technology, University of Dhaka, Dhaka, Bangladesh
Department of Chemistry, American International University of Bangladesh, Dhaka, Bangladesh

NC Dafader
Nuclear and Radiation Chemistry Division, Institute of Nuclear Science and Technology, Atomic Energy Research Establishment, Dhaka, Bangladesh

Siti Machmudah, Qifni Yasa' Ash Shiddiqi, Achmad Dwitama Kharisma, Widiyastuti and Sugeng Winardi
Department of Chemical Engineering, Sepuluh Nopember Institute of Technology, Kampus ITS Sukolilo, Surabaya 60111, Indonesia

Wahyudiono, Hideki Kanda and Motonobu Goto
Department of Chemical Engineering, Nagoya University, Furo-cho, Chikusa-ku, Nagoya 464-8603, Japan

Michael Matzen, Mahdi Alhajji and Yaşar Demirel
Department of Chemical and Biomolecular Engineering, University of Nebraska Lincoln, Lincoln NE 68588, USA

Ahamed MEH
Department of Applied Chemistry (Doornfontein Campus), University of Johannesburg, South Africa

Mbianda XY
Department of Applied Chemistry (Doornfontein Campus), University of Johannesburg, South Africa

Centre for Nanomaterial Science Research, University of Johannesburg, South Africa

Marjanovic L
Department of Chemistry, Faculty of Science, University of Johannesburg, South Africa

Jing Wang, Jiangping Guo, Yang Zhou, Qigu Huang, Yunfang Liu and Wantai Yang
State Key Laboratory of Chemical Resource Engineering, Key Laboratory of Carbon Fiber and Functional Polymers, Ministry of Education, Beijing University of Chemical Technology, People's Republic of China

Jianjun Yi, Hongming Li and Kejing Gao
Lab for Synthetic Resin Research Institution of Petrochemical Technology, China National Petroleum Corporation, People's Republic of China

Cristiane de Souza Siqueira Pereira and Fernando Luiz Pelegrini Pessoa
Technology of Chemical and Biochemical Processes, Chemistry School, Federal University of Rio de Janeiro, Brazil

Simone Mendonca and Jose Antonio de Aquino Ribeiro
Embrapa Agroenergy, Brasilia, Brazil

Marisa Fernandes Mendes
Chemical Engineering Department, Federal Rural University of Rio de Janeiro, Brazil

Joanna Liszkowska, Bogusław Czupryński and Joanna Paciorek-Sadowska
Kazimierz Wielki University, Bydgoszcz, Poland

Habiba Shehu, Edidiong Okon, Ifeyinwa Orakwe and Edward Gobina
Center for Process Integration and Membrane Technology, Robert Gordon University Aberdeen, United Kingdom

Benabid FZ, Douibi A and Benachour D
LMPMP, Faculty of Technology, Université Sétif-1, Algeria

Zouai F
Unité de Recherche Matériaux Emergents, Université Sétif-1, Algeria

Isaac Mbir Bryant and Roberta Tetteh-Narh
Department of Environmental Sciences, University of Cape Coast, Ghana

Chai YH
Chemical Engineering Department, Universiti Teknologi PETRONAS, Bandar Seri Iskandar, 31750 Tronoh, Perak, Malaysia

Suzana Yusup
Associate Professor, Chemical Engineering Department, Universiti Teknologi PETRONAS, Bandar Seri Iskandar, 31750 Tronoh, Perak, Malaysia

Chok VS
Director of Technology, Platinum Nanochem Sdn. Bhd., Lot 15-19 & PT 1409, Senawang Industrial Estate, Batu 4 Jalan Tampin, 70450 Seremban, Negeri Sembilan Darul Khusus, Malaysia

AM Sarwaruddin Chowdhury, Moshfiqur Rahman, Pinku Poddar, Kamol Dey, Noor Md Shahriar Khan and Md Ali Akba
Department of Applied Chemistry and Chemical Engineering, Faculty of Engineering and Technology, University of Dhaka, Dhaka, Bangladesh

Syed Rashedul Alam
Research and Development (R&D), Berger Paints Limited, Bangladesh

Ruhul A Khan
Nuclear and Radiation Chemistry Division, Institute of Nuclear Science and Technology, Bangladesh Atomic Energy Commission, Dhaka, Bangladesh

Zinia Nasreen
Department of Chemistry, Dhaka University of Engineering Technology (DUET), Gazipur, Bangladesh

Ishaq A Zaafarany
Chemistry Department, Faculty of Applied Sciences, Umm Al-Qura University, Makkah Al -Mukarramah 13401, Saudi Arabia

Index

www.ingramcontent.com/pod-product-compliance
Lightning Source LLC
Chambersburg PA
CBHW080637200326
41458CB00013B/4663